Lived Spaces in Late Antiquity

This volume considers "lived space" as a scholarly approach to the past, showing how spatial approaches can present innovative views of the world of Late Antiquity, integrating social, economic and cultural developments and putting centre stage this fundamental dimension of social life.

Bringing together an international group of scholars working on areas as diverse as Britain, the Iberian Peninsula, Jordan and the Horn of Africa, this book includes burgeoning fields of study such as lived spaces in the context of ships and seafaring during this period. Chapters investigate the history, function and use of different spaces in their own right and identify the social and historical logic presiding over continuity and/or change. They also explore the fluidity of lived space in both its physical and conceptual dimensions, analysing issues like agency and intentionality as well as meaning and social relations. Space is the fundamental dimension of social life, the arena where it unfolds and the stage where social values and hierarchies are represented; analysis of space allows us to understand history through different means of shaping, occupying and controlling space. Considering Late Antiquity through a spatial perspective offers a complex and stimulating picture of this pivotal period, and this volume provides avenues for the development of further research and discussion in this area.

Lived Spaces in Late Antiquity is a fascinating resource for students and scholars interested in space and spatiality in the late antique world, as well as archaeology, classical studies and late antique studies more generally.

Carlos Machado is a senior lecturer in ancient history at the University of St Andrews (UK). He has published extensively on late antique history and material culture, including *Urban Space and Aristocratic Power in Late Antique Rome* (2019) and co-edited *Poverty in Ancient Greece and Rome* (2023).

Rowan Munnery is a doctoral candidate at the University of St Andrews (UK). He focuses on utilising quantitative and network approaches in maritime archaeology to examine the maritime cultural landscape of the late antique western Mediterranean. His research interests also include the phenomenology of space aboard ships and behavioural economics in the ancient world.

Rebecca Sweetman is the director of the British School at Athens and a professor of ancient history and archaeology at the University of St Andrews (UK). She has published widely on the archaeology of Greece in the Roman and late antique periods, including *The Mosaics of Roman Crete: Art, Archaeology and Social Change* (2013). She is particularly interested in religious spaces, mobility and network analysis.

Lived Spaces in Late Antiquity

Edited by Carlos Machado,
Rowan Munnery and
Rebecca Sweetman

Routledge
Taylor & Francis Group
LONDON AND NEW YORK

First published 2024
by Routledge
4 Park Square, Milton Park, Abingdon, Oxon OX14 4RN

and by Routledge
605 Third Avenue, New York, NY 10158

Routledge is an imprint of the Taylor & Francis Group, an informa business

British Library Cataloguing-in-Publication Data
A catalogue record for this book is available from the British Library

ISBN: 978-1-138-38530-6 (hbk)
ISBN: 978-1-032-70435-7 (pbk)
ISBN: 978-0-429-42715-2 (ebk)

DOI: 10.4324/9780429427152

Typeset in Times New Roman
by Apex CoVantage, LLC

Contents

Figures

About the contributors

Michelle L. Berenfeld is the John A. McCarthy associate professor of classics at Pitzer College (Claremont, California) and a fellow of the American Academy in Rome. Her first book was *Aphrodisias XI: The Triconch House* (Reichert, 2019).

Will Bowden is a professor of Roman archaeology at the University of Nottingham (UK). He became an archaeologist through a UK government scheme for the unemployed in the 1980s before undertaking a BA at University College London and a PhD at the University of East Anglia. He has worked in Italy, the United Kingdom, Jordan and Albania.

Amelia R. Brown is a senior lecturer in Greek history and language at the University of Queensland, Australia. She will be Gertrude Smith professor at the American School in Athens in 2024 and published *Corinth in Late Antiquity: A Greek, Roman & Christian City* in 2018.

Gabriele Castiglia is an associate professor of topography of the ancient Christian world and methodology of archaeological research at the Pontifical Institute of Christian Archaeology (Rome). He has conducted archaeological excavations for 20 years, with a focus on Tuscany, Liguria, Calabria, Eritrea and Palestine. His primary research interests revolve around the dynamics of material formation and affirmation of Christianity in both urban and rural settings from the fourth century to the tenth century.

Javier Martínez Jiménez is a late antique archaeologist in the Department of Prehistory and Archaeology at the University of Granada (Spain). He specialises in the evolution of urbanism and water supplies from the late Roman into the early Umayyad period and has also worked on issues of identity and citizenship in the early Middle Ages. He has conducted field work in the Visigothic sites of Casa Herrera (Mérida) and Reccopolis.

Jason König is a professor of classics at the University of St Andrews. His work has focused especially on the Greek literature and culture of the imperial period and Late Antiquity. His recent publications include *The Folds of Olympus: Mountains in Ancient Greek and Roman Culture* (2022).

Carolyn T. La Rocco completed her PhD in ancient history at the University of St Andrews in 2023, examining the evidence for the 'Christianisation' of the Iberian Peninsula, reconsidering the role of episcopal and non-episcopal elites. She was a postdoctoral research fellow at the RomanIslam Center (Hamburg, Germany), examining urbanism in the late antique Iberia and North Africa. She is currently preparing her PhD thesis for publication.

Esen Ogus is a Harvard-trained art historian and classical archaeologist, focusing on the art, archaeology and visual culture of Asia Minor in the Roman Imperial period and Late

Antiquity. Her research has been funded by institutions including the Institute for Advanced Study (Princeton) and Alexander von Humboldt Foundation.

Rubina Raja is a professor of classical archaeology and art at Aarhus University, Denmark. She has worked extensively on urban developments, networks, iconography and religion in the Hellenistic to early medieval Near East and Mediterranean. Her latest monograph is *Pearl of the Desert. A History of Palmyra* (2022).

Christina Videbech is a part-time lecturer at the University of Bergen, Norway. She has specialised in disseminating classical archaeology to the public and worked on the Erasmus+ project "Ancient Cities", which developed a Massive Open Online Course. She is writing a popular scientific book on insanity in the Roman Empire, together with professor of psychiatry Poul Videbech titled *Sindets ild: Sindssygdom og galskab i antikken* (2024).

Nadine Viermann is an assistant professor of late antique and early Byzantine history at Durham University. Her first book was a structural analysis of the reign of Emperor Heraclius in the context of Late Roman monarchy. Currently, she works on the translation of relics in the late antique Mediterranean.

Enrico Zanini is a full professor of methodologies of archaeological research at the University of Siena, where he also teaches late antique and Byzantine archaeology. At present, he manages two archaeological fieldwork projects, in Gortys (Crete) (www.gortinabizantina.it) and Vignale (Tuscany) (www.uominiecoseavignale.it).

Acknowledgements

We would like to thank the School of Classics, University of St Andrews, as well as The Classical Association and the Institute of Classical Studies in London for their generous support for this project, including the conference that originated it. Sam Dixon, from St Andrews, provided invaluable assistance throughout. We are also grateful to all the colleagues who kindly accepted to review the manuscripts anonymously. Finally, we would like to thank Marcia Adams at Routledge, who supported this project since its inception.

1 Introduction

Lived spaces in Late Antiquity

Carlos Machado and Rebecca Sweetman

In 342 CE, the emperors Constantius II and Constans mandated a law regulating the use and preservation of temples in Rome to the urban prefect Aco Catullinus:

> Although all superstitions must be completely eradicated, nevertheless, it is Our will that the buildings of the temples situated outside the walls shall remain untouched and uninjured. For since certain plays or spectacles of the circus or contests derive their origin from some of these temples, such structures shall not be torn down, since from them is provided the regular performance of long established amusements for the Roman people.
>
> (*Cod. Theod.* 16.10.3)[1]

As the law shows, legislators at court and imperial officials on the ground were well aware that a temple was much more than just a single religious building occupying a discrete location in the city.[2] Temples embodied myths and cultural traditions, connecting different people, generations, and places through stories and rituals. Although authorities might express unease about their religious use, their continued importance in explaining and giving meaning to festivals, celebrations, and community traditions was acknowledged (Sweetman, 2015). Christian preachers were aware of this and often condemned these celebrations precisely for their religious associations (e.g. in North Africa: Van Slyke, 2005; Gassman, 2020). Temples were part of a complex of places, stories, and practices that were integrated by centuries-old religious traditions and performances. It was this broader significance that moved rulers like Constantius II and Constans to argue for their preservation.

The fate of ancient religious spaces in Late Antiquity has attracted increasing attention from specialists of different academic disciplines, working on all regions of the former Roman empire. Scholarly assessment is often contradictory, reflecting the complexity of the evidence available as much as intellectual preconceptions.[3] Whereas literary texts and especially Christian writings tend to highlight the destruction of old temples, legal codes record attempts at protecting their physical structure. Inscriptions attest to the restoration of temples but not their form of use nor their date of abandonment. Archaeological evidence is potentially more revealing, especially when excavators are interested in late and post-antique stratigraphic layers and building phases. It seems that at least in regions like Italy, Asia, Greece, and Asia Minor temples were still seen as a valuable feature of the landscape, treasured for their splendour, history, and symbolic value, and worthy of preservation wherever there was political will for it. However, as public funding declined and elites shifted their priorities to other types of cults and buildings (often inspired by their Christian preachers), temples fell into disuse and were progressively abandoned and, later, dismantled – a process that took place at different rhythms at diverse places. The spectacular destruction of a temple, like the Serapeum of Alexandria, or its direct (i.e. without an

DOI: 10.4324/9780429427152-1

intermediary phase of abandonment) conversion into a church, was as rare as it was culturally significant (Ward-Perkins, 2003; Hahn, Emmel, & Gotter, 2008; Sweetman, 2010; Schuddeboom, 2017). Whereas scholars used to see the fate of temples as a dramatic symbol of the rise of Christianity, this is now widely acknowledged as a complex and multifaceted process, determined by broader cultural and religious trends as much as by context-specific conditions.[4]

These debates about the physical condition of temples – their preservation, abandonment or conversion, and the attitudes of people in the late antique period to them – encapsulate the broader debates that have animated the field of late antique studies; between continuists or catastrophists, between advocates of the concept of decline or of transformation (Cameron, 2001, 2003). *Fora*, public buildings and even streets and their décor have been used by scholars to argue for change and continuity, providing evidence for the remarkable resilience or, alternatively, for the lamentable decline of Classical civilisation (Dey, 2014). In the case of the countryside, *villae* and other forms of residence of the elite are employed in similar debates.[5] Nowhere is this discussion more keenly seen than in the case of the cities of the eastern regions of Syria and Palestine, where the remarkable continuity of architectural forms amidst urban transformation has promoted vigorous debate about the limits of the ancient world and the transition to Islam (Kennedy, 1985, 2000; Avni, 2011, 2014). After decades of intensive investigation and considerations, the picture emerging from scholarship is one of multifaceted complexity: all around the Mediterranean, abandoned ruins lay next to restored or newly built palaces, public buildings, and Christian basilicas; *fora* and *agorai* were restored, but their appearance and forms of use were fundamentally changed; rural *villae* were partially abandoned and adapted to new uses that suggest continued economic vitality; new religious rituals gave new meaning to the urban and rural spaces, but long-running ceremonies and celebrations could still attract great crowds.[6]

Through multiple perspectives, *Lived Spaces in Late Antiquity* analyses the different ways in which the spaces where social life unfolded were transformed, and how these ever-shifting spaces in turn helped to shape social life. As we argue below – and the contributions in this book demonstrate – "space" offers a valuable perspective through which we can analyse the developments that marked Late Antiquity, in all its aspects. The period between the last decades of the third century and the beginning of the eighth century might, in fact, be seen as a time of emerging new spatialities – the ways in which space is culturally conceived, materially defined, and socially appropriated. New conceptions of sacred space were affirmed, informed by Christian stories and beliefs, drawing on past traditions about the relationship between humans and the supernatural and shaping distinctly late antique identities (Cormack, 1990). Urban architectural forms evolved from early imperial models, incorporating new cultural values and building techniques, responding to the needs and priorities of the societies that inhabited them (Dey, 2014). Traditional practices of investing in land in urban centres and in the countryside were redefined as much by the development of new economic strategies, as by the affirmation of new patrons – ones with their own priorities (Leone, 2003).

Focusing on space, rather than the more usual continuity-decline argument, highlights the need to consider how different and often contradictory historical dynamics coexisted in this process. As a category of analysis, space invites consideration of multiple processes across different regions and time periods and facilitates discussions of diverse themes such as economic, political, or religious change and of how they intersected. In the case of temples, mentioned earlier, the appreciation of religious change becomes more closely intertwined with the cultural and social elements that were involved in the development of a new attitude towards these structures, whether we are thinking of new forms of monumental building, new conceptions of sacred space, or a particularly strong attachment to historic monuments and works of art.

Therefore, considering Late Antiquity through a spatial perspective offers a nuanced, complex, and stimulating picture of this pivotal historical period.

1.1 Lived space

There are myriad definitions of space, perhaps as many as there are studies dedicated to it. Rather than obsolete, the definition of space *in itself*, as a grid where points are located, is still useful, as scholars still endeavour to identify what was where on a map. Such a positivist preoccupation, still crucial for understanding the topography of any site, does not mean that one can offer an objective and accurate description of space as it is or as it was. In fact, the very history of our disciplines is a history of how their methods, objectives, and assumptions changed over time – but the notion is still there.[7] And yet, it is safe to suggest that scholarship nowadays tends to assume a more problematic attitude towards space. For example, the influential study of Henry Lefebvre (originally from 1974) identified what he called a "conceptual triad", consisting of spatial practice, representations of space, and representational spaces (Lefebvre, 1991, 33). The contemporary scholar David Harvey distinguished absolute, relative, and relational space, only to reframe the question of "what is space?" as "how is it that different human practices create and make use of distinctive conceptualizations of space?" (Harvey, 2009 (or. ed. 1973), 13–14). Space is always qualified, even if not explicitly: metaphorical, economic, political, symbolic, urban, or religious, to name but a few (Harvey, 2004). The chapters collected in this book explore neighbourhoods, churches, ships, and even conceptual frameworks, illustrating this diversity.

Contributors to this volume approach space as a dynamic dimension of social life, both as an arena where social interactions took place and as an expression of these same interactions, whether in the form of buildings, concepts, or stories. Lived space is fluid and can change as quickly as night turns to day and as occupants and visitors come and go – or indeed as Rubina Raja illustrates in her chapter as the impact of natural disasters such as earthquakes impact on the physical history of a city. Just as in the case of temples and religious spaces that concerned imperial legislators, traditions, identities, rituals, and practices are invested in the social use and physical (re)definition of space, while they are simultaneously shaped by it. It is not our aim to present here a single definition of "space" as a sociological, philosophical, or geographical concept. Instead, this book's purpose is to bring together different perspectives to form a more nuanced understanding of the late antique period from the perspective of those who lived, constructed, and gave meaning to that space.

As an analytical perspective, "lived space" invites us to consider the challenge of integrating written and material evidence in all their forms (Manacorda, 2007). Space may be read in the archaeological record, in the distribution of settlements, the layout of a site, the architecture of a building, or the stratigraphy that records human activity therein. When considered as space, the material record can provide the starting point for our understanding of the past, both as an outcome of historical processes and as the necessary setting for them. Texts and inscriptions can help us reconstruct space: hagiographies, laws, letters, and historical accounts, as well as epitaphs, votive offerings, and monumental dedications, describe spaces and document their use and meaning. At the same time, their cultural and political significance might be more fully appreciated when their spatial context – whether described in the text or as the setting for reading or display – is integrated into our analysis. Whether we think of it as a methodological approach or as the subject of scholarly enquiry, "space" offers the opportunity to move beyond the apparent dichotomy between the "material" and the "mental", between "culture" and "reality" (Godelier, 2011; Sahlins, 1976).

"Lived space" has agency, shaping the actions of the different groups and individuals who frequent it.[8] Rather than a blank slate where patrons were free to build anew and artisans created freely, space carries the marks of previous generations. Whether these be in the form of a lavishly decorated building in a large city, tilled fields dotting the countryside, or the previously desert landscapes occupied by Christian ascetics, they bring together different pasts and temporalities. Be it real or imaginary, a concrete expression in the form of a building or a visual or written representation, space is a historical synthesis.[9] The experiences and interactions of late antique communities were thus shaped by the physical settings, established practices, and traditions inherited from previous generations, which were in turn adapted according to the priorities and conceptions of those communities. Potter (1995, 34–47) showed how the *forum* of Iol Caesarea (modern Cherchel) was adapted for new religious and commercial uses in the fifth century, including the employment of wooden structures amidst the splendour of earlier structures. A cityscape rich with marble ornaments, like Rome, offered numerous opportunities for architects and builders, who were able to reuse Classical columns and architraves in churches, or simply burn them to produce lime in the many kilns scattered throughout the city (Pensabene, 2015; Santangeli Valenzani, 2015). The study of space thus eschews the traditional opposition between catastrophist and continuist views of the period. Being confronted with the new and the old as part of the same totality, scholars are invited to consider how they are articulated, whether they are talking of temples next to churches or of workshops installed in once magnificent public buildings.

The studies collected here do not just consider the world of emperors, senators, and bishops. They also bring to the fore the actions and ambitions of workers, sailors, peasants, and monks, among other groups often neglected in our literary sources. Scholarship has recently made great progress in incorporating subaltern groups into our picture of the historical past, including in Late Antiquity (e.g. Grey, 2011; Magalhães de Oliveira, 2012). Religious processions and the pageantry of political power were an essential part of life, but it is by placing them in their spatial context, with competing groups and viewers side by side, that we can capture their full impact and significance (Latham, 2016; Falcasantos, 2020). This is not a new realisation. Already in the 1970s, Cameron had shown the importance of the circus as a space for the social, political, and cultural life of the empire – spanning from the early imperial to the late antique period (Cameron, 1976).

Space is, by definition, a totalising dimension that brings difference – social, economic, and religious – together. It does not deny inequality or conflict but rather expresses, gives meaning and engenders it. It can be felt in many domains: as workers need the tools and workshops of their patrons to make a living, as peasants visit their lords to pay their rent, and as lay worshippers attend the ritual parades of relics. Inequality and resistance are part of the same dynamic – one that is also brought together in space.[10] By focusing on its "lived" nature, we want to emphasise the multiple, complex, and oftentimes conflictual nature of space – one that resonates beyond cartographic representations and legal regulations. If the study of single monuments like churches and civic buildings is limited to a consideration of their patrons, commissioners, artists, and sometimes combined with the reactions of educated writers like Pausanias (for an earlier period) or Procopius, for example, the study of space poses the unavoidable challenge of considering society beyond its topmost strata.

By emphasising the "lived" quality of space, it is our aim to stress its multi-temporal and socially diverse aspects, while also acting to uncover the deeper and anonymous currents that defined Late Antiquity. Historians working on this period are naturally drawn to the high-profile agents, events, buildings, and processes that shaped the traditional narratives of Late Antiquity: Justinian and Gregory I, the battle of Hadrianople, the conversion of Augustine, the rise of the

Church and its splendid basilicas, the fragmentation of the empire in the West, and the development of new centres of power. While the interest and centrality of dramatic headliners cannot be denied, "lived space" allows us to reconsider them in their quotidian dimension, integrating their extraordinary character and their ordinary setting, illuminating how they impacted, and how they were impacted by the societies that lived through them.

1.2 Approaching space in late antique studies

In a broader sense, scholarly interest for late antique spaces is not new. If historians and literary scholars have only recently caught up with the need to pay proper attention to it, the late imperial phases of Classical spaces have for centuries attracted the attention of archaeologists and epigraphists. Scholars working on the nascent discipline of Christian Archaeology were more directly involved in these efforts, excavating churches and cemeteries, while attempting to trace the topographical impact of the new religion. For example, design and decoration of the Christian catacombs of late antique Rome were the subject of comprehensive investigation since the works of Antonio Bosio in the sixteenth century. This was an effort that had already reached its maturity with the publications of De Rossi in the second half of the nineteenth century – while forefronting a process that took place across the Mediterranean region. Interest for the ancient and late antique monuments of Constantinople stretches back to the sixth and seventh centuries, if not earlier, and their study was already gaining momentum in the centuries following the conquest of Mehmed II (Berger, 2013; Bassett, 2022). Much like with the broader field of late antique studies, however, attention to specifically late antique spaces was limited and narrow. In general, space was approached variously as a grid, where scholars located churches and secular structures, or as an architectural form, as a point in the map, where information (buildings, decoration, inscriptions) was retrieved. In this sense, the secular perspective of antiquarians, interested in collecting as much information about the past as possible, remained influential.

It was from the middle of the twentieth century onwards, as archaeologists became more interested in framing their research as studies of socio-economic and cultural processes, that this attitude to space began to change. The South Etruria survey (started in the 1950s) in Italy, the modern systematic exploration of Aphrodisias in Turkey (1961), and the UNESCO-sponsored excavations of Carthage (1973) in Tunisia, which paid great attention to late antique and early medieval phases, are good examples of this new perspective, revealing trends in settlement patterns and forms of spatial occupation. Combined with this was the growing adoption of stratigraphic methods in archaeological campaigns, which provided not only rich insights into the post-antique phases of a site but also invaluable information about human activity in the area. This is eloquently illustrated by the excavations in the 1980s of the Crypta Balbi in Rome. This was a project that went beyond narrow interest in specific features, instead examining the history of a whole quarter in the Campus Martius from its initial use in the last decades of the first century BCE to the present. As well as a rich succession of physical structures that occupied the site and their uses, the project provided for the first time a coherent picture of the history of Rome's material culture (or aspects of it, at least).

The so-called spatial turn, the development of a more nuanced understanding of space and of its significance for our understanding of social life, played a key role in this process. Crucially, this theoretically informed definition of space came to play a more central role in different disciplines (Withers, 2009). Partly encouraged by developments in archaeological theory, particularly in cognitive archaeology, spatial analysis has become a fundamental means for understanding the use of space over time – placing material culture within its spatial context (Seibert, 2006). Spatial analysis has been combined with quantitative methods such as GIS, network analysis, and actor-network theory, helping to marshal data, encourage wider interdisciplinary engagement, and enabling broader perspectives (Verhagen, 2018). Scholars like Luke Lavan (e.g. 2003) have focused on the notion of "activity spaces", seeking to emphasise the functional and contextual nature of space, while others have highlighted the importance of the senses in considering how dwellings and churches were experienced by contemporaries (Ellis, 2005; Harvey, 2006).

Space has increasingly been approached as multifunctional and relational, considered from the perspective of different users, including residents, visitors, officiating priests, and patrons, eliciting diverse responses. Oftentimes in dialogue with archaeologists, literary scholars and historians working on this period have also embraced space as a perspective and a subject of enquiry.[11] This has been particularly influential in the study of late antique historiography and of geographical writings, like Ammianus Marcellinus and Cosmas Indicopleustes, among others.[12] Although ancient historians working on cities have traditionally engaged with geographical and topographical issues (e.g. Pietri, 1983; Sotinel, 2005), in recent years a more theoretically informed engagement with space has been seen, whether using specific cities as context for their analyses (e.g. Shepardson, 2019; Machado, 2019) or focusing on issues like patriarchal and episcopal power (Cooper, 2007; Sessa, 2011). Whether it be in terms of theoretical elaborations, methodological developments, or scholarly interest, the time is ripe for an in-depth exploration of the world of Late Antiquity from the perspective of space and the people who inhabited, built, adapted, and gave meaning to it.

1.3 Lived spaces in Late Antiquity: aims and scope

The chapters presented here analyse the role played by different types of space, across the late antique world, in the lived experience of late antique societies. Our contributors, where possible, have taken a phenomenological approach to space rather than following architectural determinism; they approach their material in an exploratory way with a focus on agency and, in doing so, allow the unexpected to shine through. We do not seek to present an exhaustive picture of space in all its many uses, social definitions, and physical appearance over such a broad chronological and geographical span. Our chapters do not cover all the wealth of agents, developments, or locations that shaped the world of Late Antiquity. Instead, we consider "lived space" as a scholarly approach to the past, showing how spatial approaches can present innovative views of this period, integrating social, economic, and cultural developments and putting centre stage this fundamental dimension of social life.

True to the spatial turn, we as editors embraced the perspective that definitions of space are so multifarious that, rather than adopting a clear-cut definition, the best approach is to explore the lived experience of late antique societies from different theoretical standpoints and concepts. We were keen to investigate the fluidity of lived space in both its physical and conceptual dimensions, analysing issues like agency and intentionality as well as meaning and social relations. Contributors to this book have sought the original intention for a space (both practical and conceptual), while at the same time allowing for the practicalities to override the primary purposes, whether that be in moving statues or changing processional routes. They have analysed the social relations and natural catastrophes that gave form to space, as well as the cultural values and political agendas that gave meaning to it.

When we as scholars approach the study of space, it is important for us to try to understand our own role in the interpretation that we put forward. We acknowledged this in this volume by inviting contributions from colleagues of many different career stages, disciplines, and academic traditions, working on areas as diverse as Britain, the Iberian Peninsula, and Jordan, as well as the horn of Africa. The sea, the shores, and the ships that traversed them – a burgeoning field of study – deserve particular attention, as well as the city of Rome, a dramatic symbol of change in this period. Our initial plan consisted of holding a conference that would lead to the final volume but, like much else in the past few years, this was impacted by the COVID-19 pandemic. We were sorry that some of the brilliant colleagues who contributed papers to the online workshop could not submit them for the book, but the experience of living in lockdown added even greater poignancy to the need to address our subject. Rather than trying to fill perceived gaps caused by these circumstances, we opted for fostering conversations among all the authors, workshopping and circulating the papers as a way of ensuring that our plurality of views did not prevent the unity of focus.

This book is organised in four parts, defined according to the issues raised in our conversations with the different contributors. Their aim is not only to provide rich insights into crucial issues related to the analysis of lived spaces but also to invite readers to engage in the self-reflective exercise of thinking about our subjects and also about how we think about them as inseparable aspects of our analysis. Part 1, "Visualising late antique spaces", is dedicated to the issue of how modern scholars *and* their late antique sources conceptualise and conceive lived space. This issue is directly discussed in Will Bowden's "Lost in space? Finding the people in late antique archaeology", which draws on the author's own intellectual trajectory, from excavation to digital reconstruction, to consider the issue of how we can populate space as a way of giving meaning to it. Bowden raises the issue of social diversity and its (in)visibility in late antique studies, one of the recurring threads in this book. In "Suburban saints: space, place and environment in Theodoret's *Religious History*", Jason König examines the rich conceptualisation of space presented in Theodoret's work, exploring how space was conceived and represented in late antique texts. As different contributions to this volume argue, the issue of lived space is intimately connected to the issue of its scale. "Living with neighbours in late antique Rome", by Michelle L. Berenfeld, examines a specific neighbourhood of Rome, the Caelian, as a form of reconstruing the social fabric of this part of the city. Similar issues are considered by Rowan Munnery in "Ship-shape and Roman fashion: space at sea in Late Antiquity", where he reconstructs both the ship as a specific, physically defined type of space and the web of meanings and social relations formed within it.

Part 2, "Place-making and emplacement: the impact of religion", builds on the conceptions and perceptions of space discussed in the previous chapters to look at how space is created. Specifically, the impact of religion in the making of space is the focus of the chapters presented here, as Christianity played a crucial role in the development of specifically late antique spatialities. In order to examine this issue, Gabriele Castiglia, in "Space(s) in transition: the impact of early Christianity on the late antique Horn of Africa", analyses how different spatial scales were reconfigured in an area outside the bounds of the Roman Empire, where the constraints of Classical urbanism are only faintly perceptible. The issue of who is legitimately entitled to create religious space is considered by Carolyn T. La Rocco, in "Making space in the Visigothic Kingdom: church founders in sixth- and seventh-century Iberian epigraphy", who shows that, contrary to the simplistic assertions of ecclesiastical sources, different social groups could be involved in the foundation of churches. Creating religious space was not just a matter of bricks and mortar, however, but also of the ways through which it could be made effective, a topic examined in "Holy objects on the move: relics in Constantinople between city centre and urban periphery", by Nadine Viermann. The cultural significance of place-making is analysed in Rebecca Sweetman's "Space and place: late antique churches and place remaking", where issues of memory and identity are reconsidered in light of this period's context and experience of migration and displacement.

The making and remaking of spaces bring the issue of meaning to the fore, or what Lefebvre called "representational space", and this is the focus of Part 3, "Space and meaning". Here, the issue of how space is constructed is reassessed in terms of its symbolic repercussions, from three different perspectives. In "Entanglement of public space and honorific statue habit in late antique Asia Minor", Esen Ogus considers how honorific statues shaped spaces and expressed social and political hierarchies, while at the same time being adapted to suit their contexts of display. The dynamic nature of the statue habit makes it an optimal strategy to analyse long-existing cities, but we should not neglect new foundations and their ability to foster new senses of community and identity, as Javier Martínez Jiménez shows in "*Civitatem condidit.* City building and community formation in the new Visigothic urban foundations". While Esen Ogus and Javier Martínez Jiménez look at settled communities, Amelia R. Brown, in "From *Theoria* to pilgrimage: ships, shores and sacred travel around the Mediterranean sea in Late Antiquity", focuses on travellers, both pilgrims and sailors, and how their religious commitments were part of their process of giving meaning to the shipboard spaces in which they lived and moved, as well as the spaces of the harbours and sanctuaries their voyages were directed towards.

Rather than being isolated, the issues of visualising, making, and giving meaning to space cut across all chapters of this book, regardless of their specific focus. The same is true of our Part 4, "Changing spaces", as it highlights the issue of how – and to what extent – space changes and how this can be read in the evidence available. Christina Videbech looks at two of the most represented areas in the entire late antique world in "Clean death or messy resilience? *Forum Traiani* and *Forum Romanum* as activity spaces during the sixth century" using the perspective of activity spaces, exploring how spaces and their uses changed in Late Antiquity – and the implications of this fact for how we view the period. Change is a defining characteristic of cities in the East, as Rubina Raja shows in "Lived spaces in late antique and early Islamic Jerash – old questions in the light of new evidence", confronting the issue of how to identify and visualise change. The relationship between continuity and transformation is examined over the very long term in " 'Places of continuity': before, around and after the end of the Roman villas in Central Italy", by Enrico Zanini, who shows how different historical periods coexist and are rearranged in one single site. Finally, Carlos Machado analyses the relationship between space and social life in "Lived space and social change in late antique Rome: the Campus Martius", looking at how different social groups gave shape to and were shaped by the changing topography of a busy part of Rome at a time of unprecedented transformations.

We hope that this volume stimulates discussions in the study of space in Late Antiquity and helps to advance the development of approaches related to this perspective. We conceived this book as a series of deep explorations through a spatial lens into the historical diversity that marked this period; as a way of probing the limits and contributions of this type of approach; and as the contributions presented here clearly demonstrate, the picture that emerges of the Mediterranean world and its environs is as fascinating as it is original.

Notes

1 "Quamquam omnis superstitio penitus eruenda sit, tamen volumus, ut aedes templorum, quae extra muros sunt positae, intactae incorruptaeque consistant. Nam cum ex nonnulis vel ludorum vel circensium vel agonum origo fuerit exorta, non convenit ea convelli, ex quibus populo Romano praebeatur priscarum sollemnitas voluptatum" (transl. Pharr). For Catullinus, see *PLRE* I, Catullinus 2.

2 See, for a stimulating discussion of temples in the sense considered here, MacMullen (1983, 34–42). For a general discussion, with recent bibliography, see Caseau (2022).

3 See the essays collected in Lavan & Mulryan (2011), for a comprehensive overview of the bibliography and the main issues involved.

4 For the impact of Christianity, see Fowden (1978) and Saradi-Mendelovici (1990). For a more nuanced view, see Caseau (2001) and Sotinel (2010).
5 See, on cities and their spaces, Liebeschuetz (2001); Jacobs (2013); Christie & Augenti (2012). The essays collected in Christie (2004), and Bowden, Lavan, & Machado (2004), offer useful introductions for the debate on the countryside.
6 Lavan (2021), provides an essential overview of these issues and the evidence available.
7 This perspective is present in the works of Foucault, most specifically the essays collected in Crampton & Elden (2007). For ancient studies (but from a different perspective), see Coarelli (1982) and Purcell (1989).
8 Lefebvre (1991, 11) refers to the "active – operational or instrumental – role of space, as knowledge and action".
9 In the sense discussed by Koselleck in his discussion of Albrecht Altdorfer's Alexanderschlacht: Koselleck (2004, 9–25).
10 Santos (2021, 215–228); Harvey (2009); see also Yeoh (2003).
11 See, for an excellent interdisciplinary example, Grig & Kelly (2012).
12 See, on Ammianus, Sundwall (1996), for example, and the studies collected in Van Nuffelen (2019) in general; on Cosmas and his tradition, Kominko (2013) is essential. See also Belanger (2020).

Editions and abbreviations

Cod. Theod – *Codex Theodosianus*, in *Theodosiani Libri XVI cum Constitutionibus Sirmondianis*, ed. T. Mommsen & P. Meyer (1905); translated in C. Pharr (1952), *The Theodosian Code and Novels, and the Sirmondian Constitutions*.

Bibliography

Avni, G. (2011). From Polis to Madina revisited – urban change in Byzantine and early Islamic Palestine. *Journal of the Royal Asiatic Society*, *21*, 301–329.

Avni, G. (2014). *The Byzantine-Islamic Transition in Palestine. An Archaeological Approach*. Oxford University Press.

Bassett, S. (2022). Introduction. In S. Bassett (Ed.), *The Cambridge Companion to Constantinople* (pp. 1–15). Cambridge University Press.

Belanger, C. (2020). Echoes of empire: Roman imperialism in Avienus's *Descriptio orbis terrae*. *Journal of Late Antiquity*, *13*(2), 193–219.

Berger, A. (2013). *Accounts of Medieval Constantinople: The Patria*. Dumbarton Oaks Medieval Library.

Bowden, W., Lavan, L., & Machado, C. (Eds.). (2004). *Recent Research on the Late Antique Countryside* (Late Antique Archaeology 2). Brill.

Cameron, A. (1976). *Circus Factions: Blues and Greens at Rome and Byzantium*. Oxford University Press.

Cameron, A. (2001). The 'long' Late Antiquity: A twentieth-century model? In T. P. Wiseman (Ed.), *Classics in Progress* (pp. 165–191). Oxford University Press.

Cameron, A. (2003). Ideologies and agendas in late antique studies. In L. Lavan & W. Bowden (Eds.), *Theory and Practice in Late Antique Archaeology* (Late Antique Archaeology 1) (pp. 3–21). Brill.

Caseau, B. (2001). *Polemein Lithois*. La désacralisation des espaces et des objets religieux païens durant l'antiquité tardive. In M. Kaplan (Ed.), *Le sacré et son inscription dans l'espace à Byzance et en Occident* (pp. 61–123). Publications de la Sorbonne.

Caseau, B. (2022). Spaces of Roman religion and Christianity in late antiquity. In J. Halgren Kilde (Ed.), *The Oxford Handbook of Religious Space* (Chapter 20). Oxford University Press. DOI: 10.1093/oxfordhb/9780190874988.013.24

Christie, N. (Ed.). (2004). *Landscapes of Change: Rural Evolutions in Late Antiquity and the Early Middle Ages*. Ashgate.

Christie, N., & Augenti, A. (Eds.). (2012). *Urbes Extinctae. Archaeologies of Abandoned Classical Towns*. Ashgate.

Coarelli, F. (1982). Topographie antique et idéologie moderne: le *Forum* romain revisité. *Annales. Économies, Sociétés, Civilisations, 37e année, 5–6*, 724–740.

Cooper, K. (2007). *The Fall of the Roman Household*. Cambridge University Press.

Cormack, R. (1990). Byzantine Aphrodisias: Changing the symbolic map of a city. *Proceedings of the Cambridge Philological Society, 216*, 26–41.

Crampton, J., & Elden, S. (Eds.). (2007). *Space, Knowledge and Power: Foucault and Geography*. Routledge.

Dey, H. (2014). *The Afterlife of the Roman City*. Cambridge University Press.

Elden, S., & Crampton, J. (Eds.). (2007). *Space, Knowledge and Power: Foucault and Geography*. Routledge.

Ellis, S. (2005). Shedding light on late Roman housing. In L. Lavan, L. Özgenel, & A. Sarantis (Eds.), *Housing in Late Antiquity* (Late Antique Archaeology 3.2, pp. 283–302). Brill.

Falcasantos, R. (2020). *Constantinople: Ritual, Violence, and Memory in the Making of a Christian Capital*. University of California Press.

Fowden, G. (1978). Bishops and temples in the eastern Roman empire, AD 320–435. *Journal of Theological Studies*, *29*(1), 53–78.

Gassman, M. (2020). A feast in Carthage: Testing the limits of 'secularity' in Late Antiquity. *Journal of Roman Studies*, *110*, 199–219.

Godelier, M. (2011). *The Mental and the Material*. Verso Books.

Grey, C. (2011). *Constructing Communities in the Late Roman Countryside*. Cambridge University Press.

Grig, L., & Kelly, G. (Eds.). (2012). *Two Rome. Rome and Constantinople Compared*. Oxford University Press.

Hahn, J., Emmel, S., & Gotter, U. (Eds.). (2008). *From Temple to Church. Destruction and Renewal of Local Cultic Topography in Late Antiquity*. Brill.

Harvey, D. (2004). Space as a Key Word. Paper for Marx and Philosophy Conference, 29 May, Institute of Education, London. Available at https://frontdeskapparatus.com/files/harvey2004.pdf (accessed 28 August 2023).

Harvey, D. (2009). *Social Justice and the City*. Revised edition. The University of Georgia Press. (or. ed. 1973).

Harvey, S. A. (2006). *Scenting Salvation. Ancient Christianity and the Olfactory Imagination*. University of California Press.

Jacobs, I. (2013). *Aesthetic Maintenance of Civic Space. The 'Classical' City from the 4th to the 7th c AD* (Orientalia Lovaniensia Analecta 193). Peeters.

Kennedy, H. (1985). From *polis* to *madina*: Urban change in late antique and early Islamic Syria. *Past and Present*, *106*, 3–27.

Kennedy, H. (2000). Gerasa and Scythopolis: Power and patronage in the Byzantine cities of Bilad Al-Sham. *Bulletin d'études orientales*, *52*, 199–204.

Kominko, M. (2013). *The World of Kosmas. Illustrated Byzantine Codices of the 'Christian Topography'*. Cambridge University Press.

Koselleck, R. (2004). *Futures Past: On the Semantics of Historical Time*. Columbia University Press.

Latham, J. (2016). *Performance, Memory, and Processions in Ancient Rome: The 'Pompa Circensis' from the Republic to Late Antiquity*. Cambridge University Press.

Lavan, L. (2003). Late antique urban topography: From architecture to human space. In W. Bowden & L. Lavan (Eds.), *Theory and Practice in Late Antique Archaeology* (Late Antique Archaeology 1, pp. 171–195). Brill.

Lavan, L. (2021). *Public Space in the Late Antique City*, 2 vols. Brill.

Lavan, L., & Mulryan, M. (Eds.). (2011). *The Archaeology of Late Antique 'Paganism'* (Late Antique Archaeology 7). Brill.

Lefebvre, H. (1991). *The Production of Space*. Blackwell. (or. ed. 1974)

Leone, A. (2003). Topographies of production in North African cities during the Vandal and Byzantine periods. In W. Bowden & L. Lavan (Eds.), *Theory and Practice in Late Antique Archaeology* (Late Antique Archaeology 1, pp. 257–287). Brill.

Liebeschuetz, J. H. W. G. (2001). *The Decline and Fall of the Roman City*. Oxford University Press.
Machado, C. (2019). *Urban Space and Aristocratic Power in Late Antique Rome, AD 270–535*. Oxford University Press.
MacMullen, R. (1983). *Paganism in the Roman Empire*. Yale University Press.
Magalhães de Oliveira, J. C. (2012). *'Potestas Populi'. Participation populaire et action collective dans les villes de l'Afrique romaine tardive (ver 300–430 apr. J.-C.)* (Bibliothèque de l'Antiquité Tardive, 24). Brepols.
Manacorda, D. (2007). Fonti archeologiche e fonti scritte: vent'anni dopo *Le vin de l'Italie romaine* di André Tchernia. *Dimensioni e problemi della ricerca storica*, *2*, 85–100.
Pensabene, P. (2015). *Roma su Roma. Reimpiego architettonico, recupero dell'antico e trasformazioni urbane tra il III e il XIII secolo* (Monumenti di Antichità Cristiana, 2nd series, 22). Pontificio Istituto di Archeologia Cristiana.
Pietri, L. (1983). *La ville de Tours du IVe au VIe siècle. Naissance d'une cité chrétienne*. École Française de Rome.
Potter, T. (1995). *Towns in Late Antiquity: Iol Caesarea and Its Context*. University of Sheffield.
Purcell, N. (1989). Rediscovering the Roman *forum*. *Journal of Roman Archaeology*, *2*, 156–166.
Sahlins, M. (1976). *Culture and Practical Reason*. The University of Chicago Press.
Santangeli Valenzani, R. (2015). Calcare ed altre tracce di cantiere, cave e smontaggi sistematici degli edifici antichi. In A. Molinari, R. Santangeli Valenzani, & L. Spera (Eds.), *L'Archeologia della produzione a Roma (Secoli V-XV)* (pp. 335–344). École française de Rome/EDIPUGLIA.
Santos, M. (2021). *The Nature of Space*. Duke University Press.
Saradi-Mendelovici, H. (1990). Christian attitudes toward Pagan monuments in Late Antiquity and their legacy in later Byzantine centuries. *Dumbarton Oaks Papers*, *44*, 47–61.
Schuddeboom, F. (2017). The conversion of temples in Rome. *Journal of Late Antiquity*, *10*(1), 166–186.
Seibert, J. (2006). Introduction. In E. C. Robertson, J. D. Seibert, D. C. Fernandez, & M. U. Zender (Eds.), *Space and Spatial Analysis in Archaeology* (pp. xiii–xxiv). University of Calgary Press.
Sessa, K. (2011). *The Formation of Papal Authority in Late Antiquity. Roman Bishops and the Domestic Sphere*. Cambridge University Press.
Shepardson, C. (2019). *Controlling Contested Places. Late Antique Antioch and the Spatial Politics of Religious Controversy*. University of California Press.
Sotinel, C. (2005). *Identité civique et christianisme: Aquilée du IIIe au VIe siècle*. École Française de Rome.
Sotinel, C. (2010). The end of Pagan Holy places in the west: Problems and method. In C. Sotinel (Ed.), *Church and Society in Late Antique Italy and Beyond* (Chapter XI). Ashgate.
Sundwall, G. (1996). Ammianus Geographicus. *American Journal of Philology*, *117*(4), 619–643.
Sweetman, R. (2010). The Christianization of the Peloponnese: The topography and function of late antique churches. *Journal of Late Antiquity*, *3*(2), 203–261.
Sweetman, R. (2015). Memory, tradition, and Christianization of the Peloponnese. *American Journal of Archaeology*, *119*(4), 501–531.
Van Nuffelen, P. (Ed.). (2019). *Historiography and Space in Late Antiquity*. Cambridge University Press.
Van Slyke, D. (2005). The devil and his Pomps in fifth-century Carthage: Renouncing spectacula with spectacular imagery. *Dumbarton Oaks Papers*, *59*, 53–72.
Verhagen, P. (2018). Spatial analysis in archaeology: Moving into new territories. In C. Siart, M. Forbriger, & O. Bubenzer (Eds.), *Digital Geoarchaeology. Natural Science in Archaeology* (pp. 11–25). Springer.
Ward-Perkins, B. (2003). Reconfiguring sacred space: From Pagan shrines to Christian churches. In G. Brands & H.-G. Severin (Eds.), *Die spätantike Stadt und ihre Christianisierung* (pp. 285–290). Reicher Verlag.
Withers, C. (2009). Place and the "spatial turn" in geography and history. *Journal of the History of Ideas*, *70*, 637–658.
Yeoh, B. (2003). *Contesting Space in Colonial Singapore. Power Relations and the Urban Built Environment*. Singapore University Press.

Part 1

Visualising late antique spaces

2 Lost in space? Finding the people in late antique archaeology

Will Bowden

2.1 Introduction – a personal story

I have been involved with studying the archaeology of Late Antiquity for almost 30 years, starting from the moment in 1994 when I began to work at Butrint in Albania as part of the Butrint Foundation team led by Richard Hodges. My work on the town's splendid baptistery stimulated my growing interest in the changes that took place between the fourth and sixth centuries. Why were other parts of the town apparently falling into disuse at the same time as this remarkable building was being erected? At the same time, along with other members of the team, I was struck by the dramatic changes that were occurring in post-communist Albania, in which an entire political and economic system was comprehensively dismantled in the space of a few years. We were surrounded by the wreckage of Enver Hoxha's regime – abandoned factories, the ruins of collective farms and towns still dominated by concrete socialist monuments depicting the triumph of communism. It was not lost on us that we were investigating a transitional period in the Roman past while living in the middle of ongoing dramatic change. This was also an era of profound (and naïve) Western self-satisfaction. Fukuyama (1992) had told us that history had ended with the triumph of liberal democracy. We were the victors coming to Albania to harvest the spoils.

My work at Butrint led to my PhD on late antique Epirus, in which I documented the apparent changes in the towns and countryside of north-west Greece and southern Albania in the late Roman period (Bowden, 2000, 2003). Like many others in this period I was particularly fascinated by an urban change and my thinking was profoundly influenced by a number of books that came out in the 1990s (Carver, 1993; Rich, 1992; Potter, 1995; Christie & Loseby, 1996; Brogiolo & Ward-Perkins, 1999). A wave of new publications emerged in the early 2000s, in the Anglophone world driven particularly by the determination of Luke Lavan, whose *Late Antique Archaeology* conferences, volumes and now journal formed the vehicle in which a wide range of established and emerging scholars (including me) expressed new ideas about the material remains of Late Antiquity (e.g. Lavan, 2001; Lavan & Bowden, 2003).

In 2005, I began a new project at Venta Icenorum (Caistor-by-Norwich) in the United Kingdom. I was interested in taking some of the things that I had been exploring in the heart of the Mediterranean out to the barbarian fringes of the Roman Empire. How did Late Antiquity play out in Britain? Working in Britain made me look at the Mediterranean again, not only in terms of the archaeology but also in terms of disciplinary approaches. Over the past few years I have begun to suspect that late antique archaeology in the Mediterranean, particularly that relating to towns, has gone down something of a disciplinary *cul-de-sac* (Bowden, 2020a). The reasons for this lie at the heart of the whole notion of late antique archaeology and this raises questions for us as archaeologists. Archaeology as a whole also faces some significant challenges in relation

DOI: 10.4324/9780429427152-3

to inclusivity and representation, and this is also an area where debates about late antique space are found profoundly wanting. This chapter is my reflection on these challenges. Although I am deliberately setting out to question our practice as archaeologists, I do not wish to downplay the often-sophisticated work carried out by my colleagues and apologise in advance if I appear to oversimplify their thoughtful and nuanced approaches and interpretations for the sake of a rhetorical position.

2.2 A brief history of late antique archaeology

The historiography of Late Antiquity and late antique archaeology is well-trodden ground (e.g. Cameron, 2003; Saradi, 2006, 14–45; Humphries, 2019, 5–16; Dey, 2014, 5–10). It has become canonical that the negative perceptions of the late Empire emerged in the Enlightenment before being encapsulated by Gibbon and his successors who distrusted the increased intertwining of Church and State. This distrust was notoriously echoed in Jones's 'idle mouths' comment in reference to the monks and priests of late Roman Christianity (Jones, 1964, 933). This narrative was famously challenged by Brown, who reinvented the late Empire as an exciting and colourful period of Christian-inspired artistic and cultural invention, also challenging the latent orientalism implicit in many perceptions of the period (Brown, 1971).

Although with hindsight Brown's *The World of Late Antiquity* is seen as a turning point in the study of the late Empire, its effect on archaeology and archaeological practice was not immediate. Brown did not engage with archaeology and, as Fowden has pointed out, used material remains 'primarily as illustrations . . . for an argument based overwhelmingly on texts' (Bowersock et al., 1997, 44). *The World of Late Antiquity*'s use of copious images undoubtedly contributed to its impact on archaeologists in conveying the material richness of Late Antiquity. However, Brown's focus on Christianity found less traction in the secular world of Anglophone archaeology, Bryan Ward-Perkins noting his 'bewilderment' at the centrality of religion in scholarship on Late Antiquity (Ward-Perkins, 2005, 180).

In relation to late antique archaeology, Hayes's *Later Roman Pottery* (1972) arguably had more practical impact than *The World of Late Antiquity*. Hayes provided archaeologists with a robust means of dating late antique levels in the Mediterranean, where previously they had often relied on numismatic evidence, or (more commonly) often questionable associations with historically attested events such as earthquakes or barbarian incursions.

In Anglophone circles at any rate, the 1980s saw the rise of the so-called New Archaeology in which archaeologists sought to create disciplinary space between themselves and text-based historical narratives to enable archaeology (and newly emerging archaeology departments in UK universities) to carve out a new disciplinary identity. A range of exciting new theoretical approaches was adopted to enable archaeologists to use their evidence to address major questions of cultural change in ways untrammelled by narrative history. The adherents of these new methods saw archaeology as a scientific discipline in which hypotheses could be established and tested through the application of methodological rigour. Complex jargon and terminology was employed aggressively to emphasise the novelty of the approach and establish disciplinary distance between the old and the new (Johnson, 1999, 12–47). In relation to the Roman period, the influence of this 'New Archaeology' was largely restricted to field survey, which developed its own set of disciplinary mores (Bowden & Lavan, 2004, xxi–xxii).

Roman archaeology in general has never been quick to embrace change and, particularly in the Mediterranean, those encountering late Roman levels throughout the 1970s and 1980s (and up to the present day in some quarters) often continued to prioritise a restricted canon of material that could be readily accommodated within the traditional approaches of classical archaeology. Such materials (e.g. sculpture, inscriptions, mosaics, architecture, fortifications and urban infrastructure) could be treated act as an adjunct to text-based political history. Alternatively, they could be studied within an art-historical paradigm that prioritised stylistic and typological categorisation and chronologies derived from such approaches.

However, other areas of archaeology were changing. The increasing professionalisation of field archaeology, as a response to the unprecedented threats to archaeological deposits resulting from widespread development of cities and infrastructure, led to a refinement of field techniques. In Britain, Barker's Wroxeter excavations, with their pioneering fine-grained open-area approach, famously led to the identification of Wroxeter's post-Roman sequence, the nature and existence of which continues to be debated to this day (Lane, 2014; White, 2014).

The techniques employed at the Wroxeter excavations convinced British archaeologists that they had the methodologies to address questions about post-Roman urban continuity that could not be addressed by textual history. Similar developments were occurring elsewhere. In relation to the post-Roman towns of northern Italy, a fierce debate broke out between the proponents of different interpretations of the evidence for continued urban life or its absence, the interpretations often depending on the intellectual or national backgrounds of the writers in question. Ward-Perkins noted that different writers employed strikingly different descriptions of exactly the same wall excavated at Via Dante in Verona, contrasting his own description of it as a 'structure [with a] rough, mortared stone façade' with that of Carandini who described it as a 'house . . . at least in part of stone, although of hair-raising quality' (Ward-Perkins, 1997, 165–167).

Notwithstanding these developments, as Lavan (2015a) notes, the application of fine-grained stratigraphic excavation in the late antique Mediterranean has been patchy with many archaeological approaches to classical sites continuing to use techniques that prioritise the recovery of the traditional targets of classical archaeology, such as public buildings (including churches), high-status dwellings, inscriptions and sculpture. Lavan (2015a, 2–4) also noted that the disciplinary structures built around networks of academic patronage that dominate Mediterranean archaeology actively work to maintain the status quo. Many academic traditions of archaeology remain deeply hierarchical with students and post-graduates dependent on the continued patronage of their professors for career advancement. In such situations, advocating changes in methodology or research priorities or questioning an established paradigm can prove hazardous!

Alongside new excavations, a generation of scholars (including me) realised that the large-scale excavations of the twentieth century on a range of major classical urban sites in the Mediterranean had created a huge dataset that was ripe for re-examination. This approach involved detailed reappraisal of previously excavated public areas of Roman towns that allowed scholars to reassess the changing nature of late antique urban space, often examining the rather ephemeral traces of phenomena that had escaped the spades of earlier generations (Lavan, 2015b).

The advantages of this approach are obvious in that it is very low cost (involving little more than travel and subsistence expenses) and entirely non-destructive. At its simplest, it is also politically straightforward as it involves little more than wandering around sites in the manner of a particularly thoughtful tourist, unless undertaking more detailed survey. In combination with detailed trawling of earlier publications, such approaches have led to important reappraisals of sites with long excavation histories (e.g. Corinth: Brown, 2018; Sagalassos: Lavan, 2015c; Ostia: Gering, 2015; Gerasa: Raja, this volume). Similar approaches have been adopted that have focused on particular buildings or features (e.g. the study of statues and epigraphy in late antique Rome such as Chenault [2012] or Machado [2021] and Lavan's [2021] *magnum opus* on the fate of public space in Late Antiquity), or alternatively taking a broader province-wide approach to the issue (Bowden, 2003).

This work has undoubtedly driven a wholesale redefinition of our understanding of urban space in Late Antiquity. However, by its focus on reappraisal of previously (often poorly) excavated sites and the limited canon of material that earlier excavators deemed worthy of study and retention, this work has arguably become trapped within the same paradigm that shaped those earlier excavations. By this, I mean that the nature of the evidence means that archaeologists working on late antique towns inevitably concentrate on the behaviours, interests and aspirations of social and political elites, which were the foci of the earlier excavations. Equally inevitable is the comparison with the city of earlier periods. The monumental city of the late Republic and early Imperial period was the focus of research by earlier excavators and now forms the canvas on which we paint the late antique city, with the discussion framed in terms of deviation from the perceived norms of early imperial monumentality. While we have (generally) moved from discussions of decline towards ostensibly more neutral terms such as transition or transformation, the object of study – the nature of the cityscape as a vehicle of elite social expression – remains broadly the same (e.g. Dey, 2014; Machado, 2019).

I am not, of course, unique in criticising the habit of late antique archaeologists of focusing on space rather than those who inhabited that space. Similar points have been made by others (e.g. Christie, 2020; Dey, 2014, 9; Humphries, 2019, 64–65; Christina Videbech, this volume). However, it is a point worth repeating, not least because late antique archaeology seems to have entered something of a period of stasis, leaving those who work with other sources to populate the towns with the 'teeming humanity that is mute in archaeological plans, but which can be appreciated in late antique iconographic and textual depictions' (Humphries, 2019, 65). Indeed significant work has appeared in recent years that explicitly focuses on the lives and experiences of the urban poor (e.g. Magalhães de Oliveira, 2012; Carlos Machado, this volume). In general, however, it is not archaeologists producing this work but those who work predominantly with textual material.

Why does it matter that archaeologists have relinquished the role of populating the late antique town and retreated to the limited sphere of charting the changing form of the built environment and the ebb and flow of the use of decorative and epigraphic elements? To me, it matters because archaeology's core strength has always been telling the stories of those who are less represented in the textual and iconographic sources. Instead, as I have commented previously, late antique archaeology's focus seems increasingly to be identifying the changing ways in which a very small section of the population did, or didn't, spend their money (Bowden, 2020a, 26). To put it bluntly, I am not convinced that statues of local worthies occupied the day-to-day thoughts of most of the inhabitants of the Empire's towns and I think we should reflect this in our work. The fascinating furores in 2020 over the statue of Edward Colston in Bristol (UK) and the statues of confederate generals like that of Stonewall Jackson in Richmond (USA) don't disprove the point but rather show that such monuments can survive decades of relative indifference before suddenly igniting passionate disapproval or advocacy.

Finally, our models of late antique urban change are implicitly male in their viewpoint. The work of archaeologists focuses on a landscape shaped by the aspirations of men and reflecting the lives and political ambitions of men, memorialised in the public buildings, statues and inscriptions whose fates we chart so assiduously. That we do this with so little reflection on this point can no longer be defended. Our choice of subject matter is implicitly linked to assumptions and values relating to gender and we must be deeply suspicious of a disciplinary agenda that prioritises certain subject areas at the expense of others.

This is, of course, not just an issue for late antique archaeology. Eckardt's (2019) analysis of the contents of the journal *Britannia* between 1999 and 2016 noted that of the 174 papers published in that period, it was only in the study of artefacts and papers on archaeological science that the representation of women authors achieved close to parity. On papers on epigraphy/

history, excavation reports and those with a single-site focus, male authors outnumbered women authors by a margin of around 25:1, recalling Swift's (2007) comments on the 'low status' accorded to artefact studies by comparison with other areas and the implications of this in relation to underlying disciplinary ideologies and power structures.

The St Andrews conference from which these papers derive had an equal gender split between participants. Nonetheless, the subject matter itself is implicitly gendered. As Dempsey (2019) argues in relation to medieval archaeology, by focusing on issues of power representation (as the discussion of late antique space inevitably does), archaeology is writing a history that remains shaped by androcentric perspectives. This is surely no longer sustainable. If late antique archaeology is to thrive and survive during the coming decades, we must look to move beyond present goals and investigate the period in all its texture and diversity.

2.3 Different questions, different methods

The process that led me to the conclusions outlined earlier was that of publishing the Triconch Palace excavations in Butrint in southern Albania. This resulted in two volumes dealing, respectively, with the structural evidence from a major late Roman *domus* and the associated excavated sequence (Bowden & Hodges, 2011) and the non-ceramic finds (Bowden, 2020b). In essence, the first volume dealt with the use of space and the second volume with the people who inhabited that space. This was in part a result of the excavated sequence, which produced excellent deposits associated with the use of the area after the abandonment of the *domus*. With hindsight, I would have done it differently, with a chronological approach to publication rather than separating structures and materials into different volumes. However, I think the publishing sequence reflected where I was coming from at the time of the excavations with my thoughts very much engaged with the sort of spatial questions that I take issue with above. The excavated materials informed my interpretation of the space and the ways in which it changed over time, rather than being prioritised to tell a story in their own right.

In the event, it took another nine years to bring the finds to publication (with the pottery remaining outstanding). As a reviewer noted, the post-excavation recording and publication represented by *Butrint 5* is something 'that is all too rare in the discipline' (McLean, 2021). It's a fair point, although knowledge of my unpublished finds from other sites stops me from descending too far into self-congratulation. The dearth of detailed treatment of materials and bio-archaeological data from the late antique Mediterranean (and indeed the Roman Mediterranean in general) is remarkable but few archaeologists seem to be particularly perturbed by it, although this is by no means exclusive to late antique archaeology. That Roskams (2015) felt obliged to provide the readership of Lavan and Mulryan's edited volume *Field Methods and Post-Excavation Techniques in Late Antique Archaeology* with a primer to the possibilities of studying faunal remains exemplifies how far such study is from the mainstream topics of debate in late antique archaeology. Indeed, the bulk of that (very useful) volume is predominantly focused on gaining better data with which to approach the sort of spatial questions noted earlier.

The publication of the Triconch finds so long after the excavated structures and sequences forced me to look at them anew and in their own right. I have a sneaking suspicion that many archaeologists are slightly afraid of the specialist reports that lurk at the back of publications (even their own), concerned that they are likely to be boring and incomprehensible. They skip past the dense pages of graphs, tables and scatterplots in the hope of finding a few concluding paragraphs that they can understand. Perhaps the fact that these reports were going to be in a wholly separate volume for which I needed to write a meaningful introduction and conclusion made me approach them in a different way.

Several things became apparent. A number of themes began to emerge that would challenge my purely spatial approach to the site. The finds from the later phases after the abandonment of the *domus* provided a much more vivid picture of life in Late Antiquity than the architecture of the building did. They indicated the sorts of stories that we can tell about space, how it was used, and the people who used it, when we move away from picking over the carcass of twentieth-century clearance excavations. I have obviously dealt with this in the volumes themselves and also touched on this elsewhere (Bowden, 2020a), so here a brief mention of some key points will suffice.

2.3.1 High status living and subdivision in the fifth century

The Triconch Palace (Figure 2.1) was a large *domus* arranged around a central peristyle. The southern, western and northern wings had developed incrementally over time, while the eastern wing – with its huge triapsidal *triclinium* and waterfront entrance vestibule – had been added in a single act of aggrandisement *c*. 420 CE. This final phase was never completed, and the fifth century saw the southern and west wings subdivided into smaller dwellings (Figure 2.2). This process of subdivision is well known from other late antique *domus* across the Western Empire, although scholarship has moved away from using terms like 'squatting' (Uytteroeven, 2007, 45–46 with references). At Butrint, the smaller 'post-domus' dwellings were poorly preserved but seemed to occupy one or two rooms and were characterised by hearths and large storage jars or pithoi. At least some of the dwellings showed a new emphasis on second storey living (which seems to be a feature of much of the post-domus activity in the Triconch area).

Figure 2.1 Overall view of the Triconch Palace.

Source: Courtesy of Alket Islami

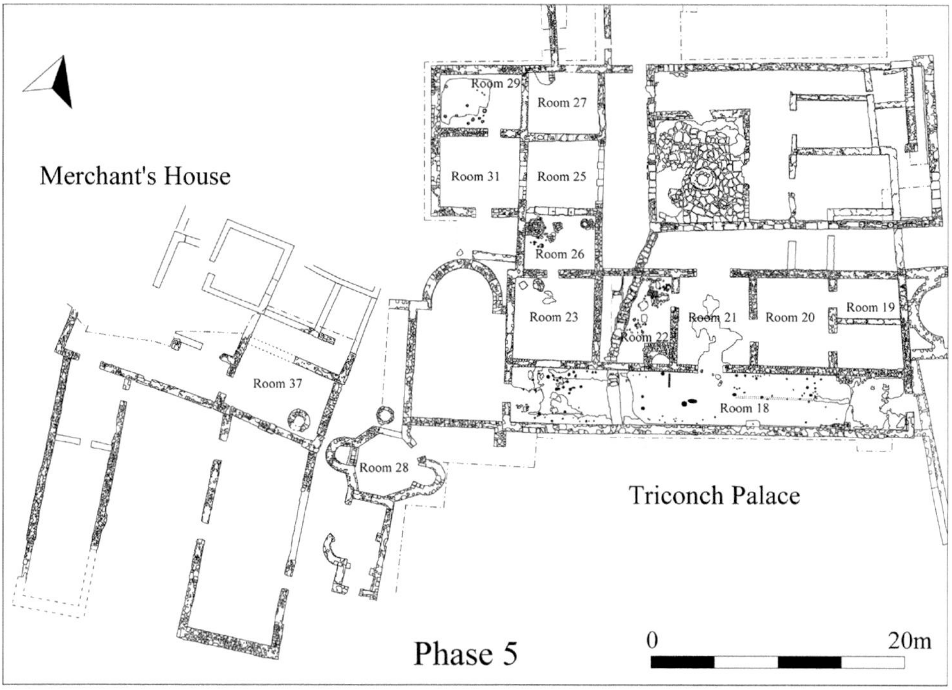

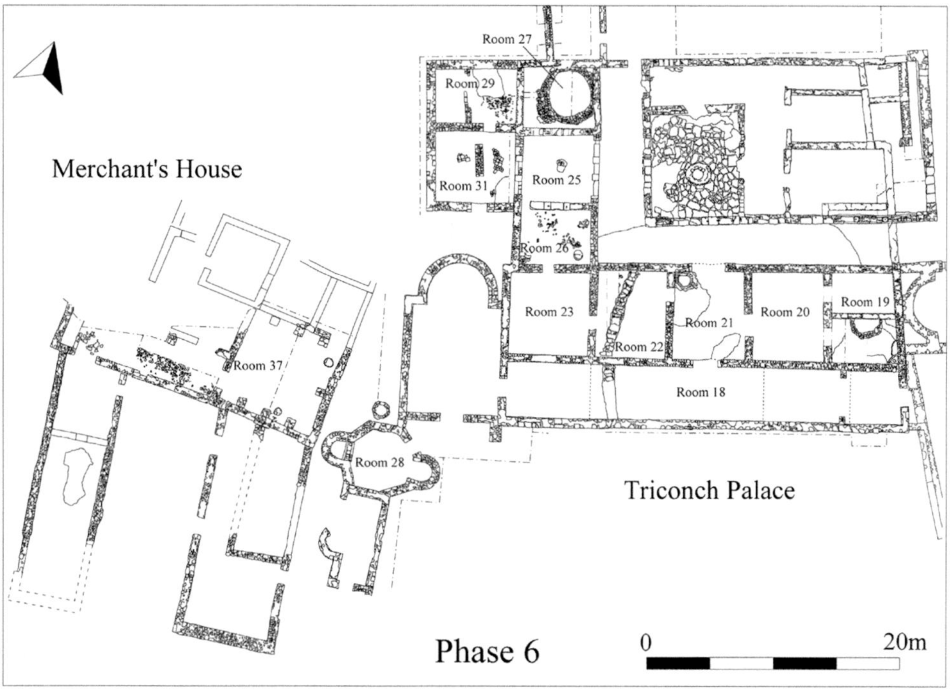

Figure 2.2 Plan of the fifth-century occupation areas of the Triconch Palace.

Source: Author's image

This sequence (outlined in Bowden & Hodges, 2011) places us in the normal territory of late antique archaeology in which elite space is redefined. However, detailed analysis of material recovered from these rooms demonstrated that the inhabitants of these fifth-century buildings clearly had access to high-quality material culture. Most of the better bone and copper alloy ornaments recovered in the excavations as well as some very high quality glass comes from this phase (Bowden, 2020c, 248–250; Mitchell, 2020; Jennings, 2020).

The inhabitants of these fifth-century dwellings also had access to considerable diversity in their diet, including access to a range of wild and domestic birds and other wild animals, suggesting participation in hunting (a high-status activity) (Bowden et al., 2020, 29–30). The wetland environment of Butrint has always provided a habitat for an abundance of water birds. The main food mammal in this period seems to be pig, continuing the trend from earlier Roman contexts elsewhere in the city (notably the forum), suggesting a recognisably urban Roman diet (Bowden et al., 2020, 39–40).

Therefore, study of the small finds, and particularly the faunal remains, changed our understanding of what appeared on the surface to be a typical picture of late antique 'subdivision' to one that showed the inhabitants using material culture and diet to express status, identity and aspiration. These were not people subsisting in the ruins of former grandeur but an urban population with access to a wide range of materials and foodstuffs.

2.3.2 ***Ruralisation, chickens and evil spirits in the sixth century***

It was the sixth century that saw a change, with much more subsistence-focused activity taking over (Bowden, 2020c, 250–253). Although the quantities involved are relatively small, material culture shows a marked change in this period, with a much greater emphasis on fishing equipment (all 14 examples coming from contexts dating to 500–650 CE) and a distinct decrease in the number and range of high-quality copper alloy and bone objects associated with dress and personal adornment. Bone items disappear entirely after 525 CE. The appearance of quernstones after 500 CE indicates food processing, with 13 of the 14 quernstone elements recovered belonging to the period 500–650 CE. Loom weights also indicate domestic textile working in the sixth century and after, with nine of the ten examples recovered coming from deposits dating to 500–650 CE. Also in these phases we can see a move away from a faunal assemblage, in which pig is the most significant constituent, to one dominated by caprines. The latter form 33 per cent of domesticates in the early to mid-sixth century, but 61 per cent of the assemblage in 525–550 CE (Bowden et al., 2020, 28–29). With the material culture and the faunal assemblage we can argue that we are moving from a recognisably urban assemblage in the fifth century to one that looks more rural in the sixth, based on a less diverse diet and much more productive/subsistence activity in the domestic sphere.

It would be uncontroversial to say that late antique archaeologists do not think about chickens very much.[1] However, the chicken data from the Triconch reveal some interesting patterns, notably a significant drop in numbers in the sixth century. Although the sample size and recovery strategies suggest that caution is required, this drop in chicken numbers is also apparent in the assemblages from other major excavations at Butrint on the extra-mural area of the Vrina Plain and may be linked to a European-wide decline in chicken numbers, possibly linked to climate change (Sykes, 2018). The chicken assemblage also shows a marked predominance of older, male birds, which is interesting as cockerels are noisy and aggressive and serve no practical purpose when you have more than one of them, unless used for cockfighting (undoubtedly common in the Roman world). However, their masculine characteristics give them a powerful apotropaic value, and representations of cockerels frequently appear in liminal contexts, notably a pair of fighting cocks that are depicted in Butrint's baptistery mosaic immediately in front of the entrance. Indeed, the struggle against supernatural forces seems to have been of significant concern to the residents of the triconch area, as they employed a wide range of amuletic devices in both decoration and portable material culture (Mitchell, 2007). Notable examples include a complex copper amulet with a range of textual and decorative devices to ward off evil (Figure 2.3) (Mitchell, 2020, 117–118), while Triconch Palace itself also included an 'all-seeing eye' symbol in its mosaic decoration (Mitchell, 2011, 261–262) and elaborate window grills with *chi-rho* decorations offering protection at vulnerable openings into the house (Mitchell, 2007, 287–289).

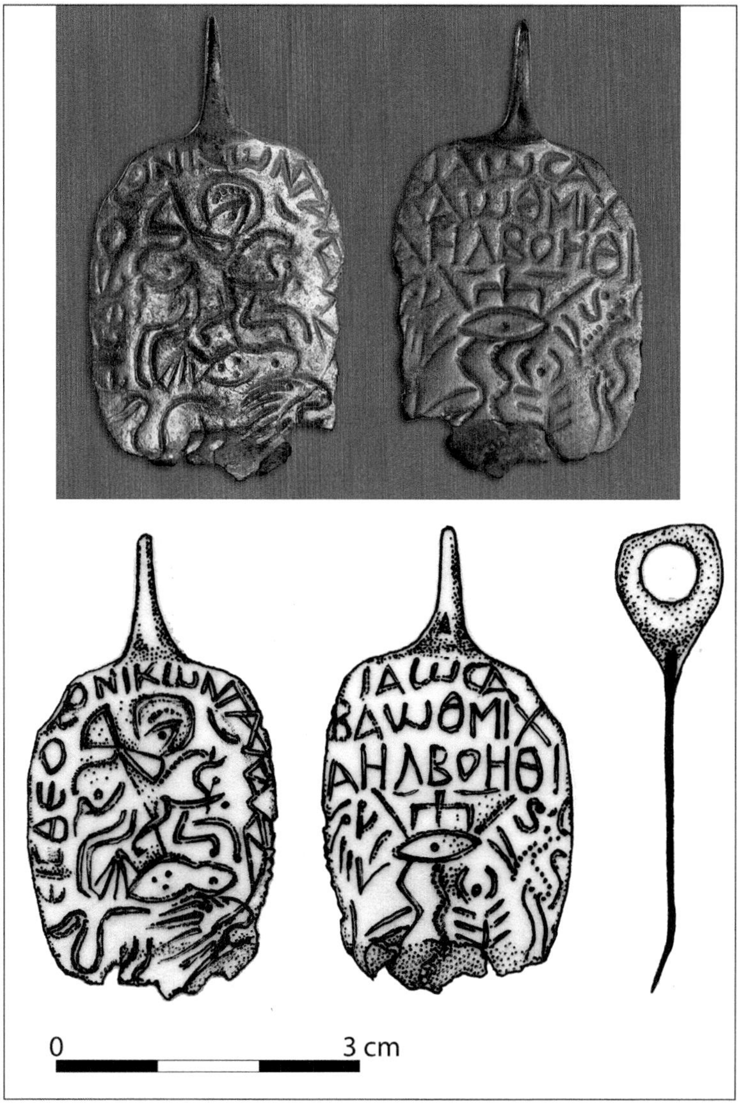

Figure 2.3 Apotropaic copper alloy pendant with rider and all-seeing eye.

Source: Courtesy of the Butrint Foundation

Sensory archaeologies have been the subject of a certain amount of discussion (an overview can be found in Skeates & Day, 2022), and the animal noise in fifth- and sixth-century Butrint must have been very striking. As well as the cockerels noted earlier, the representation of body parts at the Triconch suggests that live animals were a presence on the site (either permanently living there or brought in on the hoof for slaughter). Cats and dogs were also present, and discussions of urban space would do well to consider the four-legged denizens of late antique towns as well as those on two legs.

2.3.3 In sickness and in health

From the mid-sixth century, the Triconch area was used for burial by a section of Butrint's population, with the skeletal assemblage dominated by infants and children (Beatrice et al., 2020). The evidence (based on one adult and 14 sub-adults) suggests a very unhealthy population, with enamel hypoplasis (pits or lines in the tooth enamel) suggesting *in utero* or infant physiological stress. Others showed lesions in the eye orbits or on the cranium, symptomatic of anaemia, while some showed irregular growth of new bone symptomatic of trauma, stress or infection. The prevalence of poor health and malnutrition is striking although it is not clear as to whether this is solely local to the Triconch area. Certainly the late antique burials on the Vrina Plain (Soler et al., 2019), and those at the villa at Diaporit, do not show the same pathologies. The aDNA study of the Triconch burials also shows a population of remarkable genetic diversity, and I suggested that these burials may represent a transitory population living on the margins of the urban area (Bowden, 2020c, 256).

The collective archaeological evidence provides a clear picture of the changing lives (and arguably quality of life) of the Triconch area. In the fifth century, the Triconch area was occupied by people living in coherent dwellings living lives of relative prosperity in terms of material culture and dietary diversity. In the sixth century, we see a distinct change to a population more directly involved with subsistence, including fishing, and grinding their own grain and making textiles at a domestic level. Their diet lacked the diversity of their fifth-century predecessors, with meat consumption dominated by caprines and lacking the wild components associated with high-status activities like hunting. The apparently diminishing quality of life of those living in the Triconch area is also graphically illustrated by the poor health and high infant and child mortality of the population represented in the graves of the period after 550. To my mind, it is the combination of material culture, faunal and human remains that is important here. By moving away from the interests and self-expression of the elite, we are getting a very different picture of late antique space and how it is used.

2.4 Reconstructing Late Antiquity at Venta Icenorum

The picture that I have painted of life in the later fifth and sixth centuries in the Triconch area relied on the large-scale excavation of materially rich deposits. By contrast, the evidence on the site of life in the *domus* of the fourth and early fifth centuries was much more restricted. This is for the obvious reason that those living in high-status buildings had people to sweep their mosaics and take their rubbish somewhere else. A well-maintained domestic or civic environment will produce little archaeological evidence relating to its primary use. How do we deal with this? How do we as archaeologists look at the lives of people within our empty spaces? To explore this, I want to turn to my second example, that of the forum at Venta Icenorum (Caistor-by Norwich) (Figure 2.4).

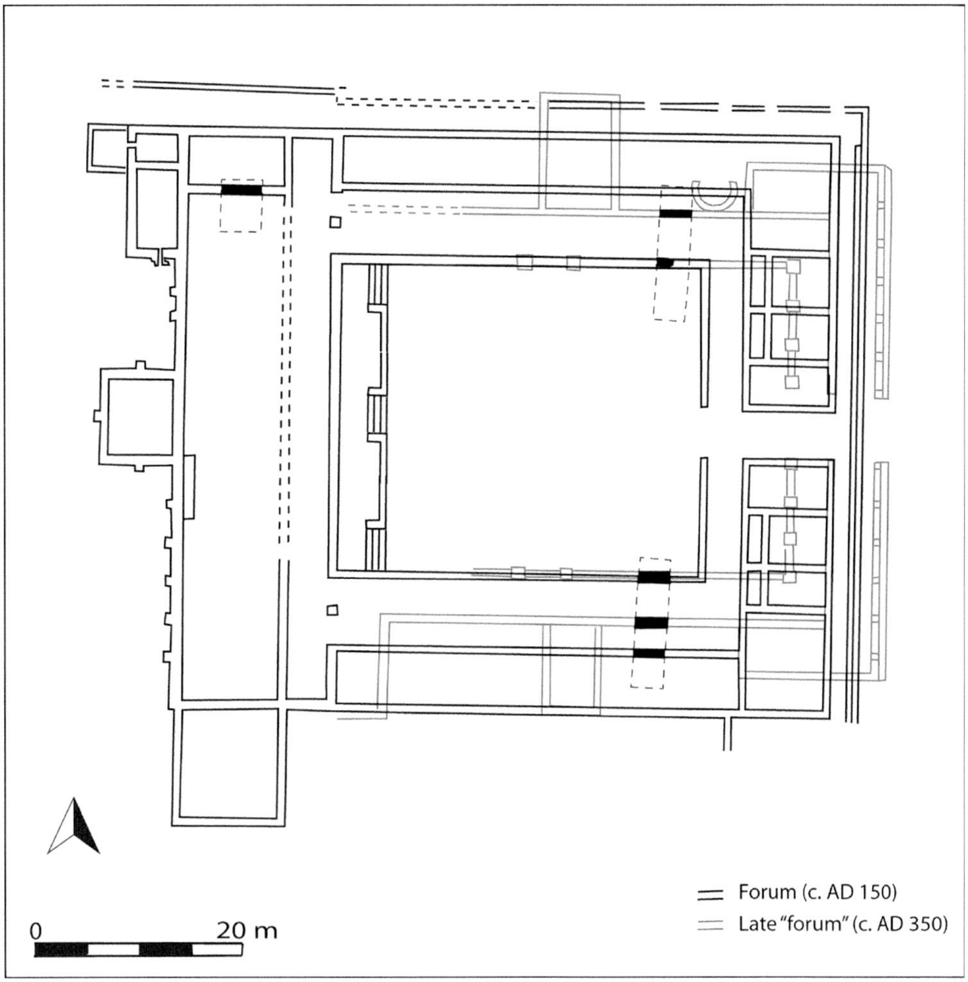

Figure 2.4 Plan of the forum at *Venta Icenorum*. Shaded walls are those revealed in the 2011 excavations.
Source: Drawn by the author after Frere (1971)

Venta Icenorum was the *civitas* capital of the Iceni and lies just to the south of modern Norwich. Unlike most Roman towns in Britain, it saw no medieval development beyond a church in its south-east corner and as a result lies almost entirely under open fields. It first drew attention in 1928 when, during a particularly dry summer, a series of spectacular aerial photographs was taken showing the street grid and public buildings as parch-marks in the barley crop. Subsequent publication of these images led to significant public and scholarly interest in the site, which resulted in the first campaigns of excavation, carried out by Donald Atkinson between 1929 and 1935. I have been working in and around the town since 2006 in a major project to look at the development and ultimate disappearance of the Roman town (see Bowden, 2013a and 2013b, for summary and a full bibliography on the site). The forum was excavated by Atkinson in 1931 and 1933 (Frere, 1971) and I directed some further excavations there in 2011 (Bowden, 2013b) The excavated sequence of the forum is undoubtedly problematic. Like many earlier excavators of Roman towns, Atkinson prioritised the recovery of the architectural plan of the buildings, following the lines of walls and expanding his trenches outwards at particular points of interest. The eastern wing of the complex was excavated in its entirety in 1931, alongside more limited trenching of the north and south wings. The basilica, which formed the western end of the complex, was explored in a fairly piecemeal fashion in 1933. Atkinson did not publish his work, and it was left to Sheppard Frere to stitch together a report from Atkinson's notes and plans (Frere, 1971). Frere encountered significant problems, not least that Atkinson's composite plan of the two seasons of excavation added an extra 30 feet to the east-west dimensions of the building by comparison with the plan derived from the aerial photographs. Frere ultimately arrived at a best-guess solution to its size and plan by what he described as a process of 'exasperating trial and error' (Frere, 1971, 3).

Little material was recorded from the forum excavations, with the exception of coins and Samian ware that could be used to date the structural phases. Atkinson identified a masonry forum (Forum I), that he dated to the late first or second century CE, according to lecture notes summarised by Frere (1971, 8). He suggested that this had been destroyed by fire, necessitating a rebuild in the later second century before a further fire destroyed the complex again. Frere, by contrast, argued that the first masonry forum was no earlier than 150–160 CE and that the evidence for both fires was limited. Both Atkinson and Frere, however, agreed that the forum had subsequently been rebuilt on an entirely different plan (Forum II), with Frere arguing that this took place around 270–290 CE.

Our excavations in 2011 involved small trenches across the northern and southern wings and a trench in the north part of the basilica. In the lower levels of these trenches we recovered traces of an earlier timber building (possibly a forum) that had clearly been entirely destroyed by fire prior to the construction of the first masonry forum in the mid-second century. The masonry forum seems to have lost its civic function in the very late third or early fourth century, with both the north and south wings used for the deposition of refuse. This was followed by the creation of a major new building (the so-called Forum II) which (*contra* Frere's dates) cannot be much earlier than the 340s, based on coin finds from underlying levels. Although the fourth-century building is described as a forum in most secondary literature, it seems unlikely that it functioned in the same way as the earlier public space. *Fora* in Britain seem to be abandoned quite early, with many being turned over to secondary uses such as metalworking in the third and early fourth centuries (Rogers, 2011, 130–135). Given that late Roman Venta occupied an increasingly strategic position in eastern *Britannia* at the head of a riverine system protected by the so-called Saxon Shore forts at Burgh Castle and Caister-on-Sea, the second 'forum' seems more likely to be a building associated with a more militarised authority that deliberately referenced the earlier civic structure. The basilica, which belongs to the mid-second-century masonry forum, seems to have stayed in use throughout (or at least lacked the thick deposits of rubbish that filled the north and south porticos), with the latest internal deposits belonging to the late fourth or early fifth century.

As will be apparent from the above, the nature of human life within this building is hard to extrapolate from the archaeological sequence. This created some challenges when we decided to use the forum to trial a new form of immersive digital reconstruction (Bowden et al., 2020; Brundell et al., 2022).

The possibilities of using reconstructions to address the issue of archaeologists' focus on the built environment have been noted by others (Humphries, 2019, 89–90; Christie, 2020), although it is noteworthy that the examples that Christie illustrates are hand-painted rather than digital, while Humphries's suggestion is one of aspiration rather than application. Indeed, although Virtual Reality (VR) reconstructions have quite a long history in relation to Roman towns, they have tended to represent urban environments as largely empty landscapes of monumental buildings (see, e.g. the long-running Rome Reborn project [Frischer & Massey, 2022]). There are several reasons for this. First, the buildings are the easiest elements to create and utilise within a VR environment. Put very crudely, on a practical level, simple rectilinear shapes use up much less processing power than uneven shapes and require fewer textures to render them. People and animals require more shapes, textures and memory for a VR model to run smoothly. Second, however, I suspect that Roman archaeologists have tended to reconstruct the things that we are most interested in – monumental buildings and public spaces. If we were really that interested in the people, we would put them centre stage.

The focus on the built environment continues to drive the use of VR in relation to Roman towns, even in those projects that ostensibly focus on human experience. For example, Frischer and Massey (2022) explore how people would have viewed the buildings of the fourth-century *Forum Romanum*. However, their people are small, distant figures, gazing in awe at the towering, monumental, buildings that dwarf them. Similarly, recent work exploring the oft-cited ideas of optical axiality in the Pompeian house did so in the context of a reconstruction that was empty of human life (Campanaro & Landeschi, 2022). Within their own terms, these projects represent pioneering uses of new technologies, but they are also pursuing the traditional interests of classical archaeology, creating reconstructions that are unconscious reflections of the long-standing intellectual frameworks that prioritise the built environment over the people who used it.

At Caistor, we have a long history of using VR reconstructions. The first iteration was created by my friend and Norwich resident Danny Voisey in the early 2000s, as an exercise to learn how to use 3D Studio Max software (Figure 2.5). He based it on a beautiful painted reconstruction of the town in the early fourth century by Sue Walker-White (commissioned by the Norfolk

Figure 2.5 VR of *Venta Icenorum* (*c.* 2006).

Source: Courtesy of D. Voisey

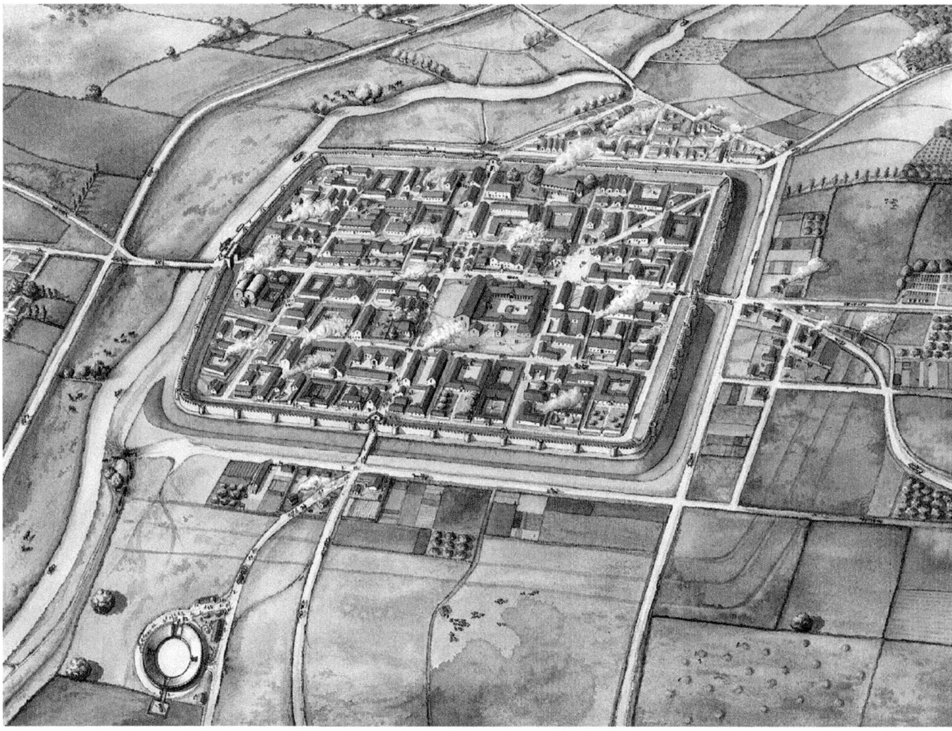

Figure 2.6 Reconstruction of *Venta Icenorum* as envisaged in 1991.
Source: Copyright 1991 by Sue Walker-White. Reprinted with permission

Archaeological Trust) (Figure 2.6). When I began to work at the site in 2006, Danny let me use his reconstruction with its fly-through animation for public outreach and it eventually became part of the permanent display of the site within Norwich Castle Museum's Boudica gallery.

Walker-White's reconstruction was based on aerial photographs and Atkinson's excavations as well as earlier reconstructions of more fully understood sites such as Colchester. It depicted the town with its walls (suggested by Frere [2005, 324] to date to around 275–280 CE) and the second-phase forum. The space within the walls was shown as an ordered grid-plan settlement with white-painted houses and red-tiled roofs. The town was shown surrounded by regular field systems, dotted with occasional orchards, with boats plying a River Tas that was depicted as much wider than it is today (in accordance with the prevailing view of the site at the time). It can be argued that this representation of Venta Icenorum, along with the published plans of the site, was an unconscious evocation of the ideas of Rome as a civilising influence that remained current in Britain until the early 1990s (Bowden & Bescoby, 2008, 333).

By 2010, when Channel 4 approached me to make a programme for their *Time Team* special strand, we had a very different picture of the town derived from large-scale geophysical survey, showing large open spaces within the walled area. My enthusiasm for doing *Boudica's Lost Tribe: A Time Team Special* was driven in part by the fact that they would fund new VR reconstructions of the site. On the basis of my understanding of the changing nature of late antique towns and the evidence of the short-lived nature of Roman public spaces in Britain, I decided to consciously challenge the image of the Roman town as a clean and ordered urban space. My new Caistor would be much messier. It would have animals grazing and cultivation within the walls. Steaming heaps of manure was prominent in the foreground. Buildings were thatched as well as tiled (Figure 2.7).

Figure 2.7 Reconstruction of *Venta Icenorum* for *Time Team* (2011).
Source: Copyright 2011 by W. Bowden and Channel 4. Reprinted with permission

In addition to breaking down the regimented order of the Roman town, I deliberately made the weather gloomy. Weather is perhaps an under-discussed aspect of reconstructions. Alan Sorrell, 'the man who created Roman Britain', in a series of hugely important reconstruction paintings of sites commissioned by the Ministry of Works, used weather as a key component of his images (Sorrell & Sorrell, 2017, 2). He depicted the forts of Hadrian's Wall hunkered down beneath the Northumbrian rain in greys, blacks and a muted palette of washed-out pastels. This incurred the displeasure of his clients at the Ministry of Works, with an internal memo noting caustically that 'Like Mr Piper, Mr Sorrell appears to have had [no] luck with the weather' (Sorrell & Sorrell, 2017, 174). Sorrell, however, was a gifted artist who used the weather to evoke an image of the forts existing in a hostile environment. Walker-White's Caistor had been drenched in sunshine, perhaps unconsciously conjuring an image of the halcyon days of the height of Romano-British life. Mine was under a leaden sky, with the red tiles muted and the white plaster of the walls reduced to a dirty grey, a deliberate ploy to provoke a different image of Roman Britain in the eyes of the 2.5 million people who watched the programme on its first showing.

In 2015, I had the chance to create a new iteration of the reconstruction. The Norfolk Archaeological Trust, funded by Natural England, wanted to renew the interpretation panels of the site, driven in part by the changing understanding of the town that had emerged from our research. This time, however, we would use augmented reality (AR) to allow visitors to view the town from different angles via tablets or mobiles. By moving their devices, they would be able to see the town laid out before them. AR is in many ways an ideal technology for Caistor, which is otherwise a hard site for visitors to understand, with nothing now visible apart from the wall circuit. Jam Creative Studios in Cardiff were commissioned to create the *Caistor AR* app (Brundell et al., 2022)

Mobile-based AR in 2015 presented some significant technical challenges. Channel 4's CGI only had to work from one angle viewed as an animation on a TV screen. The new version had to work in 360 degrees and had to respond to the user's movements. In computer gaming, it appears to the user that they are moving smoothly through a virtual environment but this movement in fact requires the environment to be constantly rebuilt and its surface textures re-rendered by the computer. For this to work with the limited processing power of a mobile phone, the designers must reduce the number of textures present and the structural complexity of a virtual model. This is one of the key reasons that VR models of ancient buildings have traditionally tended to feature sparsely furnished clean spaces and plain rectilinear structures. This meant that to include the sort of detail we wanted in the 2015 *Caistor AR* app, we had to abandon photorealistic textures to ensure that most phones would be able to run the app to allow the user to scan views of the townscape (Figure 2.8).

The 2015 app also included archaeological finds that the user could 'collect' as they went round the visitor trail. The finds were 3D-scanned to allow people to turn and manipulate them on screen. All the content of the app was triggered by codes hidden in the design of static interpretation panels placed at different points of interest around the site.

The 2011 and 2015 iterations of Caistor depicted the town in the fourth century and showed the wall circuit and the rebuilt 'forum'. They also showed other public buildings including the two second-century temples, a bathhouse and a theatre – the latter hypothesised from the geophysical survey (Bowden & Bescoby, 2008). We do not know if these buildings were extant or functioning at this time, but they were key features of the town and, in terms of creating a visitor experience, it made sense to include them. The technical limitations, however, meant that including people and animals was difficult. In 2018, inspired in part by the work on the Triconch Palace materials described earlier and also by debates on representation of people in Roman Britain that were playing out in sections of the press,[2] I applied for funding from the Arts and Humanities Research Council's Next Generation of Immersive Experiences strand to create an AR experience that would centre the inhabitants of the Roman town rather than its buildings. In addition, we wanted to look at some of the technical challenges of using immersive AR in a field with no infrastructure (Bowden et al., 2020).[3]

The solution we adopted for this project, which was christened *Virtual Venta*, was to create a version of the forum that users could move through in real time and space via their mobile

Figure 2.8 Reconstruction of *Venta Icenorum* (2015).

Source: Produced by Jam Creative Studios and Norfolk Archaeological Trust

phones. We chose the forum as it was an enclosed space, which made it easier to create an immersive virtual environment and created a cast of characters that people would interact with as they moved through the space. We chose to reconstruct the first masonry forum but depicted it as we imagined it to be *c.* 300 CE, in a state of decrepitude with peeling plaster and a range of different activities taking place (Figure 2.9). The aim was to challenge the viewer's perception of Roman Britain as a clean, ordered and militarised space, and indeed the lack of soldiers was commented on by users (Bowden et al., 2020, 17).

Like the Triconch Palace in its primary phase as a high-status dwelling – and like most of the spaces that we deal with in late antique towns – we have little evidence of the actual activities that occurred in Caistor's forum, although what we do have suggests the presence of metalworking and butchery in the vicinity and rubbish building up in the porticoes. Therefore, we decided to populate it with a range of characters, participating in a range of activities based wholly on archaeological evidence from the town as a whole. It required the use of imagination, but in a way that was intellectually defensible. This set up a whole series of challenges and questions that, I would argue, lay well beyond the questions we normally ask in relation to the late antique town. What would our people wear? What hairstyles would they have? What colour would their skin be? What languages would they speak?

We came up with a number of solutions for these issues. The answers were based mainly on small finds evidence, isotope analysis and pathologies of skeletal remains, as well as parallels from elsewhere in Britain. The language aspect was particularly intriguing, and Alex Mullen came up with a script in which people spoke a British fourth-century Latin based on her work on the spread of Latin across Europe as part of the European Research Council-funded LatinNow project. We had people from other parts of the Empire (evidenced by isotope data from cemetery finds) and the first two characters that the user encountered were women, as we

Figure 2.9 The forum, as reconstructed for the *Virtual Venta* app.

Source: Produced by Jam Creative Studios

Figure 2.10 Characters from the *Virtual Venta* app.

Source: Produced by Jam Creative Studios

wanted to challenge people's perception of Roman Britain as white and male (Figure 2.10). The third character was old and severely disabled (again based on burial evidence). The characters guided the user through a story arc (based around buying an eye salve from a Syrian healer). The story arc directly related to materials found at the site, including a so-called oculist's stamp and a curse tablet dedicated to Neptune. After they had completed the story, the user was left to explore the forum as they wished. The space was dirty, the forum pavement hidden beneath dung and straw, and chickens scratched around the entrance. The action took place against a raucous aural backdrop of animal noise, the hammering of a blacksmith and shouts in multiple languages. The latter included 'Pretty girl, give me beer' in Gaulish, based on a graffito on a spindle-whorl from Autun!

To make the experience work on a mobile phone, we had to make a range of compromises in terms of what we would include and what we could leave out. We tried to find a compromise between maximising the information that we could give to the user without including obvious anachronisms. The chronology of 'c. 300 CE' was deliberately vague, and indeed it could be argued that the forum was entirely abandoned by this time rather than just a little run-down. The disabled character that we included was based on a second-century skeleton, whereas the evidence of a non-local population comes from the fourth century.

It could be argued that this is just public-facing interpretation – a politically correct gamification of archaeology – and not a serious intellectual endeavour. I would dispute this. I don't see that considering what breeds of chicken to include in a reconstruction is any less valid (or less evidence-based) than trying to date a subdivision built of *spolia* or deciding when a particular statue was moved. It is simply that the latter sits more comfortably within the long-standing intellectual traditions in which we work.

2.5 Conclusion

The interpretation of the Triconch Palace data and the Caistor forum AR reconstruction are very different, but they both prioritised people over the built environment. The Triconch material was the result of years of excavation and post-excavation, whereas the Caistor forum was the result of setting ourselves an interpretative task that forced us to seek answers to particular questions. They are two very different approaches to late antique space, but some similar themes emerge from both of them.

1. Both projects sought deliberate distance from the datasets, methodologies and questions that shaped Classical archaeology in the twentieth century and that were focused on the built environment.
2. Both used a much greater range of archaeological data than is often employed in late antique archaeology. In the case of *Virtual Venta*, this sometimes involved making significant interpretative decisions (e.g. in relation to language), but in many aspects (e.g. in relation to hairstyles) it simply involved making better use of data that were actually quite extensive.
3. Both addressed the lived experience of a broader cross section of society than is normally considered in relation to the late antique town.

Does this have any broader implications for the ways in which we think about urban space? As I noted in the introduction to this chapter, the power of archaeology lies in its ability to give voice to those who do not appear in textual sources or whose experience is mediated through the invariably male aristocratic world view of the texts. Actively seeking the experience of the people who inhabited our late antique townscapes must surely change our targets for study and the methodologies that we apply to those targets (not least causing us to question the value of further reassessment of change in public areas). As archaeologists, we need to continually, and critically, evaluate what we are doing and why we are doing it. In relation to late antique space, I don't think it's enough to nuance and add detail to a picture that we are broadly happy with. We now have a range of scientific and technological tools available for research and dissemination that even ten years ago were the stuff of science fiction, yet we continue to ask questions that are shaped by the methodologies, ideologies and data of the twentieth century.

While we may justifiably claim to have discredited decline as a paradigm, in many ways our work continues to be framed by it and shaped by a world view that prioritises elite representation above all else. To challenge the status quo, we do not all have to embrace VR, but we do need to emerge more from the disciplinary silo that late antique archaeology finds itself in. If we do not, we may find that when we do come out the world has moved on and left us behind.

Notes

1 Here I should acknowledge the input of Naomi Sykes and Holly Miller, who made me think about chickens a great deal.
2 *Daily Mail* (7 August 2017): www.dailymail.co.uk/news/article-4766466/Mary-Beard-online-row-black-Roman-soldier.html.
3 *Virtual Venta* was funded by the Arts and Humanities Research Council. I am grateful to my colleagues on the project – Pat Brundell, Natasha Harlow, Adam Martin-Jones and Alex Mullen.

Bibliography

Beatrice, J., Fenton, T., Hurst, C., Jenny, L., Wankmiller, J., Mutolo, M., Rauzi, C., & Foran, D. (2020). The human skeletons from the Triconch Palace and the Merchant's house. In W. Bowden (Ed.), *Butrint 5: Life and Death at a Mediterranean Port. The Non-Ceramic Finds from the Triconch Palace* (pp. 42–67). Oxbow.

Bowden, W. (2000). *Town and Country in Late Antique Epirus Vetus* (unpublished PhD thesis, University of East Anglia).

Bowden, W. (2003). *Epirus Vetus: The Archaeology of a Late Antique Province*. Duckworth.

Bowden, W. (2013a). The urban plan of Venta Icenorum and its relationship with the Boudican revolt. *Britannia*, *44*, 145–169.

Bowden, W. (2013b). Townscape and identity at Caistor-by-Norwich. In H. Eckardt & S. Rippon (Eds.), *Living and Working in the Roman World. Essays in Honour of Michael Fulford on his 65th Birthday* (pp. 47–62, JRA Suppl. Ser. 95). Journal of Roman Archaeology.

Bowden, W. (2020a). Approaches to late antique urban transformation: Looking beyond the archaeology of public space. In A. Carneiro, N. Christie, & P. Diarte-Blasco (Eds.), *Urban Transformations in the Late Antique West: Materials, Agents, and Models* (pp. 23–43). Coimbra University Press.

Bowden, W. (Ed.). (2020b). *Butrint 5: Life and Death at a Mediterranean Port. The Non-Ceramic Finds from the Triconch Palace*. Oxbow.

Bowden, W. (2020c). The Triconch Palace and Merchant's house as lived environments in Late Antiquity. In W. Bowden (Ed.), *Butrint 5: Life and Death at a Mediterranean Port. The Non-Ceramic Finds from the Triconch Palace* (pp. 246–259). Oxbow.

Bowden, W., & Bescoby, D. (2008). Mapping Venta Icenorum: New geophysical survey at Caistor-by-Norwich. *Journal of Roman Archaeology*, *21*, 325–334.

Bowden, W., Brundell, P., Harlow, N., Martin-Jones, A., & Mullen, A. (2020). *Developing immersive experience at Caistor Roman Town Final report on AHRC Next Generation of Immersive Experiences Project (2018)*. https://nottingham-repository.worktribe.com/OutputFile/5366659.

Bowden, W., & Hodges, R. (Eds.). (2011). *Butrint 3: Excavations at the Triconch Palace*. Oxbow.

Bowden, W., Knapp, Z., Powell, A., & Westoby, J. (2020). The faunal remains. In W. Bowden (Ed.), *Butrint 5: Life and Death at a Mediterranean Port. The Non-Ceramic Finds from the Triconch Palace* (pp. 25–41). Oxbow.

Bowden, W., & Lavan, L. (2004). The late antique countryside: An introduction. In W. Bowden, L. Lavan, & C. Machado (Eds.), *Recent Research on the Late Antique Countryside* (pp. xvii–xxvi). Brill.

Bowersock, G. W., Cameron, A., Clark, E. A., Dihle, A., Fowden, G., Heather, P., Rousseau, P., Rousselle, A., Torp, H., & Wood, I. (1997). Comments. *Symbolae Osloenses*, *72*(1), 31–69.

Brogiolo, G.-P., & Ward-Perkins, B. (Eds.). (1999). *The Idea and Ideal of the Town Between Late Antiquity and the Early Middle Ages*. Brill.

Brown, A. R. (2018). *Corinth in Late Antiquity. A Greek, Roman and Christian City*. I.B. Tauris.

Brown, P. (1971). *The World of Late Antiquity*. Thames and Hudson.

Brundell, P. R., Harlow, N., Cameron, H., & Bowden, W. (2022). Rome is where the heart is: Designing immersive experiences to augment and build personal connections to outdoor environments. In *Academic Mindtrek '22: Proceedings of the 25th International Academic Mindtrek Conference November 2022* (pp. 318–323). Association for Computing Machinery.

Cameron, A. (2003). Ideologies and agendas in late antique studies. In L. Lavan & W. Bowden (Eds.), *Theory and Practice in Late Antique Archaeology* (pp. 3–21). Brill.

Campanaro, D., & Landeschi, G. (2022). Re-viewing Pompeian domestic space through combined virtual reality-based eye tracking and 3D GIS. *Antiquity*, *96*(386), 479–486.

Carver, M. (1993). *Arguments in Stone: Archaeological Research and the European Town in the First Millennium*. Oxbow.

Chenault, R. (2012). Statues of senators in the forum of Trajan and the Roman forum in Late Antiquity. *Journal of Roman Studies*, *102*, 103–132.

Christie, N. (2020). From royalty to refugees: Looking for the people in reconstructing urban change in late antique Italy. In A. Carneiro, N. Christie, & P. Diarte-Blasco (Eds.), *Urban Transformations in the Late Antique West: Materials, Agents, and Models* (pp. 323–349). Coimbra University Press.

Christie, N., & Loseby, S. (Eds.). (1996). *Towns in Transition: Urban Evolution in Late Antiquity and the Early Middle Ages*. Scolar Press.

Dempsey, K. (2019). Gender and medieval archaeology: Storming the castle. *Antiquity*, *93*(369), 772–788.

Dey, H. (2014). *The Afterlife of the Roman City: Architecture and Ceremony in Late Antiquity and the Early Middle Ages*. Cambridge University Press.

Eckardt, H. (2019). Editorial. *Britannia*, *50*, 1–11.

Frere, S. S. (1971). The forum and baths at Caistor by Norwich. *Britannia*, *2*, 1–26.

Frere, S. S. (2005). The south gate and defences of Venta Icenorum: Professor Atkinson's excavations, 1930 and 1934. *Britannia*, *36*, 311–327.

Frischer, B., & Massey, D. (2022). 3D urban models as tools for research and discovery: Two case studies of the *rostra* in the Roman forum utilizing Rome Reborn. In K. Gartski (Ed.), *Critical Archaeology in the Digital Age: Proceedings of the 12th IEMA Visiting Scholar's Conference* (pp. 23–47). The Cotsen Institute of Archaeology Press.

Fukuyama, F. (1992). *The End of History and the Last Man*. Hamish Hamilton.

Gering, A. (2015). Ruins, rubbish dumps and encroachment. Resurveying late antique Ostia. In L. Lavan & M. Mulryan (Eds.), *Field Methods and Post-Excavation Techniques in Late Antique Archaeology* (pp. 249–288). Brill.

Hayes, J. W. (1972). *Later Roman Pottery*. British School at Rome.

Humphries, M. (2019). *Cities and the Meaning of Late Antiquity*. Brill.

Jennings, S. (with contributions from W. Bowden & K. Stark). (2020). The vessel glass from the Triconch Palace: A catalogue. In W. Bowden (Ed.), *Butrint 5: Life and Death at a Mediterranean Port. The Non-Ceramic Finds from the Triconch Palace* (pp. 218–245). Oxbow.

Johnson, M. (1999). *Archaeological Theory: An Introduction*. Blackwell Publishers.

Jones, A. H. M. (1964). *The Later Roman Empire 284–602: A Social, Economic and Administrative Survey*. Basil Blackwell.

Lane, A. (2014). Wroxeter and the end of Roman Britain. *Antiquity*, *88*, 501–515.

Lavan, L. (Ed.). (2001). *Recent Research in Late Antique Urbanism*. Journal of Roman Archaeology.

Lavan, L. (2015a). Field methods and post excavation Techniques in late antique archaeology: Anyone for discussion? In L. Lavan & M. Mulryan (Eds.), *Field Methods and Post-Excavation Techniques in Late Antique Archaeology* (pp. 1–13). Brill.

Lavan, L. (2015b). Distinctive field methods for Late Antiquity: Some suggestions. In L. Lavan & M. Mulryan (Eds.), *Field Methods and Post-Excavation Techniques in Late Antique Archaeology* (pp. 51–90). Brill.

Lavan, L. (2015c). The agorai of Sagalassos in Late Antiquity: An interpretive study. In L. Lavan & M. Mulryan (Eds.), *Field Methods and Post-Excavation Techniques in Late Antique Archaeology* (pp. 289–353). Brill.

Lavan, L. (2021). *Public Space in the Late Antique City* (2 vols). Brill.

Lavan, L., & Bowden, W. (Eds.). (2003). *Theory and Practice in Late Antique Archaeology*. Brill.

Machado, C. (2019). *Urban Space and Aristocratic Power in Late Antique Rome: AD 270–535*. Oxford University Press.

Machado, C. (2021). Statue habit and statue culture in late antique Rome. *Journal of Roman Archaeology, 34*(2), 632–666.

Magalhães de Oliveira, J. C. (2012). *Potestas populi: Participation populaire et action collective dans les villes de l'Afrique romaine tardive (vers 300–430 apr. J.-C.).* (Bibliothèque de l'Antiquité Tardive, 24). Brepols.

McLean, A. (2021). *Review of Butrint 5: Life and death at a Mediterranean port: The non-ceramic finds from the Triconch Palace, Bryn Mawr Classical Review 2021.04.16.* https://bmcr.brynmawr.edu/2021/2021.04.16/.

Mitchell, J. (2007). Keeping the demons out of the house: The archaeology of apotropaic strategy and practice in late antique Butrint and Antigoneia. In L. Lavan, E. Swift, & T. Putzeys (Eds.), *Objects in Context, Objects in Use: Material Spatiality in Late Antiquity* (pp. 273–311). Brill.

Mitchell, J. (2011). The mosaic pavements and painted walls of the *domus*. In W. Bowden & R. Hodges (Eds.), *Butrint 3: Excavations at the Triconch Palace* (pp. 230–276). Oxbow.

Mitchell, J. (2020). The small finds. In W. Bowden (Ed.), *Butrint 5: Life and Death at a Mediterranean Port. The Non-Ceramic Finds from the Triconch Palace* (pp. 106–217). Oxbow.

Potter, T. W. (1995). *Towns in Late Antiquity: Iol Caesarea and its Context.* Ian Sanders Memorial Committee.

Rich, J. (Ed.). (1992). *The City in Late Antiquity.* Routledge.

Rogers, A. (2011). *Late Roman Towns in Britain: Rethinking Change and Decline.* Cambridge University Press.

Roskams, S. (2015). Food for thought: The potential and problems of faunal evidence for interpreting late antique society. In L. Lavan & M. Mulryan (Eds.), *Field Methods and Post-Excavation Techniques in Late Antique Archaeology* (pp. 513–552). Brill.

Saradi, H. (2006). *The Byzantine City in the Sixth Century.* Society of Messenian Archaeological Studies.

Skeates, R., & Day, J. (Eds.). (2022). *The Routledge Handbook of Sensory Archaeology.* Routledge.

Soler, A., Isaac, C. V., Beatrice, J., & Fenton, T. W. (2019). The human skeletons from the Vrina Plain. In S. Greenslade (Ed.), *Butrint 6: Excavations on the Vrina Plain. The Finds* (Vol. 2, pp. 59–76). Oxbow.

Sorrell, J., & Sorrell, M. (2017). *Alan Sorrell: The Man Who Created Roman Britain.* Oxbow.

Swift, E. (2007). Small objects – small questions? Perceptions of finds research in the academic community. In R. Hingley & S. Willis (Eds.), *Roman Finds: Context and Theory* (pp. 18–28). Oxbow.

Sykes, N. J. (2018). Biogeography. In S. L. López Varela (Ed.), *The SAS Encyclopedia of Archaeological Sciences.* Wiley-Blackwell.

Uytteroeven, I. (2007). Housing in Late Antiquity: Thematic perspectives. In L. Lavan, L. Özgenel, & A. Sarantis (Eds.), *Housing in Late Antiquity: From Palaces to Shops* (pp. 25–66). Brill.

Ward-Perkins, B. (1997). Continuitists, catastrophists and the towns of post-Roman northern Italy. *Papers of the British School at Rome, 65*, 157–176.

Ward-Perkins, B. (2005). *The Fall of Rome and the End of Civilization.* Oxford University Press.

White, R. (2014). A brave new world? The archaeology of Western Britain in the fifth and sixth centuries. In F. K. Haarer (Ed.), *AD 410: The History and Archaeology of Late and Post-Roman Britain* (pp. 155–164). Society for the Promotion of Roman Studies.

3 Suburban saints

Space, place and environment in Theodoret's *Religious History*

Jason König

My aim in this chapter is to chart some of the features that make Theodoret's representation of space in the *Religious History* (written in the 440s CE) both typical of fifth-century hagiography more broadly, in its portrayal of the tension between wild landscape and human culture, and in some respects quite distinctive and unusual (cf. Urbainczyk, 2002, 40–51, for general discussion of the text's relationship with earlier hagiography). The first section surveys the text's representation of the relationship between saints and cities, with reference to Theodoret's manipulation of the long tradition of representing mountains as spaces on the edges of urban territory, both within and beyond human control. The second half of the chapter examines Theodoret's representation of human–environment relations from an ecocritical perspective, arguing that there is a closely related tension between anthropocentric and ecocentric visions of ascetic practice running through the text: we need to take account of both of those strands and their interrelationship if we want to understand Theodoret's treatment of that theme in full.

3.1 Saints, cities, mountains

One major achievement from the new wave of research on the saints' lives over the past few decades has been to show that the image of solitary ascetics inhabiting wild spaces often conceals a more complicated reality, where monastic communities were closely integrated into the communities around them. That has become increasingly clear from the archaeological evidence (e.g. see Brooks Hedstrom, 2017). It is also often visible in the texts themselves, if we pause to look at some of the less spectacular details behind their idealising portraits of ascetic solitariness (see Rousseau, 1978, 33–67, 1985, 149–173; Goehring, 1999, especially 45–52; also Brown, 2008, li–liv). Much of the recent work on that subject has focused on Egypt, as the setting for the first ancient works of hagiography, but some researchers have also pointed out that the integration of saints into urban culture is even more pronounced for the Syrian ascetics as Theodoret represents them (see Price, 1985, xx: 'significant too is their location on frequented routes; contact with the outside world seems to have been cultivated rather than shunned'; see also Urbainczyk, 2002, 36, 80–88, who discusses the tension between isolation and engagement in Theodoret, and the way in which one of Theodoret's goals is to present Syrian asceticism as a serious rival to its Egyptian equivalents). The same goes for their counterparts in Palestine, whose monastic communities were close to and closely linked with Jerusalem (Binns, 1994, especially 79–147).

Of course, that is partly a matter of geographical reality. Theodoret is acutely aware of what makes the geography of Syria distinctive. For example, the cultural entanglements of Theodoret's saints are, not surprisingly, quite different from what we see for the Egyptian ascetics in other texts. Theodoret regularly looks eastward to Persia and beyond. Theodoret's first subject, James, comes from Nisibis, which Theodoret introduces as 'a city on the boundary of

DOI: 10.4324/9780429427152-4

the Roman and Persian empires' (Theodoret, *Religious History* 1.1), and James spends time in Persia. In a later chapter, Theodoret tells us that 'Macedonius is known to all people – Phoenicians, Syrians, and Cilicians, and the neighbouring people who border on them know him too' (13.1) (Simeon Stylites similarly has many followers from the east, but in some ways he is an exception, since Theodoret also stresses that he is known to peoples in other parts of the world too; e.g. the Iberians in the west: see especially 26.1, 26.11 and 26.13, with Burrus, 2019, 129). In addition, the distinctive relief map of Syria, with high ground often very close to cities, and a large number of villages at relatively high altitude, goes a long way towards shaping the close relationships between the saints and the communities around them that Theodoret presents (for good introductory discussion of the geography and religious culture of late antique Syria, see Brown, 1971, 83, drawing a contrast with the deserts of Egypt; Canivet, 1977, 147–205; Millar, 1993, 236–241; also Schor, 2011, 5–6, with extensive further bibliography). The lack of any firm dividing line between city and wilderness in Theodoret's work might also be explained by the fact that some 'urban' settlements in late antique Syria were relatively small and quite rural in character. The prominence of Antioch in Theodoret's account suggests, however, that this is not the only explanation: Antioch was in this period one of the largest cities of the Roman Empire, but Theodoret makes it clear that the city was even so closely integrated with the marginal, mountainous territories traditionally associated with ascetic habitation. More accurate might be to say that Theodoret is responding to the wide range of different settlement sizes that characterised late antique Syria: he is interested in the way in which ascetic–community relations work in strikingly similar ways across that full range, with Antioch at one end of the spectrum, then much smaller cities like Rhosus (10.1, 11.1) and Gabala ('a small and charming city', 28.1), and Theodoret's own home town of Cyrrhus (3.1, 14.1, 17.2), then a series of very much smaller habitations described by Theodoret explicitly as 'villages' (e.g. Capersana at 19.1, Omeros at 20.1, Tillima at 22.1).

Theodoret's text thus reflects some of the realities of Syrian geography (for the distinctive features of Syrian monasticism in its geographical setting, including discussion of the evidence from Theodoret, see Canivet, 1977, especially 207–233; Finn, 2009, 136–143). At the same time, it is also clear that his portrayal is shaped by his own distinctive literary vision, albeit using imaginative resources of geographical representation that have a long pre-Christian history. Mountains in earlier Greek and Latin literature had often been represented as having an ambiguous relationship with city culture. We see that even as early as the eighth century BCE, in the *Iliad*: some of the mountains in that text, especially the mountains of the similes, are represented as spaces beyond human control or even, in the case of Mt Olympus, as dwelling places of the gods, utterly inaccessible to human presence. Others, by contrast, are represented as having a close relationship with the economy and identity of particular cities, in a way which reflects the legal status and economic significance of mountain territories across many centuries of Mediterranean history. We see that repeatedly for Mt Ida, which stands next to the city of Troy: this is where the Trojans pasture their flocks, in times of peace, or gather wood. There are also many examples in the Catalogue of the Ships in *Iliad* Book 2 ('those who inhabited Arkadia beneath the high mountain of Kyllene by the tomb of Aipytos, where the men are good at fighting hand-to-hand' [2.603–2.604]; 'next was Odysseus, leading the great-hearted Cephallenians, who held Ithaca and Mt. Neriton with its windswept woods' [2.631–2.632]). Those different conceptions of city–mountain relations stand in tension with each other throughout the poem (König, 2022a, 30–35). There are many parallels in later literature. Dio Chrysostom's *Euboian Oration* (dating from around 100 CE) contrasts an idealised image of isolated rustic existence in the mountains of Euboia with a series of what are represented as unjust and inept attempts by the nearby city dwellers to assert ownership over that land in the mountains and to make it economically productive (König, 2019, 2022a, 267–282). We see a version of those tensions in other hagiographical writing too, for example in Jerome's *Life of Hilarion* (late fourth century CE), which shows us urban culture repeatedly breaking in on Hilarion's idealised mountain retreats (König, 2017, 2022a, 300–304). Theodoret's representation of the entanglement between mountains and cities thus has a long heritage lying behind it. It is also, however, unusually complex and sustained: there are not many other hagiographical texts that present us with so many different versions of that kind of city–mountain relationship side by side with each other (on mountains and late antique hagiography, see König, 2022a, 283–304, including brief discussion of Theodoret at 290–292, and Whalin, 2021; also della Dora, 2016, 147–175 on the Byzantine material).

Many of Theodoret's subjects oscillate between positions outside and inside the cities of Syria. Some start off outside civilisation early on in their lives. To take the example of Theodoret's first subject once again, James of Nisibis 'embraced the isolated and quiet life, and gaining the summits of the highest mountains he lived among them. In spring, summer, and autumn he made use of thickets for shelter, having the sky for a roof; in winter a cave received him and provided sparse protection' (1.2). We will see other examples later. Theodoret's life of Eusebius of Teleda, however, opens by signalling a transition from the world of the desert, in the first three lives, to the 'inhabited land' (τὴν οἰκουμένην) (4.1), with the intention of showing that more populated terrain is no hindrance to virtue (but see Urbainczyk, 2002, 72, pointing out that that claim ignores the fact that James himself had later in his career been bishop of Nisibis). There is also a repeated pattern whereby many of the saints become absorbed within cities later in their lives, despite their reluctance. Macedonius is deceived into coming down from his mountain, at 13.4, and ordained by trickery (see also 13.7, where Macedonius comes down voluntarily to intervene against a pair of Roman generals who have ordered the destruction of the city). James of Cyrrhestica is brought down to the city when he is sick and unconscious (21.9). In other cases the bodies of ascetics are brought into the city after their deaths, or in some cases disputed over by competing cities or villages (e.g. see *Religious History* 15.5 and 16.4; also see 19.3 on Salamanes, who is carried backwards and forwards between two rival villages while still alive; Shepardson, 2014, 189–190; Jerome, *Life of Hilarion* 32.7 for an earlier parallel). Theodosius is a good example. Theodoret tells us that he was born to an illustrious family in Antioch. He spends much of his life far outside the city living simply with a small community around him, but later the leaders of the church are afraid that he will be taken hostage for ransom by barbarians and persuade him to return to the city. When he dies, his body is carried 'through the middle of the town' in procession and buried in the shrine of the holy martyrs, 'escorted by all, including those entrusted with the major offices' (10.8) (cf. Theodoret, *Ecclesiastical History* 1.32 for similar scenes on the death of Constantine; also again Jerome, *Life of Hilarion* 32.7 for a good hagiographical parallel; also Nadine Viermann and Esen Ogus in this volume for examples of the kinds of civic, processional place-making that Theodoret is responding to here).

The precise place of habitation is often important for these effects. Many of the ascetics inhabit mountains that stand just a few miles outside the cities. James of Cyrrhestica lives on a mountain which is just 30 stades from the town of Cyrrhus (21.4). Thalelaeus lives on a hill 20 stades from Gabala (28.1) (for other examples, see 4.2, 10.1, 12.2, 18.1, 22.1, 22.1, 23.2). In some cases these choices of habitation are represented as attempts to take over spaces connected with Greco-Roman religion, especially in examples where Theodoret's ascetics drive out demons from pagan shrines (see Shepardson, 2014, 196–198; see also Kleinkopf, 2017 on Christianisation of the landscape, involving occupation of pagan spaces by the saints) (cf. Rebecca Sweetman in this volume: these incidents are less formal than the church foundation processes she studies, in the sense that Theodoret's ascetics always maintain at least a semi-nomadic character, but they are nevertheless closely related, as examples of a less settled but not for that reason less significant variety of place-making). The 'hill' (γήλοφος) inhabited by Thalelaeus was home to a 'sanctuary dedicated to demons and long honoured by the ungodly with many sacrifices' (28.1); Thalelaeus drives them away eventually by his piety (cf. other examples at 4.2, 6.4, 9.10, 16.14). There are parallels for that motif in earlier Christian literature, for example in Jerome's *Life of Hilarion* (especially *Life of Hilarion* 31.4–31.5), but Theodoret uses it with unusual frequency. Others have rather different reasons for inhabiting these spaces. For example, Julian Saba is persuaded late in his life to come back to Antioch from the desert and ends up living even closer to urban territory, on Mt Silpius (cf. 11.1 on Romanus for another ascetic on Mt Silpius). Mt Silpius is a good example of the proximity between mountains and cities that is so distinctive to the geography of Syria: the mountains stands on the very edge of the city of Antioch, by contrast with the mountains of Egypt, which are located much farther from human habitation, and where the words mountain (*oros*) and desert (*erêmos*) are used almost interchangeably (see König, 2022a, 283). Theodoret tells us that he was greeted by a great crowd and settled 'in the caves at the foot of the mountain (ἐν τῇ ὑπωρείᾳ), where they say that the holy apostle, the blessed Paul, settled and was hidden' (2.18). In that case, his return to the city is a response to theological conflict. Theodoret tells us that Bishop Meletius, whom he represents positively as a fellow adherent to orthodox Nicene doctrine, had been expelled from the city (in 365–369 CE) and that his followers had taken to assembling precisely in public spaces just outside the city:

> now occupying the foot of the mountain (τήν ὑπώρειαν), they would hold in that place their sacred assemblies; now they would make the bank of the river a place of prayer; now they would do the same for the military training ground situated in front of the north gate; for the enemy did not allow the pious to settle in one location.
>
> (2.15)

A rumour circulates that Julian has embraced Arian doctrines, and he is persuaded to come back to Antioch to disprove it (Shepardson, 2014, 24–25). His decision to settle just outside the city on Mt Silpius is among other things a way of enacting his theological position in sympathy with Meletius.

Others inhabit spaces which could almost be described as suburban, in the sense that they are represented as being on the edges of the city's built space. Domnina, for example, sets up a small hut made of millet stalks in her mother's garden (30.1): that location involves an unusual combination of an urban setting and a rustic style of dwelling. Others live in tombs, for example Peter the Galatian in Antioch (9.3). Sometimes we see both of these features combined, in the case of ascetics who live shut up in sealed, tomb-like dwellings within or just outside the city boundaries. Aphrahat moves to the city of Edessa, which Theodoret describes as 'very large and well-populated and exceptionally brilliant in its piety'; there he finds 'a small house outside the city-walls (ἔξω τῶν περιβόλων) and confines himself' (8.1). Later he returns to Antioch, where he settles 'in front of the town (πρὸ τοῦ ἄστεως)' (8.2). Marana and Cyra, two female ascetics from Beroea who were still alive at the time Theodoret was writing, 'acquired a small place in front of the town (πρὸ τοῦ ἄστεως), and going inside it they walled up the door with clay and stones' (29.2). Salamanes (19.1) and Maris (20.1) live immured in huts in their own villages. The occupation of burial places, of course, has plenty of parallels in earlier Christian culture – the *Life of Antony* is an obvious example (Athanasius, *Life of Antony* 8–11) – but that motif in Theodoret's case gains additional force through being part of this series of examples where the ascetics inhabit marginal, suburban spaces. In an Antiochene context, cemeteries also have special associations as locations for the *martyria* that were so important for the Christian communities of the city, as reflected in the writings of John Chrysostom in the fourth century (Shepardson, 2014, 25). In occupying tombs or tomb-like spaces on the edges of the cities, Theodoret's ascetics make their dwelling places equivalent to living martyr shrines.

We have seen, then, that Theodoret is working with familiar concepts and literary resources; also, that he is reflecting some of the topographical realities of the space around Antioch and around the cities of Syria more broadly. At the same time, he manipulates those resources and realities in challenging and creative ways. His technique of viewing multiple saints and multiple locations side by side with each other allows him to map out a huge range of possibilities for marginal, suburban or extra-urban locations. In that sense, Theodoret offers his readers (perhaps in part because of his own role in mediating between urban and rural Christian communities within his own career, discussed further later) an unusually multifaceted portrayal of the way in which ascetic isolation could be in tension with various kinds of communal engagement and urban location.

We might even view this as a text that anticipates – like other saints' lives too – modern accounts that deconstruct the notion of 'wilderness'. In practice, the spaces we tend to idealise as wildernesses in modern culture tend to be profoundly shaped by human culture and human imagination (see Cronon, 1995; cf. König, 2022a, 267, for further discussion, with reference also to scholarship on the European pastoral tradition, e.g. Williams, 1973, which similarly explores the artificiality of idealised, nostalgic images of separation between rural and urban). Recent ecocritical writing has increasingly tended to be interested in hybrid spaces of nature that are marked by the inevitable presence and debris of the Anthropocene, which can prompt us to acknowledge the inevitability of entanglement between human and environment (see Clark, 2019, 31–37, for discussion of the increasing volume of work on these themes). Theodoret's version of that tension is of course quite different in some ways from its modern equivalents, but for him too there is no such thing as a space of pristine and lasting ascetic isolation: urban culture always impinges, even on the mountains of Syria as the *Religious History* represents them. Some of the ascetics seem to accept the inevitability of that impingement by their choice of hybrid, suburban spaces just on the edges of civilised habitation (cf. Burrus, 2019, 106 on the *Life of Antony*: 'distinctions between urban and rural and artificial and natural break down'). Even tombs and other tomb-like habitations tend to be open to intrusion from urban visitors (e.g. Peter the Galatian's tomb, at 9.3, which has an upper level with a ladder to make it open to visitors; or Marana and Cyra's walled-up hut: at 29.5, Theodoret describes visiting regularly and being invited to dig through the door in order to go inside).

Theodoret's erosion of the boundary between city and wilderness – which involves making the wilderness open to urban influences and contacts, but also in some respects making urban communities 'wilder' by their absorption of ascetic figures who would usually be defined by their hostility to urban civilisation – also has important functions for Theodoret's self-positioning in relation to the theological controversies that dominated the Syrian church in the fourth and fifth centuries CE. One of his goals is to challenge standard perceptions of the Syrian holy men as inhabitants of the wild places outside human culture and to show that they are ultimately under clerical control, in an attempt to bolster his own authority in the theological struggles between Antiochene and Alexandrian Christianity (on the wider context of Theodoret's patronage and friendship networks in Syria in the light of his involvement in Christological disputes, see Schor, 2011). Theresa Urbainczyk (2002, 9) has argued along similar lines: 'Theodoret wanted to show that he was an extraordinary bishop with an army of remarkable monks behind him, the equal of any Egyptian. To the upper classes of these big cities these were wild men from the Syrian mountains. To Theodoret they were devoted and loyal' (also more detail at Urbainczyk, 2002, 115–129; cf. Schor, 2011, 119: 'all of this served as an apologetic to outsiders.

By displaying Syria's holy people, Theodoret lent support to their clerical associates. Indirectly he countered suspicions of heresy. Just as important, the *Religious History* reinforced relationships within the Antiochene network. He featured cooperative ascetics, which made them more attractive to clerics. And he featured ascetic-minded clerics, which made them more attractive to the monks'). That is especially the case in the second half of the work, where he moves from talking about the ascetics of previous generations to those he has known himself, emphasising his own intimacy with them (see Schor, 2011, 118–120 for the way in which Theodoret's letters reinforce the picture of cooperation between Syrian clerics and ascetics that we find in the *Religious History*). But it is also an important theme in the first half of the work, where he describes the ascetics of fourth-century Syria. Many of the incidents we have looked at already, involving ascetics returning to the cities, were interventions in theological conflict. We have already seen the example of Julian Saba's return to Antioch in support of Meletius, at 2.18. There are several other examples in the *Religious History*, and also several mentions of these developments in Theodoret's later work the *Ecclesiastical History*, which, despite its more exclusively urban focus, does give us occasional glimpses of how the mountain ascetics could exercise theological and ecclesiastical authority at moments of crisis (see also Schor, 2011, for discussion of the involvement of ascetics in Theodoret's networks, e.g. at 47, 62, 72–75). In *Ecclesiastical History* 1.6, for example, Theodoret describes the attendance of James of Nisibis at the Council of Nicaea, with an explicit cross-reference to the account of James's life in the *Religious History*; at 3.19, he retells the story of Julianus Saba's return to Antioch, again with cross-reference to the *Religious History*. Those incidents act as precedents for the picture of cooperation between ascetics and clerics that Theodoret paints for his own adult lifetime in the second half of the *Religious History* (cf. Schor, 2011, 119 for the way in which parts of the *Religious History* 'drew a lineage of mentorship from founding fathers (like Marcian and Julian Saba) to contemporaries').

It was thus in Theodoret's interest to draw attention to the permeability of the dividing lines between city and mountain. In doing so, he makes it clear that he is dealing with a context quite distinctive to fifth-century Syria and quite different from the patterns of monastic separation that his readers would be more familiar with from earlier hagiography with an Egyptian context. It is important to emphasise, however, that Theodoret's attempts to bring his ascetics into the ambit of the urban church society he was a part of had its limits (see Urbainczyk, 2002, 115–129, for a survey of relations between ascetics and the church in the text), and that Theodoret would not anyway have wanted to dilute too much the rural positioning that gave the ascetics so much of their power. Ultimately, in other words, the geographical vision of the *Religious History* is very different from the vision of theological and geographical interconnection that we find in Theodoret's *Ecclesiastical History* and also in his letters, which conjure up a picture of a cosmopolitan network of correspondents in which Theodoret plays a central role. The networks in which the saints of the *Religious History* are involved are for the most part very different in character, founded on local links with both notable and ordinary people in nearby towns and villages, as we have seen repeatedly throughout this section, for example in the repeated scenes where Theodoret mentions his own personal visits to the individuals he describes (e.g. see 29.5, already discussed earlier, for his visits to Marana and Cyra; also among others 21.33, 22.3, 24.4, 28.4; and detailed discussion with a full list by Price, 1985, xiii). In some respects the difference between the *Religious History* and Theodoret's other works reflects the dual nature of his role as bishop, involved in the wider church, but also with local duties to rural communities (on Theodoret's role as bishop and his relations with rural society, see Schor, 2011, 6–7). It must be right, in other words, that Theodoret has his wider theological networks and debates in mind even in a work like the *Religious History* where they are not discussed explicitly very much, and it must be right that his goal is to enhance his own authority by showing his own connection to the Syrian ascetics. It is still striking, however, that he keeps them separate from that wider world in many respects. In that sense, the word 'suburban' again seems like the right one to use: they are on the edge of urban concerns, but not at the heart of them; they always maintain a certain distance from the elite intercity networks that are so central to Theodoret's activity and identity in other contexts.

3.2 Human and environment in the *Religious History*

I have argued that Theodoret's *Religious History* makes use of some standard patterns of spatial representation from ancient literature – especially the tension between images of mountains within and outside urban culture, and the tension between solitary and communal models of asceticism – reshaping them and intensifying them in distinctive ways. My argument in the second half of the chapter is that Theodoret's portrayal of human–environment relations is thematically connected with those elements of his text.

The role of environmental themes in the early saints' lives has gained renewed prominence recently, thanks to Virginia Burrus's book *Ancient Christian Ecopoetics*, published in 2019. Her research responds among other things to long-standing assumptions about the anthropocentrism of early Christianity, which has been taken as having a formative influence on the devaluation of the natural world in Western culture whose consequences we are living with today. That view was shaped by a very brief but still influential article by Lynn White Jr from the 1960s (White, 1967). Burrus by contrast argues that the saints' lives repeatedly subvert anthropocentric assumptions, representing the ascetics as figures on the very edge of human identity (see especially Burrus, 2019, 2–4 on responses to White, 1967). She breaks new ground (if that is not an inappropriate image for a work that explores alternatives to human-centred images of dominance over nature) in applying ecocritical, new materialist and posthumanist approaches to late antique hagiography; there has been an increasing volume of recent work along similar lines in other areas of classical scholarship too (for new materialist and posthumanist approaches within Classics, see among others Chesi & Spiegel, 2019; Bosak-Schroeder, 2020; also Schliephake, 2017, 2020 for wide-ranging accounts of the relevance of ecocriticism for Classics; and for an introductory account of ecocriticism as a discipline more broadly, see Clark, 2019, especially 111–136 on 'material ecocriticism', including discussion of new materialist approaches and extensive further bibliography). Burrus's wonderful reading of Theodoret's life of Simeon Stylites (123–135) is one of the centrepieces of her argument.

Burrus's (2019) book clearly has the potential to transform the way in which these issues are discussed. Even she, however, chooses not to take the further step of tracing these themes in their full complexity across Theodoret's work. One of the first things that become clear in making that attempt is that there are many other sections of the work that give powerful support to Burrus's claim that the text offers a posthuman vision of ascetic practice (cf. Kleinkopf, 2017, for extensive discussion along similar lines on the way in which the saints of Theodoret's text embed themselves in the landscapes they inhabit). At 21.3, we hear that James of Cyrrhestica

> has the sky for his roof, and receives all the competing impacts of the open air, now deluged by violent rain, now frozen by frost and snow, at other times burned and set on fire by the rays of the sun, and yet persevering in the face of all these things.
>
> (cf. 1.2 and 23.1 for other good examples)

Here, the surfaces of the human body seem to be vulnerable to environmental forces. In that sense, Theodoret's representation of James, like many other representations of hagiographic exposure, anticipates modern ideas about the immersive quality of our experiences of landscape and environment, for example in Tim Ingold's concept of the 'weather world':

> Rather than thinking of ourselves only as observers, picking our way around the objects lying about on the ground of a ready-formed world, we must imagine ourselves in the first place as participants, each immersed with the whole of our being in the currents of a world-in-formation: in the sunlight we see in, the rain we hear in and the wind we feel in.
>
> (Ingold, 2011, 129)

There are many other passages along similar lines in the *Religious History* that emphasise the vulnerability of ascetic bodies to the elements. For example, the hut of the ascetic Maris 'received much rain from the nearby mountain; in winter it even spouted out streams of water' (20.1). Sometimes this closeness to the elements gives the ascetics a quasi-animal identity: for example, Acepsimas is mistaken for a wolf, at 15.2 (cf. Burrus, 2019 for regular discussion of ascetics represented in animal terms, especially 106–110 on Athanasius's *Life of Antony*, Jerome's *Life of Paul of Thebes* and Sophronius's *Life of Mary*).

One of the most persistent manifestations of that broad theme is in scenes where the ascetic body is in close contact with the ground. There is a particular cluster of these scenes in the second half of the text. At 17.2, for example, the ascetic Abraham travels to a village that has a reputation for impiety, together with his companions; the inhabitants hear them singing psalms: 'Men, children and women gathered together and walled up the doors from the outside' – mirroring the kind of immurement that Theodoret's ascetics often impose on themselves freely – 'and heaping up a great pile of earth they poured it down from the roof above'. But far more often, this is a voluntary act. James of Cyrrhestica's exposure to the weather, already discussed earlier, causes him to be immersed in snow:

> Often when snow has fallen for three days and the same number of nights, he has been covered by it, while lying face downwards and praying to God, to such an extent that not even a small piece of the rags that he wears can be seen. And often his neighbours, using spades and forks, and in this way shovelling aside the snow that covers him, pull him out and revive him as he lies there.
>
> (21.13)

In many cases, the saints are brought close to the ground by iron, which they bind to themselves as an act of discipline. That is true for Simeon Stylites, as Burrus has noted. At 26.10, he attaches himself to a rock with iron and insects torment his skin beneath the bonds (see Burrus, 2019, 128; earlier, at 26.7, Simeon had spent some of his time immersed in the ground in a waterless cistern). Cyra, at 29.4, weighs herself down with iron: she is 'bent down to the ground, and not at all able to straighten her body'. Others find other ways of producing the same effects without the same dependence on the products of human industry: Polychronius uses the heavy root of an oak tree to weigh himself down (24.6). There is a particularly intriguing example at 4.6, where Eusebius of Teleda is distracted from a reading of the scriptures by the sight of some men ploughing the land in the plains down below. In his shame he makes a rule for himself that his eyes will never take similar delight again from looking at the plains and the heavens: 'binding his waist with an iron belt, and placing a very heavy collar around his neck, with another piece of iron he attached the belt to the neck-collar in order that bent over in this way he would be forced to stoop down and look at the ground continually' (see König, 2022a, 292, for discussion of the way in which this passage relates to ancient ideas about looking from mountains and hillsides). One of the many fascinating things about this scene is the fact that the original act of looking is directed at 'farm labourers' (γηπόνων δέ τινων) who are themselves interacting with the ground, in a way that is productive in conventionally anthropocentric terms. Eusebius's response is thus represented as a rejection not just of the pleasures of viewing but also of human industry in relation to nature. His relationship with the ground could hardly be more different.

When we look at the text in full, however, it becomes clear that Theodoret's understanding of the relative value of different types of human–environment relations is more complicated than these images of human immersion in the landscape suggest. Just as the ideal of a wilderness setting for ascetic habitation is often subverted or at least qualified, among other things by the regular movement of ascetics back into the cities, so these images of the human body open to the air and to the ground stand in tension with other more anthropocentric portrayals of human control over the natural world (see also Cox Miller, 1994, for a rather different kind of tension, between the earthbound nature of the saints' bodies and the angelic identity they attain through their practices of asceticism). Theodoret's account of Simeon contains some of the most prominent examples in the text of the human body merged with its environments, as Burrus (see especially 2019, 31–33) makes clear, but it is worth noting that it is in some respects far from typical of the work as a whole (e.g. Urbainczyk, 2002, not mentioned by Burrus, has emphasised the way in which Theodoret's account of Simeon is far from typical, and in some respects disapproving of his excesses).

The importance of anthropocentric models for the *Religious History* becomes clear from the very first sentence of the work after the end of the preface, which tells us that 'Moses, the divine law-giver – who laid bare the sea bed, inundated the moistureless desert with water and performed all the other miracles – wrote down the way of life of the holy men of the past' (1.1) (see Krueger, 2004, 15–32 for Theodoret's unusually extensive use of biblical typology in the *Religious History*; especially 30–31, pointing out that this passage also compares Moses's inspiration with Theodoret's own; cf. Rapp, 1998, 290–291, n. 63). James is then compared implicitly with Moses in the first miracle described in the text, which represents a less tolerant rewriting of the story of Moses's defence of the daughters of Jethro at the well at Exodus 1:16–17. James sees some girls washing clothes at a spring behaving immodestly; he curses the spring, which dries up, and also the girls, whose hair instantly turns grey. The local people plead with him to soften his anger: 'without even a moment's delay he addressed his supplication to the Lord, and ordered the springs to pour forth once again, and immediately they appeared again out of their underground reservoirs' (1.4) (he offers to restore the girls' original hair colour too but then decides to let that punishment stand when they do not appear) (see Rapp, 1998, 291, comparing this with passages from other authors where a holy man is described as a 'new Moses'). A remarkable third passage in this series then makes it clear that James's powers – or at least the divine powers that James channels – outdo the kinds of military control over landscape that are descended ultimately from the stories of the Persian king Xerxes and his hybristic acts of landscape alteration, in the whipping of the Hellespont and his digging of the peninsula at Mt Athos (for those traditions, see König, 2022b, which explores the tensions between positive and negative representations of human mastery in ancient historiography and geography). The new Persian king Sapor has besieged James's home city of Nisibis. He builds towers and mounds and attempts to undermine the city walls. When he has no success with those techniques, he dams up the river, then releases the dam so that the rush of water breaks through the walls. They postpone their assault because of the depth of the water; overnight, the inhabitants build up the walls as far as they can. They also call James for help:

> he obeyed and went up, and having seen the many tens of thousands of the enemy, he prayed to God to send on them a cloud of gnats and mosquitoes. He spoke, and God sent, persuaded as he was by Moses. And the men were wounded by the divine missiles; and the horses and elephants, breaking their bonds, ran away scattering in all directions, unable to bear those arrows.
>
> (1.11)

The Persians give up and return home (see Theodoret, *Ecclesiastical History* 2.26, for repetition of the story of James and the Persian siege of Nisibis). James's Moses-like ability to call up divine help against an impious enemy is thus one of several passages that represent him as exercising control over the natural world, much more effectively so than the Persian king with his earthworks. It stands in tension with the passages earlier in the text which represent him living patiently immersed in his environment, 'having the sky for a roof' and 'gathering the natural fruits of the wild trees . . . refusing the use of fire' (1.2).

There are other passages too where Theodoret's ascetics exercise control over the natural world. The best match for these waterworks comes in 10.7, where Theodosius calls forth water from a cliff face by prayer, having previously built a conduit to bring the water into the monastery below. This passage may not be relevant to an account of Theodoret's own design, since it comes in a passage whose authenticity has been generally discounted (see Canivet & Leroy-Molinghen, 1977–1979, I, 449), but even so, it suggests a later reader or editor who is alert to the importance of that theme elsewhere in the text. Julian Saba too is compared with Moses, in much the same way as for James of Nisibis, whose life immediately precedes his. At 2.7–2.8, for example, he brings forth a spring of water:

> The spring has lasted even until now, bearing witness to the Moses-like prayer of the divinely inspired old man; for just as Moses once long ago by striking that sterile rock with his stick flooded it by the generation of river waters, in order to satisfy those many tens of thousands of thirsty people, so this holy man, by watering with his tears that very dry sand, drew forth streams of spring-water, in order to slake the thirst not of many tens of thousands but of a single youth.
>
> (2.8) (see Rapp, 1998, 289–290, n. 56)

There are other similar incidents later narrated without explicit mention of Moses but nevertheless involving control over the kinds of natural forces that we find in the Exodus stories. At 26.19, Simeon foretells a drought and also a plague of locusts, with reassurances that they will not damage human food supply. At 7.14, Aphrahat drives away a plague of locusts with holy water. At 24.7, Polychronius produces oil for a suppliant at a time of drought.

The text also includes an intriguing cluster of scenes involving deforestation of various kinds (which is not, to my knowledge, a common theme in other fourth- and fifth-century hagiography). As we have seen already, Thalelaeus, at 28.1, settles on a hill that has a shrine dedicated to demons. They try to frighten him away, but with no success:

> Therefore, filled with rage and madness, they launched an attack against the trees that were planted there: there happened to be many flourishing fig trees and olive trees on this hill. Of

> these they say that more than five hundred were suddenly pulled up; I heard this story told by the nearby country-people.

The mention of demons' rage and frenzy invites us to see this as an act of mindless destruction, a senseless manifestation of their emotions of frustration. The mention of the country-people hints, but without making it explicit, at a negative impact on the surrounding population. This is represented, in other words, as a negative, destructive act of landscape alteration, standing in contrast with the more positive versions of ascetic control over the natural world already noted earlier.

Oddly, however, there are also two scenes where the saints themselves are associated with the cutting down of trees. The first is in the account of Acepsimas, who lives in a state of seclusion; at one point, a visitor climbs a plane tree near to his enclosure in an attempt to find out what the saint is doing and is immediately paralysed: Acepsimas 'predicted that his health would be restored by the felling of the plane-tree: for he ordered the tree to be cut down at once, in order to prevent anyone else from doing the same thing and suffering the same consequence' (15.3). Here, the desire to maintain ascetic seclusion is represented as an admirable justification for destruction of natural resources. Similarly, at 23.1, John cuts down a tree that is making him too comfortable: 'When some earnest person planted an almond right by his bed, and then in time it became a tree and provided him with shade and feasted his eyes, he ordered it to be cut down, in order to avoid getting any pleasure from it' (cf. Burrus, 2019, 100, for interesting discussion of trees in hagiographical *locus amoenus* scenes, pointing out that these passages often involve trees that are relatively sparse and unimpressive by comparison with their more luxuriant equivalents in the Greco-Roman *locus amoenus* tradition, with special reference to *Life of Antony* 49–50). We could choose to view this as an example of John rejecting anthropocentric alteration of the natural environment (i.e. tree planting), but it is hard to suppress entirely the impression that John's own activity is an anthropocentric act of landscape alteration in itself: his own ascetic identity seems to matter more than any ideal of passive coexistence with non-human entities.

Moreover, in Theodoret's use of metaphor, the posthuman ideals we find on display in some sections of the text are in tension with heavily anthropocentric imagery. The predominant metaphors for ascetic practice, as in other hagiographical writing, are human-centred, often urban or agricultural. In the former category are the ubiquitous images of athletic struggle (Urbainczyk, 2002, 85). In the latter category are images of cultivation and pasturing. The summary to Theodoret's account of Eusebius of Teleda is typical of both:

> Such victorious competitors did the holy Eusebius bring before God, as coach and athletic trainer of all these contests. And he has trained very many others of this kind and sent them to be teachers in other wresting schools, and they have filled that whole holy mountain with these divine and sweet-smelling meadows.
>
> (4.13)

Aphrahat finds a hut outside the city walls of Edessa and immures himself; there 'he tended his own soul, like some excellent farmer pulling out by the roots the thorns of his passions and cleansing the divine crop, and offering to the Lord the seasonal fruits from the seeds of the Gospel' (8.1). By contrast, natural-force imagery tends to be associated with external dangers. For example, Aphrahat goes soon afterwards to Antioch, which 'was being seriously stirred up by the storm of heresy' (8.2). The accession of Emperor Valens is a time when 'storms and hurricanes stirred up the sea around us, and a dangerous swell rose up and from all sides groups of waves dashed against the ship', and Valens himself is compared with a wild beast scattering the 'flock' of the pious (8.5) (there are good parallels in Theodoret's own work, e.g. in the very opening metaphor of *Ecclesiastical History* 1.1; and in other late antique historiography, e.g. in Ammianus Marcellinus, in the opening sections of the work as it survives, at *Res gestae* 14.1.1 and 14.1.10). Of course, there may be a paradox implied: that their metaphorical cultivation of their souls means that the ascetics are less involved in attempts at literal mastery of the external, physical world. But images like these nevertheless extend the impression, which as we have seen is common in other parts of the text too, that the ascetic association with non-human kinds of identity and practice oscillate with heavily anthropocentric assumptions. That tension is not unique to Theodoret – there are plenty of parallels for both of these strands from other hagiographical texts from the late fourth and early fifth centuries (see Burrus, 2019, 81–144 for the wider context; see Jerome, *Life of Hilarion* 40, for another good example, describing Hilarion's Moses-like control over a tsunami; and Gregory the Great, *Dialogues*, for a good sixth-century example of a text that includes repeated images both of human immersion in the environment and ascetic control over rocks and water, e.g. at 2.5) – but his exploration of it plays an unusually prominent role in his design of the text as a whole.

3.3 Conclusion

My argument in this chapter has been that if we want to understand Theodoret's representation of human–environment relations in the *Religious History* we need to read the text from end to end, with attention to the productive inconsistencies that are embedded in it. As Virginia Burrus (2019, especially 123–135) has pointed out, the posthuman resonances of much early hagiography, with their interest in the fluid dividing lines between culture and nature, between human and animal, reveal the limitations of traditional ideas about the anthropocentrism of early Christianity. It is also clear, however, that anthropocentric imagery can nevertheless play an important role in these texts. Theodoret's *Religious History* is a good example. What we find in that text, as for so many other examples of ancient environmental thinking too, is a spectrum of different possibilities for envisaging human control over or immersion in the natural world. Theodoret has a sophisticated awareness of the way in which entanglement between human saints and the more-than-human environment, which is one of the defining features of ascetic identity in early Christian hagiography, can never be fully insulated from more conventional images of anthropocentric manipulation of the natural world. Those insights are thematically intertwined with Theodoret's interest in the way in which his ascetic subjects, however much they try to immerse themselves in wild, mountain places cut off from human society, nevertheless repeatedly end up being brought into the ambit of urban culture.

Editions and abbreviations

Athanasius, *Life of Anthony – Athanase d'Alexandre, Vie d'Antoine* (Sources chrétiennes 400), ed. G. J. M. Bartelink (1994). Cerf.

Gregory the Great, *Dialogues – Grégoire le Grand, Dialogues* (Sources chrétiennes 251, 260, 265), ed. A. de Vogüé (1978–1980). Cerf.

Jerome, *Jérôme: Trois vies de moines (Paul, Malchus, Hilarion)* (Sources chrétiennes 508), eds. E. M. Morales & P. Leclerc (2007). Cerf.

Theodoret, *Religious History – Théodoret de Cyr, Histoire des moines de Syrie* (Sources chrétiennes 234, 257), eds. P. Canivet & A. Leroy-Molinghen (1977–1979). Cerf.

Theodoret, *Ecclesiastical History – Theodoret de Cyr, Histoire ecclésiastique* (Sources chrétiennes 501, 530), eds. L. Parmentier, G. C. Hansen, J. Bouffartigue, A. Martin, & J. Canivet (2006–2009). Cerf.

Bibliography

Binns, J. (1994). *Ascetics and ambassadors of Christ: The monasteries of Palestine, 314–631*. Oxford University Press.

Bosak-Schroeder, C. (2020). *Other natures: Environmental encounters with ancient Greek ethnography*. University of California Press.

Brooks Hedstrom, D. L. (2017). *The monastic landscape of late antique Egypt: An archaeological reconstruction*. Cambridge University Press.

Brown, P. (1971). The rise and function of the holy man in Late Antiquity. *Journal of Roman Studies*, *61*, 80–101.

Brown, P. (2008). *The body and society: Men, women, and sexual renunciation in early Christianity* (Revised ed.). Columbia University Press. (original work published 1988).

Burrus, V. (2019). *Ancient Christian ecopoetics: Cosmologies, saints, things*. University of Pennsylvania Press.

Canivet, P. (1977). *Le monachisme syrien selon Théodoret de Cyr* (*Théologie historique*, 42). Éditions Beauchesne.

Chesi, G. M., & Spiegel, F. (Eds.). (2019). *Classical literature and posthumanism*. Bloomsbury.

Clark, T. (2019). *The value of ecocriticism*. Cambridge University Press.

Cox Miller, P. (1994). Desert asceticism and the body from nowhere. *Journal of Early Christian Studies*, *2*, 137–153.

Cronon, W. (1995). The trouble with wilderness: Or getting back to the wrong nature. In W. Cronon (Ed.), *Uncommon ground: Rethinking the human place in nature* (pp. 69–90). W.W. Norton & Company.

della Dora, V. (2016). *Landscape, nature and the sacred*. Cambridge University Press.

Finn, R. (2009). *Asceticism in the Graeco-Roman world*. Cambridge University Press.

Goehring, J. E. (1999). *Ascetics, society, and the desert: Studies in early Egyptian monasticism*. Trinity Press International.

Ingold, T. (2011). *Being alive: Essays on movement, knowledge and description*. Routledge.

Kleinkopf, K. (2017). A landscape of bodies: Exploring the role of ascetics in Theodoret's *Historia Religiosa*. *Studia Patristica*, *96*, 283–292.

König, J. (2017). Rethinking landscape in ancient fiction: Mountains in Apuleius and Jerome. In M. P. Futre Pinheiro, D. Konstan, & B. D. MacQueen (Eds.), *Cultural crossroads in the ancient novel* (pp. 275–288). De Gruyter.

König, J. (2019). Mountain and city in Dio Chrysostom's *Euboicus*. In M. Taufer (Ed.), *La montagna nell'antichità – Berge in der Antike – Mountains in Antiquity* (pp. 327–345). Rombach Verlag.

König, J. (2022a). *The folds of Olympus: Mountains in ancient Greek and Roman culture*. Princeton University Press.

König, J. (2022b). Ecocritical readings in late Hellenistic literature: Landscape alteration and hybris in Strabo and Diodorus. In J. König & N. Wiater (Eds.), *Late Hellenistic Gree literature in dialogue* (pp. 119–148). Cambridge University Press.

Krueger, D. (2004). *Writing and holiness: The practice of authorship in the early Christian east*. University of Pennsylvania Press.

Millar, F. (1993). *The Roman near east, 31 B.C.–A.D.* 337. Harvard University Press.

Price, R. M. (1985). *A history of the monks of Syria, by Theodoret of Cyrrhus*. Cistercian Publications.

Rapp, C. (1998). Comparison, paradigm and the case of Moses in panegyric and hagiography. In M. Whitby (Ed.), *The propaganda of power: The role of panegyric in Late Antiquity* (pp. 277–298). Brill.

Rousseau, P. (1978). *Ascetics, authority and the church in the age of Jerome and Cassian*. Oxford University Press.

Rousseau, P. (1985). *Pachomius: The making of a community in fourth-century Egypt*. University of California Press.

Schliephake, C. (2017). *Ecocriticism, ecology, and the cultures of antiquity*. Lexington Books.

Schliephake, C. (2020). *The environmental humanities and the ancient world*. Cambridge University Press.

Schor, A. M. (2011). *Theodoret's people: Social networks and religious conflict in late Roman Syria*. University of California Press.

Shepardson, C. C. (2014). *Controlling contested spaces: Late antique Antioch and the spatial politics of religious controversy*. University of California Press.

Urbainczyk, T. (2002). *Theodoret of Cyrrhus: The bishop and the holy man*. University of Michigan Press.

Whalin, D. (2021). Mountains and the holy in Late Antiquity. In D. Hollis & J. König (Eds.), *Mountain dialogues from antiquity to modernity* (pp. 89–107). Bloomsbury.

White, L., Jr. (1967). The historical roots of our ecologic crisis. *Science*, *155*, 1203–1207.

Williams, R. (1973). *The country and the city*. Chatto & Windus.

4 Living with neighbours in late antique Rome[1]

Michelle L. Berenfeld

All urban dwellers live in multiple cities at once. Each of us inhabits a version of our city that is different from that of our neighbours. 'Our' city, moreover, changes depending on what we are doing, with whom and when. It is a commonplace to say that a city is a palimpsest, made up of fragmentary layers and traces, but in any given moment, every city is also a whole. Ancient Rome, like any other urban environment, contained countless versions of itself – multiple, simultaneous cities with joins and points of overlap that were not always visible, similarly perceived or fixed in the minds of its occupants. These multiple cities coexisted in the same physical topography while at the same time being constantly constructed and reimagined as residents and visitors walked through the streets noticing and remembering different things, or not, and encountered each other. This chapter explores these concepts on the Caelian hill in Rome in the period from Constantine's arrival in 312 CE to Alaric's sack of the city in 410 CE and its aftermath. It seeks to better understand the Caelian as a neighbourhood in which changes in the physical fabric interacted with social habits and religious practices of its residents and visitors.

The Caelian is roughly bounded by the Colosseum and the Temple of Divine Claudius to the west and the Aurelian wall to the east (Figure 4.1). The Lateran Cathedral and what is now the church of Santo Stefano Rotondo anchor its north-east and south-east regions, respectively. These parameters broadly coincide with the limits of *Regio II Caelimontium* as reconstructed based on the Regionary Catalogues of the fourth century (Valentini & Zucchetti, 1940–1942, 92–94; Coarelli, 2007, 213–214) but extend beyond the eastern limit of *Regio II* to include the Lateran and slightly into *Regio III* to the north, in the valley that extends east from the Colosseum and separates the Caelian from the Oppian and the Esquiline.

The Caelian has been the object of exploration and excavation for centuries. The current study builds upon and is indebted to decades of work of other scholars and their resulting publications. The first modern account of the archaeology of the Caelian as a distinct topographical area by A.M. Colini (1944) remains foundational.[2] Colini's work was revisited and updated in a study by Carlo Pavolini (2006), whose publications over the past three decades (Giuliani & Pavolini, 1999; Palazzo & Pavolini, 2013; Pavolini, 1993a, 1993b, 2000, 2007, 2013; Pavolini et al., 1993) have presented the results of archaeological investigations of many major structures on the hill, most of which are now largely inaccessible. The Lateran Cathedral and its surroundings are most recently the focus of the ongoing *Rome Transformed* project, which promises to dramatically expand our understanding of that part of the city.[3]

The Caelian is best known in late antique studies for two things: the archaeologically fragmentary but huge *domūs* that occupy much of its upper plateau (Barbera et al., 2008; Brenk, 1999, 2003, 113–121; Carignani, 2000; Guidobaldi, 1986; Machado, 2012b, 2019, 230–258; Spinola, 2000) and the fourth- and fifth-century churches preserved near its base, which were built out of or on top of smaller but still luxurious private houses (Alberti et al., 2014; Brandenburg, 2005;

DOI: 10.4324/9780429427152-5

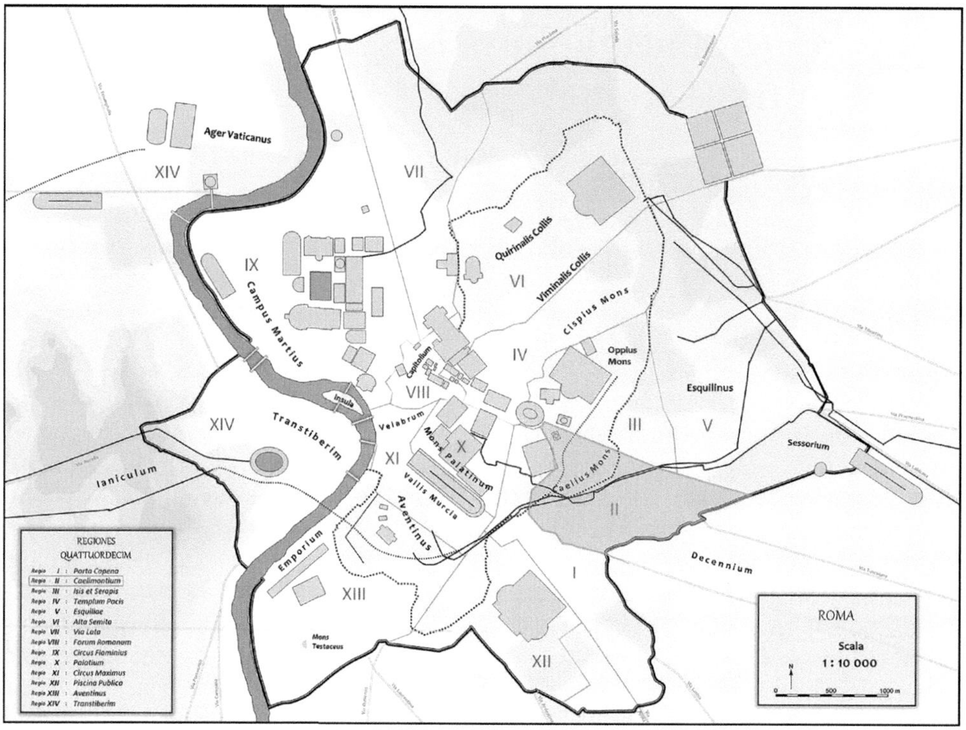

Figure 4.1 Map of Rome with *Regio II Caelimontium* highlighted.

Source: Adapted from 'RegioIIRome2.png' by Cassius Ahenobarbus (Wikimedia Commons). CC BY-SA 4.0.

Brenk, 2003, 82–113; Guidobaldi, 1989; Heikonen, 2017; Pietri, 1976; Reekmans, 1989; Spera, 2013, 121–127). Excavations have also revealed remains of roads, *tabernae*, pagan shrines, and infrastructure connected to the Flavian amphitheatre (Colini & Cozza, 1962; Coates-Stephens, 2001; Heikonen, 2017; Palazzo & Pavolini, 2013; Schuddeboom, 2016).

The history of visual representation of the Caelian closely parallels the history of scholarship and is relevant to this study. Colini (1944) and Pavolini (2006) both provide documentation of archaeological remains across the entire Caelian that give a sense of the overall topography of the hill and what is preserved. Colini (1944, Figure 2) published a reconstructed plan of the imperial-period Caelian that restores the shapes of major structures (amphitheatres, the Temple of Divine Claudius, barracks of the *equites singulares*, city walls) and streets but renders other remains as fragmentary. More recent publications by Pavolini and collaborators (cited earlier) provide updated state plans of excavated structures and present additional details about phasing of individual buildings. Palazzo and Pavolini (2013) include a plan of archaeological remains from a large segment of the Caelian overlaid with modern topography.

As is true in studies of late antique houses more broadly, other scholarly considerations of the *domūs* on the Caelian usually reproduce plans in which the houses are isolated from their surroundings (Brenk, 1999, 2003; Guidobaldi, 1986; Machado, 2019); this is also the case with the late antique churches (Alberti et al., 2014; Brandenburg, 2005; Heikonen, 2017). In plans of the whole ancient city, in contrast, scale requires the illustration of major monuments and roads as simplified masses and lines. These necessarily leave swaths of the city and often the Caelian largely blank when they were in fact filled with houses, shops, and the rest of the fabric of daily

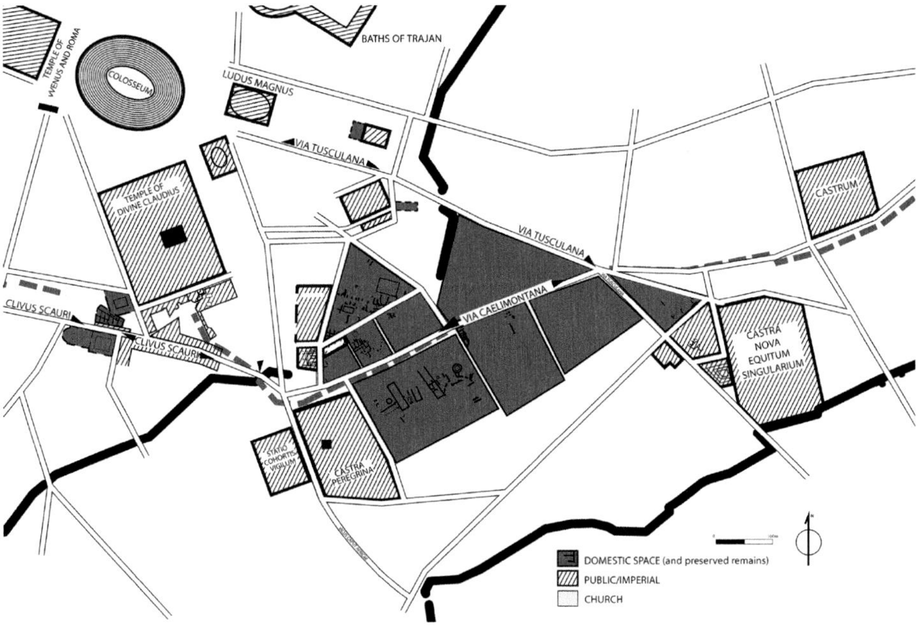

Figure 4.2 Plan of Caelian neighbourhood, *c.* 312.

Source: Author's image

life. Those plans also tend to omit, often for lack of evidence, smaller streets, access points, and routes people took to get from one place to another. The houses of the elite are almost always absent in such city plans as well, which can be misleading. On the Caelian, for example, aristocratic houses collectively occupied a far larger portion of the neighbourhood than any other type of building or public space (see Figure 4.2). In a related phenomenon, in plans illustrating the development of Christian Rome – such as Richard Krautheimer's often reproduced rendering of the city *c.* 500 CE (Figure 4.3) – the city seems to be covered by a constellation of churches represented by symbols, while the dense fabric of the still extant imperial city largely disappears.

4.1 Aims and methods

The Caelian is referred to in ancient sources as a district associated with elite dwellings (Martial, *Epig.* 12.18; SHA, *Marc. Aurel.* 4.1; *Commod.* 7.16; Symm., *Ep.* 3.12, 3.88, 7.18). It is conceived here as a large neighbourhood, broadly defined as a spatially distinct urban area in which physical proximity of structures and institutions as well as regular face-to-face encounters shaped the social lives of its inhabitants (Garrioch & Peel, 2006; Keith, 2003; Smith, 2010; Suttles, 1972; York et al., 2011). Such encounters included regular daily movements, informal gatherings, and holidays, as well as unusual or unexpected events. While the Caelian had a strong residential character and a notable concentration of elite domestic buildings, like other Roman neighbourhoods it also contained spaces dedicated to a range of public and private activities (Andrews, 2014; Guilhembet & Royo, 2008; Laurence, 1994, 38–50; Lott, 2004; Mignone, 2016; Moralee, 2018). Its residents and visitors were not just members of the cosmopolitan Roman elite but also practitioners of traditional polytheistic religion, adherents of a newly legitimized Christianity, foreigners, workers, and slaves, and these categories intersected in multiple ways. They were also members of 'imagined communities' (Anderson, 2016; Mol & Versluys, 2015), with

Figure 4.3 Map of Rome, *c*. 500.

Source: Krautheimer (2000). Copyright © 2000 by Princeton University Press. Reprinted with permission

ties to real and invented pasts as well as other parts of the city and the empire (Calvino, 1974; Graeber & Wengrow, 2021, 276–282). The built environment of their neighbourhood may be understood as a slowly evolving manifestation of what they decided to preserve or change, even if not all of those decisions were made consciously or forcefully.

This study begins by putting together elements of the late antique topography of the Caelian that are often kept apart, in both narrative and visual terms – private and public spaces, still-functioning imperial monuments and late antique structures, Christian and pagan shrines. Relying largely on published reports and documentation, as well as personal autopsy, it presents an architectonic and phenomenological approach (Steiner & Sternberg, 2015) to reconstructing the social and visual experience of this neighbourhood at a human scale – both spatially and chronologically – through four roughly generational intervals from 312 to around 410 CE.

The first 'generation' coincides with the completion of the *Basilica Constantiniana* (the Lateran) in the first quarter of the fourth century CE. The second focuses on the mid-fourth century and polytheistic religious activity that took place alongside Christian celebrations on the Caelian. The third looks at the era when Damasus was bishop of Rome and the development of early *titulus* churches near the base of the hill. For the fourth generation, I consider the social world of some of the city's wealthiest aristocrats near the end of the fourth century and as Alaric bore down on the city in the early fifth century. In this period, the mass of elite property on top of the hill continued to serve as a collective representation of elite power in the city, even as some aristocratic residents became Christian and others did not. Finally, I discuss how that representation may have been redistributed and redefined by an emerging Christian elite after Alaric's sack.

The series of neighbourhood plans (Figures 4.2, 4.4–4.6) created to accompany this chapter correspond to these generational intervals. On the basis of published reports, they bring together structures that are usually represented separately and illustrate as much as possible their urban footprints and topographical relationships and how they changed over time. By rendering phases as generational intervals rather than tying them to chronological sequences of particular buildings, my intention is to view the structures and institutions on the Caelian as parts of an always interacting whole. The plans support a narrative analysis that draws on archaeological and literary evidence to reconstruct aspects of social and religious experience on the Caelian in Late Antiquity and argues that the streets and human movement through them were at least as important to the Christianization of the neighbourhood as the buildings. One of those routes was the *via Tusculana* (see Figure 4.2), which ran east/south-east from the Flavian amphitheatre and along the north ridge of the Caelian hill towards the Aurelian Wall. The other route started at the base of the *Clivus Scauri* and continued uphill to the *via Caelimontana*, which extended north-east through a concentration of elite houses. The two routes intersected not far from the site of the Lateran Cathedral but existed long before its construction.

Between 312 and 410 CE, life on the Caelian underwent significant changes, but even with the construction of the Lateran church, its topography was fairly consistent for much of that time. The houses of the aristocracy maintained a significant presence on top of the hill; the amphitheatre and other enormous imperial structures maintained theirs at the bottom. The neighbourhood's main streets, lined with more modest dwellings, taverns, and shops, continued to form the connective tissue that tied them all together. By the mid-fifth century, however, the Caelian was quite changed. That physical transformation did not happen overnight or solely as a result of Alaric's sack. By exploring the Caelian at the scale of a community with multiple, overlapping experiences over generations in Late Antiquity, this study attempts to view the fourth and early fifth centuries in Rome as an evolving present, in which its occupants had differing understandings of the past and were actively shaping an unknowable future.

4.2 Generation one: Constantine and the Caelian

Constantine had no personal attachment to the ancient capital and little or no experience of its streets and neighbourhoods. Through his building projects, however, he not only asserted his dominance over Maxentius but also shaped his imperial image in Rome. He capitalized on the city's symbolic value to legitimize and strengthen his own position within the still shifting ranks of the imperial college (Omissi, 2018, 103–116). In doing so, Constantine demonstrated a commitment to the city and its glorious past while also looking towards a radically different Christian future. These principles were reflected in his first monumental contributions to the city – his triumphal arch and his church – which bracketed the Caelian neighbourhood from the west and east, respectively.

After his victory at the Milvian Bridge, Constantine returned to Rome two more times that we know of: in 316 CE, for the celebration of his *decennalia* (Eus. *VC* 1.48), which would have followed closely upon the dedication of his triumphal arch along the *via Sacra* near the Colosseum and the Temple of Divine Claudius (Barnes, 1981, 47; Marlowe, 2006; Rose, 2021, 1–3), and for his *vicennalia* in 326 CE – probably his last visit (Eus. *VC* 3.15; Zos. 2.29.5; Curran, 2000, 178). By then, the first imperially sponsored Christian church, the *Basilica Constantiniana*, was complete (Figure 4.4, No. 1).[4] It stood at the far east end of the Caelian hill. Like the triumphal arch, the church took a traditional Roman architectural form. The purpose of the basilica was entirely new, however, and its location near the edge of the city unusual.

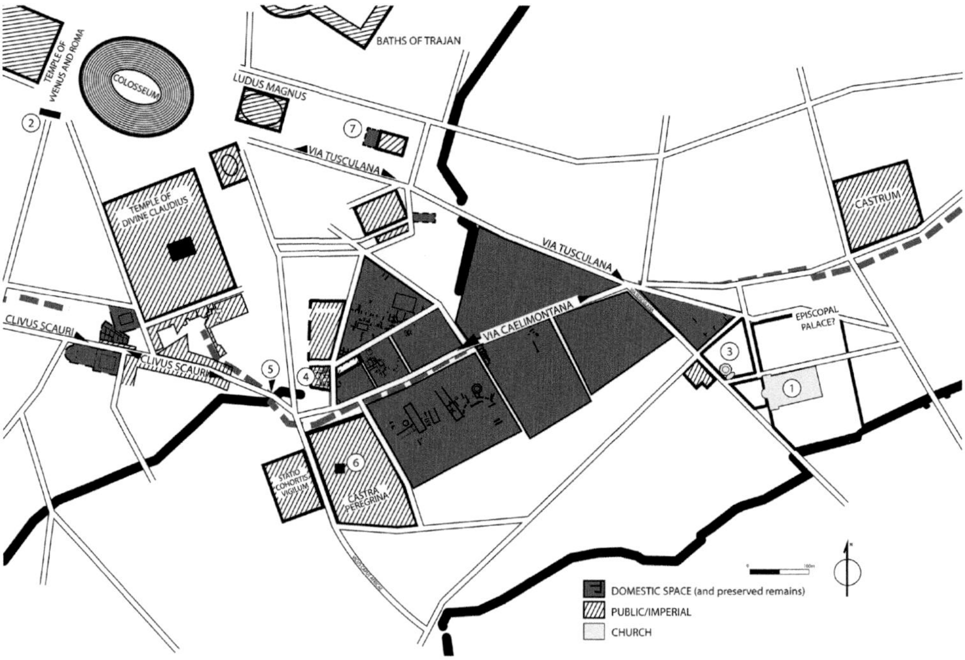

Figure 4.4 Plan of Caelian neighbourhood, *c*. 326.

Source: Author's image

As is well known, Constantine's church was built on the site of the barracks of the *equites singulares*, the military backers and guardians of Maxentius, and not far from the Porta Asinaria in the Aurelian wall (Brandenburg, 2005, 20–36; Curran, 2000, 94–95; Haynes et al., 2013, 2014; Pietri, 1976, 4–11). The replacement of their headquarters with a church was not just a statement of Constantine's allegiance to the Christian God but also a symbol of the defeat of his enemy. By the time of his *vicennalia* in 326, Constantine had no visible enemies left (Haynes et al., 2020, 4; Bosman, 2020, 168). The barracks had stood in an area long occupied by the military and surrounded by imperial property, all now owned by Constantine (Kinney, 2001, 126–133; Haynes & Liverani, 2020; Reekmans, 1989, 865–866). He was therefore able to build quickly and without having to buy or confiscate any land.

Constantine did not choose this location near the edge of town in order to be polite to pagan aristocrats as is sometimes suggested.[5] If he was worried about the local elite, he did not show it when he paraded the head of a leader many of them supported through the streets of the city impaled on a pike (Barnes, 1981, 44–45; Salzman, 2021, 43; *Pan. Lat.* 12(9).18.3; Zos. 2.17.1). Nor did he have much reason to be concerned, given the immediate support offered to him by the Senate in the aftermath of that victory (Barnes, 2011, 83; Salzman, 2021, 43–55). Moreover, the location of Constantine's Basilica did not avoid the elite at all. It is true that it did not interfere with revered public monuments and temples – there was little space in the city centre for it to do so – but Constantine was not hiding from the elite or the city. His church was a prominent addition to what was, at the time, a very fancy neighbourhood (Curran, 2000, 95–96; Chin, 2017, 24–25) and was both highly visible and easily accessible from other parts of town.

To Romans, the basilica as a building type had long associations with civic activities and imperial projects. Thus far, they were usually located in or alongside a forum and Constantine's Basilica might have seemed strangely misplaced at the edge of town. Its size and luxurious materials, however, were in keeping with its patron and the aristocratic landscape that surrounded it. As would become standard in the churches that came later, the exterior of Constantine's Basilica was plain, but the huge sanctuary inside was lined with coloured-marble columns leading to an altar framed by a gilded gate and decorated with silver angels (Brandenburg, 2005, 23–26; Bosman, 2020). The height and position of the basilica was such that it would have been visible from other parts of Rome (Kinney, 2001, 126, 129–131), maybe even from the Capitoline.[6]

The construction of the church would have had some immediate effects on life in the Caelian neighbourhood, whose residents and visitors would have seen clear evidence of this major imperial building project in their daily lives. The razing of the barracks and their replacement with a basilica was a massive undertaking that took several years, including quarrying bricks and stone decoration from the demolished castrum and transporting a large number of coloured-marble columns drawn from local stockpiles of *spolia* (Bosman, 2020). Some building materials and supplies, along with workers and animals, would have been brought to the construction site from outside the walls, but some portion of them would have likely travelled on the streets of the neighbourhood (DeLaine, 2000).

Once the *Basilica Constantiniana* was complete, the *via Tusculana* became a major connective artery between the church and the ancient city centre (see Figure 4.4).[7] Before then, that street would have been used by those heading to the military barracks or the elite properties around it and those leaving town, including those visiting cemeteries outside the walls. Indeed, because the basilica was (and is) near the edge of the walled city, a number of major streets led right to it. Constantine's church would have attracted new people to the area, sometimes in crowds and on a regular basis. From the Forum Romanum and the amphitheatre the *via Tusculana* offered a clear route to the church and its major ceremonial entrance on the north side as well as the baptistery (Figure 4.4, No. 3; Brandenburg, 2005, 36; Kinney, 2007, 148, Figure 4). The other major path to the Lateran was along the *Clivus Scauri* and the *via Caelimontana*; they connected the church to the Circus Maximus, the Palatine, and the Aventine.[8] Neither of these routes was particularly long or difficult and both ran past the wealthiest houses on the Caelian on the way to Constantine's Basilica. Today, one can walk at a leisurely pace from the Colosseum to the Lateran along the rough equivalent of *via Tusculana* in less than 20 minutes. The walk is only about five minutes longer for someone coming from the east end of the Circus Maximus and using the *Clivus Scauri* and *via Caelimontana*.

The built environment along these routes was largely unchanged in this first generation after Constantine's victory over Maxentius, but the walk along them would have been experienced differently by devout Christians on their way to the Lateran than by pagan visitors heading to their patrons' houses or on their way home from the amphitheatre and at different times of either the day or year.[9] By the time Constantine's basilica was completed, there were certainly Christians living and working in the area and throughout the city. As has been articulated by Rebillard (2012, 2020), however, their daily lives continued to be guided by forces and motives that lay outside the purview of their religion. Christian Romans attended games in the arena, spent afternoons at the baths, frequented market stalls, and attended legal proceedings and meetings of their *collegia*; those not fortunate enough to be members of the aristocracy got up every day to go to work (Dossey, 2019, 709–710, 715; Tran, 2020; Mol & Versluys, 2015, 453).

Among the familiar daily social patterns that remained in place in this period was the *salutatio* – the long-standing practice of powerful men in Rome receiving less powerful visitors in their houses each morning (Ellis, 1991; Machado, 2019, 215–217). The Caelian would have bustled with this activity each day, as people ascended the hill towards the homes of the super-rich, waited at their doorways, and socialised in the streets.[10] Those walking to *salutatio* visits each morning would have seen and passed – but perhaps not recognized as such – Christians who had additional social obligations in the early hours. Since at least the second-century

Christians would ideally start their day with a morning meeting held at someone's house (Rebillard, 2012, 15; Burns & Jensen, 2014, 242). Many Roman Christians, including members of the upper classes, continued to participate in *salutatio* as well (Maier, 1995, 233).

While active throughout the day with people going about the daily business of urban life, in the late afternoon and evening, some streets of the Caelian would also fill with people on their way to and from dinners, with some going to houses owned by wealthy pagans and others attending evening meals in the homes of well-off Christians (McGowan, 2004; Meeks, 1993, 97). Gatherings in dining rooms of private houses were important means of affirming Christians' sense of community and occasions for religious instruction (Acts 2:42, 2:46–47; Hallbäck, 1998; Meeks, 2003, 75–78, 157–160). In earlier days, the Eucharist was an actual meal. Although the Eucharist later became part of the liturgy performed in churches, the communal meal remained important to Christian social life long afterwards (McGowan, 2004; Meeks, 1993, 37–38, 45; White, 1998, 178–181).

The Lateran created a new and important node in the Christian topography of the Caelian, but even after its construction, life in the neighbourhood remained rooted in the long-standing daily rhythms of its residents and the cultural norms that had sustained them for centuries. People might have been able to identify who was Christian and who was pagan in the streets by their daily movements or perhaps the occasional greeting or gesture (Rebillard, 2012, 13–20), but little else. For visitors from lower social strata, the class and rank of the owners of elite houses on the Caelian hill would have been far more powerfully advertised than their religion. That difference in rank and class was displayed not only through the size and sumptuousness of individual properties but also by their collective presence in the neighbourhood. While the precise numbers of individual houses is not known, aristocratic *domūs* had since at least the second century CE occupied an enormous urban footprint on the Caelian plateau. Like Roman houses of any era, even these large estates presented relatively plain façades to the street and external signs of property divisions may have been difficult to discern.[11] Ephemeral elements like lamps, garlands, open doors, and the presence of servants would have marked the sometimes grand entrances to elite houses (Machado, 2019, 244–245) more clearly in the past than they do in the archaeological record, but to many visitors a cluster of enormous mansions like the one on the Caelian would have appeared as an imposing and largely undifferentiated mass of wealthy property. Finding one's way to a particular house, especially at night, would have required some guidance or prior knowledge of the layout of the neighbourhood. Those who lived or regularly spent time there would not only know their way around but also recognize each other on the street. Newcomers and outsiders would be noticed. One such newcomer in this generation was Constantine's mother, Helena, who moved into the Sessorian palace northeast of the Lateran. Her presence there would have attracted aristocratic visitors and clergy and offered a connection to an otherwise largely absent imperial family (Curran, 2000, 96, 100–101; Salzman, 2021, 65–66).

While Constantine's victory and the construction of the Lateran Basilica were major developments in this generation, the familiar topographical signifiers of class and wealth in this neighbourhood remained much as they had been since the imperial period. Christians and pagans passed each other on the street as part of their daily social practice, in which their timing often coincided. These daily interactions and observations of each other's comings and goings continued to reinforce long-standing class and social relationships as much or more than religious affiliations at this time. The uppermost strata of the elite lived on the plateau at the top of the hill. Houses of the less well off, but still prosperous, dotted other parts of the neighbourhood, such as those identified along the lower parts of the *via Tusculana* and the *Clivus Scauri*. Most such houses are known to us now because they were later transformed into churches, but as houses their designs and decoration were similar regardless of the religious affiliations of their owners. They were outfitted with apsidal dining rooms and courtyards and decorated with marbles, mosaics, paintings, and silver to the degree that their owners could afford. Whether they hosted Christian morning meetings or dinner parties for cultivated pagans, they were venues for display of material wealth, cultural sophistication, and education steeped in the traditions of *paideia* (Cameron, 2011, 357–360, 384 and *passim*, among many others).

4.3 Generation two: gods in the streets

The topography of the Caelian remained largely stable for the next generation as well, who maintained the daily social customs and neighbourly behaviour of their parents. Public religious experience on the Caelian did shift, however, with an increasingly confident and growing Christian population moving through the streets on their way to and from the Lateran Basilica. These shifts would have had visible impacts on the broader time frames that governed urban life, such as the annual religious calendar. In the mid-fourth century, the Roman calendar was packed with celebrations of Christian and polytheistic cults (Salzman, 1990; Fraschetti, 1999), which played out in urban neighbourhoods. For this generation, the triumph of Christianity and disappearance of the old gods were by no means inevitable and the latter unimaginable to most. Festivals dedicated to polytheistic cults continued to take place in the open (Latham, 2020; Machado, 2019, 162–163). Previously modest or hidden Christian celebrations became more visible in urban life but had not yet achieved the organization or scale of polytheistic and major civic processions (Latham, 2014). Still other practices associated with traditional Roman religion became more ambiguous in their public significance (Markus, 1990, 107–124; Rothaus, 1996). While some polytheistic and Christian rituals necessarily took place indoors and out of view of the uninitiated, they also frequently implicated public urban space, such as the open areas around shrines, where worshippers gathered, and city streets, through which they moved on the way to and as part of their celebrations. Contrasting ritual actions of adherents to the old gods and the Christian one charged these spaces with new meanings. An examination of one part of the Caelian illustrates how the physical surroundings of a neighbourhood could remain unchanged as far as we can see in the archaeological record and yet be transformed by the actions and perceptions of people participating in rituals and festivals – in this case, the rites of Magna Mater and Easter (Latham, 2020; Salzman, 1990, 85, 128, 165–166), which brought processions of both pagan and Christian worshippers to the *via Caelimontana* in the spring.

The building on the Caelian most closely associated with Magna Mater/Cybele and her consort Attis was the headquarters of the *dendrophoroi*, the 'bearers of the tree' of Attis (Carignani in Pavolini et al., 1993, 713; Palazzo & Pavolini, 2013; Van Haeperen, 2012); the building is now known as the Basilica Hilariana (see Figure 4.4, No. 4). It was located near a major intersection of the *via Caelimontana* and another street leading north towards the city centre.[12] The Basilica Hilariana is securely identified through inscriptions. It was named for its patron – a pearl merchant called M(anius) Poblicius Hilarus – who paid for an important phase of its construction, probably in the second century (Colini, 1944, 48; Palazzo & Pavolini, 2013, 22, 25; Pavolini, 1993a, 42). The building is a relatively modest structure in comparison with the large *domūs* that were its neighbours to the east, but it was well appointed for its purposes. The significance of the association of the *dendrophoroi* with the Caelian is attested by the use of the term *arbor sancta* – a reference to their responsibility for the sacred tree of Attis – as a topographical feature of the *Regio II Caelimontium* in the fourth-century Regionary Catalogues (Colini, 1944, 49; Diosono, 2016, 264; Valentini & Zucchetti, 1940–1942, 93).

The *arbor intrat* ritual began on 22 March every year (Latham, 2020; Salzman, 1990, 85, 128, 165–166), when the *dendrophoroi* brought the sacred pine tree of Attis, festooned with garlands, out of their headquarters and carried it to the Temple of Magna Mater on the Palatine (Pavolini, 2006, 79–81, 2007, 315–317; Pensabene, 2008; Van Haeperen, 2012). From the Basilica

Hilariana they turned right (west) down *via Caelimontana* towards and through the 'arch of Dolabella' (Figure 4.4, No. 5; Colini, 1944, 215–219; Pavolini, 1993a). From there, they followed the ancient route of the *Clivus Scauri* to the *via Sacra* and the Palatine. The arch of Dolabella was in fact a gate in the ancient Servian wall and seems to have been the threshold through which Attis was considered to have crossed the *pomerium* into the city, hence the *intrat* (Colini, 1944, 33; Diosono, 2016, 264; Pavolini, 2007, 318–319; Pavolini, 1993a, 32). A second procession, possibly also involving the *dendrophoroi*, would have brought a representation of Magna Mater (probably a stone) to the Almo, a Tiber tributary, for a ritual washing known as the *lavatio* a few days later on 27 March (Latham, 2020, 268, 271, 290–292; Pensabene, 2008, 25). For the rest of the year, the Basilica Hilariana would have been used for *collegium* meetings and social gatherings of the *dendrophoroi* (Tran, 2020, 208–212). Members entered via a flight of steps from the street on its east side, just a few metres away from private homes that had surrounded it since its foundation. Most of the time the regular comings and goings of the *dendrophoroi* were probably little remarked upon by the surrounding community, but in March their presence would have been hard to miss.

As Latham (2020) recently argued, worship of Magna Mater attracted substantial negative attention from literary upper-class Christians in Late Antiquity, who not only decried it as pagan but also sought to portray it as incompatible with the classical culture they so valued.[13] Nonetheless, despite objections from Christians and misgivings even among some pagans, the practices associated with Magna Mater – both the March events described earlier and the *ludi Megalensis* in honour of the goddess in April – were highly visible and persistent in Rome (Latham, 2020; Machado, 2019, 177, 180; Pensabene, 2008, 23; Salzman, 1990, 87–88).

The other eastern god whose death and resurrection was commemorated on the Caelian each spring was Jesus Christ. By at least the mid-fourth century, the *Basilica Constantiniana* had been established as the official locus of Easter and Holy Week observances presided over by the bishop of Rome. These included the mass baptism of the catechumens (Saxer, 1989, 948). In spring, Roman Christians would have streamed up the Caelian hill, many travelling on the *via Caelimontana* or *via Tusculana*, past the fine houses near the bottom of the hill and through the neighbourhood of enormous mansions at the top on their way to and from the church.

The weeks-long festivals of Magna Mater/Cybele and Attis and the lead up to and celebration of Easter sometimes overlapped.[14] When they did, bands of men carrying a tree symbolizing an eastern god who came back from the dead walked in one direction across the Caelian hill at the same time as groups of Christians moved in the opposite direction to worship their own resurrected eastern deity who had carried his own cross. Each spring, the via *Caelimontana* and the *Clivus Scauri* were simultaneously Christian and pagan festival routes as well as stages for the performance of devotion to different gods and competition between them.

It should be noted that Magna Mater, Attis, and Christ were not the only foreign divinities whose presence were felt on the Caelian in Late Antiquity. Archaeologists have also identified two shrines of Mithras that were still functioning in the mid-fourth century. One constructed inside the *castra peregrina* (see Figure 4.4, No. 6) was expanded in the late third century and remained in use until at least the early fifth century (Lissi, 1986; Pavolini, 2006, 55–60; Schuddeboom, 2016, 227–229, 235–236). The second *mithraeum* was part of a house that would become the site of the church of San Clemente (Figure 4.4, No. 7); it, too, was still functioning in the second half of the fourth century (Brandenburg, 2005, 142; Heikonen, 2017; Schuddeboom, 2016, 232).

In addition to Mithras, Pavolini (2007, 312–313) suggests that several objects related to Isis found in the south-west region of the hill support the location of a shrine to the Egyptian goddess in that area.[15] Ovid (*Fasti* 3.835–8) describes an ancient (already in his time) shrine to Minerva Capta as being located on the lower slopes of the Caelian where the ground becomes almost level. It has not been identified in the archaeological record, but Colini (1944, 39) noted that as a foreign deity her shrine would have to be outside the Servian walls and argued for a location on the *via Tusculana* near the later church of Santi Quattro Coronati, where the wall crossed the hill at a low point. Festivals of Minerva were also celebrated in March, with one connected to Mars on 19 March and the *Natalis Minerves* on 22 March, which is thought to have been celebrated somewhere on the Aventine, the Esquiline, or the Caelian (Salzman, 1990, 161–163).

While the Caelian neighbourhood had no major temples to the ancient Roman gods, it was clearly not lacking in divinities who continued to be worshipped through the fourth century in both public and private contexts. Even as some residents turned their souls over to the Christian God, others continued to entrust theirs to Mithras, make offerings to Isis, and join in the celebrations of Minerva's birthday and the rituals of Attis and Magna Mater. Still others may have dabbled in the new religion of the emperors while still holding onto some of the old ways. In this first generation after Constantine, and as Christianity gained increasing visibility in Rome, residents of the Caelian could witness, share in, support, ignore, or condemn the religious practices of their neighbours. We might also imagine some corresponding shifts in social relationships – from the Christian senator whose colleague and frequent dinner companion refused to abandon the old gods to the craftsman and initiate of Mithras whose aristocratic patron insisted that he did – along with individuals' sense of their roles within their communities.

4.4 Generation three: Damasus and friends

In the third quarter of the fourth century, Rome's transformation from a symbolic political capital to a Christian one began in earnest. The late fourth-century bishop Damasus was instrumental in that process and clearly understood that the physical topography of the city could play a crucial role in the spiritual transformation of its residents. He helped create what Jaś Elsner (2003, 72) called a Christian 'alternative sacred topography' that would come to coexist alongside and within Rome's traditional civic, religious, and economic ones. Damasus's efforts on this front are visible in several ways. Most famously, he composed Vergilian epigrams for martyrs' tombs on the outskirts of town, applying classical cultural tradition to Christian devotion (Cameron, 2011, 350–351; Trout, 2005), while also establishing a visible, monumental Christian itinerary in the landscape around Rome (Sághy, 2012; Trout, 2015). Damasus was almost 60 when he took his seat as the bishop of Rome in 366 CE and had never experienced religious persecution (Krautheimer, 2009, 316), yet stories of martyrdom loomed large in his imagination and were important to his efforts to emphasize the 'Roman-ness' of Christianity (Löx, 2013, 43; Sághy, 2012; Trout, 2005) and the 'holy ancestors' of his contemporary Roman Christian community (Brown, 2012, 253; Grig, 2004, 215; Trout, 2015, especially 11, 39, 42–45). Inside the city, he celebrated liturgies and hosted synods at the Lateran, drawing crowds to the Caelian and making the Church ever more visible to his neighbours. He also turned his attention to private property, encouraging the burgeoning practice of elite Christians donating their houses as *tituli* to the Church (Curran, 2000, 142–146; Löx, 2013, 45–58). These activities had repercussions for the Caelian and the experience of walking along the major routes up and across the hill.

4.4.1 Damasus at the Lateran

Damasus officially began his episcopate on the Caelian in the midst of religious conflict that had nothing to do with pagans but rather a dispute between Christians over the selection of the bishop of Rome (Curran, 2000, 137–140; Maier, 1995, 234, 243). Damasus secured his position only after a bitter struggle with a rival, Ursinus, in which factions of Roman Christians engaged in physical violence in support of their preferred candidate (Amm. Marc. 27.3.12–27.3.13; *Coll. Avell.* 1.5–1.14; Rufin., *HE* 11.10; Krautheimer, 2009, 316–317; Latham, 2014, 133–135; Trout, 2015, 6–7). Damasus succeeded in defeating Ursinus but not before his followers slaughtered a huge number of his rival's supporters and subjected the city to the spectacle of violent Christians attacking each other in churches and raging in the streets of Rome.[16] The culminating episode in that conflict took place at the Lateran and would have been alarmingly visible to the neighbourhood around it. Damasan partisans, which according to a hostile source (*Coll. Avell.* 1.6) included 'perjurers and gladiators' to complement the charioteers who participated in the massacre, marched to the Lateran Basilica and occupied it for several days, refusing to leave until Damasus was anointed bishop (Latham, 2014, 133–135). The conflict was only resolved when the Urban Prefect and another city official offered their support to Damasus and agreed to exile Ursinus. This was a victory for Damasus but also a sign of his relative weakness when compared with the senatorial elite at the time (Krautheimer, 2009, 317; Machado, 2019, 171–172; Salzman, 2021, 29).

Despite the brief and tidy account in the *Liber Pontificalis* (Life 39.1–39.3), this bloody conflict clearly had an impact on Damasus's reputation and cast a pall on his episcopate (Amm. Marc. 27.13; Brown, 2012, 251–252; Krautheimer, 2009, 317; Trout, 2015, 8–9).[17] After this rocky start, Damasus needed to demonstrate that he could bring together a fractious Christian community under his authority while integrating himself and his office into the broader power matrix of Roman civil society. He also sought to establish the primacy of the bishop of Rome over the other episcopal sees in the empire (Trout, 2015, 15). Achievement of these goals required the cultivation of relationships, both in face-to-face encounters with potential allies in the city and in conceptual connections with imagined Christian communities, past and future, which he helped establish through his interventions in the urban landscape, signs of which can be traced on the Caelian.

At the Lateran basilica, Damasus presided over crowds of Christian worshippers especially in the weeks surrounding Easter. As had now been true for decades, many made their way there from other parts of the city along the *via Tusculana* and the *via Caelimontana*. As the Christian population grew, the gatherings of the catechumens, mass baptisms, and other Christian celebrations would have made Damasus and his brethren even more visible to the surrounding neighbourhood (Liverani, 2020, 16–17).[18] In addition to these regular ceremonies, as part of his effort to establish his authority over the other episcopal sees, Damasus called a synod in Rome on the anniversary of his ordination in 371/2 and another one in 381 (Löx, 2013, 30–31), thereby bringing the spectacle of high-ranking clergy from around the empire and their entourages to the Caelian.

4.4.2 Damasus and the neighbourhood elite

Damasus was also, as Krautheimer (2009, 326) put it, 'a political animal' and he needed the support of the local urban elite (Brown, 2012, 248–253; Löx, 2013, 25–30). According to the rules of the day, as bishop he had senatorial rank (Krautheimer, 2009, 318), but like most bishops in this era (Lizzi Testa, 2009, 533) he was not born into great wealth or a noble aristocratic family. As bishop, however, he was at least officially a resident of the Caelian. While the precise location of the fourth-century episcopal palace is the subject of debate, it was close to the Lateran Basilica and therefore the grand aristocratic properties nearby. Scholars had long identified the late antique episcopal palace with a certain Domus Faustae, a house offered by Constantine in 313 as a venue for an episcopal council and residence of the then bishop of Rome, Miltiades (Curran, 2000, 94, citing Opt. *De Don.* 1.23; Krautheimer & Trachtenberg, 2000, 21). The physical identification of the Domus Faustae has been questioned in recent years and Paolo Liverani (2020, especially 18–22) now suggests that the late antique episcopal palace lay in the area of the Scala Sanctum, just north-east of the Lateran.

In this era, the daily rhythms of neighbourhood life were still dictated in large part by the social needs and practices of the rich and powerful, but aristocratic pagans had increasingly fewer venues for public service and benefactions that reinforced their positions as the century wore on (Machado, 2019; Latham, 2020). Christian numbers were growing among the upper classes, although they were probably not yet a majority at the start of Damasus's episcopate (Salzman, 2021, 37).[19] While Damasus surely cultivated relationships with some of his aristocratic neighbours and was sometimes derided for doing so (Amm. Marc. 27.3.14; *Coll. Avell.* 1.9; Krautheimer, 2009, 318; Löx, 2013, 28–30), he also paid particular attention to the lesser elite – those who were wealthy, but not that wealthy, or rich but not from traditionally noble Roman stock. Economically, Damasus and the Church itself would have counted among that lesser elite (Brown, 2012, 248–253). The Church could not begin to compete with the upper echelons of the Roman aristocracy in terms of cash or land holdings, either in or outside the city (Latham, 2012, 302–303).[20] Christianity, however, offered new ways for the lower ranking and less wealthy to assert themselves in Rome (Lizzi Testa, 2009, 530–535). Importantly, for neighbourhoods like the Caelian, those without aristocratic pedigrees or with smaller fortunes could become revered patrons of the Church through the donation of private property in the form of *tituli* (Curran, 2000, 142–146; Guidobaldi, 1989, 395–396). The *titulus* church is a category unique to Rome that designated, among other things, a structure built on or converted from private property and donated to the church (Brandt, 2009, 163–167; Guidobaldi, 1989; Hillner, 2006; Kinney, 2017, 68–71; Reekmans, 1989). *Titulus* churches also offered alternatives to the Vatican and the Lateran where Christians were allowed to gather and experience the liturgy with the approval of the bishop (Kinney, 2017, 71–72).

By the end of the fourth century there were at least ten *titulus* churches in Rome, which are known from literary and epigraphic sources (Guidobaldi, 1989; Löx, 2013, 45–46; Reekmans, 1989, 866–874). The precise sequence and chronologies of Roman *tituli* are not secure, but there is scholarly consensus that few existed before the mid-fourth century and that their numbers increased during the episcopate of Damasus, who encouraged their establishment as part of a parish-like network within the city (Löx, 2013, 45–49; Reekmans, 1989, 868).[21] *Tituli* were predominantly built on the site of and sometimes preserved the fabric of elite *domūs* in which the characteristic late antique apsidal dining room could be converted into the sanctuary (Guidobaldi, 1989).

At least three of the *tituli* recorded in the synod decree of 499 were established on the lower slopes of the Caelian alongside major streets that led up to the top of the hill. Of those San Clemente (Figure 4.5, No. 2; *titulus Clementis;* Brandenburg, 2005, 142–147; Heikonen, 2017) and what would become Santi Quattro Coronati (Figure 4.5, No. 3; *titulus Aemilianae)* were former *domūs* near the *via Tusculana* (Brandenburg, 2005, 195–196; Pavolini, 2006, 104–105).[22] The third, a *titulus Pammachius*, was located at the base of the *Clivus Scauri* (Figure 4.5, No. 4; now Santi Giovanni e Paolo). It was built on top of a rare example of a house containing a clearly Christian private chapel (Brenk, 2003, 98–111); it was donated as a *titulus* in the late fourth century. The church, however, may not have been completed until after Alaric's sack (see below).

San Clemente and the house that would become the church of Santi Quattro Coronati were mutually intervisible and connected to each other and the Lateran by proximity to the *via Tusculana*. They also stood in the shadow of the Flavian amphitheatre, which towered over that part of town (and still does). They were within earshot of the training grounds in the *Ludus Magnus* and other facilities that supported the games that still took place in the amphitheatre in this period (Heikonen, 2017, 35–36; Latham, 2014, 130; Parlato, 2017, 167). Clergy and congregants could have capitalized on the location of these two churches to advertise their piety to fellow citizens headed to and from the amphitheatre – including Christians who did so in defiance of pastoral instruction – to see games celebrating the accomplishments of and occasionally paid for by some of Rome's most prominent citizens (Machado, 2010, 301–305; Dossey, 2019, 709–710).[23]

Of the *tituli* on the Caelian, San Clemente is the most closely associated with Damasus himself (Curran, 2000, 146; Löx, 2013, 46; Trout, 2015, 16), but his direct involvement with this or other specific *tituli* is less significant than the growing number of them in this era and their impact upon the social and material landscape of Rome (Curran, 2000, 142–147; Löx, 2013, 45–49). By transforming their private houses into churches, owners incorporated their personal

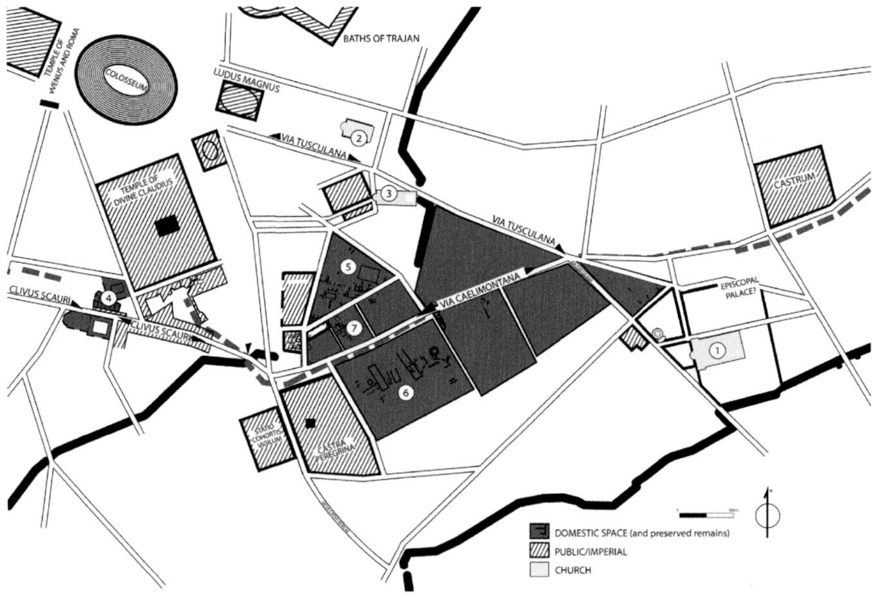

Figure 4.5 Plan of Caelian neighbourhood, *c*. 375.

Source: Author's image

property and local social worlds into an increasingly layered urban landscape. These houses that became churches and churches that looked like houses may have been visually ambiguous in the urban landscape – recognizable as Christian to those in the know but simply looking like nice houses to those who were not, and with shifting significance for different people who passed by them at different times (Elm, 2014, 172–174).

Inside, their congregations gathered in spaces of similar scale and appearance as those they had used for centuries. This embodied experience of performing ancient rituals in familiar spaces could have reinforced the sense of deep ancestry and continuity that Damasus was so keen to instil in Roman Christians. For their donors, *tituli* were visible demonstrations of piety and opportunities for those who may have been positioned at a lower rung of the late Roman social ladder to switch to another ladder entirely – one that relied on different measures and on which they could perhaps climb higher.[24]

The installation of *tituli* churches at the base of the hill added an additional layer of social practice to the Caelian neighbourhood – the actions of clergy. Their movements through the streets would have helped make Christian ritual more regularly visible in the neighbourhood. For instance, only the bishop could consecrate the Eucharist, so priests or their representatives would have been required to walk up to the Lateran to retrieve it (Saxer, 1989, 924–927; *LP* Life 33.2, 40.2). Every Sunday, and on other days in the period leading up to Easter especially, members of the Christian clergy would have walked up the *via Tusculana* (and other streets) towards the Lateran in the morning alongside those heading to *salutatio* visits in mansions up the hill. They later retraced their steps carrying sacks of consecrated bread. Those priests and their neighbours might also pass another type of Christian who frequented aristocratic neighbourhoods in the later fourth century – ascetics and monks who sought to influence and instruct the growing Christian elite. Their movements would have been less regular and less predictable, but the ascetics, especially, would have visibly represented a side of Christianity that was not easy to reconcile with the lifestyles of the people they were visiting (Curran, 2000, 293–5) and to whom we now turn.

4.5 Generation four: aristocratic neighbours in the shadow of invasion

In the last decades of the fourth century, the Roman aristocracy remained in possession of enormous wealth and still occupied a massive bloc of domestic property on top of the Caelian hill (Chin, 2017, 21–23; Machado, 2012a, 2019, 241–252; Pavolini, 2000, 2006). Historically, along with their privilege came responsibility to their fellow citizens – they were patrons, public benefactors, and employers of everyone else. Some members of the aristocratic class would break that social contract when Alaric threatened and ultimately sacked the city. Their abandonment of those responsibilities and of Rome itself – some fled to their estates in North Africa, others to different cities (*Vita Mel. Gr.* 20–21; Jerome, *Ep.* 130.4–130.7; Pavolini, 2013, 179) – created openings for those who remained, including Christians among the lesser elite, to establish themselves in an emerging new social order (Salzman, 2021, 101–102).

Two famous aristocratic personalities from this period – Quintus Aurelius Symmachus and Melania the Younger – are associated with houses on the Caelian and may serve to illustrate some social and religious developments in this aristocratic neighbourhood. Symmachus and Melania were both members of great Roman families, the Symmachi and the Valerii, respectively. Their lineage and enormous landholdings placed them at the very top of the aristocratic ranks and their lives are traceable in some detail through the literary record (Cameron, 2011; Chin & Schroeder, 2017). Symmachus was a prominent senator, prolific writer, and pagan patrician, best known for his defence of Rome's traditional institutions and deities (Cameron, 2011; Ebbeler, 2007).[25] Melania the Younger was a woman, a devout Christian, and became famous for her renunciations of worldly wealth and devotion to asceticism (Chin & Schroeder, 2017; Clark, 1984). Their paths diverged radically, but they had much in common and, like other aristocratic pagans and Christians in Rome, may have been neighbours.

Symmachus owned at least three houses in Rome and in his letters he mentions a *domus* on the Caelian, which he refers to as his *lar* (*Ep.* 7.18).[26] The remains of the house identified with him (*domus Symmachorum*, see Figure 4.5, No. 5) occupy a roughly triangular space and cover some 6,500–8,500 square metres, making it one of the larger, but not largest, houses on the plateau (Pavolini et al., 1993, 483–502; Carignani, 2000; Machado, 2019, 243–246; Pavolini, 2006, 84).[27] Like other *domūs* nearby, it was the product of expansion of an earlier house, probably built in the late second or third century CE (Carignani, 2000, 150). Both before and after the expansion, the *domus* was oriented roughly north-east–south-west, probably with an entrance from the street on its north-east side (not preserved), which connected it to the lower slopes of the hill, the amphitheatre, and the city centre. Machado (2019, 245) identifies an additional entrance on the south-east side of the house. Visitors using either route were led to the large interior courtyard and a spacious apsidal hall (Pavolini et al., 1993, 483–488; Carignani, 2000).[28] The house was close to the *via Tusculana* and connected to the *via Caelimontana* through cross streets but did not face directly onto either of them.

Melania the Younger was descended from the *gens Valeria* through her grandfather and the *gens Antonia* through her grandmother and namesake (Chin & Schroeder, 2017, 4–5). The colossal structure known as the *domus Valeriorum* (Figure 4.5, No. 6) is located south of the *domus Symmachorum* on the *via Caelimontana*. The remains of the house lie under the site of the modern Ospedale dell'Addoloratta (Brenk, 1999; Pavolini, 1994–1995, 91; Pavolini, 2006, 110–111). Modern exploration of the *domus Valeriorum* began as early as the sixteenth century and excavations in the nineteenth century and in recent decades have revealed at least one colonnaded courtyard, the remains of a private bath, a possible fountain, a large corridor decorated with wall frescoes, fragments of opus sectile and mosaics, as well as a number of statues and other precious objects (Barbera et al., 2008; Brenk, 1999, 2003; Guidobaldi, 1986; Pavolini, 2006). Like the house of the Symmachi, the *domus Valeriorum* is identified with the Valerii by inscriptions honouring members of the family and also by a large bronze lamp in the shape of a boat carrying the apostles Peter and Paul and bearing an inscription commemorating the baptism of Valerius Severus, the name of Melania's father-in-law (Brenk, 1999, 2003; Guidobaldi, 1986, 186–188; Xanthopoulou, 2010, cat. LA 16.001).[29] The association of the *domus Valeriorum* with Melania is also drawn from a story recounted in the *Life of Melania the Younger*, written by an acolyte, Gerontius, of her departure from Rome (Chin, 2017, 20–23; Chin & Schroeder, 2017; *Vita Mel. Gr.*).[30] It can be summarised as follows: Melania and her husband, Valerius Pinianus, were both born to Christian parents and married as teenagers, probably around 400 CE. After their two children died young, they took a vow of chastity and turned to a life of asceticism. Without heirs, the question of the fate of their properties – which the couple vowed to liquidate – became a point of contention in their families. Sometime around 408 CE, Melania and Pinianus tried to sell their enormous house in Rome but found no buyers who could afford it (*Vita Mel. Gr.* 14). As the hagiographical story goes, they left Rome to pursue their ascetic lifestyle, leaving their grand estate behind. While they were gone, the city was sacked by Alaric's forces and their house was burned. They later sold it for a song. The size and opulence of the *domus Valeriorum* on the Caelian are suited to this description, but while archaeological remains show signs of damage by fire, evidence for the precise date and extent of fire remains elusive (Hillner, 2003, 143; Machado, 2012a, 118–119, 2019, 250; Pavolini, 1993b, 53, 2013, 163).

While the archaeological evidence is not sufficient to confirm that Symmachus and Melania themselves lived in these houses, residence by members of their families is possible and by people very much like them in terms of wealth, class, and religious affinities, assured. In the late fourth century, members of their social stratum still navigated a shared social and built environment. The ongoing interest among the upper classes in classical literary culture is well attested and much discussed (Brown, 2012; Cameron, 2011; Kaldellis, 2009, 45–46; Watts, 2006), but classical *visual* culture was pervasive and arguably had a wider reach. Classical imagery continued to dominate domestic interiors of both pagans and Christians and therefore remained part of the daily experience of all manner of people – clients and friends, women and children, employees and slaves. In domestic settings, visitors would have to look quite closely to discern the religion of a house owner based on interior décor and, even then, one's conclusions might well be wrong. For instance, the religious affiliations of one Gaudentius (Figure 4.5, No. 7), a next door neighbour of the *domus Symmachorum*, can only be guessed at. Among potentially misleading visual cues is a larger than life-sized Hadrianic statue of Antinous in the guise of Dionysus. It is one of the finer objects associated with the house of Gaudentius and was displayed in the triclinium, where it would have been seen by visitors (Moltesen & Hast, 2004, 114–115; Spinola, 2000, 152–153). It is certainly possible that Gaudentius was a pagan, but even this large and prominently displayed statue is not sufficient to support that identification, given the ubiquity of Dionysian imagery in this era. The statue could have been as or more important as a sign of its owner's cultural sophistication and connection to classical artistic traditions than to the god himself. Moreover, if Gaudentius was poorer than his neighbours, as seems likely, given the relatively small size of his house, his possession of a high-quality imperial-period statue of Antinous might have suggested to his guests that he came from a family that had been prominent since the time of Hadrian, which would have been worth advertising (Cameron, 2011, 11; Weisweiler, 2015, 17–20). Even his potential friendship with Symmachus is of no help. Symmachus compliments one Gaudentius in his letters, but Symmachus said nice things about a great many people, including Christians (Ebbeler, 2007).

Christian images in houses are probably more reliable identifiers of the beliefs of their owners than pagan ones, but they did not dominate domestic spaces. The aforementioned bronze lamp from the house of the Valerii may have been one of the several Christian objects visible in the late fourth-century house, but its ancient distinction within this sprawling estate filled with precious things – including statues of mythological figures, herms, and portraits – has surely been overblown by our modern interest in it (Barbera et al., 2008; Brenk, 1999, 2003). Its shape of a boat and sailors would be unlikely to merit much attention among people who had encountered countless lamps in the form of human heads, animals, and male genitalia. Christians in this era certainly owned fine objects of silver or painted glass with specifically Christian imagery that they displayed or showed to guests in their homes, but the material and style of these objects were still quite similar to those decorated with traditional non-Christian imagery (Grig, 2004).[31] As with the Peter and Paul lamp, a literate Christian visitor to an elite house might well note the devotional significance of such objects, but pagan guests or illiterate servants could easily miss their message. Moreover, the surroundings of late Roman elite society were not made up of isolated images of pagan gods and Christian symbols but were the product of centuries of accumulation, deterioration, and small adjustments that made up their world. Just as the imperial monuments of Rome continued to stand and function in Late Antiquity, so did the frescoes, mosaics, and decorative objects that filled the houses of the urban elite. Visitors to these houses would have registered the luxury value of the materials more easily than their specific iconographies (few of which were specifically Christian). Images from classical mythology and allusions to noble ancestors would have been for many Christians and pagans merely background, signifying wealth and status above all.

What may have differentiated elite houses among their neighbours who were inclined to notice were the comings and goings of aristocratic residents and the sorts of visitors who arrived at their houses each morning or left each night. When senatorial aristocrats like Symmachus left the house dressed in a toga and surrounded by an entourage to go to the Forum or attend a public event, they projected an image of traditional Roman aristocratic values and practices, whether they were Christian or not (Machado, 2010, especially 298–299).[32] At the same time, aristocratic participation in public religious ritual may have been a rarer sight in the later fourth century. Pagans had fewer opportunities for such things while Christians at the highest levels of the upper classes had the option to avoid the large, undifferentiated gatherings where new converts professed their faith (Aug. *Conf.* 8.2; Brown, 2012, 46–49; Kinney, 2017, 72). When a woman like Melania, however, walked out of her colossal, luxurious estate wearing the plain, rough clothes of an ascetic but carrying piles of precious objects to go visit the empress (*Vita Mel. Gr.* 8, 11–12), she would have made quite an impression. We hear of a number of Christian aristocrats who turned to 'house asceticism', dedicating themselves to celibacy and prayer at home in a sort of visible seclusion (Jerome, *Ep.* 127.15; Curran, 2000, 264–279; Trout, 1999, 121–123). Their neighbours might have noticed – as his enemies did when Damasus called on noble women (Amm. Marc. 27.3.14; *Coll. Avell.* 1.9) – when members of the clergy or even the bishop visited them (*Vita S. Ambrosii* 10.1; Curran, 2000, 269–280; Maier, 1995; Sessa, 2007, 199–201) or when they ventured out to participate in such rituals as the rites of the Reconciliation of the Penitents, wearing sackcloth and mortifying themselves as part of a crowd in front of the Lateran (Jerome, *Ep.* 77; Romano, 2020, 405–411). Over time, such shifts in elite practices, large and small, brought with them corresponding changes in the comings and goings of clergy and lower-ranking Romans and in the perceptions of their peers. Together these slowly reshaped conceptions of time and space in the neighbourhood, inscribing its topography with new meanings even without radically affecting its physical appearance.

On a broader scale, conversion or resistance to Christianity among aristocrats affected the fortunes of their dependents and everyone else. At least some of the apartment buildings, shops, and storehouses that lined the streets and lower slopes of the Caelian hill were likely owned and patronized by the types of families who had houses at the top (Garnsey & Whittaker, 1998, 318–323, 328–335; Pavolini, 2013, 180; *CTh.* 14.3.4, 364; *CTh.* 13.9.4, 391; *CTh.* 13.5.20, 392). Control of urban property meant control over the livelihoods of the people who lived in and operated them (Guilhembet & Royo, 2008, 205–207). Together with long traditions of public civic benefaction and newer ones of Christian charity, aristocratic property and power were entangled in every facet of daily life among every social class (Brown, 2012, 53–90 and *passim;* Garnsey & Whittaker, 1998, 331–332). The Symmachi and the Valerii, for instance, owned property around the city as well as vast expanses of agricultural land in North Africa and elsewhere in the provinces. Through their estates, they controlled the lives of thousands of people, both free and enslaved, who worked for and depended on them (Brown, 2012, 296–297). These were political families – who literally had more money than God – and their decisions to turn to Christianity, or not, had ripple effects well beyond their households. The situation is exemplified by Augustine's oft-repeated statement in which he claimed that if a certain powerful nobleman converted to Christianity 'there would be no pagans left' (Aug. *Enarr. in Ps*. 54.13, cited by Cameron, 1999, 109).

With all of this in mind, we may return to the hagiographical story of Melania, Pinianus, and their gigantic house. Even if their house was not actually the *domus Valeriorum* on the Caelian, the tale is illustrative. The story was meant to show the couple's lack of interest in worldly wealth and status, but their neighbours might have understood their actions quite differently. For at least two years before the sack of 410 CE, Alaric had been extorting Rome. Its wealthiest citizens were called upon to contribute funds and help assure its protection. Melania and Pinianus refused. When, in 409, the Urban Prefect Pompeianus unsuccessfully tried to confiscate their property in the city, which, remarkably, would have been sufficient to satisfy Alaric's demands (*Via Mel. Gr*. 15, 19; Brown, 2012, 294–296, 299), the couple left town without relinquishing the property or offering funds to the city, choosing to use their still enormous fortune elsewhere and as they saw fit. These acts that were later framed in terms of ascetic Christian virtue and charity would have looked to many of their Roman neighbours more like a stark rejection of aristocratic responsibility to the city and its people (*Vita Mel. Gr*. 10–20; Mathisen, 2013, 96–97). Melania and Pinianus were not the only ones among the wealthiest Romans to abandon the city or to see their fortunes diminished or carted off by the Visigoths (Brown, 2012, 300–303; Krautheimer & Trachtenberg, 2000, 45). Those who stayed would shape the Christian city the others left behind.

4.6 There goes the neighbourhood: towards a Christian capital

At least four *domūs* at the base of the Caelian hill would be converted into churches, but none of the grand mansions on the top ever were. Even at the turn of the fifth century, the topography of the Caelian was dominated not by churches but by the houses of the city's wealthiest families. That elite neighbourhood was not destroyed in Alaric's sack, but it was damaged and depopulated. Over the course of the fifth century, the enormous houses that had so long covered the Caelian plateau fell out of use (Pavolini, 2013; Pavolini et al., 1993, 453–454), opening up more space for the Church and the Christian elite to make their mark on the city.

The post-Alaric years also saw the demise of pagan institutions in the neighbourhood. The activities of the *dendrophoroi* were outlawed by imperial decree in 415 CE (*CTh*. 16.10.20.2; Diosono, 2016, 262–263). The Basilica Hilariana's argue that it ceased to serve as the seat of the *dendrophoroi*, with parts of it abandoned and filled with refuse by the mid-fifth century (Palazzo & Pavolini, 2013, 89–93; Pavolini, 1993b, 610, 2013, 178). If the *collegium* of the *dendrophoroi* was at least partially financially dependent on its rich neighbours (Spinola, 2000, 152), elite abandonment of the Caelian may have hastened its decline. The Temple of Magna Mater on the Palatine was also abandoned by the mid-fifth century (Machado, 2019, 163–164), which suggests that the cult of Cybele and Attis was by that time severely compromised if not defunct. Across the *via Caelimontana* from the Basilica Hilariana, the *mithraeum* that had been inside the *castra peregrina*, which had survived Alaric, was violently destroyed in the early fifth century presumably by Christians (Lissi, 1986; Schuddeboom, 2016, 235–236; Pavolini, 2013, 167). At the time of that destruction, a number of Mithraic objects and a head of Isis were stored there (Lissi, 1986; Schuddeboom, 2016, 235–236; Pavolini, 2006, 59–60, 2007, 312). Pavolini suggests that those objects could have been brought to the still functioning *mithraeum* from other shrines that had by then gone out of use (Pavolini, 2006, 59–60). It is also possible that they were collected to be smashed along with other sacred objects in that process (Schuddeboom, 2016, 228–229). The large church of Santo Stefano Rotondo (Figure 4.6, No. 2) was not built on the site until at least the mid-fifth century; it would become an anchor point in a papal processional circuit – which included the *via Caelimontana* – through the city for centuries to come (Brandenburg, 2005, 201–204, 213; Pavolini, 2006, 60; Schuddeboom, 2016, 231).

Although looted by the Visigoths, the Lateran remained intact (Bauer, 2013, 259–261). By the mid-fifth century, the once solitary basilica stood at the top of a sort of ecclesiastical pyramid. The elite houses on the upper reaches of the hill fell into disrepair over the course of the fifth century. The reception halls in the house of Gaudentius were abandoned around the mid-fifth century, with some continued utilitarian activity after that, while the sewer system and exterior streets around the *domus Symmachorum* fell out of use, as did the house not long afterwards; the *domus Valeriorum* seems to have followed a similar trajectory (Pavolini, 2013). Meanwhile, the churches built from the houses on the lower slopes were maintained and elaborated, presumably with funds from those who stayed in town (or perhaps moved in) after Alaric (Spera, 2013, 125–126), and at least one, maybe two were built at the base of the route of the *Clivus Scauri*. The aforementioned *titulus Pammachius* was on the north side of the street (Figure 4.6, No. 4; now Santi Giovanni e Paolo). Both the precise chronology and identification

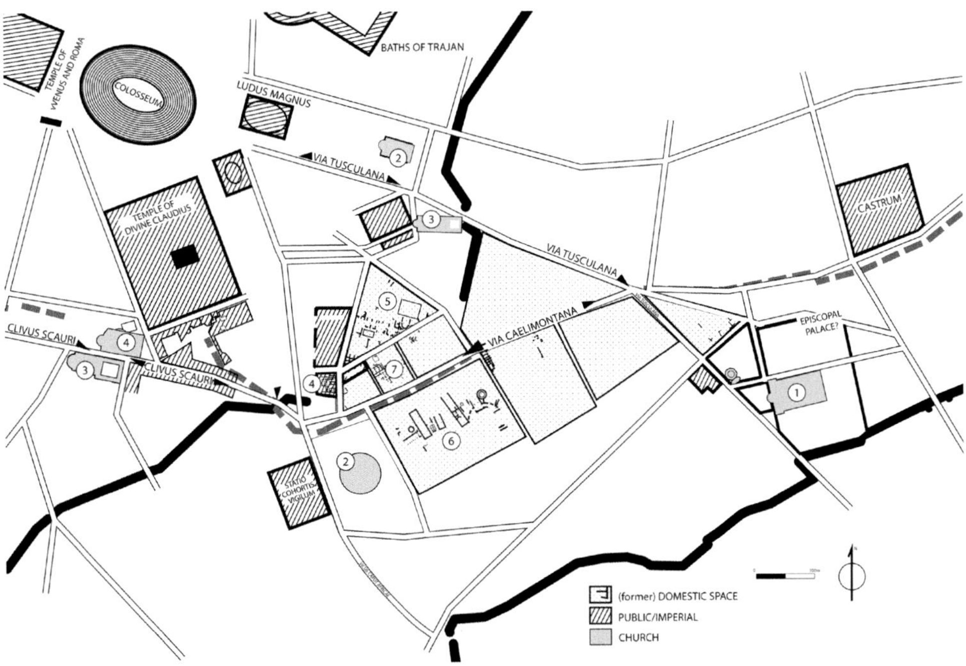

Figure 4.6 Plan of Caelian neighbourhood, *c.* 475.

Source: Author's image

of the donors of this church are matters of debate, but as mentioned earlier, the property was likely donated to the church before Alaric's sack. The archaeological evidence suggests that the basilica was constructed in the early fifth century and that it covered the earlier house and its chapel (Brenk, 2003, 82–112). It is not clear whether the church building was completed before or after the sack (Bauer, 2013, 263–265; Brandenburg, 2005, 155–161).[33] The eponymous Pammachius has long been identified with a proconsul of 396 and friend of Jerome who died in the siege of Alaric (Bauer, 2013, 263–265; Pavolini, 2006, 32–35), but this identification has been questioned (Brenk, 2003, 109–112; Hillner, 2006, 62–65; Leyser, 2007).[34] The church came to be associated with the martyrs John and Paul by the late sixth century (Leyser, 2007).

The location of Santi Giovanni e Paolo alongside the *Clivus Scauri* was, as Brandenburg (2005, 156) put it, 'prominent, even strategic'. It was even more so once it was paired with another church across the street (Reekmans, 1989, 873). Sometimes referred to by the misleading term 'library of Agapito' (Figure 4.6, No. 3), it was a fourth-century *domus* with an apsidal hall and courtyard that became a church in the fifth century (Guidobaldi, 1986, 199–202). The property may have been owned by the Anicii family (Giuliani & Pavolini, 1999, 95; Brandenburg, 2005, 155; Pavolini, 2000, 148). Together, these two churches stood astride the *Clivus Scauri*, long a key access point to the Caelian neighbourhood and the *via Caelimontana*. They also stood at the base of the still massive and imposing platform of the Temple of Divine Claudius. These two buildings, their apses projecting towards anyone approaching from the Palatine and the *via Sacra*, now flanked the entrance to a neighbourhood of the city that had long been associated with some of the most powerful people in the world. For the patrons of these churches and others like them, whatever their prior status in earlier generations, such associations were elevating. Whoever the patron of the *titulus Pammachius* actually was, he could not do much better than the Anicii, one of the most important noble families in Rome, as an ecclesiastical neighbour (Brown, 2000, 341, 2012, 286; Machado, 2011, 501).

In earlier generations, these buildings had been private locations for small Christian groups to gather, steeped in the visual culture of the Roman elite, but unable to compete with the luxury or size of houses belonging to people like Symmachus or Melania. After 410, with the absence or diminishment of aristocratic fortunes like theirs, those who stayed were able to stake their claim as patrons of the Christian city and reshape neighbourhoods like the Caelian. Alongside the pope and with his support, they helped create a Christian capital using an architectural and visual language closely tied to the material world of the traditional ruling class.

By the mid-fifth century, churches stood alongside both of the two major routes leading up the Caelian hill from the ancient city centre. These streets no longer welcomed rowdy worshippers of Magna Mater or fans of gladiators. They traced the same routes they always had, but now they were part of a circuit that connected them to the Cathedral of Rome and formed the backbone of a new Christian topography of the Caelian. As domestic buildings, they were modest compared to their neighbours up the hill, but those neighbours were now gone. As churches, they became revered and ancient. Their form and locations along these specific routes embedded them with associations based on generations of social and religious practice. While felt and remembered differently, those associations powerfully anchored their patrons and the Church in the neighbourhood's elite past as well as its Christian future.

Notes

1 I wish to thank Carlos Machado and Rebecca Sweetman for inviting me to participate in the conference that led to this volume and for their guidance in developing this essay. I thank Rowan Munnery for his patience and organization throughout. Support for this research was provided by the American Academy in Rome and the American Council on Learned Societies. I am very grateful to

the anonymous reviewers for their constructive and insightful comments. All remaining errors are my own.

2 Lanciani (1897) covers the Caelian in 335–357; see more recently Fatucci (2012, 342–358) in *The Atlas of Ancient Rome*.

3 Ian Haynes, principal investigator. In addition to interim reports published in the *Papers of the British School at Rome,* updates on the *Rome Transformed* project are presented in public lectures at the British School at Rome (attended by the author in April and December 2022) and the project website (with bibliography): https://research.ncl.ac.uk/rometrans/. For the Lateran itself, see Bosman et al. (2020). For previous documentation of the eastern Caelian, see Consalvi (2009).

4 I refer here to the church as the *Basilica Constantiniana* or the Lateran. Brandenburg (2005, 20) gives the ancient names of the building as *Basilica Laterensis*; *Basilica Constantiniana,* and *Basilica Salvatoris*. See also Liverani (2020, 14).

5 The notion that Constantine sought to avoid upsetting prominent Roman pagans appears in a range of discussions of his church-building programme, including textbooks about Roman art: see for example Kleiner (2010, 299), which states that putting his churches on the outskirts of town 'permitted Constantine . . . to avoid any confrontation between Rome's Christian and pagan populations'. A statement in Krautheimer and Trachtenberg (2000, 21) that it was 'politically inadvisable to set up Christian buildings in or near the center of Rome with its temples and administrative buildings dominated by the pagan Senate' has been influential. An example from a historical overview is Davis (*Liber Pontificalis*, xxxii): 'At the date of the [Constantinian] foundations it is likely that not even an emperor would take the risk of upsetting pagan sensibilities by disturbing the monumental centre of Rome'. Johnson (2006, 283) mentions this reasoning but emphasizes that the location of the church was 'also very much a practical and logistical decision'.

6 The roof of the Lateran is visible from the Capitoline now. It is not clear if this was true in the fourth century or, if so, from where on the Capitoline one would have been able to see it. The *Rome Transformed* project analyses of ancient ground levels may answer this question definitively in the coming years.

7 *Via Tusculana* is roughly aligned with the modern Via dei Santi Quattro Coronati. See Heikonen (2017, 38), with references.

8 *Via Caelimontana* is roughly contiguous with the modern via Santo Stefano Rotondo.

9 I use the term pagan here for convenience to denote adherents of traditional polytheistic cults and practices who did not consider themselves Christian or Jewish, following a formulation and rationale in Watts (2015, 1, n. 2).

10 Martial *Epig*. 12.18 imagines Juvenal as just such a visitor. See Kelly (2018, 168–169, 174–175).

11 Property divisions are certainly difficult to discern now (Saliou, 1994, 1). As is evident from better-preserved sites like Pompeii, however, even fancy houses presented largely featureless walls to the street punctuated by doorways. While it is possible that divisions between properties were indicated in now lost painted borders or similar, where exterior walls of elite urban houses are preserved they often appear to run together.

12 The street is sometimes referred to as the *vicus Caput Africae*.

13 Latham argues that literary Christians in Late Antiquity drew on examples in classical authors (i.e. Verg. *Aen*. 9.614–20; Ovid *Fasti* 4.183, 243–44; Mart. 2.45, 2.57) that represented worshippers of the goddess as foreign, effeminate, or sexually deviant in order to inveigh against the cult and the goddess herself. He (2020, 278) also argues that this was somewhat disingenuous, as by the fourth century 'the carefully constructed difference between the "Roman" and the "foreign" in the cult of Magna Mater had largely disappeared in practice' and that '[o]nce "deviant" rites were now [in the fourth century] publicly acceptable'.

14 Latham (2020, 271) notes that the calendar of Polemius Silvius says Easter of 449 fell on 27 March and referred to the date as what the 'ancients used to call the *lavatio*'.

15 The *Historia Augusta* mentions an *Iseum Metellinum* (SHA, *Tyr. Trig*. 25.4) near sacred groves on the Caelian; it could conceivably refer to this.

16 Ammianus (27.12) says 137 supporters of Ursinus were killed by Damasan partisans in the Basilica Sicininus; Rufinus (*HE* 11.10) cites the same basilica and says the place of prayer ran with human blood; the *Collectio Avellana* offers a slightly different sequence of events and says 160 people were killed. The location of that massacre is debated; a long-accepted association with the site of Santa Maria Maggiore is now considered incorrect (Krautheimer, 2009, 317, n. 43; Latham, 2014, 134).

17 At least one more violent episode followed two years later, when Damasus and his followers broke up what Richard Krautheimer (2009, 317; citing *Coll. Avell*. 8–9) calls a 'sit-in' at the cemetery of Sant'Agnese in Nomentana.

18 Liverani (2020) argues that the catechumens who flocked to the Lateran as described in Prud. *c. Symm.* 1.585–1.586 were headed to the episcopal palace.

19 Salzman reiterates her claim that Christians did not constitute a majority among the Roman elite until the late fourth century, noting that others argue for a majority by mid-century (2021, n. 2). Machado (2019, 164) also sees major conversions of aristocrats taking place only in the second half of the fourth century.

20 Latham notes that Q.A. Symmachus spent nearly five times the annual income of the bishop of Rome on the praetorian games in honour of his son in 401 (2012, n. 14, 15 for references and calculations). See also Latham (2014, 127), Krautheimer (2009, 323) and Kinney (2001).

21 He seems to have led by example with the *titulus Damasi* (San Lorenzo in Damaso). Löx (2013, 46, 49–58) provides a description of the archaeology of the building and discusses literary sources. He argues that while the site was probably the property of Damasus's family it was not, as previously thought, his family's *domus*. The titulus is mentioned in *Liber Pontificalis* 39.4 (see Davis' Introduction, p. xxxvii).

22 As with most *tituli*, the dates of donation and conversion of these *domūs* to churches are not securely established and open to some debate. The Quattro Coronati martyrs were associated with the *titulus Aemilianae* by the late fifth century (Brandenburg, 2005, 195) and its apse was retained from the mid-fourth century *domus* that preceded it (Brandenburg, 2005; Pavolini, 2006, 105; Guidobaldi, 1986, 296). The phasing of San Clemente is complicated, but it was also a *domus* by the fourth century and at least some of its fabric was retained in the church (for detailed discussion of phases, see most recently, Heikonen, 2017, especially 33–37).

23 Dossey discusses John Chrysostom's disdain for urban activities that took place at dusk or in the evening, including crowds in the streets after games let out (citing *De Anna serm.* 3.5, with refs) and how Christian sober vigils could stand in contrast to rowdier night-time gatherings (2019, 720).

24 Hillner (2006, 66) argues that most *tituli* donors were not among the senatorial aristocracy.

25 Cameron (2011) argues that the characterization of Symmachus and his peers as leaders of a vociferous pagan resistance to Christianity (i.e. his pleas for the restoration of the Altar of Victory in the Roman Curia and the restoration of funds to some pagan cults) is misleading. See especially Cameron (2011, 37, 353–398).

26 Hillner (2003, 135–136) argues that the term *lar* indicated the house where the *paterfamilias* held power, and that therefore the Caelian property held particular significance for him, even if he did not necessarily live there. She suggests that Symmachus acquired this house when he got married, before his father died. Her point is about inheritance, that the house that holds the *lar* is not necessarily the one that was inherited through the family line. Two statue bases with honorific inscriptions are associated with the structure, both erected by Q. Aurelius Symmachus's son. On these, see also Carignani (2000, 149) and Cameron (2011, 629).

27 The limits of most of the *domūs* on the Caelian are not securely identified, but the *domus Symmachorum* appears to have been smaller than the nearby *domus Valeriorum* (estimated at least 12,000 square metres based on published drawings), but much larger than the neighbouring *domus Gaudentius* (estimated at 1,250 square metres based on published drawings). The dimensions of the *domus Gaudentius* are more typical of late antique urban houses in Rome (Guidobaldi, 1986, 218–219) and other cities in the empire (see, e.g. Baldini Lippolis, 2005; Bowes, 2010).

28 Dunbabin (1991) discusses apsidal halls in relation to dining and other social practices in late antique houses.

29 The inscription: *Dominus legem dat Valerio Severo Eutropi vivas.*

30 The *vita* by Gerontius exists in Greek and Latin versions and there is debate about which text is primary. Elizabeth Clark (1984) supports a Greek original. Passages in the *vita* referred to here are from her translation in that volume. Julia Hillner's argument (2003, especially 141–142) that there is no reason to believe that the *domus Valeriorum* belonged to Melania and Pinianus or even consistently to the Valerii has been influential. Barbera et al. (2008, 86) and Pavolini (2006, 109–111) offer evidence for association with the family in the fourth century; the argument in Brenk (1999, 2003) is tied to the hypothesis that the family lived there. See Machado (2019, 250) for discussion.

31 Brenk (1999, 2003) discusses two silver amulets and a small pot with images and text related to Peter and Paul said to have been found in the *domus Valeriorum* in eighteenth-century excavations. He dates them to the second half of the fourth century and suggests that they along with the boat-shaped lamp could have been used in a private Christian chapel in the house (his argument is closely tied to his understanding of the house as belonging to Melania). No other evidence of a chapel has been identified.

32 Machado cites testimony (Symm *Ep.* 1.120; *De cos. Prob. et Olyb. 226–233*) for consular inaugurations in which the honorees were 'paraded' (2010, 298) from their house to the Forum accompanied by numbers of people and other such displays.

33 Bauer argues that the church could have been completed before 410; Brenk (2003, 105–106) argues that it was built in the fifth century (based in part on evidence of amphorae in one of the rooms of the house). Kinney (2017, 70) says only that the building demonstrates architectural refinements of 'the turn of the fifth century'.

34 This identification of Pammachius is supported by Bauer (2013, 263), but Hillner (2006, 62–63) and especially Brenk (2003, 109–110) argue against it. The association of the *titulus Pammachius* with the current church of Santi Giovanni e Paolo is based on an inscription in the vestibule dated to the episcopate of Leo I (440–461), which mentions Pammachius and the dedication to the two martyrs (Hillner, 2006). The secondary association of the property with *titulus Vyzantis* is apparently based on a mention of the two *tituli* next to each other in the council acts of 499, but Brenk (2003, 111) argues that this is a mistake. The *Liber Pontificalis* credits Pope Symmachus (498–514), with building steps behind the apse of a church of Saints John and Paul (Life 53.9). A senator named Vyzantis and his son, Pammachius, feature in John and Paul in the martyrs' passion story as protectors of their relics, but that story was only later retrojected onto the church of Santi Giovanni e Paolo. The *passio* said that the martyrs were executed during the reign of Julian and their relics buried in their own house, but all of this took place in Antioch; the martyrs, their house, and this story have no historical connection with Rome or the building on the *Clivus Scauri* (Brenk, 2003, 111–112; Leyser, 2007).

Editions and abbreviations

Amm. Marc. – *Ammianus Marcellinus. History, Volume III: Books 27–31. Excerpta Valesiana* (Loeb Classical Library), trans. J. C. Rolfe (1939). Harvard University Press.

Aug. *Conf.* – Augustine, *Confessions, Volume 1: Books 1–8* (Loeb Classical Library), trans. C. J.-B. Hammond (2014). Harvard University Press.

Coll. Avell. – *Epistolae Imperatorum Pontificum Aliorum Inde ab a. CCCLXVII usque DLIII datae Avellana Quae Dicitur Collectio* (CSEL 35), ed. O. Guenther (1895). Tempsky.

CTh. – *Codex Theodosianus*. In *Theodosiani Libri XVI cum Constitutionibus Sirmondianis*, ed. T. Mommsen & P. Meyer (1905); trans. C. Pharr (1952), *The Theodosian Code and Novels, and the Sirmondian Constitutions*.

Eus. *VC* – *Eusebius. Life of Constantine*, with translation and commentary by A. Cameron & S. G. Hall (1999). Oxford University Press.

Jerome, *Ep.* – *Sancti Eusebii Hieronymi Epistulae*, vol. 3: Ep. 121–154 (CSEL 56), ed. I. Hillberg (1918). Tempsky.

Liber Pontificalis – *The Book of Pontiffs* (Liber pontificalis): *The ancient biographies of the first ninety Roman bishops to AD 715 (Translated Texts for Historians)*, trans. R. Davis (2010). Liverpool University Press.

Martial, *Epig.* – *Martial, Epigrams, Vol. 3: Books 11–14* (Loeb Classical Library), ed. and trans. D. R. Shackleton Bailey (1993). Harvard University Press.

Pan. Lat. – *In Praise of Later Roman Emperors: The Panegyrici Latini*, trans. C. E. V. Nixon & B. S. Rodgers (1994). University of California Press.

Rufin., *HE* – *History of the Church*, trans. P. R. Amidon (2016). Catholic University of America Press.

SHA – Scriptores Historiae Augustae. In *Historia Augusta, Volume 1* (Loeb Classical Library), trans. D. Magie, revised by D. Rohrbacher (2022). Harvard University Press.

Symm., *Ep.* – Symmachus, *Epistulae*. In *Symmaque. Correspondence*, 4 vols., ed. and trans. J.-P. Callu (1972–2002). Les Belles Lettres.

Vita Mel. Gr. – *The Life of Melania the Younger*, with introduction, translation, and commentary by E. A. Clark (1984). The Edwin Mellen Press.

Vita S. Ambrosii – Paulinus, *Vita S. Ambrosii*. In *Ambrose*, trans. B. Ramsey (1997). Routledge.

Zos.– *Zosimus: New History* (Byzantina Australiensia), trans. R. T. Ridley (1982). Brill.

Bibliography

Alberti, B., Englen, A., Filetici, M. G., & Palazzo, P. (Eds.). (2014). *Caelius II. Pars inferior. Le case romane sotto la basilica dei Santi Giovanni e Paolo*. L'Erma di Bretschneider.

Anderson, B. (2016 [1983]). *Imagined Communities: Reflections on the Origin and Spread of Nationalism*. Verso Books.

Andrews, M. M. (2014). A Domus in the Subura of Rome from the Republic through Late Antiquity. *American Journal of Archaeology*, *118*, 61–90.

Baldini Lippolis, I. (2005). *L'architettura residenzale nelle città tardoantiche*. Carocci.

Barbera, M., Palladino, S., & Paterna, C. (2008). La domus dei Valerii sul Celio alla luce delle recenti scoperte. *Papers of the British School at Rome*, *76*, 75–98, 349–354.

Barnes, T. D. (1981). *Constantine and Eusebius*. Harvard University Press.

Barnes, T. D. (2011). *Constantine. Dynasty, religion and power in the later Roman empire*. Wiley-Blackwell.

Bauer, F. A. (2013). Saccheggi e distruzioni nell'anno 410? Nuove problematiche e nuove prospettive. In J. Lipps, C. Machado, & P. von Rummel (Eds.), *The Sack of Rome in 410 AD. The Event, Its Context and Its Impact. Proceedings of the Conference Held at the German Archaeological Institute at Rome, 4–6 November 2010. Palilia, Vol. 28* (pp. 259–275). Reichert.

Bosman, L. (2020). Constantine's *Spolia:* A set of columns for San Giovanni in Laterano and the arch of Constantine in Rome. In L. Bosman, I. P. Haynes, & P. Liverani (Eds.), *The Basilica of Saint John Lateran to 1600* (pp. 168–196). Cambridge University Press.

Bosman, L., Haynes, I. P., & Liverani, P. (Eds.). (2020). *The Basilica of Saint John Lateran to 1600*. Cambridge University Press.

Bowes, K. (2010). *Houses and Society in the Later Roman Empire*. Duckworth.

Brandenburg, H. (2005). *Ancient Churches of Rome From the Fourth to the Seventh Century. The Dawn of Christian Architecture in the West* (A. Kropp, Trans.). Brepols. (or. ed. 2004).

Brandt, O. (2009). The archaeological record: Problems of interpretation. In P. Rousseau (Ed.), *A Companion to Late Antiquity* (pp. 156–169). Wiley-Blackwell.

Brenk, B. (1999). La cristianizzazione della Domus dei Valerii sul Celio. In W. V. Harris (Ed.), *The Transformations of Urbs Roma in Late Antiquity* (pp. 69–84). Journal of Roman Archaeology.

Brenk, B. (2003). *Die Christianisierung der spätrömischen Welt: Stadt, Land, Haus, Kirche und Kloster in frühchristlicher Zeit*. Reichert.

Brown, P. (2000). The study of elites in Late Antiquity. *Arethusa*, *33*(3), 321–346.

Brown, P. (2012). *Through the Eye of a Needle: Wealth, the Fall of Rome, and the Making of Christianity in the West, 350–550 AD*. Princeton University Press.

Burns, J. P., & Jensen, R. M. (2014). *Christianity in Roman Africa: The Development of Its Practices and Beliefs*. Eerdmans Press.

Calvino, I. (1974). *Invisible Cities* (W. Weaver, Trans.). Harcourt Brace. (or. ed. 1972).

Cameron, A. (1999). The last Pagans of Rome. In W. V. Harris (Ed.), *The Transformations of Urbs Roma in Late Antiquity* (pp. 109–121). Journal Roman Archaeology (Suppl Series No. 33).

Cameron, A. (2011). *The Last Pagans of Rome*. Oxford University Press.

Carignani, A. (2000). La *domus* 'dei Simmaci'. In S. Ensoli & E. La Rocca (Eds.). *Aurea Roma: Dalla città pagana alla città cristiana* (pp. 149–151). L'Erma di Bretschneider.

Chin, C. M. (2017). Apostles and Aristocrats. In C. M. Chin & C. T. Schroeder (Eds.), *Melania: Early Christianity Through the Life of One Family* (pp. 19–33). University of California Press.

Chin, C. M., & Schroeder, C. T. (2017). Introduction. In C. M. Chin & C. T. Schroeder (Eds.), *Melania: Early Christianity Through the Life of One Family* (pp. 1–13). University of California Press.

Clark, E. A. (1984). *The Life of Melania the Younger. Introduction, Translation, and Commentary*. The Edwin Mellen Press.

Coarelli, F. (2007). *Rome and Environs. An Archaeological Guide*. University of California Press.

Coates-Stephens, R. (2001). Muri dei bassi secoli in Rome: Observations on the re-use of statuary in walls found on the Esquiline and Caelian after 1870. *Journal of Roman Archaeology*, *14*, 217–238.

Colini, A. M. (1944). *Storia e topografia del Celio nell'antichità*. Tipografia Poliglotta Vaticana.

Colini, A. M., & Cozza, L. (1962). *Ludus Magnus*. "L'Erma" di Bretschneider.

Consalvi, F. (2009). *Il Celio Orientale. Contributi alla Carta Archeologica di Roma. Tavola VI settore H*. Instituto Nazionale di Studi Romani – ONLUS.

Curran, J. (2000). *Pagan City and Christian Capital: Rome in the Fourth Century*. Oxford University Press.

DeLaine, J. (2000). Building the eternal city: The construction industry of imperial Rome. In J. Coulston & H. Dodge (Eds.), *Ancient Rome: The Archaeology of the Eternal City* (pp. 119–141). Oxford University School of Archaeology.

Diosono, F. (2016). *Professiones gentiliciae:* The *collegia* of Rome between paganism and Christianity. In M. Salzman, M. Sághy, & R. Lizzi Testa (Eds.), *Pagans and Christians in Late Antique Rome* (pp. 251–270). Cambridge University Press.

Dossey, L. (2019). Night in the big city: Temporal patterns in Antioch and Constantinople as revealed by Chrysostom's sermons. In C. de Wet & W. Mayer (Eds.), *Revisioning John Chrysostom: New Approaches, New Perspectives* (pp. 698–732). Brill.

Dunbabin, K. M. D. (1991). *Triclinium* and *stibadium*. In W. J. Slater (Ed.), *Dining in a Classical Context* (pp. 121–148). University of Michigan Press.

Ebbeler, J. V. (2007). Religious identity and the politics of patronage: Symmachus and Augustine. *Historia. Zeitschrift für Alte Geschichte*, *56*, 230–242.

Ellis, S. (1991). Power, architecture, and decor: How the late Roman aristocrat appeared to his guests. In E. Gazda (Ed.), *Roman Art in the Private Sphere* (pp. 117–134). University of Michigan Press.

Elm, S. (2014). Church – festival – temple: Reimagining civic topography in Late Antiquity. In C. Rapp & H. A. Drake (Eds.), *The City in the Classical and Post-Classical World: Changing Contexts of Power and Identity* (pp. 167–182). Cambridge University Press.

Elsner, J. (2003). Inventing Christian Rome: The Rome of early Christian art. In C. Edwards & G. Woolf (Eds.), *Rome the Cosmopolis* (pp. 71–99). Cambridge University Press.

Fatucci, G. (2012). Region II. Caelimontium. In A. Carandini with P. Carafa (Eds.), *The Atlas of Ancient Rome. Biography and Portraits of the City. Vol. I. Text and Images* (pp. 342–358). Princeton University Press (or. ed. 2012).

Fraschetti, A. (1999). *La conversione da Roma pagana a Roma cristiana*. Laterza.

Garnsey, P., & Whittaker, C. R. (1998). Trade, industry and the urban economy. In A. Cameron & P. Garnsey (Eds.), *The Cambridge Ancient History, Vol. XIII. The Late Empire, A.D. 337–425* (pp. 312–337). Cambridge University Press.

Garrioch, D., & Peel, M. (2006). Introduction. The social history of urban neighborhoods. *Journal of Urban History*, *32*, 663–676.

Giuliani, E., & Pavolini, C. (1999). La 'Biblioteca di Agapito' e la Basilica di S. Agnese. In W. V. Harris (Ed.), *The Transformations of Urbs Roma in Late Antiquity* (pp. 69–84). Journal of Roman Archaeology.

Graeber, D., & Wengrow, D. (2021). *The Dawn of Everything: A New History of Humanity*. Farrar, Straus & Giroux.

Grig, L. (2004). Portraits, pontiffs and the Christianizaiton of fourth-century Rome. *Papers of the British School at Rome*, *72*, 203–300.

Guidobaldi, F. (1986). L'edilizia abitativa unifamiliare nella Roma tardoantica. In A. Giardina (Ed.), *Società romana e impero tardoantico, vol. 2: Roma, politica, economia, paesaggio urbano* (pp. 165–237). Laterza.

Guidobaldi, F. (1989). L'inserimento delle chiese titolari di Roma nel tessuto urbano preesistente: osservazioni ed implicazioni. In P. Pergola & F. Bisconti (Eds.), *Quaeritur inventus colitur: miscellanea in onore di padre U.M. Fasola* (pp. 381–96). Pontificio istituto di archeologia cristiana.

Guidobaldi, F. (2000). Le abitazioni private e l'urbanistica. In A. Giardina (Ed.), *Storia di Roma dall'antichità a oggi. Roma Antica* (pp. 133–161). Editori Laterza.

Guilhembet, J.-P., & Royo, M. (2008). L'aristocratie en ses quartiers (IIe s. avant J.-C. – IIe s. après J.-C.). In M. Royo, E. Hubert, & A. Bérenger (Eds.), *Rome des quartiers: Des vici aux rioni. Cadres institutionnels, pratiques sociales et requalifications entre antiquité et époque moderne, Université de Paris IV: Paris-Sorbonne* (pp. 93–227). De Boccard.

Hallbäck, G. (1998). Sacred meal and social meeting: Paul's argument in 1 Cor. 11.17–34. In I. Nielsen & H. S. Nielsen (Eds.), *Meals in a Social Context: Aspects of the Communal Meal in the Hellenistic and Roman World* (pp. 166–176). Aarhus University Press.

Haynes, I., & Liverani, P. (2020). The Castra Nova and the Severan transformation of Rome. In L. Bosman, I. P. Haynes, & P. Liverani (Eds.), *The Basilica of Saint John Lateran to 1600* (pp. 91–113). Cambridge University Press.

Haynes, I., Liverani, P., Piro, S., & Spinola, G. (2013). The Lateran project: Interim report on the July 2012 and January 2013 seasons (Rome). *Papers of the British School at Rome*, *81*, 360–363.

Haynes, I., Liverani, P., Piro, S., Spinola, G., & Turner, A. (2014). The Lateran project: Interim report for the January 2014 season (Rome). *Papers of the British School at Rome*, *82*, 331–335.

Heikonen, J. (2017). *San Clemente in Rome: A New Reconstruction of the Early Fifth Century Basilica and Its Origins*. (Ser. Aalto University publication series. Doctoral dissertations, 60/2017). School of Arts, Design and Architecture.

Hillner, J. (2003). Domus, family, and inheritance: The senatorial family house in late antique Rome. *The Journal of Roman Studies*, *93*, 129–145.

Hillner, J. (2006). Clerics, property and patronage: The case of the Roman titular churches. *Antiquité Tardive*, *14*, 59–68.

Johnson, M. J. (2006). Architecture of empire. In N. Lenski (Ed.), *The Cambridge Companion to the Age of Constantine* (pp. 278–297). Cambridge University Press.

Kaldellis, A. (2009). *The Christian Parthenon*. Cambridge University Press.

Keith, K. (2003). The Spatial patterns of everyday life in old Babylonian neighborhoods. In M. L. Smith (Ed.), *The Social Construction of Ancient Cities* (pp. 56–80). Smithsonian Books.

Kelly, G. (2018). From Martial to Juvenal (Epigrams 12.18). In A. König & C. Whitton (Eds.), *Roman Literature Under Nerva, Trajan and Hadrian: Literary Interactions, AD 96–138* (pp. 160–180). Cambridge University Press.

Kinney, D. (2001). The church basilica. In J. Rasmus Brandt & O. Steen (Eds.), *Imperial Art as Christian Art. Christian Art as Imperial Art. Expression and Meaning in Art and Architecture From Constantine to Justinian. Acta ad archaeologicam et artum historiam pertinentia 15* (pp. 115–135). Bardi.

Kinney, D. (2007). Bearers of meaning. *Jahrbuch für Antike und Christentum*, *50*, 139–153.

Kinney, D. (2017). Expanding the Christian footprint: Church building in the city and the *suburbium*. In I. Foletti & M. Gianandrea (Eds.), *The Fifth Century in Rome: Art, Liturgy, Patronage* (pp. 65–97). Viella.

Kleiner, F. S. (2010 [2007]). *A History of Roman Art*. Wadsworth.

Krautheimer, R. (2009). Damasus in his time. In A. Paolucci & F. Buranelli (Eds.), *L'Antica Basilica di San Lorenzo in Damaso. Indagini archeologiche nel Palazzo della Cancelleria (1988–1993), Vol. 1. Gli Scavi* (pp. 313–327). Musei Vaticani/Bibliotheca Hertziana – Istituto Max Planck per la Storia dell'Arte.

Krautheimer, R., & Trachtenberg, M. (2000). *Rome: Profile of a City, 312–1308*. Princeton University Press.

Lanciani, R. (1897). *The Ruins and Excavations of Ancient Rome*. Houghton, Mifflin and Co.

Latham, J. (2012). From literal to spiritual soldiers of Christ: Disputed episcopal elections and the advent of Christian processions in late antique Rome. *Church History*, *81*, 298–327.

Latham, J. (2014). Battling bishops, the Roman aristocracy, and the contestation of civic space in late antique Rome. In J. D. Rosenblum, L. C. Vuong, & N. P. DesRosiers (Eds.), *Religious Competition in the Third Century CE: Jews, Christians, and the Greco-Roman World* (pp. 126–137). Vandenhoeck & Ruprecht.

Latham, J. (2020). Re-defining aristocratic distinction: Christian verse invective, classical culture, and the cult of Mater Magna in late antique Rome. *Memoirs of the American Academy in Rome*, *65*, 261–306.

Laurence, R. (1994). *Roman Pompeii: Space and Society*. Routledge.

Leyser, C. (2007). 'A church in the house of the saints:' Property and power in the *Passion of John and Paul*. In K. Cooper & J. Hillner (Eds.), *Religion, Dynasty, and Patronage in Early Christian Rome, 300–900* (pp. 140–162). Cambridge University Press.

Lissi, E. (1986). *Il Mitreo dei Castra peregrinorum* (S. Stefano Rotondo). Brill.

Liverani, P. (2020). The evolution of the Lateran: From the *Domus* to the episcopal complex. In L. Bosman, I. P. Haynes, & P. Liverani (Eds.), *The Basilica of Saint John Lateran to 1600* (pp. 6–24). Cambridge University Press.

Lizzi Testa, R. (2009). The late antique bishop: Image and reality. In P. Rousseau (Ed.), *A Companion to Late Antiquity* (pp. 525–538). Blackwell Publishing.

Lott, J. B. (2004). *The Neighborhoods of Augustan Rome*. Cambridge University Press.
Löx, M. (2013). *Monumenta Sanctorum. Rom und Mailund als Zentren des frühen Christentums: Märtyrerkult und Kirchen bau unter den Bischöfen Damasus und Ambrosius*. (Ser. Spätantike – Frühes Christentum – Byzanz. Kunst im ersten Jahrtausend, Reihe B: Studien und Perspektiven, Band 39). Reichert Verlag.
Machado, C. (2010). The city as stage: Aristocratic commemorations in late antique Rome. In C. Sotinel & É. Rebillard (Eds.), *Les frontières du profane dans l'antiquité tardive* (pp. 287–317). École française de Rome.
Machado, C. (2011). Roman aristocrats and the christianization of Rome. In P. Brown & R. Lizzi Testa (Eds.), *Pagans and Christians in the Roman Empire: The Breaking of a Dialogue* (pp. 493–516). Lit Verlag.
Machado, C. (2012a). Between memory and oblivion: The end of the Roman *domus*. In R. Behrwald & C. Witschel (Eds.), *Rom in der Spätantike: Historische Erinnerung im städtischen Raum* (pp. 111–138). Franz Steiner.
Machado, C. (2012b). Aristocratic houses and the making of late antique Rome and Constantinople. In L. Grig & G. Kelly (Eds.), *Two Romes: Rome and Constantinople in Late Antiquity* (pp. 136–158). Oxford University Press.
Machado, C. (2019). *Urban Space and Aristocratic Power in Late Antique Rome, AD 270–535*. Oxford University Press.
Maier, H. O. (1995). The topography of heresy and dissent in late-fourth-century Rome. *Historia: Zeitschrift für Alte Geschichte*, *44*, 232–249.
Markus, R. (1990). *The End of Ancient Christianity*. Cambridge University Press.
Marlowe, E. (2006). Framing the sun: The arch of Constantine and the Roman cityscape. *The Art Bulletin*, *88*, 223–242.
Mathisen, R. W. (2013). Roma a Gothis Alarico duce capta est. Ancient accounts of the sack of Rome in 410 CE. In J. Lipps, C. Machado, & P. von Rummel (Eds.), *The Sack of Rome in 410 AD. The Event, Its Context and Its Impact. Proceedings of the Conference Held at the German Archaeological Institute at Rome, 4–6 November 2010* (pp. 87–102). Reichert.
McGowan, A. (2004). Rethinking Agape and Eucharist in early north African Christianity. *Studia Liturgica*, *34*, 165–176.
Meeks, W. A. (1993). *The Origins of Christian Morality: The First Two Centuries*. Yale University Press.
Meeks, W. A. (2003). *The First Urban Christians: The Social World of the Apostle Paul* (2nd ed.). Yale University Press.
Mignone, L. M. (2016). *The Republican Aventine and Rome's Social Order*. University of Michigan Press.
Mol, E., & Versluys, M. J. (2015). Material culture and imagined communities in the Roman world. In R. Raja & J. Rüpke (Eds.), *A Companion to the Archaeology of Religion in the Ancient World* (pp. 451–461). Wiley.
Moltesen, M., with Hast, R. (2004). The Antinous Casali in the Ny Carlsberg Glyptotek. *Analecta Romana Instituti Danici*, *30*, 101–117.
Moralee, J. (2018). *Rome's Holy Mountain. The Capitoline Hill in Late Antiquity* (Oxford Studies in Late Antiquity). Oxford University Press.
Omissi, A. (2018). *Emperors and Usurpers in the Later Roman Empire. Civil War, Panegyric, and the Construction of Legitimacy*. Oxford University Press.
Palazzo, P., & Pavolini, C. (Eds.). (2013). *Gli dèi propizi. La Basilica Hilariana nel contesto dello scavo dell'Ospedale Militare Celio (1987–2000)*. Quasar.
Parlato, E. (2017). Colosseo al nero: distopie del rudere. In R. Rea, S. Romano, & R. Santangeli Valenzani (Eds.), *Colosseo* (pp. 166–181). Electa.
Pavolini, C. (Ed.). (1993a). *Caput Africae I. Indagini archeologiche a Piazza Celimontana (1984–1988). La storia, lo scavo, l'ambiente*. Istituto Poligrafico e Zecca dello Stato, Libreria dello Stato, Soprintendenza Archeologica di Roma.
Pavolini, C. (1993b). L'area del Celio fra l'antichità e il medioevo alla luce delle recenti indagini archeologiche. In L. Paoli & P. Delogu (Eds.), *La storia economica di Roma nell'alto Medioevo alla luce dei recenti scavi archeologici* (pp. 53–70). Edizioni All'Insegna del Giglio.

Pavolini, C. (1994–1995). Nuovi contributi alla topografia del Celio da rinvenimenti casuali di scavo. *Bullettino della Commissione Archeologica Comunale di Roma*, *96*, 71–94.
Pavolini, C. (2000). Le *domus* del Celio. In S. Ensoli & E. La Rocca (Eds.), *Aurea Roma: Dalla città pagana alla città Cristiana* (pp. 147–148). L'Erma di Bretschneider.
Pavolini, C. (2006). *Lexicon topographicum urbis Romae, Supplementum III. Archeologia e topografia della regione II (Celio). Un aggiornamento sessanti'anni dopo Colini*. Quasar.
Pavolini, C. (2007). L' 'Agrippina-Orante' di Villa Casali e la politica religiosa degli imperatori sul Celio. In A. Leone, D. Palombi, & S. Walker (Eds.), *Res bene gestae. Richerche di storia urbana su Roma antica in onore di Eva Margareta Steinby* (pp. 309–334). Quasar.
Pavolini, C. (2013). Le consequenze del sacco di Alarico sul Celio. Alcune ipotesi dagli scavi recenti. In J. Lipps, C. Machado, & P. von Rummel (Eds.), *The Sack of Rome in 410 AD. The Event, Its Context and Its Impact. Proceedings of the Conference Held at the German Archaeological Institute at Rome, 4–6 November 2010* (pp. 163–183). Reichert.
Pavolini, C., Carignani, A., Pacetti, F., Spinola, G., & Vitti, M. (1993). La topografia antica della sommità del Celio. Gli scavi nell'Ospedale Militare (1987–1992). *Mitteilungen des Deutschen Archaeologischen Instituts Roemische Abteilung/Bullettino dell'Istituto Archeologico Germanico sezione Romana*, *100*, 443–505.
Pensabene, P. (2008). Il culto di Cibele e la topografia del sacro a Roma. In B. P. Venetucci (Ed.), *Culti orientali tra scavo e collezionismo* (pp. 21–39). Artemide.
Pietri, C. (1976). *Roma Christiana. Recherches sur l'église de Rome, son organisation, sa politique, son idéologie de Miltiade à Sixte III (311–440)*. École française de Rome.
Rebillard, É. (2012). *Christians and Their Many Identities in Late Antiquity, North Africa, 200–450 CE*. Cornell University Press.
Rebillard, É. (2020). Approaching 'religious identity' in Late Antiquity. In R. Flower & M. Ludlow (Eds.), *Rhetoric and Religious Identity in Late Antiquity* (pp. 15–27). Oxford University Press.
Reekmans, L. (1989). L'implantation monumentale chrétienne dans le paysage urbain de Rome 300 à 850. In N. Duval, F. Bartiel, & P. Pergola (Eds.), *Actes du XIe congrès international d'archéologie chrétienne, Lyon, Vienne, Grenoble, Genève, Aoste, 21–28 septembre 1986* (pp. 861–916). École française de Rome.
Romano, J. (2020). The rite of the reconciliation of penitents at the Lateran Basilica. In L. Bosman, I. P. Haynes, & P. Liverani (Eds.), *The Basilica of Saint John Lateran to 1600*. (pp. 400–427). Cambridge University Press.
Rose, C. B. (2021). Reconsidering the frieze on the Arch of Constantine. *Journal of Roman Archaeology*, *34*, 1–36.
Rothaus, R. (1996). Christianization and de-paganization: The late antique creation of a conceptual Frontier. In R. W. Mathisen & H. S. Sivan (Eds.), *Shifting Frontiers in Late Antiquity* (pp. 299–308). Variorum/Ashgate.
Sághy, M. (2012). *Renovatio Memoriae*: Pope Damasus and the Martyrs of Rome. In R. Behrwald & C. Witchel (Eds.), *Rom in der spätantike: Historische erinnerung im städtischen raum* (pp. 251–265). Franz Steiner Verlag.
Saliou, C. (1994). *Les lois des bâtiments: voisinage et habitat urbain dans l'Empire romain: recherches sure les rapports entre le droit et la construction privée du siècle d'Auguste au siècle de Justinien* (Bibliothèque archéologique et historique, no. 116). Institute Français d'archéologie du Proche-Orient.
Salzman, M. R. (1990). *On Roman Time. The Codex-Calendar of 354 and the Rhythms of Urban Life in Late Antiquity*. University of California Press.
Salzman, M. R. (2021). *The Falls of Rome: Crises, Resilience, and Resurgence in Late Antiquity*. Cambridge University Press.
Saxer, V. (1989). L'utilisation par la liturgie de l'espace urbain et suburbain: l'exemple de Rome dans l'Antiquité et le Haut Moyen Âge. In N. Duval (Ed.), *Actes du XIe congrès international d'archéologie chrétienne. Lyon, Vienne, Grenoble, Genéve, Aoste, 21–28 septembre 1986* (pp. 917–1033). École Française de Rome.
Schuddeboom, F. L. (2016). The decline and fall of the *mithraea* of Rome. *Babesch*, *91*, 225–245.

Sessa, K. (2007). Christianity and the *Cubiculum*: Spiritual politics and domestic space in late antique Rome. *Journal of Early Christian Studies*, *15*, 171–204.

Smith, M. (2010). The archaeological study of neighborhoods and districts in ancient cities. *Journal of Anthropological Archaeology*, *29*, 137–154.

Spera, L. (2013). Characteristics of the Christianisation of space in late antique Rome: New considerations a generation after Charles' Pietri's *Roma Christiana*. In T. Kaiser, A. Leone, E. Thomas, & R. Witcher (Eds.), *Cities and Gods: Religious Space in Transition* (pp. 121–142). Peeters.

Spinola, G. (2000). La *domus* di Gaudentius. In S. Ensoli & E. La Rocca (Eds.), *Aurea Roma: Dalla città pagana alla città cristiana* (pp. 152–155). L'Erma di Bretschneider.

Steiner, H., & Sternberg, M. (Eds.). (2015). *Phenomenologies of the City. Studies in the History and Philosophy of Architecture* (Ashgate Studies in Architecture). Ashgate.

Suttles, G. D. (1972). *The Social Construction of Communities*. University of Chicago Press.

Tran, N. (2020). The meeting places of associations in the city of Rome. In M. L. Caldelli & C. Ricci (Eds.), *City of Encounters: Public Spaces and Social Interaction in Ancient Rome* (pp. 197–223). Quasar.

Trout, D. (1999). *Paulinus of Nola*. University of California Press.

Trout, D. (2005). Damasus and the invention of early Christian Rome. In D. B. Martin & P. Cox Miller (Eds.), *The Cultural Turn in Late Ancient Studies: Gender, Asceticism, and Historiography* (pp. 298–315). Duke University Press.

Trout, D. (2015). *Damasus of Rome. The Epigraphic Poetry*. Oxford University Press.

Valentini, R., & Zucchetti, G. (1940–1942). *Codice topografico della città di Roma*. Fonti per la Storia d'Italia pubblicate dal R. Istituto Storico Italiano per il Medio Evo.

Van Haeperen, F. (2012). Collèges de dendrophores et autorités locales et romaines. In M. Dondin-Payre & N. Tran (Eds.), *Collegia. Le phénomène associatif dans l'Occident romain* (pp. 47–62). Boccard.

Watts, E. J. (2006). *City and School in Late Antique Athens and Alexandria*. University of California Press.

Watts, E. J. (2015). *The Last Pagan Generation*. University of California Press.

Weisweiler, J. (2015). Domesticating the senatorial elite. Universal monarchy and transregional aristocracy in the fourth century AD. In J. Wienand (Ed.), *Contested Monarchy: Integrating the Roman Empire in the Fourth Century AD* (pp. 17–41). Oxford University Press.

White, L. M. (1998). Regulating fellowship in the communal meal: Early Jewish and Christian evidence. In I. Nielsen & H. S. Nielsen (Eds.), *Meals in a Social Context: Aspects of the Communal Meal in the Hellenistic and Roman World* (pp. 177–205). Aarhus University Press.

Xanthopoulou, M. (2010). *Les lampes en bronze à l'époque paléochrétienne*. Brepols.

York, A. M., Smith, M. E., Stanley, B. W., Stark, B. L., Novic, J., Harlan, S. L., Cowgill, G. L., & Boone, C. G. (2011). Ethnic and class clustering through the ages: A transdisciplinary approach to urban neighbourhood social patterns. *Urban Studies*, *48*, 2399–2415.

5 Shipshape and Roman fashion

Space at sea in Late Antiquity

Rowan Munnery

The humble merchant ship was a ubiquitous sight in the world of Late Antiquity, acting as the workhorse of the vast engine of maritime exchange, the commonplace backdrop for thousands of sailors, and an inescapable presence in the dockyards of any Mediterranean harbour city. Their sheer commonality meant that the specifics of everyday life aboard hardly merited much time or effort being wasted in description by the vast majority of ancient writers, being both essentially taken for granted and crewed by the lower rungs of society.

However, this issue of overlooking the seemingly inconsequential actions of routine existence aboard ship is not one that is solely confined to the casual snobbery or unthinking familiarity of late antique literati. It is also an oft-repeated blind spot in the way that maritime archaeological scholarship considers the material remains of long-lost vessels, creating a situation where the excavation and survey of ships is 'more concerned with underwater technology than with historical results' (McGhee, 1998). With the history of seafaring frequently being interpreted through a purely mechanistic lens, a 'spatial turn' in maritime archaeology towards life aboard ship has a huge potential to provide new insights into the human experience through an understanding of how the everyday shaped, and was shaped by, the space in which it occurred.

This lived space, or the *espace vécu* of Minkowski (2013) and Lefebvre (1974), is intimately connected with the material reality of the everyday and we can leverage a range of insights into routine practices, architecture, and sociality in order to link it with our knowledge of late antique craft. By turning the focus on the ship as the 'largest and most complex machine' (Muckelroy, 1978, 3) of pre-industrial societies to new ends through concentrating on how its architecture and artefacts shaped social interaction and reflected maritime culture, we can reclaim the ship in abstract as a vessel for human action rather than just human ingenuity.

In the words of de Certeau, 'space is a practiced place' (1984, 117), and as such this chapter will focus on how the routines of everyday life (see Goodwin, 2002, 188–190) such as sleeping, cooking, eating, and praying, were mediated and controlled by the material realities of ships in Late Antiquity. This material approach, combined with artistic representations and literary extracts from the period, will allow us to create a more complete reconstruction of the ship as a 'complex cultural unit' (Gilroy, 1992, 193). By understanding how 'everyday space' (Upton, 2002) was organised on these vessels, we illuminate how patterns of behaviour on the part of the people on board may have been affected by their interaction with it.

Beginning with a consideration of how we can apply the concept of lived space to our knowledge of late antique ships, this chapter will move on to outline an archaeological case study of a well-documented wreck and compare it with ancient representations of life at sea to form a picture of how both architecture and social space were arranged on most vessels. With this, it will examine a series of contexts, including religious practices, domestic arrangements, social interactions, and foodways in order to understand the relationship the routines that comprised

DOI: 10.4324/9780429427152-6

them had with everyday space. In all, this chapter aims to provide new insights into how the social structures of shipborne residents may have been shaped by the spaces they inhabited, as well as how these may have reflected the maritime society in which they lived.

5.1 Space is the place

The term 'lived space' itself exposes one of the most interesting elements of how the human experience of space differs from space as an abstract geometric entity. Simply saying that a person is aboard a ship is little more than an acknowledgement that it acts as a container within which they are physically located. This is ultimately accurate, but in itself tells us little about the consequential impact of being within on said person's life from their own point of view. Instead, for lived space to be 'lived', it requires the active participation of the individual in the process of existing within it.

This perspective is shared at the base level in de Certeau's (1984, 117) insistence on practice being required to enact space; Lefebvre's (1991, 31) determination that every society produces its own space; Bollnow's (2011, 24) assertion that space is fixed in relation to man, and that man forms space; or Ingold's (2000, 143) argument that the world is constantly formed around the inhabitant. Put together, this suggests that our departure into examining lived space aboard late antique ships must necessarily start from a consideration as to how the minds of their inhabitants would perceive them.

In the phenomenological view of space espoused by Bollnow, the key aspect of the mind's perception of space is that it both dwells within the body and the immediate space that surrounds it (Bollnow, 2011, 267–288). This derives from the work of Heidegger (1927, 1997) and Merleau-Ponty (1945), who advance the argument that 'dwelling' is the core of one's relationship with the world – and to dwell is to ground oneself in a place in such a way as to find peace, stability, and security within it. In all, we can see that one's relationship to a space is heavily influenced by the restrictions that it places upon the way it can be experienced.

One of the most important of these is the opportunity to travel through it, the perception of which Bollnow labels 'space as elbow-room' (2011, 37). This space follows oneself as one moves and is the space around the body that the mind dwells within. To do so, however, it is the natural reaction of people to secure and protect their elbow-room in order to maintain the peace and stability that is synonymous with dwelling. This protection is most notable when confronted by other people and results in the creation of a possession space that allows for the possibility of unhindered activity and is often established by physical markers (Bollnow, 2011, 37). In such a way, 'making space' and 'clearing space' can be seen as fundamental elements of human organisation (Bollnow, 2011, 38).

However, a corollary to this must obviously be that the power to control how people move through space, and thus how they relate with it, is necessarily the power to shape the society which that space engenders (see Lefebvre & Levich, 1987, 7–11). At sea this is often expressed through the – still highly relevant – observation of Aubert and Arner (1959, 202) that 'the ecological location of cabins has always symbolized the distinction in rank', reflecting the frequently regimented spaces found aboard ship. The positioning of certain possession spaces within a ship (be those cabins, messes, lockers, and so forth) is the visible territorial order that manifests from the imposition of structured organisation upon those subjected to it through the strategic division of space (de Certeau, 1984, 35–48).

Altogether, we can see that the everyday practice of movement is a core component of the formation of space, identity, and meaning through the inclusion or exclusion of people from particular places (see also Clifford, 1997, 17–46; Lee & Ingold, 2006). By observing where the

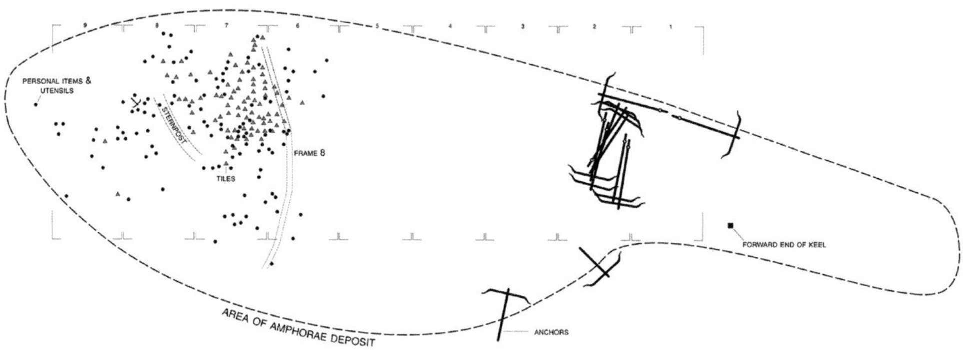

Figure 5.1 Site plan of the seventh-century *Yassıada* wreck.

Source: Drawn by the author after Bass & van Doorninck (1982), Wreck Plans III, IV, and V

actions of the everyday took place and how that affected the routines that were formed, we can gain a much deeper appreciation of why spaces gained and maintained particular functions as well as how they may have reflected and shaped relationships between individuals. In this, the ship as an entity is particularly amenable to our approach, as our major source of information comes from the archaeological afterlives of vessels that slipped beneath the waves on the course of their journeys (Muckelroy, 1978). While there is no brushing over the limitations of any archaeological investigation, they still provide a rich tableau of information about the everyday lives of their crew and passengers (Gibbins & Adams, 2001).

After all, shipwrecks present a fairly pristine snapshot of a specific moment in the life of a ship and its crew, with the archaeological remains maintaining a certain 'time capsule' like quality (Pearson, 2014, vii). The fact that this is a final moment does mean that we need to remember that we are seeing a typically organised space in a state of chaotic disorganisation (Adams, 2001, 297–298) and that taphonomic processes mean that no wreck is truly pristine (Kery & Stauffer, 2015). However, their relatively closed and static nature is an important asset. Due to this, shipwrecks allow for fine-tooth investigation and the exciting potential for obtaining a surfeit of highly specific data which can be seen at the scale of an individual mariner (Bowens, 2008, 18).

Quite luckily, our first application of spatial thinking to a material assemblage in this chapter can utilise one of the best studied and preserved wrecks of Late Antiquity. This can be found in the seventh-century Byzantine merchant vessel of *Yassıada*, off Bodrum in Turkey, whose excavation was carried out between 1961 and 1964 under the direction of George Bass (Figure 5.1). Considered to be a landmark study in maritime archaeology, it is still hailed as one of the finest excavation reports of a Mediterranean ship due to its careful recording of spatial context and the extent of its material corpus (Carlson & Leidwanger, 2015, xv).

5.2 Under the sea

Having sunk after hitting a dangerous reef off the small island for which it is named, the *Yassıada* ship settled to the seabed in a remarkably intact state. This resulted in a site with a clearly defined axial layout that allowed the excavation team to straightforwardly determine the vessel's orientation and saw the bow, hold, galley, and stern being easy to distinguish from one another (van Doorninck, 1982a, 32–34). Furthermore, this allowed the location of the surviving artefacts to their appropriate place of use or storage prior to the wrecking event to be a relatively simple process as they had largely remained clustered around their initial resting places (van Doorninck, 1982b, 87–89).

Anchors are required to be at the bow for use, and *Yassıada* displayed a large concretion of iron anchors at one extremity of the visible shipwreck (Figure 5.1: Sector 1/2), with no others apart from a smaller grapnel located at the opposite end of the site (van Doorninck, 1982c, 121; Figure 5.1: S. 8). From the anchors back, there was the remains of a cargo that comprised around 800–900 transport amphorae, naturally concentrated around the metacentre of the ship in order to preserve stability (Bass, 1982, 155–188). Finally, after the eighth transverse frame of the ship (Figure 5.1: S. 6/7), there was a concentrated cluster of personal items such as a lead seal and bronze belt-buckle, tableware, kitchenware, equipment, and tools – some of which had slumped forward onto the cargo (Bass & van Doorninck, 1982, Wreck Plan Va and Vb).

More than simply being clumped together at the stern, the concentrations were precise enough that we can see there must have been a locker for the ship's carpenter in this area, as their tools were found tightly grouped together. Furthermore, the objects' positions revealed that storage vessels and the water jar were stored at starboard in a below-deck enclosed area where religious objects such as a bronze censer were also found. Finally, valuables seem to have been kept in a forrard locker in this area, and the boatswain's equipment and fishing gear were in a separate aftward locker in the stern (Bass & van Doorninck, 1982, 314–315). The cramped below-deck space measured around 3 metres by 1.5 metres and is believed to have been a galley as well as a storage area (van Doorninck, 1982b, 89–90).

Inside, there is evidence of bricks and a grate that would have been used for a hearth, something of an upgrade to the small caboose known as a *foculus* that was often in use aboard ships in this period (Beltrame, 2015, 63–66). Along with this we have additional evidence for a tiled roof (Figure 5.1: centred on S. 7), which would have been used to protect the structure from the chimney exhaust, and an extensive set of ceramic utensils (van Doorninck, 1982b, 87–92). These included at least 21 cooking pots, 16 jars, 18 pitchers or jugs, and four or five settings of tableware, which included fineware red slip bowls, plates, and dishes, as well as cups (van Doorninck, 2015, 206).

This galley appears to have been the only below-deck area aside from the main hold, which is fairly unsurprising, given that the ship was not particularly large – with a length of 19 metres and a beam of only 5.22 metres (Bass & van Doorninck, 1982, 312–313), it sits neatly within the standard range of modern racing sail yachts. However, between the galley's deckhouse and the mainmast there would have likely been a lightweight deck shelter used both for helming the ship and for eating and sleeping, although this is now totally deteriorated with nothing remaining (van Doorninck, 1982a, 53). This type of structure is known to have existed on other Roman ships (see Casson, 1971, Figure 154) and a group of tacks from sector 5 of the *Yassıada* ship (see Bass & van Doorninck, 1982, Wreck Plan II; Figure 5.1: S. 5) in an area where there were no artefacts found in storage were considered to be probable evidence for its existence (van Doorninck, 1982a, 57).

This would have posed little problem for accommodating the ship's crew, which likely numbered no more than a dozen members, given its dimensions (Whitewright, 2016, 882). The most common numbers seem to have been around five or six in similarly sized vessels (Beltrame, 2002, 48–50), with experiments using the *Kyrenia II* replica ship showing that a crew of six was perfectly capable of navigating and sailing the vessel over two voyages between Zea and Cyprus (Katzev, 1987). Considering that at least some would have had to be on watch at any given time, with *Kyrenia II* having three or four crew manning it at any one moment (Katzev, 1987), the amount of living space required was probably quite small. What is more, the ad hoc nature of maritime passenger travel meant that there was little incentive to create extensive facilities for their use (McCormick, 2001, 404–410).

This recounting provides evidence of a shipboard layout that saw a division of material between heavy equipment at the bow, cargo amidships, and everyday items at the stern. What is more, this appears to have been a very strong division. All everyday items were found in quite tight proximity to one another, heavily suggesting that nearly all activities, save for specific job-related tasks, involving navigating or securing the ship, took place at the stern. Rather than those aboard being confronted with a series of demarcated areas, they would have consistently been within a singular overlapping space and the majority of everyday activities would have been carried out together with others.

5.3 Holy diver

What can we tell, then, about the way the routines of the everyday were shaped by the way space on board the *Yassıada* ship was organised and used? One seemingly minor detail we can immediately pick up on is that the main piece of religious equipment found on the ship, a bronze censer with a cross-shaped finial that was seemingly based on an Egyptian design, was stored and likely used in the stern (Bass & van Doorninck, 1982, Wreck Plan Vc). Censers like this would have been used for a variety of purposes, including as part of common prayers in a domestic setting, as part of specific rituals offering prayers of thanks, and as a 'sanctifying' accompaniment to legal oaths (Katzev, 1982, 266–268).

Religious practices were extremely important in the lives of late antique sailors (see Amelia R. Brown, this volume), so its existence is no surprise, but its position is intriguing in relation to both ancient and more modern parallels. It seems to have been common in antiquity to represent religious ceremonies on board ship as occurring at the stern, often in association with altars of some sort in thanks for safe passage (see Torlonia Port Relief: Testaguzza, 1970, 171; Casson, 1971, 182; *figs.* 125, 130). Similarly during the nineteenth century, Greek sailors would often utilise an icon of St Nicholas in the poop for the same purpose (Tzamtis, 1972, 59). Indeed, it is a general trend among excavations of ancient ships to find ritual objects associated with libations and sacrifices mostly in the stern (Atkins, 2009, 82).

This enduring association of the stern with the *axis mundi* might seem somewhat arbitrary but begins to become quite logical when we consider the types of activities that took place in this area. One of the obvious ones we can also relate it to is the common association of the helm and the helmsman as a tool of divine action, both of which would have been located at the stern. Prayers for safe passage, divine warnings, and entreaties for aid in dangerous situations, all make sense to have occurred in close proximity to the helm, as the helmsman was the most active participant in extracting the ship from potentially deadly circumstances. From Pindar's (Pin. *Pyth.* 4.191–4.196) description of a libation to Zeus upon reaching port to Callimachus's (Callim. *Aet.* 4.103) 'Hero at the Stern' worshiped at Phalerum, or Vergil's (Verg. *Aen.* 3.521) portrayal of Anchises's post-embarkation prayers, communication, and requests for guidance between man and the divine were consistently coupled to these aspects.

Furthermore, more specifically in the case of the *Yassıada* ship, we can see the obvious relationship between the domestic setting of the stern in opposition to the heavily job-focused nature of the remainder of the ship and relate it back to the usage of incense braziers in the home as a conduit for prayer (Katzev, 1982, 267). Despite their similarities, one might make the objection that you cannot simply assume that a religious practice would be ported over between these spaces. However, there are additional pieces of evidence that suggest a relationship between ships and household ritual practices. One of the best known of these is the practice of putting a coin in the socket of the mast-step, which is known to have occurred on a wide range of vessels (Carlson, 2007).

The shipwreck site of Port-Vendres A is a good late antique example of this, where a bronze coin of Constantine was found in this position, with the ship's estimated date of sinking occurring around three quarters of a century later than the coin was struck (Jézégou et al., 2017). This strongly implies that the coin was considered to be of long-term ritual significance, and the practice has been linked to the custom of placing coins in building foundations as votive objects. These foundation coins acted as sacrifices to consecrate a construction or ask for divine protection for the building (Donderer, 1984, 181). This ritual usage became increasingly associated with the *domus* in the Roman world, and it seems likely that the spatial part-identity of the ship as a domestic setting is displayed in this practice (Carlson, 2007, 322).

In order to add to this identification of specific religious connotations being applied to particular spaces in the ship due to the types of activity taking place, we can also look to the bow. While the front of the *Yassıada* ship is unfortunately the most decayed segment due to its position after sinking (van Doorninck, 1982a, 33–34), the large number of anchors in this section neatly illustrates the most common activity that would have taken place. This would have been the centre of practices intended to prevent damage to the vessel. Anchors needed to be let out from the bow both to prevent the ship from running aground while in harbour, but also as a safety measure in storms, where a sea anchor would be let out to stabilise it against the waves (as is recounted in *Acts* 27:13 about Paul's voyage; Oleson, 2008; Beresford, 2014).

Furthermore, in all periods, sounding leads would have also been used from the bow to prevent running aground (*Acts* 27:27–30; Ritchie, 2000; Galili et al., 2010), as well as to provide navigational information in conditions of obscured visibility or out of sight of land (Oleson, 2000; Beresford, 2014). Therefore, as the area was associated with human activity to ward off danger, so too would it likely have gathered analogous sacred connotations. This suggestion is given force through the strong association of the bow of ancient ships with apotropaic elements intended to assist the ship such as painted eyes or totemic animal horns (Brody, 1998; Atkins, 2009, 123–124). These are not just found in artistic representations either, as elements such as marble eyes nailed to the bow (Carlson, 2009) or protective symbols such as Priapus's phallus acting as figureheads (Neilson, 2002) have been uncovered on ancient wrecks.

As such, sacred space seems to map quite closely to function, with the repeated routines of the steersman and the lookout, of anchoring and navigating, and the separation between the domestic and the professional, combining together in layers of overlapping meaning to produce the axial distinctions that we have been discussing. This type of space is not unique but, nevertheless, uncommon as an entity in comparison to our usual conceptions of lived space on land. This is, as Crawford's (1999, 9) discussion of everyday urbanism suggests, because lived space usually possesses a fragmentary nature which is split between separately definable 'realms' such as the home, workplace, or institution.

The ship, on the other hand, is a total institution, a space that is 'closed in on itself' in the words of Foucault (1984, 49). It sees a complete superimposition of all of these elements within the same space, turning the classical ideal of the Hippodamian city – neatly segmented into public, private, and sacred space (Aris. *Pol.* 1267b.22–1267b.38) – on its head by producing a heterogeneous space which is forced to reconcile irreconcilable spaces all at once (Casarino, 2002, 34). As such, the assumption of spatial significance by particular areas through the routines of the everyday is uncommonly entangled in ships – with the creation of spatial dichotomies being deeply affected by elements outside their own traditional spheres.

5.4 Where did you sleep last night?

Even more than ritual, we can probably be safe in saying that one of the fundamental routines of human existence is sleeping. Sleep plays a particularly important role in our relationship to space due to the way it is inextricably bound up with the notions of dwelling and home – as well as the way that control and security are key elements of both (Kellet & Moore, 2003; Bollnow, 2011). In order to fall asleep in any manner aside from a collapse in exhaustion, a feeling of security and the trust that it will remain in the absence of conscious observation are indispensable, which leaves where one sleeps as something of the *ne plus ultra* of possession spaces (Bollnow, 2011, 175–176).

As previously stated, the separation of sleeping areas is one of the strongest markers of social stratification in ships, as the division between ranks both physically separates different elements of the crew and conceptually separates them through isolating them into different spheres of trust (see also Moloney, 2018). Sociological studies of modern ships provide support to this theory, arguing that the formation of informal social groups and the reinforcement of status is heavily impacted by the way that the physical architecture of ships divides their inhabitants and that the 'democratisation' of social life was dependent upon the integration of space (Lezaun, 2013). With this in mind, how this manifests on late antique ships has a considerable potential impact on our understanding of their social structure.

Casting our mind back to the *Yassıada* ship, one of the striking elements of it is the total lack of spatial differentiation in where those on board would have slept. With no internal spaces to lie down upon, save for the floor of the small galley, it seems most probable that the crew and any passengers aboard all sheltered beneath the small deckhouse. Indeed, sleeping on deck certainly was not unknown during Late Antiquity, as literary and legal sources are quite clear on the common usage of animal skins as a combination of sleeping bag and foul weather gear for those who had no other protection from the elements (*Sea Law* 3.11; Amalarius, *Versus marini*, 428.65). Certainly, the size of the vessel would have made spatial distinctions difficult to construct, but the fact that its inhabitants would have been forced to dwell in a conceptually co-equal space might mean that the social dynamics were more open than we would otherwise have cause to suspect.

Nevertheless, the *Yassıada* ship is only one example and to come to any firmer conclusions we ought to cast a wider net. In doing so, one of the first inferences we can make when examining the wider scope of late antique vessels is that the practice of all those aboard sleeping on deck was likely quite common. The *Yassıada* ship is a good example of the vast majority of late antique ships, which sat around the 50–60 gross tonnage mark and consequently would have measured in at around 15–21 metres in length (Gertwagen, 2014, 159). As such, the same dimensional constraints, which left the vessel with little space for more than a below-decks galley after discounting the hold, would still apply – and give good reason for the frequent usage of animal skins.

However, at the same time, it is not as if internal spaces for rest were unknown. The fourth-century statesman Themistius indicates that there must have been liveable spaces below the stern deck of enough ships to make them familiar, as his oration to Theodosius employs the imagery of a helmsman climbing from the belly of the ship to the bridge to take command as an unremarkable allusion (Them. *Or*. 15.195a). Similarly, literature from Late Antiquity such as Bishop Synesius's description of a voyage along the North African coast also suggests the presence of relatively large habitable spaces, with the approximately 50 passengers separated by gender through the presence of a jury-rigged sailcloth curtain running along the middle of a room (Syn. *Ep*. V.5).

Together with textual evidence, there are also visual representations of ships with above-deck cabins from the second and third centuries CE. The Torlonia relief shows a ship that has a substantial built-up aft section, with a fully enclosed cabin of rectangular shape that possessed a flat roof, windows, and doors. This extends forward to the mainmast, meaning that it must have had a substantial interior space, quite possibly split up into a series of individual cabins (Rougé, 1984, 237). A series of other depictions such as a mosaic in the Capitoline Museum, which shows the stern of a ship which appears to have a two-level cabin descending into the main structure of the ship (Rougé, 1984, 238), or the Rimini mosaic, in which the rightmost vessel also has a discernible cabin aft that includes a visible roof, windows, and portico, also suggests their presence (Gentili, 1979).

Unfortunately, an issue we face is that that the archaeology does not provide much in the way of an outline of what these spaces may have looked like. Due to decomposition, there are no wooden wrecks from Late Antiquity that have even partially complete upperworks – even the most intact such as the Byzantine wreck of *Sinop D*, which was substantially preserved by the anoxic waters of the Black Sea floor, only have fragments (Ballard et al., 2001; Brennan et al., 2011). Nevertheless, we can once again say that the descriptions and depictions of internal habitable areas seem to suggest that they were confined to the stern. This means that, even if they did contribute to an increased privatisation of sleep, they still would have required their inhabitants to share their elbow-room to a substantial extent. Moreover, it is just as likely that they simply represented more enclosed shelters, with an open-plan or only partially enclosed berths, similar to the common arrangements found in early-modern merchant vessels of which we have more detailed knowledge (Eriksson, 2014).

Therefore, I would argue that the crew aboard a majority of late antique ships not only had little in the way of private space but also saw little in the way of architectural differentiation in their social status. Hierarchy aboard must have been expressed in the social sphere, as a mutual agreement between those on board, rather than being imposed by their surroundings. What is more, it would seem as if the majority of captains did not see a need to separate themselves from the crew. This may have been because they felt as if there was no reason to claim a status beyond what their rank and ability provided them to ensure obedience. Or, just as likely, it was considered more important to keep a close eye on potential grumblings in order to guarantee good behaviour.

Certainly, we are aware that there were differences in rank and wage on late antique vessels – and cash in hand has always been a source of social division. From the Rhodian Sea Law we know that ships of this period notionally had a variety of dedicated ranks, including the captain, helmsman, lookout, carpenter, boatswain, able seamen, and cooks, all of whom were paid different shares of the profits of a voyage (*Sea Law* 1.57–1.59; captain with a double share, officers with one-and-a-half, seamen with one, and the cook with a half). There even seems to have been additional gradations within these broad ranks, with different grades of captains and regular seamen being attested (Rougé, 1966, 216). However, it is difficult to say how closely this idealised layout was held to on the majority of ships, as well as what the exact impact might have been on social relations, as this information was simply never recorded.

What we can say is that there are some indications that late antique crews were relatively convivial. A measure of this can be found in Synesius's travelogue, in which he describes part of the crew as 'pass[ing] their time in jesting one with another, accosting their comrades not by their real names, but by distinguishing marks of their misfortunes' (*Epistles* IV, 4), which does suggest a camaraderie and casual familiarity among them. While descriptions of interactions among late antique crews are frustratingly rare, we do get a sense that a distinct maritime community of sorts not only existed but was relatively tight-knit (Rauh et al., 2008; Wade, 2014). Seafarers were not cut off from interactions with greater society, but they were perceived as a group unto themselves which drew from a large and diverse collection of individuals that spread across the ancient Mediterranean, throwing together people of many lands, faiths, and languages (Wade, 2014).

This went as far as creating spatially distinct maritime neighbourhoods in many large ports (Magdalino, 2000) which seemed to be the home of a type of maritime counterculture, which often expressed itself through opposition to overbearing authority in moments of mass unrest (Rauh et al., 2008). This has many parallels to the trans-Atlantic maritime society of the early modern period – known about in far greater detail – where sailors are noted to have possessed a strong collectivist streak and were a major source of labour unrest against oppressive conditions (Rediker, 1987). This, combined with the impression of a relatively open social environment that we have been given, might well mean that the crews of Late Antiquity were largely made up of individuals that had a strong group identity and had worked with each other in the past – much like the documented crews of many early modern merchant ships (Van Royen, 1992, 154).

5.5 Red red wine

Naturally, alongside sleep, one of the other fundamental routines of human life is eating. Food and its consumption is an area which touches everything in the realm of social interaction and social space (Counihan & Van Esterik, 1997). In particular, commensality, a necessity aboard ship due to limitations on cooking space, can be viewed as a social arena in microcosm where the structure of social relationships are negotiated and defined (Fahlander, 2010, 37–38; Fischler, 2011). The inclusion or exclusion of groups at the table, the placement of settings, the determination of what tableware one is allowed to use, or what food one is allowed to eat, all form part of the spatial setting of mealtimes, with the actions involved impacting on the social space that all participants inhabit.

From a psychological and sociological perspective, it has long been recognised that eating together prompts a sense of belonging among the participants and can act to strengthen interpersonal bonds and feelings of intimacy (Marte, 2007; Fahlander, 2010, 37; Fischler, 2011). Conversely, the spatial layout of the diners and the way they are served can also be used to reinforce social barriers. This appears to have often been the case in formal dining throughout Roman antiquity, as the positioning of guests around the table highlighted who was particularly in favour with the host (Hudson, 2010a; Stephenson, 2016). Indeed, this gradation of status seems to have become even more fine-grained in Late Antiquity, as the seating from right to left around the table followed a strict social hierarchy from highest to lowest (Stephenson, 2016, 65).

Along with this type of seating arrangement, we also have discrete spatial organisations that manifest depending on what sort of service is being used. The two main categories of this are generally labelled individual service and collective service dining, differing from each other in whether the diner is provided with individual tableware in which the main dish of the meal is served or whether they take food from a dish shared in common with their neighbours (Hudson, 2010a). The ideal of individual service, according to Pliny (Plin. *Ep.* 2.6), was the enforcement of equality through ensuring that all diners received their fair share. However, as complaints from authors such as Martial (Mart. 3.60) and Lucian (Luc. *Merc. Cond.* 15) show, it could just as easily be used to reinforce hierarchies through the provision of different qualities of food or the usage of arcane rituals which worked to separate the truly favoured from the rest.

On the other hand, shared service dining obviates many of these potential methods of hierarchical display through its basic functions. As all diners are sharing from a small group of large vessels, it not only becomes quite difficult to create inequality through serving different dishes to different people but also largely forces closer interaction between everyone at the table, highlighting personal connections and breaking down barriers by imposing a more intimate type of spatial association (Smith, 2006; Marte, 2007; Stephenson, 2016).

The way that we can differentiate the type of service in an archaeological context is by looking at the size and shape of the primary tableware in an assemblage. From compiling the dimensions of tableware across a wide range of cultures, it appears to be a general rule that shared vessels are about twice the diameter and three times the volume of those which are intended for individual service (Henrickson & McDonald, 1983, 632; Juhl, 1995, 35). The diameter tends to be the easiest measure in practice, and the raw numbers we are talking about are an average of 24.6 centimetres for shared vessels and 14 centimetres for individual vessels (Henrickson & McDonald, 1983).

Hudson (2008, 2010a) contends that Late Antiquity saw an accelerating divide between elite and sub-elite serving practices, with silver tableware associated with status dining found at sites such as Aphrosidias or Augusta Raurica maintaining the characteristics of vessels intended for individual service. Conversely, assemblages of common ceramic tableware from both cities such as Alexandria and smaller urban centres such as Sagalassos increasingly shifted to shared service forms (Hudson, 2010a, 687–690). The argument that he presents is that this was a conscious reaction by the non-elite segments of society to their increasing disenfranchisement, along with the increasing spread of religions such as Christianity which put considerable emphasis on communal and egalitarian aspects of everyday life (Hudson, 2010a; also see Brown, 2012; Maxwell, 2021). Given what we have already discussed concerning the maritime community, does this narrative seem to be replicated on late antique ships?

I would argue that the answer appears to be yes – at least in the west and central Mediterranean. In a survey of reasonably well-recorded late antique wreck sites that I have studied in this area, we can see a clear pattern in the types of tableware that are present. The most common materials are ARS dishes of the Hayes 59 (e.g. Kamenjak: Pešić, 2019; Packoštane: Boetto et al., 2012) and Hayes 61 (e.g. Olbia: D'Oriano, 2002; Sobra: Parker, 1992, no. 1100) types. These are both large, flat-based, dishes with a 22–41-centimetre diameter, tending towards the upper end of this range on the referenced sites (Hayes, 1972, 96–102).

Along with them, Hayes 50 and 58 (e.g. Cape Zanavin: Kaleb & Pešić, 2018), Hayes 67 (e.g. Veliki Piruzi: Feiderling, 2019), 88, and 104 (e.g. Palud A: Long & Volpe, 1998) type dishes or bowls also frequently appear in the archaeological record which, together with Forms 59 and 61, make up a plurality of extant table ceramics on late antique shipwrecks in the west and central Mediterranean. These are respectively a large dish with 20–40 centimetres in diameter, a large dish with 22–42 centimetres in diameter, a large bowl with 20–45 centimetres in diameter, a large dish with approximately 40 centimetres in diameter and another large dish with 32–45 centimetres in diameter (Hayes, 1972, 69–160). Altogether, it seems clear that shared vessels played the main role aboard these late antique ships.

This picture also seems to neatly match up with the material uncovered from the *Yassıada* ship. The original report posited that the plates, bowls, and dishes, which were found with the rest of the tableware in the galley represented several individual settings for officers or important passengers due to being fineware (Bass, 1982, 188). However, their dimensions closely match the range seen on the sites just discussed, with four large plates of about 30 centimetres in diameter and at least five dishes that are 22 to 26 centimetres in diameter (Bass, 1982, 167). Together with these, there are around five smaller bowls which seem to fall within the range of size and functionality displayed by side dishes in other assemblages (Hudson, 2010b). This leaves the whole set as appearing more similar to a collection of serving plates, shared vessels, and side dishes intended for a shared service layout than a collection of individual sets (Hudson, 2010a) – with the more co-equal nature of this arrangement tying in tidily with our discussion of the ship's architecture and the maritime community.

What is more, staying with the *Yassıada* ship, the spatial elements of eating – and its attendant interrelated systems of food acquisition, preparation, and distribution known collectively as foodways (see Deetz, 1977, 50; Landon, 2002, 220–221; Fahlander, 2010) – are not only limited to what goes on at the table. The movement space created by the way that food travels from storage areas to preparation areas, and onwards to the plate, can also illuminate different types of social structures. As previously mentioned, dining probably took place under a light deck shelter behind the mainmast (van Doorninck, 1982a, 53), while cooking was done in the internal galley whose hatch and associated deckhouse was positioned slightly further aft (van Doorninck, 1982b, 89–90).

In this case, cooking was not separated by anything more than a single nearby opening from the space used for eating and sleeping on the ship. This is notable due to the theory that spatial contexts which see a separation and privatisation of the architecture of sleeping, eating, and cooking, reflect and reinforce an individualised society that is consciously or subconsciously creating hierarchical divisions between differing social strata (see Johnson, 1996, 174–178). Logically, the inverse of this would suggest that the lack of these architectural boundaries act as factors which reduce hard-edged individuality and prompt a psychological association of familiarity with the others who share in these circumstances (see also Upton, 2002).

What is particularly intriguing is that this general colocation of spaces for sleeping, eating, and cooking does not seem to have been forced upon the architects of ancient ships by inherent limitations of technology. In contrast to the *Yassıada* ship, the second century BCE wreck site at La Chrétienne off Cap Dramont shows a design with a structure on the foredeck that was roofed with tiles and contained the remains of food storage vessels and kitchen items like a mortarium (Joncheray, 1975, 31–34). This does not provide definite proof of it being a galley, but in light of the fact that similar evidence concerning the colocation of artefacts and tiled rooves on other vessels have usually been taken as evidence for the existence of a galley (Beltrame, 2015, 65), it is highly suggestive. What is more, there was also a living area in the stern (Parker, 1992, no. 304), apparently indicating a sharp spatial separation between the two spheres.

The visible transport of food out onto the open foredeck and likely towards an eating area at the stern would have been suboptimal from a practical point of view, being subject to wind, spray, and rolling – all of which could easily lead to a ruined dinner. Yet by creating an almost performative sense of movement, it would have clearly highlighted a division between the servers and the served (Eriksson, 2014; see also Lee & Ingold, 2006 on movement). This would have potentially reinforced a hierarchical structure aboard by controlling peoples' movements and, intrinsically, making a bid to control their social identity (Upton, 2002, 719). As such, I would argue that the fact that it is the *Yassıada* style of spatial organisation which appears to be replicated on most other shipwrecks in Late Antiquity for which this type of spatial information is available (see, e.g. Leidwanger et al., 2021; Long & Volpe, 1998; Llabrés et al., 2022) is significant in this context, rather than simply being in vogue among shipwrights.

5.6 Conclusion

Our aim here has been to see how the routines of the everyday both created space aboard ships in Late Antiquity and what that space might tell us about the way life was organised on them. Across this chapter, the way people aboard ships utilised certain areas has been shown to accord them particular connotations in both the sacred and the social realms, which mediated and manoeuvred specific human behaviours. We have also seen how the majority of everyday routines were carried out in a shared space that would have placed those involved in fairly close quarters with one another. The lived space, and living space, of these ships did not seem to put much spatial emphasis on distinctions of hierarchy, with inhabitants both high and low being arranged in a way that limited the distance between them. As such, we can conclude that understanding spatial organisation on late antique ships can give us a variety of insights into social organisation aboard.

This is particularly important, given that we know very little about late antique mariners in their own words. As such, the fact that the way in which space was created on their ships seems to point towards a social climate and social organisation that was, at least somewhat, egalitarian is extremely valuable. Naturally, we should be cautious about pushing this conclusion too far, as we can never be sure exactly how influential immaterial aspects were to social relations. However, the picture that seems to have largely been revealed is of a small group of individuals who would have likely been familiar with one another, who were lodged in space which distinguished little between them. Furthermore, their other routines, such as eating, reinforce this image with both the architectural layout of the whole process, and the table settings in particular, serving to encourage interpersonal bonds and to reduce the perception of hierarchical differences.

This chapter has covered only a few of the ways in which spatial theory can help us utilise archaeology in combination with other historical evidence to provide more complex and nuanced portraits of life aboard late antique ships. After all, one could easily further discuss the spatial implications of a wide range of personal and professional effects and equipment that archaeological research has turned up from shipwreck sites. A quick perusal of works like Parker (1992) or Beltrame (2002) will show examples of tools for carpentry and sailmaking, jewellery, toiletries, arms, armour, fishing equipment, clothing, medical equipment, marlin spikes, axes, mallets, knives, and many more, which have been recovered and categorised from a wide range of vessels across the Mediterranean and beyond. By paying more attention to previously overlooked aspects of these assemblages, we can unlock new perspectives on the social conditions of shipboard life.

One of the main benefits, as well as one of the main drawbacks, of examining shipboard society is that it is not a 'passive reflection' (Adams, 2001, 306) of what the wider society in which it was situated must have been like, but an active application of that society's attitudes to a particular setting. In this context, understanding the different ways that lived space was experienced by those at sea can provide us not only with a better comprehension of the peculiarities of late antique maritime communities but also, so long as we are careful, new insights into a largely undocumented section of society. In doing so, we can take a deeper dive into a relatively neglected field of inquiry and provide a different story to our instinctive assumptions. We cannot say beyond a shadow of a doubt that late antique ships are reflections of an egalitarian maritime society, but the evidence shown here makes a compelling argument that this is a serious possibility that should be explored further.

Editions and abbreviations

Acts – *Acts of the Apostles: New International Version*. (2011). Zondervan.
Arist. *Pol.* – Aristotle. *Politics* (H. Rackham, Trans.) (1944). Harvard University Press.
Callim. *Aet.* – Callimachus. *Aetia* (D. Clayman, Ed.) (2022). Loeb Classical Library.
Luc. *Merc. Cond.* = Lucian. *De mercede conductis* (A Harmon, Trans.) (1921). Macmillan.
Mart. = Martial. *Epigrammata* (J. Borovskij, Ed.) (1925). Teubner.

Pind. *Pyth.* = Pindar. *Pythian Odes* (D. Svarlien, Trans.) (1990). Perseus.
Plin. *Ep.* = Pliny. *Epistulae* (R. Mynors, Ed.) (1966). Clarendon.
Sea Law = *Nomos Rodiōn nautikos* (W. Ashburner, Trans.) (1909). Clarendon Press.
Syn. *Ep.* = Synesius of Cyrene. *The letters of Synesius of Cyrene* (A. Fitzgerald, Trans.) (1926). Oxford University Press.
Them. *Or.* = Themistius. *Oration 15, 'To Theodosius, or the Most Royal of the Virtues'* (P. Heather & D. Moncur, Trans.) (2001). Liverpool University Press.
Verg. *Aen.* = Virgil. *Aeniad* (R. Mynors, Ed.) (1969). Oxford University Press.
Versus marini = Amalarius. *Versus marini* (J.-P. Migne, Ed.) (1844). J.-P. Migne.

Bibliography

Adams, J. (2001). Ships and Boats as Archaeological Source Material. *World Archaeology*, *32*(3), 292–310.
Atkins, C. (2009). *More Than a Hull: Religious Ritual and Sacred Space on Board the Ancient Ship*. Texas A&M University.
Aubert, V., & Arner, O. (1959). On the Social Structure of the Ship. *Acta Sociologica*, *3*(4), 200–219.
Ballard, R. D., Hiebert, F., Coleman, D., Ward, C., Smith, J., Willis, K., Foley, B., Croff, K., Major, C., & Torre, F. (2001). Deepwater Archaeology of the Black Sea: The 2000 Season at Sinop, Turkey. *American Journal of Archaeology*, *105*, 607–623.
Bass, G. F. (1982). The Pottery. In G. F. Bass & F. van Doorninck Jr. (Eds.), *Yassi Ada – Volume I: A Seventh-Century Byzantine Shipwreck* (pp. 155–188). Texas A&M University Press.
Bass, G. F., & van Doorninck, F. Jr. (Eds.). (1982). *Yassi Ada – Volume I: A Seventh-Century Byzantine Shipwreck*. Texas A&M University Press.
Beltrame, C. (2002). *Vita di bordo in età romana*. Istituto Poligrafico e Zecca dello Stato.
Beltrame, C. (2015). The Contribution of the *Yassıada* Shipwreck Excavation to the Knowledge of Life Aboard Ancient Ships. In D. Carlson, S. Kampbell, J. Leidwanger, & G. F. Bass (Eds.), *Maritime Studies in the Wake of the Byzantine Shipwreck at Yassiada, Turkey* (pp. 205–216). Texas A&M University Press.
Beresford, J. (2014). Oil on Troubled Waters: A Reappraisal of the Storm Tactics Described in Acts 27. *The International Journal of Maritime History*, *26*(4), 752–761.
Boetto, G., Radić Rossi, I., Marlier, S., & Brusić, Z. (2012). L'épave de Pakoštane, Croatie (fin IVe – Début Ve siècles apr. J.-C.). *Archaeonautica*, *17*, 105–151.
Bollnow, O. (2011). *Human Space* (C. Shuttleworth, Trans.). Hyphen Press.
Bowens, A. (Ed.). (2008). *Underwater Archaeology: The NAS Guide to Principles and Practices*. Wiley-Blackwell.
Brennan, M., Ballard, R. D., Croff Bell, K., & Piechota, D. (2011). Archaeological oceanography and environmental characterization of shipwrecks in the Black Sea. *The Geological Society of America Special Papers*, *473*, 179–188.
Brody, A. (1998). *Each Man Cried Out to His God: The Specialised Religion of Canaanite and Phoenician Seafarers*. Brill.
Brown, (2012). *Through the Eye of a Needle: Wealth, the Fall of Rome, and the Making of Christianity in the West, 350–550 AD*. Princeton University Press.
Carlson, D. (2007). Mast-Step Coins among the Romans. *The International Journal of Nautical Archaeology*, *36*(2), 317–324.
Carlson, D. (2009). Seeing the Sea: Ships' Eyes in Classical Greece. *Hesperia: The Journal of the American School of Classical Studies at Athens*, *78*(3), 347–365.
Carlson, D., & Leidwanger, J. (2015). Introduction. In D. Carlson, J. Leidwanger, & S. Kampbell (Eds.), *Maritime Studies in the Wake of the Byzantine Shipwreck at Yassiada, Turkey*. Texas A&M University Press.
Casarino, C. (2002). *Modernity at Sea: Melville, Marx, Conrad in Crisis*. University of Minnesota Press.
Casson, L. (1971). *Ships and Seamanship in the Ancient World*. Princeton University Press.
Clifford, J. (1997). *Routes: Travel and Translation in the Late Twentieth Century*. Harvard University Press.

Counihan, C., & van Esterik, P. (1997). *Food and Culture: A Reader*. Routledge.
Crawford, M. (1999). Introduction. In J. Chase, M. Crawford, & J. Kaliski (Eds.), *Everyday Urbanism* (pp. 1–20). Monacelli Press.
de Certeau, M. (1984). *The Practice of Everyday Life* (S. Rendell, Trans.). University of California Press.
Deetz, J. (1977). *In Small Things Forgotten – The Archaeology of Early American Life*. Knopf.
Donderer, M. (1984). Münzen als Bauopfer in römischen Privathäusern. *Bonner Jahrbucher*, *184*, 177–187.
D'Oriano, R. (2002). Relitti di storia: Lo scavo del porto di Olbia. In *L'Africa romana: Atti del 14. Convegno di studio, 7–10 dicembre 2000, Sassari, Italia* (pp. 1249–1262). Carocci editore.
Eriksson, N. (2014). *Urbanism Under Sail*. Södertörns högskola.
Fahlander, F. (2010). The Nose, the Eye, the Mouth and the Gut: Social Dimensions of Food-Cravings and Commensality. In F. Fahlander & A. Kjellström (Eds.), *Making Sense of Things: Archaeologies of Sensory Perception* (pp. 35–50). TMG Sthlm.
Feiderling, M. (2019). *Brodolom Veliki Piruzi*. International Centre for Underwater Archaeology in Zadar.
Fischler, C. (2011). Commensality, Society and Culture. *Social Science Information*, *50*(3), 528–548.
Foucault, M. (1984). Des Espaces Autres. *Architecture, Mouvement, Continuité*, *5*, 46–49.
Galili, E., Oleson, J., & Rosen, B. (2010). A Group of Exceptionally Heavy Ancient Sounding Leads: New Data Concerning Deep-Water Navigation in the Roman Mediterranean. *The Mariner's Mirror*, *96*(2), 136–148.
Gentili, G. (1979). Il mosaico dell'Hercules bibax o del porto canale tra i mosaici di una domus adrianea di Rimini. *Bollettino d'Arte*, *64*, 49–56.
Gertwagen, R. (2014). Nautical Technology. In P. Horden & S. Kinoshita (Eds.), *A Companion to Mediterranean History* (pp. 154–169). Wiley-Blackwell.
Gibbins, D., & Adams, J. (2001). Shipwrecks and Maritime Archaeology. *World Archaeology*, *32*(3), 279–291.
Gilroy, P. (1992). *The Black Atlantic*. Harvard University Press.
Goodwin, L. (2002). Everyday Life. In C. Orser (Ed.), *Encyclopedia of Historical Archaeology* (pp. 188–190). Taylor & Francis.
Hayes, J. (1972). *Late Roman Pottery*. British School at Rome.
Heidegger, M. (1927). *Sein und Zeit*. Halle a. d. Saale.
Heidegger, M. (1997). Building, Dwelling, Thinking (N. Leach, Trans.). In N. Leach (Ed.), *Rethinking Architecture: A Reader in Cultural Theory* (pp. 100–108). Routledge.
Henrickson, E., & McDonald, M. (1983). Ceramic Form and Function: An Ethnographic Search and an Archeological Application. *American Anthropologist*, *85*(3), 630–643.
Hudson, N. (2008). Three Centuries of Late Roman Pottery. In *Aphrodisias Papers 4: New Research on the City and Its Monuments* (pp. 319–345). Journal of Roman Archaeology.
Hudson, N. (2010a). Changing Places: The Archaeology of the Roman "Convivium". *American Journal of Archaeology*, *114*(4), 663–695.
Hudson, N. (2010b). The Dining Assemblage Unpacked. *American Journal of Archaeology*, *114*. https://doi.org/10.3764/ajaonline1144.Hudson.suppl
Ingold, T. (2000). *The Perception of the Environment: Essays on Livelihood, Dwelling and Skill*. Routledge.
Jézégou, M.-P., Huet, N., Gassend, J.-M., Le Corre, G., Musard, O., Magenti, V., Salvat, M., Luc Bonnardot, M., Villevielle, S., & Paul, F. (2017). Heur et malheur d'une épave d'époque romaine découverte à Port-Vendres (Pyrénées-Orientales). *Patrimoines Du Sud [En Ligne]*, *6*. http://journals.openedition.org/pds/1846.
Johnson, M. (1996). *An Archaeology of Capitalism*. Blackwell.
Joncheray, J.-P. (1975). *L'Épave C de la Chrétienne*. Cahiers d'archéologie subaquatique.
Juhl, K. (1995). *The Relation Between Vessel Form and Function. A Methodological Study*. Arkeologisk museum Stavanger.
Kaleb, M., & Pešić, M. (2018). Remnants of an Antique Wreck of Rivanj Island's Cape Zanavin. *Submerged History*, *8*, 17–21.
Katzev, M. (1982). Iron Objects. In G. F. Bass & F. van Doorninck Jr. (Eds.), *Yassi Ada – Volume I: A Seventh-Century Byzantine Shipwreck* (pp. 231–265). Texas A&M University Press.

Katzev, M. (1987). An Analysis of the Experimental Voyages of Kyrenia II. In H. Tzalas (Ed.), *Proceedings of the 2nd International Symposium on Ship Construction in Antiquity* (pp. 245–256). Hellenic Institute for the Preservation of Nautical Tradition.

Kellett, P., & Moore, J. (2003). Routes to Home: Homelessness and Home-Making in Contrasting Societies. *Habitat International*, *27*, 123–141.

Kery, S., & Stauffer, J. (2015). *Hydrodynamics Related to Shipwreck Taphonomy*. 1–21. www.researchgate.net/publication/315921125_Hydrodynamics_related_to_shipwreck_taphonomy

Landon, D. (2002). Food and Foodways. In C. Orser (Ed.), *Encyclopedia of Historical Archaeology* (pp. 220–221). Routledge.

Lee, J., & Ingold, T. (2006). Fieldwork on Foot: Perceiving, Routing, Socializing. In S. Coleman & P. Collins (Eds.), *Locating the Field: Space, Place and Context in Anthropology* (pp. 67–85). Routledge.

Lefebvre, H. (1974). *La production de l'espace*. Anthropos.

Lefebvre, H. (1991). *The Production of Space* (D. Hicholson-Smith, Trans.). Blackwell.

Lefebvre, H., & Levich, C. (1987). The Everyday and Everydayness. *Yale French Studies*, *73*, 7–11.

Leidwanger, J., Greene, E. S., & Donnelly, A. (2021). The Sixth-Century CE Shipwreck at Marzamemi. *American Journal of Archaeology*, *125*(2), 283–317.

Lezaun, J. (2013). Ergonomics of Democracy. In K. Cupers (Ed.), *Use Matters: An Alternative History of Architecture* (pp. 215–232). Routledge.

Llabrés, S., Cardell, J., de Juan, C., Cau, M., Bernal-Casasola, D., Picornell, L., & Garcia Riaza, E. (2022). Ses Fontanelles Shipwreck (Mallorca, Balearic Islands): An Exceptional Late Roman Vessel and Its Cargo. *Journal of Maritime Archaeology*, *17*, 487–505.

Long, L., & Volpe, G. (1998). Le chargement de l'épave de la Palud (VIe s.) à Port-Cros (Var). Note préliminaire. In M. Bonifay, M.-B. Carre, & Y. Rigoir (Eds.), *Fouilles à Marseille, Les mobiliers (Ier-VIIe s.)* (pp. 317–342). Errance.

Magdalino, (2000). The Maritime Neighborhoods of Constantinople: Commercial and Residential Functions, Sixth to Twelfth Centuries. *Dumbarton Oaks Papers*, *54*, 209–226.

Marte, L. (2007). Foodmaps: Tracing the Boundaries of Home Through Food Relations. *Food and Foodways*, *15*, 261–289.

Maxwell, J. (2021). *Simplicity and Humility in Late Antique Christian Thought: Elites and the Challenges of Apostolic Life*. Cambridge University Press.

McCormick, M. (2001). *Origins of the European Economy*. Cambridge University Press.

McGhee, F. (1998). Towards a Postcolonial Nautical Archaeology. *Assemblage – The Sheffield Graduate Journal of Archaeology*, *3*. https://archaeologydataservice.ac.uk/archives/view/assemblage/html/3/3mcghee.html

Merleau-Ponty, M. (1945). *Phenomenologie de la perception*. Gallimard.

Minkowski, E. (2013 [1936]). *Le temps vécu: Etudes phenomenologiques et psycho-pathologiques* (3rd ed.). Presses Universitaires de France.

Moloney, M. (2018). Re-Imagining Shipboard Societies: A Spatial Approach to Analysing Ships of the British Royal Navy During the Eighteenth and Nineteenth Centuries. *The International Journal of Maritime History*, *30*(2), 315–342.

Muckelroy, K. (1978). *Maritime Archaeology*. Cambridge University Press.

Neilson, H. (2002). A Terracotta Phallus From Pisa Ship E: More Evidence for the Priapus Deity as Protector of Greek and Roman Navigators. *The International Journal of Nautical Archaeology*, *31*(2), 248–253.

Oleson, J. (2000). Ancient Sounding-Weights: A Contribution to the History of Mediterranean Navigation. *Journal of Roman Archaeology*, *13*, 293–310.

Oleson, J. (2008). Testing the Waters: The Role of Sounding Weights in Ancient Mediterranean Navigation. In R. L. Hohlfelder (Ed.), *The Maritime World of Ancient Rome* (pp. 119–176). University of Michigan Press.

Parker, A. J. (1992). *Ancient Shipwrecks of the Mediterranean and the Roman Provinces*. Tempus Reparatum.

Pearson, C. (2014). *Conservation of Marine Archaeological Objects*. Butterworths.

Pešić, M. (2019). *Afrički import iz hrvatskog podmorja od I. do V. stoljeća* [Doctoral thesis]. University of Zadar.

Rauh, N., Dillon, M., & Davina-McClain, T. (2008). Ochlos Nautikos: Leisure Culture and Underclass Discontent in the Roman Maritime World. In R. Hohlfelder (Ed.), *The Maritime World of Ancient Rome*. (pp. 197–242). University of Michigan Press.

Rediker, M. (1987). *Between the Devil and the Deep Blue Sea: Merchant Seamen, Pirates, and the Anglo-American Maritime World, 1700–1750*. Cambridge University Press.

Ritchie, S. (2000). A Brief History of 19th Century Deep Sea Sounding. *International Hydrographic Review*, *1*, 6–14.

Rougé, J. (1966). *Recherches sur l'organisation du commerce maritime en Méditerranée sous l'Empire romain*. S.E.V.P.E.N.

Rougé, J. (1984). Le confort des passagers à bord des navires antiques. *Archaeonautica*, *4*(1), 223–242.

Smith, M. (2006). The Archaeology of Food Preference. *American Anthropologist*, *108*, 480–493.

Stephenson, J. (2016). Dining as Spectacle in Late Roman Houses. *Bulletin of the Institute of Classical Studies*, *59*(1), 54–71.

Testaguzza, O. (1970). *Portus: Illustrazione dei Porti di Claudio e Traiano*. Julia Editrice.

Tzamtis, A. (1972). Ships, Ports and Sailors. In S. Papadoupoulos (Ed.), *The Greek Merchant Marine (1453–1850)*. National Bank of Greece.

Upton, D. (2002). Architecture in Everyday Life. *New Literary History*, *33*(4), 707–723.

van Doorninck, F., Jr. (1982a). The Anchors. In G. F. Bass & F. van Doorninck Jr. (Eds.), *Yassi Ada – Volume I: A Seventh-Century Byzantine Shipwreck* (pp. 121–141). Texas A&M University Press.

van Doorninck, F., Jr. (1982b). The Galley. In G. F. Bass & F. van Doorninck Jr. (Eds.), *Yassi Ada – Volume I: A Seventh-Century Byzantine Shipwreck* (pp. 87–120). Texas A&M University Press.

van Doorninck, F., Jr. (1982c). The Hull Remains. In G. F. Bass & F. van Doorninck Jr. (Eds.), *Yassi Ada – Volume I: A Seventh-Century Byzantine Shipwreck* (pp. 32–64). Texas A&M University Press.

van Doorninck, F., Jr. (2015). The Seventh-Century Byzantine Ship at *Yassıada* and Her Final Voyage: Present Thoughts. In D. Carlson, S. Kampbell, J. Leidwanger, & G. F. Bass (Eds.), *Maritime Studies in the Wake of the Byzantine Shipwreck at Yassiada, Turkey* (pp. 205–216). Texas A&M University Press.

van Royen, P. (1992). Seamen and the Merchant Service, 1650–1830. In R. Gardiner (Ed.), *The Heyday of Sail: The Merchant Sailing Ship 1650–1830* (pp. 152–159). Conway Maritime Press.

Wade, J. (2014). Sailors, merchants and the maritime cults that sailed into the ports (and streets) of early Byzantium. In D. Dzino & K. Parry (Eds.), *Byzantium, Its Neighbours and Its Cultures* (pp. 269–288). Brill.

Whitewright, J. (2016). Ships and Boats. In G. Irby (Ed.), *A Companion to Science, Technology, and Medicine in Ancient Greece and Rome* (pp. 870–888). Wiley-Blackwell.

Part 2

Place-making and emplacement

The impact of religion

6 Space(s) in transition

The impact of early Christianity on the late antique Horn of Africa

Gabriele Castiglia

6.1 Introduction: space(s) and interpretative problems

The aim of this chapter is to analyse the changes that affected the spaces of late antique urbanism in the Horn of Africa, by using the impact (both physical and conceptual) of Christianity as the main (but not exclusive) guideline.[1] This is somewhat of a forced choice because – as we will see – the archaeological agenda in the Horn of Africa has been largely focused on the excavation of churches, which thus became the main indicators of urban and material changes (Phillipson, 2009). With this, I do not mean that they are privileged indicators, but only that very often they are the only ones that we have. This is problematic, since the exclusive use of the church evidence might lead to some bias in the interpretative process. At the same time, churches stand as the main source of information on topographic developments and the use of space in the *longue durée*, given the general lack of other excavated structures. By default, churches have become the key source for understanding processes of spatial evolution in the context of religious change.

Further to this limitation, the late antique Horn of Africa does not benefit a *corpus* of written or epigraphic sources comparable to those of other contemporary areas (especially) in the Mediterranean basin. This means that we have to rely on other datasets to try to analyse the evolution of these geographical and historical contexts over a long period. The analysis of space can be one of these datasets. Space, in fact, can help us understand how specific territories and cities were impacted by historical, political, religious, and social changes over the centuries. As we will see, the management and organization of urban spaces were deeply affected and modified by the emerging power of Christianity, which imposed radically new topographic arrangements in the Aksumite towns.

In order to deal with these difficulties, our analysis will proceed at different levels, using the churches as a way of understanding the dynamics of spatial relationships and "social" perceptions in the urban context. In this sense, the case of Adulis, and more generally all the Christian architecture of the Aksumite Kingdom, can be considered exemplary, since the basilicas were built on top of large podiums (contextual to the creation of the worship halls) that significantly raised them above the normal level of city life. In this way, therefore, the churches imposed themselves as new driving forces in the ways urban space was perceived. At the same time, it is more complex to use the churches as tools for decoding spaces on a wider scale, especially because there is, in fact, a lack of knowledge of the rural church network, which obviously makes the transformation of extra-urban spaces in relation to Christianity somehow elusive.

Bearing these limitations in mind, we must in the first place clarify what we mean by space, and then ask whether we can really understand it relying only on the available data. Answering these questions is not simple, and it would be particularly difficult to offer unique and definitive

DOI: 10.4324/9780429427152-8

answers to them. The analysis of the internal spaces of Christian basilicas – in the case where they are well excavated and preserved – can undoubtedly offer important information about their articulation, their liturgical use, and, in some cases, also their internal patterns of movement. On the other hand, however, churches alone cannot provide us much-detailed information about their broader spatial contexts. This is particularly true in a case like Adulis, where there is a lack of real topographic context of religious buildings, since much of the city still has to be excavated.

A useful way of debating "space" is to start with the considerations of Luke Lavan, especially his *Public Space in the Late Antique City* (Lavan, 2021; also 2003a, 2003b). As he emphasized, we need to move beyond the classical approaches to the study of space in the late antique city, looking at space as a "simple location of points within a physical grid" (Lavan, 2003a, 172) and, more importantly, trying to incorporate into our analyses not only the new buildings but also the pre-existences and the neglected structures. Lavan states that "studies of architecture tend to concentrate on new buildings built in Late Antiquity, such as churches. In so doing they neglect older structures, where important traditional activities may or may not still be continuing, such as civil basilicas" (Lavan, 2003a, 178–179). In this regard, I may add that in the Aksumite context (and especially in Adulis) the focus on earlier structures (indeed often neglected) will be the key to better understand the evolution of space in the *longue durée*. Lavan proposes an approach that should concentrate more on exploring cities "as historical townscapes, full of people, rather than as lifeless ruins" (Lavan, 2003a, 184). In this perspective, he also underlines the necessity to fully integrate archaeology with written and epigraphic sources.

While this is of course the ideal approach, it is not always possible given the vagaries of the survival of the evidence. In fact, the Horn of Africa in the late antique period (for this area, it refers to what is defined as the *Middle Aksumite Period*, i.e. 380–580 CE) provides a good archaeological dataset for the understanding of urban spaces, yet it is only scarcely documented by written evidence. Moreover, research on the urban landscapes of the Horn of Africa in the late antique period has focused (with few exceptions) mostly on monumental buildings such as palaces and churches, neglecting more mundane and "ordinary" structures like houses, *tabernae*, *horrea*, and workshops. Our archaeological record is thus limited and provides only a partial view of the past. We also need caution when trying to apply Roman and Mediterranean models to the study of urbanism in the Horn of Africa. While tempting and somewhat irresistible, we must keep in mind that this is an area that, even though it had many contacts with the Mediterranean world, was never under Roman rule. Although there are parallel developments in the Mediterranean, the Horn of Africa evolved independently.

6.2 The context: geography, topography, and early Christianity

From the first to the early eighth centuries CE, the Horn of Africa was the seat of a great and powerful polity, the Aksumite Kingdom, named after the capital city of Aksum, today in Ethiopia (Munro-Hay, 1991; Phillipson, 1998, 2012). The kingdom extended through the territories of northern Ethiopia (mostly in present-day Tigray), Eritrea, southern Sudan, Djibouti, and, for a short period (first half of the sixth century), also of Yemen and Saudi Arabia (Figure 6.1).

Written sources like the anonymous *Periplus Maris Erythraei* (Casson, 1989), Pliny the Elder, or the *Topographia Christiana* by Cosmas Indicopleustes (Wolska-Conus, 1968–1973) throw light on the kingdom's long distance trade contacts with the Mediterranean and the Indian Peninsula. This is confirmed by numismatic data, as Aksum was the only potentate of sub-Saharan Africa to mint its own coins (Munro-Hay, 1984). A real turning point in the history of the kingdom was its conversion to Christianity at the beginning of the fourth century. In his *Historia*

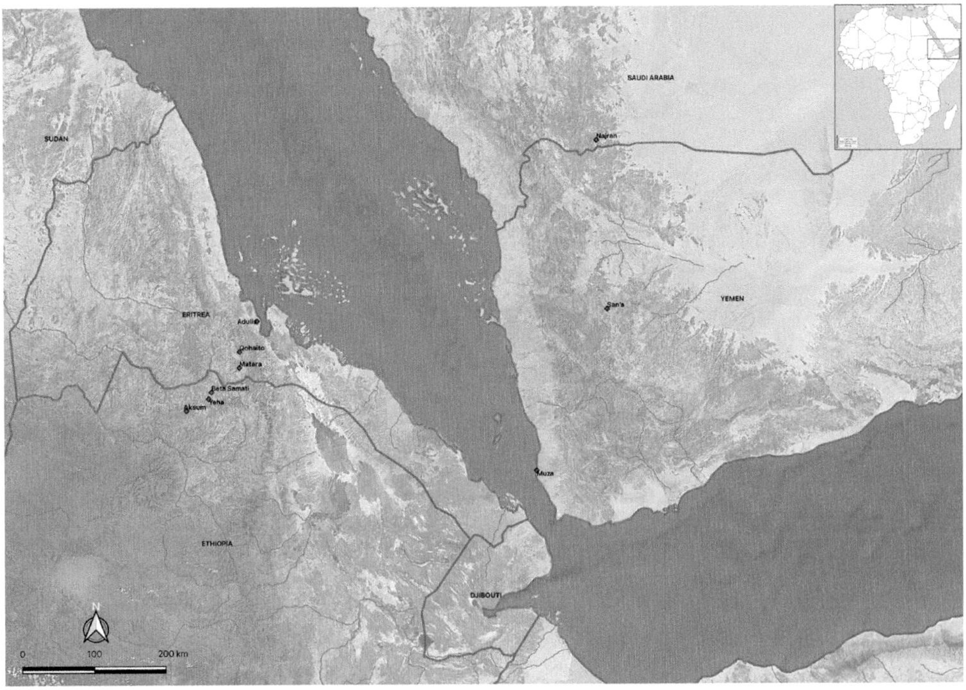

Figure 6.1 Cities of the Aksumite Kingdom at its furthest extent.

Source: Author's image

Ecclesiastica, Rufinus of Concordia notes that in the early decades of the century two Christian Syrian brothers, Frumentius and Aedesius, were captured by pirates in the Red Sea and sent to the royal court of Aksum as gifts (Amidon, 1997). There, they quickly converted Ezana, the young heir to the throne, to Christianity (Kaplan, 1982; Munro-Hay, 1988). Once Frumentius was freed, he moved to Egypt, and in Alexandria he told the patriarch Athanasius how Christianity was growing in Eastern Africa. As a response, Athanasius appointed Frumentius as the first bishop of Aksum. The same episode is recounted also in other sources and later chronicles, such as Socrates Scholasticus (*HE* 1.19), Sozomen (*HE* 2.24), Theodeoretus (*HE* 1.23), Nicephorus Callistus (*HE* 8.35), and John of Nikiou (*Chron*. 77.111f, see Zotenberg, 1883) (for an overview, see Fiaccadori, 2010). The consolidation of the new religion is also indicated by the official inscriptions of King Ezana, in which he is called "servant of Christ" and references are made to the Holy Trinity (*DAE*, inscr. 11), as well as by the coins, where the disc and the crescent symbol that had appeared above the effigies of the Kings until then were replaced by the cross (Munro-Hay, 1984, 75–85).

The sixth century marked at least two pivotal events in the history of the Aksumite Kingdom: the visit of Cosmas Indicopleustes and the military campaigns against Himyar in the Arabic Peninsula. Cosmas was an Egyptian geographer and merchant who visited Adulis and the Aksumite Kingdom in the first half of the sixth century, according to his own account titled *Topographia Christiana*. He was sent to Adulis by the Aksumite king, Ellaztbaas, who required Cosmas to copy the inscriptions placed in the so-called Throne of Adulis. The king planned to use them to legitimize his forthcoming military campaigns against the south Arabian kingdom of Himyar, mostly located in present-day Yemen, back then partially under the control of Aksum (Wolska-Conus, 1968–1973, II-56). Although impossible to prove, the ideological and propagandistic intent of the king in having the inscriptions copied and publicized is also suggested by another key source for this period, the *Martyrium of Saint Arethas*, dated in its original nucleus to the seventh and eighth centuries CE (Detoraki, 2007). In the first part of this work, the anonymous author recounts how king Ellaztbaas was about to invade the kingdom of Himyar, where the dominant Jewish government had promoted persecutions and massacres against the local Christian communities (Detoraki, 2007, 182–185). The inscriptions carved in the throne of Adulis – which recorded earlier military expeditions to the Arabic Peninsula – might have represented a sign of good omen for the forthcoming campaings of Ellaztbaas (see also Kominko, 2013, 25). Through these campaigns, King Ellaztbaas intended to punish the new Hymyarite king, Yusuf As'ar Yath'ar (better known as Joseph), who had suddenly revolted against the Aksumite protectorate, persecuting the small Christian enclaves in the southern part of the Arabic Peninsula, mostly in the oasis of Najran, where many Christians were slaughtered. This episode marked the beginning of a period in which Byzantium showed greater interest in the Arabian Peninsula and the Red Sea trade routes (Greatrex, 1998, 225–240). In fact, the sixth century marked a period of close relationship between Byzantium and Aksum, due to the fact that Justinian wanted to have the Aksumite Kingdom as an ally against Persia for the control of silk market (Greatrex, 1998).

The growing interest of Byzantium in Aksumite policies and, therefore, the desire to initiate a series of diplomatic contacts with the kingdom of the Horn of Africa in relation to strategies for controlling the silk market is clearly visible in the diplomatic missions promoted by Justinian right after the episode of Najran. Although various scholars have explored the philological and factual details of the accounts of the different embassies sent by the emperor to *Aksum* (Shahîd, 1960; Greatrex, 1998; Beaucamp, 2010; Robin, 2012, 2015; for an overview on Himyar, see Gajda, 2009), it is worth considering these narratives here. As Greatrex (1998) showed, Justinian's interest in this area was tightly linked to the silk trade in the context of the great Persian wars. Until then, the Persians had monopolized the silk trade, and the Christian emperor wanted to convince the Aksumites and the Himyarites to help Byzantium in the Persian wars. This would have given fruitful advantages to all the parties, as recounted by Procopius: the Aksumites and the Himyarites would have gained direct control of the silk trade and Byzantium would no longer need to buy silk from their Persian enemies.

> In this way, the Ethiopians, by purchasing silk from India and selling it to the Romans, would themselves gain much money while causing the Romans to profit in only one way, namely, that they would no longer be forced to pay over their money to their enemy.
>
> (Procop., *Wars* 1.20.9)

We have three different sources for these discussions, all of them differing from each other on specific points: Procopius's *Wars*, Malalas's *Chronicle*, and Photios's *Bibliotheca* (see Beaucamp, 2010). By meticulously analysing the three of them, Irfan Shahîd has determined that these three sources are actually accounts of the same embassy (Shahîd, 1960). Greatrex and Beaucamp, in their recent reviews of this interpretation, have concluded that – despite all Justinian's efforts – Byzantium did not manage to be fully involved in the Red Sea commercial and political affairs.

The archaeology of this area has benefitted in the last decades of very important syntheses, which have helped to better frame the chronological evolution of the history of Aksum in different periods: *Pre-Aksumite* (800–360 BCE), *Proto-Aksumite* (360–80 BCE), *Early Aksumite* (80 BCE–160 CE), *Classic Aksumite* (160–380 CE), *Middle Aksumite* (380–580 CE), *Late Aksumite*

(580–825 CE), and *Post Aksumite* (825–900 CE) (Bard et al., 2014). Archaeology testifies to the *Middle Aksumite* period as the heyday of the kingdom, a moment where it is possible to witness, indeed, the rise of a great urban civilization, mostly in the capital city – Aksum – but also in other towns such as Adulis, Qohaito, and Matara (Phillipson, 2012, 119–132). The main archaeological testimonies of this urban revolution are represented by great elite structures, like the palace of Ta'akha Maryam in Aksum (Phillipson, 1998, 84–85). Churches also stood as crucial elements of the expanding cityscapes, mostly from the sixth century onwards (Phillipson, 2009; Di Salvo, 2017). The Christian basilicas, with their peculiar architecture characterized by high podia that imposed them way above the level ground of the cities (Di Salvo, 2017), likely represented the most innovative elements of this period.

Another pivotal and distinctive element of Aksuite archaeology are the great funerary stelae that marked the tombs of the kings, concentrated in the capital city, even though they had been in use only until the middle of the fourth century, coeval to the arrival of Christianity in the kingdom (Phillipson, 1994; see also *infra*). In this regard, the timing of the Aksumite conversion to Christianity can be explained in a number of ways: in fact, it could be seen as a process that might have had an immediate material outcome on the landscapes of the kingdom or, on the contrary, as a result of the involvement of Byzantium in Aksumite policy, so only from the early sixth century onwards. I will now try to answer these questions, analysing the spaces of the whole kingdom and, more specifically, those of Adulis, at different levels (macro, meso-macro, and micro spaces).

6.3 Playing with scales: macro and meso-macro spaces

In the different categories of spaces that I aim to analyse, the one of "macro-space" is without doubt the most complex and opaque. Macro space can be defined as the territory of the whole kingdom, where the actual relationship between towns and countryside plays a key role in our ability to understand the changes of spaces. With this in mind, it is necessary to consider whether the arrival of Christianity had an effective impact at this level or not. As I have already mentioned, the heyday of church construction within the kingdom was in the sixth century, with so far only one possible exception, that of the recently discovered church in Beta Samati, whose first phase is dated by radiocarbon analyses to the fourth century (Harrower et al., 2019). This seems indeed to be an exceptional case, especially if we consider the fact that the cathedral of Maryam Ts'iyon in Aksum, the capital city, where Christianity had taken its roots in the very first stages, can be reasonably dated in its first phase to the sixth century, albeit only through architectural comparisons with other structures (Phillipson, 1997, 169–178). Moreover, according to the archaeological and topographical data known so far, Beta Samati is likely to be of a limited size in comparison to other urban spaces, so we can consider it to be more a rural settlement than a town. It is possible that Christianity made more of an impact on this rural centre before the capital city because of the geographical position of Beta Samati. It was located in the proximity of Aksum (and therefore close to the newly converted court and to the bishopric), something that may have resulted in an earlier need for gathering places like churches for the new Christian communities that were developing throughout the territory.

If we look at the centres where Christianity left material traces (mostly churches), we can easily see how it spread through those towns standing along the main communication system of the kingdom, the ridge connecting Aksum to Adulis (Figure 6.1). It is also important to take into consideration that the only access to the capital city in the whole kingdom, due to the geography and orography of the area, was through Adulis. This means that the Adulis–Aksum route was fundamentally the only one that guaranteed contacts over long overseas distances:

trade, commodities, culture, and religion passed through this axis. From this perspective, it can be reasonably assumed that Christianity did not change the macro space but rather adapted to a former reality. With this, I mean that in such a wide spatial perspective, the "new" religion did not cause any change to the perception or the management of spaces. It is not possible to identify how this macro-scale impacted social reality and daily perceptions of space, which is why we need to consider these issues differently.

Meso-macro spaces refer to spatial frames that have a closer focus on the territory, such as the topography of the cities and of their suburbia. Again, of course, this categorization is arbitrary, but what matters is the information that these frameworks can give us about the changes in the management and perception of spaces. Going down the hierarchical scale of spaces and, therefore, focusing on urban structures and, even more so, on the internal developments (and, where possible, on the elevation) of churches allows for a more detailed and therefore more reliable dataset. In this category, I would like to focus mostly on the city of Adulis.

As seen earlier, Cosmas Indicopleustes represents a key source in this perspective. His work offers crucial information about the kingdom of Aksum, which he personally visited. Cosmas, moreover, provides a series of illustrations in support of his text, very useful for an analysis of spaces (see Kominko, 2013). Among these, there stands a representation of Adulis and its environs, where we can also see Aksum and the oversized depiction of the so-called Throne of Adulis (Figure 6.2). In the left part of the image, we can see the Red Sea shores and three sites: Gabaza, Adulis, and Samidi. Even though these three sites are represented as separate (as they actually are in the topography of the area), they were part of the same urban and suburban territory of Adulis. The first, Gabaza, to the south, was the proper port of the city: back in Aksumite times, it was formed by at least three islands, shaping a natural harbour. Due to the withdrawal of the coastlines, now Gabaza stands as a complex of three hills (known today as Galala Hills). Surveys on the hills have revealed an astonishing amount of surface pottery, mostly amphorae, and signs of architecture thought to be related to port facilities (Peacock & Blue, 2007, 33–37). Four kilometres to the north stands Adulis itself, a great town of almost 40 hectares, with no defensive walls and with a densely settled pattern. This was a strategic site for the whole kingdom, being its main and probably unique access to the sea. Adulis also represents a crucial spot for understanding the material setting of Christianity, as we will see below. Samidi, the last site mentioned by Cosmas, is situated at the northern end. Surveys carried out *in situ* seem to point out, with very few doubts, that this was a funerary centre, characterized by both monumental and simple tombs. In particular, there stands a great structure (visible only in small part, since it has never been excavated) that might possibly be a church, arguably a cemeterial sanctuary, surrounded by tombs. The tombs – also not excavated – are mainly of circular shape and made up of schist slabs; by considering the peculiarities of the visible structures and the pottery found in the ground surface of the area, it is likely that Samidi was coeval to the late antique phase of Adulis (Peacock & Blue, 2007).

The characteristics of these spaces have some distinguishing features, while also having elements in common with Mediterranean towns: the late antique Mediterranean, in fact, had a Roman background, which means that new spaces and buildings that emerged during Late Antiquity had to adapt themselves to former spaces and structures. The same partially happens in the Horn of Africa, but the background was something completely different, mainly in its architectural features. Moreover, churches – as we have seen – were built on brand new high podia and never reused ancient structures, adopting completely new building solutions. Furthermore, it is possible to trace analogies and differences with the Mediterranean also in funerary customs. Cemeteries of Mediterranean cities, in fact, are usually extra-urban, often in the

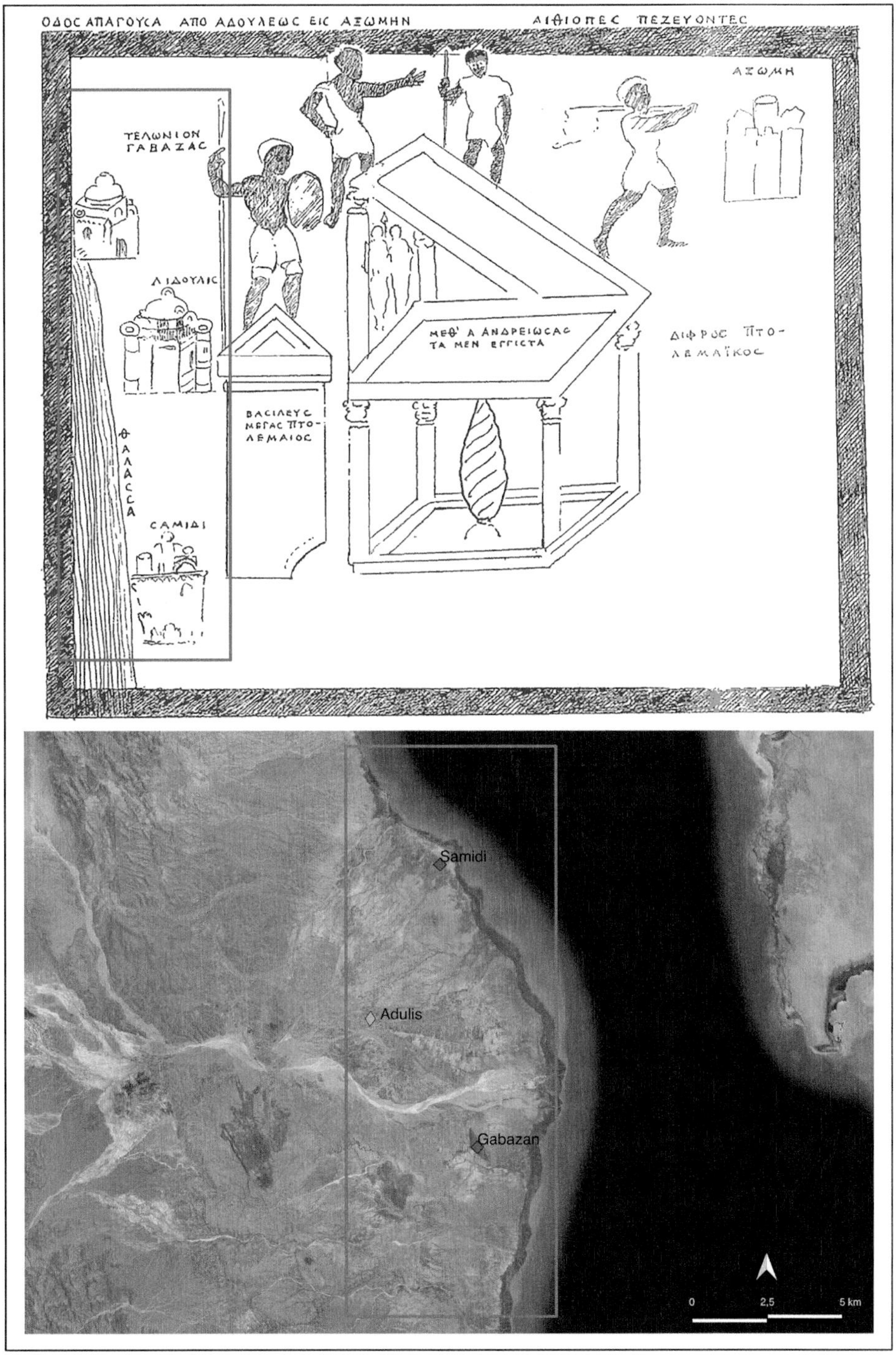

Figure 6.2 Illustration of Adulis, Gabaza, and Samidi in Cosmas Indicopleustes's *Christian Topography*.

Source: Author's image

suburbs. In Adulis, there are no traces of funerary areas in the immediate proximity of the town, with the cemeterial nucleus of Samidi standing almost eight kilometres north of the city, further than one would expect in a Greco-Roman settlement.

The port is six kilometres away from the town and thus topographically and physically detached from it. This planned topographical organization may recall, from my point of view, a selection of "activity spaces", as Lavan (2003a) called them: we have the commercial spot (Gabaza), the space of the living and of religious cult (Adulis), and the space of the dead (Samidi). Although most of what we know about urban topography for this period is related to churches, recent excavations in the south-western part of Adulis and north of the episcopal complex indicate a long continuity in the use of residential and commercial buildings. In south-western Adulis, the stratigraphy of structures identified as domestic revealed the presence of beaten earth floorings, fireplaces, and almost exclusively locally produced ceramics; to the north of the episcopal complex, the masonry techniques and the greater heterogeneity of material culture (including many imported ceramics) led to the identification of these environments as pertaining to residences of an affluent class. The stratigraphies highlighted in these structures, moreover, have demonstrated how their functions remained basically the same from the second to the seventh centuries CE (Zazzaro et al., 2014). Another outstanding issue in our knowledge of the urban topography is represented by the likely existence of a defensive circuit: so far, it has not been possible to identify any evidence of urban walls. This is not surprising, since these kind of structures are completely absent (or, at least, not documented) in all the other Aksumite towns (Phillipson, 2012).

The most visible impact of Christianity on the evolution of space can be seen in the urban fabric of Adulis. As noted earlier, it is impossible to form a precise understanding of the broader topography of the town due to the traditional archaeological focus on monumental complexes. This is starting to change, however, thanks to campaigns of excavation and survey that have also considered other typologies of structures: this is the case, for example, of the residential structures excavated by the university "L'Orientale" of Naples in the south-western part of the city and, again, of the structures identified north of the cathedral (see Zazzaro et al., 2014). As these findings suggest, there is a clear contrast between the urban plan of the late antique city and the alignment of earlier structures (Figure 6.3). This is particularly evident if we compare the positioning of the churches in relation to the orientation of the earlier structures revealed through magnetometer surveys (that can be seen in the grids), which can be fairly interpreted as private houses or storage rooms. It is possible to document the same trend also by analysing the archaeological sequence of the structures excavated immediately around the churches. These structures exhibit entirely different orientations compared to later Christian basilicas, many of which have either cut or, in some cases, partially obliterated the pre-existing buildings. This shows that the need to orientate the churches east-west, with the apse pointing East, was a greater driving force in their construction than the former spatial organization of the town. A comparison of chronologies of the central eastern church – most probably the cathedral of Adulis – and those of the surrounding (pre-existing) buildings is revealing in this sense. The first phase of the annex buildings can be dated to the second century CE (Zazzaro et al., 2014), while the church has a foundation date of the late fifth/early sixth, as indicated by radiocarbon analyses and the stunning amount of material culture, mostly decorated marble and alabasters (Castiglia et al., 2020, 2021). Nevertheless, the building activities of the church did not fully affect the continuity of life of the former structures, likely to be private houses: some of them were indeed erased by the Christian building, but others testify to an uninterrupted continuity of use until the late seventh century.

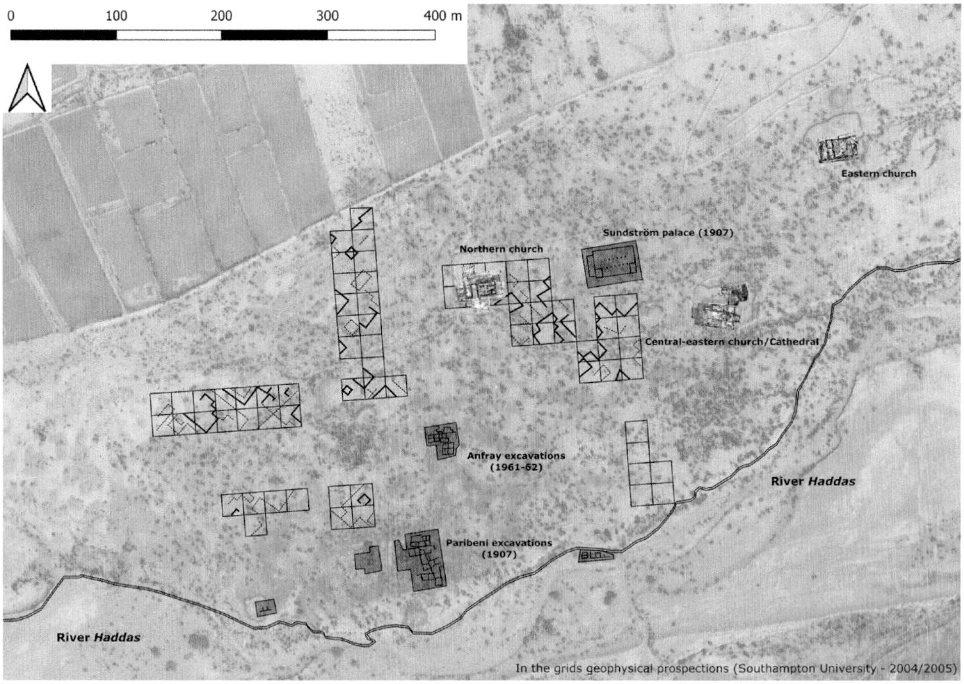

Figure 6.3 Excavation plan of Adulis.
Source: Author's image

Another monumental structure, standing a few meters north of the church and of (probably) close date as indicated by the similar building technique, has been tentatively identified as a *palatium*. The identity of this building cannot be proved, since the structures are not visible anymore and all we have is Sundström's report, but it is suggested by the building's massive size, much larger than any of the churches (Sundström, 1907; see also Giostra, 2017). Substantial walls, in various states of repair, were identified during a survey around the church-palace complex and can be identified as elements of a defensive system based on their size and typology (1.20 metres thick and constructed with basalt and schist blocks significantly bigger than those of the other buildings). If this interpretation is correct, it is possible that this represents a fortified episcopal complex within the urban perimeter. This was a common feature of late antique urbanism in different parts of the Mediterranean, as for example Salona (present-day Croatia) and Luni (Italy), where the episcopal complexes were fortified by massive defensive walls, creating a sort of citadel within the city (for Salona, see Turkovič & Marakovič, 2017; for Luni, see Durante, 2003). The presence of this complex indicates a conscious and planned strategy, on the part of ecclesiastical authorities and patrons, of changing the urban space in a very clear way, in what can be described as a policy of discontinuity in continuity. The discontinuity is indeed marked by the change in function of this part of the city, while this change assured the continued relevance and use of this space (something that Enrico Zanini has called "structural resilience": see Enrico Zanini, this volume). This raises important questions, in particular, concerning the relationship between function and space. It is possible that this complex functioned as the *episcopium*, or it may also have had mixed functions as a bilateral "activity space", used by both secular and spiritual powers (Carolyn T. La Rocco has suggested a similar dynamic for Visigothic Spain, defining the synergy between religious and secular entities in strategies related to new foundations and the redefinition of urban spaces as a "collaborative foundation": see Carolyn T. La Rocco, this volume). These two different hypotheses reflect the debate about the acropolis of Caričin Grad/Iustiniana Prima in present-day Serbia. There, in fact, the upper part of the city – which was fortified – hosted the episcopal complex and other structures that some scholars have interpreted as the residence of the bishop (Duval, 1996), while others see as the administrative neighbourhood of the city (Caillet, 2012).

From a bird's-eye view, the position of the three churches suggests that they were built in order to create a sort of east–west processional route dividing the city, possibly linked to a stational liturgy as well (again Figure 6.3). Unfortunately, as there are no traces of a road connecting the three churches, this remains just a hypothesis. Nevertheless, it is quite evident that this "church-line" indeed cut the spatial perception of the city in two, with the Christian element becoming the new defining character of the late antique space and mindscape. It is by taking another step through our spatial categories, however, that we can appreciate the depth of the impact of Christianity on the lives of contemporaries.

6.4 Micro space(s)

At the lowest level of enquiry, micro spaces require the analysis of the architecture of the churches and their liturgy. The first element that must be highlighted, in this sense, is the vertical development of space. One of the main peculiarities of the first Christian churches in Adulis is that they were constructed on podia that made them considerably higher than the ground level of the city and the surrounding buildings; this is also the case in other Aksumite cities such as Matara, Argula, Qohaito, and Beta Samati (for an overview of the peculiarities of Aksumite architecture, see Breton, 2015 and Di Salvo, 2017). Most of the surviving podia were just the base of the churches, so we have to imagine also the elevation of the basilicas themselves. This aspect indeed testifies to a spectacular impact of the novel Christian architecture on the perception that people might have had of the new religion. It is likely that this was a strategy of communication planned directly by the ecclesiastical hierarchies, in order to impose a message of innovation, as well as of power and control, compared to previous religious creeds and the general management of space. With these changes, the space becomes – more than before – religious, with the imposition of its own materiality as a tool of control over the daily life and mindscapes of people.

As much as their height, the horizontal architecture of churches is an important element for the interpretation of space at the micro level. An analysis of the plan of the churches might reveal the heterogeneous prism of cultural influences felt in the late antique Horn of Africa, and perhaps even some of its liturgical developments. This requires caution because, as Lavan pointed out, the absence of liturgical texts often implies that the interpretation of church floor plans is largely speculative (Lavan, 2003a, 176). Nonetheless, it is still worth trying to understand the organization of internal spaces of the churches and their roles. I will use the church that we have recently excavated (Figure 6.4) – the possible cathedral – as a model (again Castiglia et al., 2020, 2021).

Thirty metres long and almost 20 wide, the so-called Cathedral Church has at least seven different building phases (Figure 6.5). In the first phase, securely dated to the late fifth and early sixth centuries, thanks to radiocarbon analysis, the church had a narthex to the West, which led to the *quadratum populi* via a monumental entrance. The *quadratum populi* was divided into three naves by two rows of pillars; it was located before the presbyterial area, which was marked

Figure 6.4 Photogrammetry of the "Cathedral Church".

by a Y-shaped chancel that divided the central nave from the apsidal area. The apse was flanked by two *pastaphoria*, with the southern one most likely used to host a baptistery. This interpretation derives, at first, from the fact that this room was paved with hydraulic mortar, the only case known in Adulis apart from the baptistery in the Eastern church (all the other pavements in the city are, in fact, made up of schist slabs). Although only preserved in approximately half of the room, the pavement bears the clear trace of a curving edge, possibly relating to a baptismal font that was positioned off-centre and was lost when the floor collapsed. In this phase, the church was decorated by a significant number of carved pieces of marble and alabaster (over 500 fragments found so far): some of these, like fragments of *plutei* and columns belonging to the presbyterial chancel, recall an early Byzantine manufacture, typical of the sixth century. It is no wonder, then, that in the ninth century, the Arabic historian Al-Azraki recounted how Justinian, three centuries earlier, had sent marble-workers to South Arabia and Aksum in order to help the local builders in the construction of churches (Finster & Schmidt, 1994; for the edition of Al-Azraki's account, see Wüstenfeld, 1857–1861). In the second phase, likely in the last decades of the seventh century, a chapel was created in the southern part of the narthex and was paved with great schist slabs and, in the centre, a quadrangular base. Their levels of preservation make the first two phases the most interesting and reliable for trying to understand the management of the micro space by the Christian community.

The general architecture of this church (which is basically the same for the churches of the whole kingdom) recalls what is known as the "Syriac model", whose main peculiarity is represented by the rectilinear perimeter and by the two rooms siding the apse, something that Mathews defined also as the "triple-sanctuary" (Mathews, 1971, 106; see also Mulholland, 2014). Krautheimer believed the "Syriac model" to have been translated from Syria directly

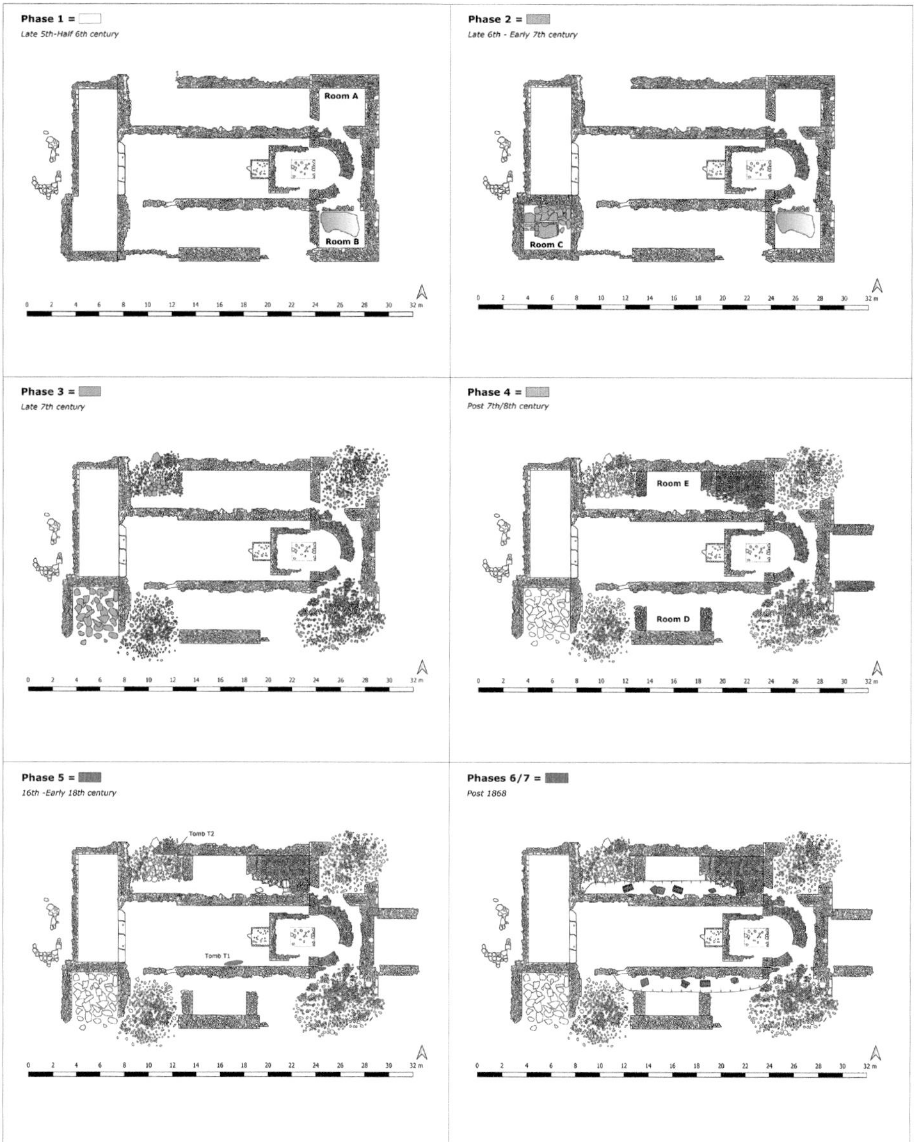

Figure 6.5 Phases of construction and use.

Source: Author's image

to Egypt (Krautheimer, 1965, 307), which is a plausible interpretation. In fact, we know that Frumentius had been appointed as the first bishop of Aksum directly by the Egyptian patriarch of Alexandria, Athanasius. This event marked the first episode of a long-lasting link between the Egyptian and Ethiopian churches and it is likely that this architectural and liturgical model was subsequently translated from Egypt to the Horn of Africa (see Munro-Hay, 1997).

If we try to identify the hypothetical internal routes that the clergy and the faithful might have followed, that of the first phase seems to be simpler, with the baptistery and the presbytery as a fulcrum (Figure 6.6). Since the excavations have revealed the presence of a passage between the eastern wall of the church and the external part of the apse and that the room hosting the baptistery was directly linked to this passage, it is fair to hypothesize the presence of a direct link between the northern *pastophorion* and the baptistery. If this interpretation is correct, we may then assume that the catechumens, after entering the church, went to the northern nave, then to the northern *pastophorion*, where they would have attended the rites of preparation and, then, finally, to the baptistery. The northern *pastophorion* might have had a double function, serving as a sacristy when not in use as a baptistery, as it was typical of the Byzantine rite (Mathews, 1971, 138; Fritsch, 2007, 2012).

The second phase of the church seems to have been marked by the presence of a quadrangular base in the centre of the room added in the southern part of the narthex. This base might have functioned as an altar, where the faithful left their offerings before entering the church, something that Mulholland has recently framed as a paraliturgical function (2014, 3). This interpretation, moreover, fits well with Taft's suggestion that leaving the offerings in the narthex was one of the main elements in Byzantine liturgy (Taft, 1995; also Mathews, 1962). Because of the poor preservation of the elevation of the walls in this area, it is difficult to ascertain if there was a direct access to the church directly from the chapel or if, on the contrary, there was a sort of "circular" path that required worshippers to return to the narthex before proceeding to the *quadratum populi*. The surviving height of the walls indicates that the quadrangular base might have been part of a staircase leading to an upper level, to galleries for example, as it is common in the churches of nearby Sudan (Obłuski, 2016).

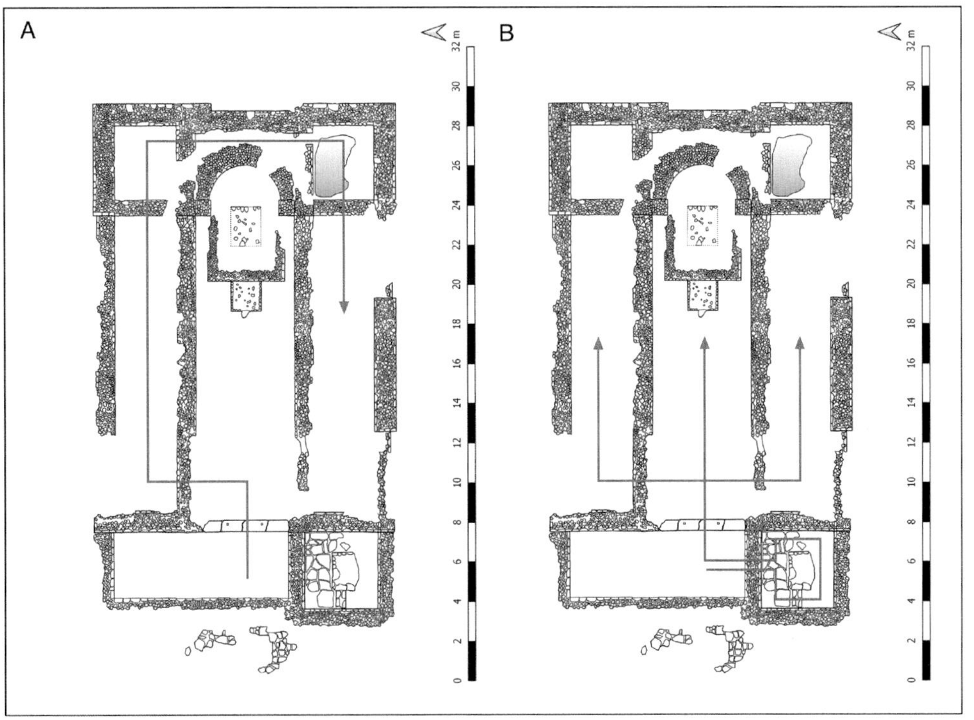

Figure 6.6 Avenues of movement.

Source: Author's image

An examination of the later phases of the building sheds light on the management and perception of space, albeit later and in a completely different context. Our excavations have shown that the building ceased to function as a church in the eighth century. Nevertheless, this did not represent the end of use of the structure. We have in fact identified two burials, one cut in the abandonment layer in the central nave of the church and another in the robbed out part of where the northern stylobate was (again Figure 6.5). Both the burials have been processed with radiocarbon analysis but, unluckily, we could extract the collagen from one of them only. The analysis carried out testified to a chronology to be framed most likely in the sixteenth century. Both the burials are single inhumations, primary deposition, lying in left lateral decubitus, with the head facing Mecca. These burials are clearly indicative of funerary customs typical of the Islamic community, which according to accepted written sources, had invaded the Horn of Africa from the early eighth century, even though the first archaeological attestations are quite later, from the twelfth century onwards (Insoll, 2003, 36–85). We can put these two burials in the frame of the eve of the Ottoman conquest of the Gulf of Zula, where Adulis stands (Chekroun & Hirsch, 2020). The choice by the Islamic community of a ruined church for the creation of a small cemetery in the late fifteenth or early sixteenth century may well represent, on the one hand, the desire for a conceptual reappropriation of a sacred place, by generating a multi-layered context of religious expression "overlaying rather than erasing pre-existing cultural strata" (Loiseau et al., 2021, 509). On the other hand, this material process may also introduce the concept of hybridization into the religious transition: the choice by the Islamic community was indeed symbolic and may be seen as a strategy for the creation of memory, where the Early Christian past became a sort of vernacular heritage, with the superimposition of a new cosmopolitan element represented by Islam (on the concept of "vernacular cosmopolitism" applied to archaeology, see González-Ruibal, 2021; on "cosmopolitan archaeology", see Meskell, 2009; again on the concept of cosmopolitanism and religious transition in Adulis, see Castiglia, 2022). An analogous process, testifying in this case to the occupation of polytheistic spaces by the Christian community, can be found not far from the Ethiopic area, in the Egyptian town of Philae. This is one of the cases where the religious transition can be well interpreted as a top–down and programmatic event, where the former structures had not been abandoned for a long time but, on the contrary, were reconverted to the Christian liturgy in only few decades (first half of the sixth century) (see Dijkstra, 2011).

6.5 Concluding remarks

All the data presented earlier concur to some general considerations about the evolution of space(s) in the late antique Horn of Africa. As I noted earlier, we have to be aware of the fact that our dataset is not as complete as we would hope, being mostly limited to churches and monumental buildings. The *corpus* of contemporary written sources is quite limited with regard to the description and perception of both urban and rural topographies; at the same time, we also lack contemporaneous liturgical texts. There might be just an exception, the so-called Aksumite Liturgical Collection, which is believed by Bausi (2006) to be a product of the Aksumite Kingdom and datable to between the fourth and the seventh centuries. Nevertheless, the collection bears many later corruptions that may have heavily invalidated the reliability of the text, which only survives in a fourteenth-century manuscript. Another significant challenge is posed by differences in the quality of the excavations. As we have seen, there is no excavation of a town in its entirety, since the archaeological agenda has always privileged the middle-Aksumite phases. This means that we have to rely basically on these markers, with Early Christianity, like it or not, to be considered as a "privileged" observation point.

The analysis carried out at different dimensional levels of space shows how Christianity provoked significant changes in the management of space and in the perception that people had of it. It is harder to understand these changes at a macro level, as we have seen, while they are more evident at a lower (micro) levels of analyses; it is unfortunate, in this respect, that we know so little about the people who used these spaces, also due to the lack of epigraphic data. In a semi-macro perspective of space, it is possible to see how the material expressions of early Christianity in great part reshaped the mindscapes of towns, with the churches becoming the new main elements in late antique topographies. The stunning architecture of the Aksumite churches and, most of all, their dominating position in the urban setting, may indeed be indicative of a new role of the bishops, who likely became the new civic leaders of the cities.

The rise of episcopal power does not point to a conflict between religious and secular authorities, however. Indeed, the ecclesiastical hierarchy gained a new leading place in the cities at the same time that monumental secular palaces occupied a prominent position in urban topography. This is the case of the capital, Aksum, and as it is well demonstrated by the relationship between the cathedral and the great palace nearby, most probably enclosed within the walls of a citadel as discussed earlier. I rather believe that church and secular government gradually evolved towards a shared power, even though the church seemed to be the main actor, at least in the sixth century. This is clearer in Adulis where, as we have seen, active and lively churches were built, replacing part of the former structures in order to make room for the Christian buildings themselves. This development suggests that the influence of episcopal authority was considerable even among secular governors, who allowed the church to conquer urban space. Unfortunately, there are very few traces of the pre-Christian religion venerated in the early Aksumite times. The only significant structure that survives is the great temple at Yeha, close to Aksum, which was most probably dedicated to the South-Arabian moon god *Almaqah* (Phillipson, 2012, 24). We don't have clear stratigraphic data about the chronology of abandonment of the temple as a polytheistic site, but we know that it was converted into a church with a baptistery in the sixth century. Since we do not know the exact chronological range between the abandonment of the building as a temple and its re-occupation as a church, we are not allowed to hypothesize a strategy of conflict imposed by the ecclesiastical hierarchies over the pagan past. Nevertheless, we can testify to a re-occupation of a religious space by a different creed, similar to what happened, even though in different modalities and chronologies, with the small Islamic cemetery overlapping the ruins of the cathedral of Adulis in the fifteenth and sixteenth centuries (see, e.g. the fundamental Hahn et al., 2008). We may never know if this was an ideological imposition by the ecclesiastical hierarchy or if it was just a utilitarian solution, but what it is significant is the definitive change of space and function not only of a building but probably of an entire area of the town as well.

All these elements lead to the key point of conversion that, due to the scarcity of written and epigraphic sources, needs to rely mainly on the topography of space to be interpreted. As we have seen, written documents and inscriptions are mainly useful for understanding the chronology of the formal adoption of Christianity by the royal court, which took place in the early decades of the fourth century. Nevertheless, the archaeological and topographic dataset points out to a widespread rooting of Christianity that took over only from the sixth century onwards, with the "explosion" of church-building throughout the kingdom (for a synthesis, see Castiglia, 2021). There is, then, a sort of discrepancy between the approach to the new religious creed by the rulers and the one implemented by the majority of the people: the chronological hiatus between them (fourth to sixth centuries) seems to reveal an initial recalcitrance by the latter to immediately accept Christianity, something that can find significant comparisons also in some areas of the Mediterranean basin, such as in Corsica (Castiglia & Pergola, 2019) and Greece, where conversion had never been a top–down process but, rather, a "strategic change" (Sweetman, 2015; for the conversion processes in Adulis, see Castiglia, 2022).

Although challenging, the evolution of funerary spaces is also worth examining. One of the most remarkable and symbolic expressions of the Aksumite civilization are the great stelae that marked the tombs of the kings until the times of Ezana who, as far as we know, was the last Aksumite ruler to have a stela to mark his own tomb. It seems clear that the use of stelae ceased with the arrival of Christianity in the first half of the fourth century. But this was a gradual process of transition as seen in the use of the monumental stelae in kings' burials in Adulis (Phillipson, 1994), since the first Christian king of Aksum, Ezana, was still being represented through them (Fattovich, 1987; Poissonier, 2012). The transitional character of this period is also suggested by the fact that the boom of church construction only took place in the sixth century, almost two centuries after the official conversion of the royal court. Nevertheless, it is not clear why the Christian kings took the decision to change the appearance of their funerary monuments. One possible answer might be that they were keen to adopt a more sober lifestyle. While this may be a predictable answer (applied widely in Christian scholarship), it is too simplistic – in part because the material expression of the Aksumite churches was everything but sober and simple (see, for further examples of this contradiction, Brown, 2012). It is possible to hypothesize that funerary customs had undergone a profound change by this period, which would help explain why the data on burial practices became scarce from the arrival of Christianity onwards. A plausible interpretation could be that the evidence relating to Christian necropolises has not yet been identified, especially with regard to the less affluent classes of the population. Unfortunately, therefore, the evolution of funerary practices in the long term (and especially in the moment of religious transition from non-Christian to Christian status) is destined, given the current state of our knowledge, to remain a significant question mark and one of the biggest challenges for the future research agenda. Of course, to better frame this phenomenon and, most of all, to understand where to look for the material traces of the necropolises, it will perhaps be crucial to try to abandon the Mediterranean-centric interpretive paradigms that we are used to. Nevertheless, it goes without any doubt that Christianity acted – as in the Mediterranean world – as one of the leading forces that literally overturned space at different levels: territorial, topographic, architectural, liturgical, cultural, and even conceptual.

Note

1 This choice is due at first to the fact that, since 2017, I am the co-director for the Pontificio Istituto di Archeologia Cristiana (PIAC) of Rome of the excavations of the Early Christian churches in the Aksumite town of Adulis, in present-day Eritrea, in the frame of a wider project on the Origins of Early Christianity in the Horn of Africa (see Castiglia, 2019; Castiglia et al., 2020, 2021).

Editions and abbreviations

Cosmas Indicopleustes, *Topographia Christiana* – Wolska-Conus, W. (1968–1973). *Cosmas Indicopleustes, Topographie chrétienne. Introduction, texte critique, illustration, traduction et notes*. Les éditions du Cerf.

DAE = Littmann, E. (1913). *Deutsche Aksum-Expedition (1906)*. G. Reimer.

John of Nikiu, Chron. – Zotenberg, H. (1883). *Chronique de Jean, Evêque de Nikiou. Texte éthiopien*. Imprimerie Nationale.

Martyrium of Saint Arethas – Detoraki, M. (2007). *Le martyre de Saint Aréthas et de ses compagnons: (BHG 166)*. Collège de France – CNRS.

Nicephorus Callistus, *HE* – Migne, J. P. (1865). *Nicephori Callisti. Ecclesiasticae Historiae*. Patrologiae Cursus Completus, Series Graeca 145.

Periplus Maris Erythraei – Casson, L. (1989). *The Periplus Maris Erythraei. Text with Introduction, Translation and Commentary*. Princeton University Press.
Procop, *Wars* – Dewing, H. B. (2014). *The Wars of Justinian. Prokopios* (revised and modernized, with an Introduction and Notes, by A. Kaldellis). Hackett Publishing.
Rufinus, *HE* – Amidon, R. (1997). *Rufinus of Aquileia. History of the Church*. CAU Press (reprint 2016).
Socrates, *HE* – Hansen, G. C. (1995). *Sokrates Kirchengeschichte*. Akademie Verlag.
Soz, *HE* – Hansen, G. C. (2004). *Sozomenos Historia ecclesiastica – Kirchengeschichte*. Turnhout.
Theodor, *HE* – Parmentier, L. (1998). *Theodoret. Kirchengeschichte*. Akademie Verlag.

Bibliography

Amidon, P. R. (1997). *Rufinus of Aquileia. History of the Church*. Oxford University Press (reprint 2016).
Bard, K. A., Fattovich, R., Manzo, A., & Perlingieri, C. (2014). The chronology of Aksum (Tigrai, Ethiopia): A view from Bieta Giyorgis. *Azania*, *49*, 285–316.
Bausi, A. (2006). La collezione aksumita canonico-liturgica. *Adamantius*, *12*, 43–70.
Beaucamp, J. (2010). Le rôle de Byzance en Mer Rouge sous le règne de Justin: mythe ou réalité? In Beaucamp, J., Briquel-Chatonnet, F., & Robin, C. J. (Eds.), *Juifs et chrétiens en Arabie aux Ve et VIe siècles: regards croisés sur les sources* (pp. 197–218). CNRS.
Breton, J.-F. (2015). *Les bâtisseurs des deux rives de la mer Rouge*. Éditions de Boccard.
Brown, P. (2012). *Through the Eye of a Needle: Wealth, the Fall of Rome, and the Making of Christianity in the West, 350–550 AD*. Princeton University Press.
Caillet, J. P. (2012). Le cas de Caričin Grad (Serbie) et le problème de l'identification de certains "palais épiscopaux" de l'Antiquité tardive. In Balcon-Berry, S., Baratte, F., & Sandron, D. (Eds.), *Des domus ecclesiae aux palais épiscopaux, actes du colloque d'Autun (26–28 novembre 2009)* (pp. 147–162). Brepols.
Castiglia, G. (2019). *In Adule, Aethiopum Urbs Maritima*. L'impatto monumentale del Cristianesimo ad Adulis e nel Corno d'Africa in età tardo antica. *Antiquité Tardive*, *26*, 327–348.
Castiglia, G. (2021). Between Aksum and Byzantium. The rise of early Christianity in the Horn of Africa: Sources and new archaeological data. *Vetera Christianorum*, *57*, 85–105.
Castiglia, G. (2022). An archaeology of conversion? Evidence from Adulis for early Christianity and religious transition in the Horn of Africa. *Antiquity*, *390*, 1555–1573.
Castiglia, G., & Pergola, P. (2019). Between change and resilience: Urban and rural settlement patterns in late antique Corsica. In Cau Ontiveros, M. A., & Mas Florit, C. (Eds.), *Change and Resilience: The Occupation of Mediterranean Islands in Late Antiquity* (pp. 25–49). Oxbow Books.
Castiglia, G., Pergola, Ph., Ciliberti, M., Larentis, O., Maletić, B., & Pola, M. (2020). Le campagne del Pontificio Istituto di Archeologia Cristiana ad Adulis (Eritrea) – Scavi 2018 e 2019. *Rivista di Archeologia Cristiana*, *96*, 221–274.
Castiglia, G., Pergola, Ph., Ciliberti, M., Larentis, O., Maletić, B., & Pola, M. (2021). For an archaeology of religious identity in Adulis and the Horn of Africa: Sources, architecture, and recent archeological excavations. *Journal of African Archaeology*, *19*, 1–32.
Chekroun, A., & Hirsch, B. (2020). The Muslim-Christian wars and the Oromo expansion: Transformations at the end of the middle ages (ca. 1500–ca. 1560). In Kelly, S. (Ed.), *A Companion to Medieval Ethiopia and Eritrea* (pp. 454–476). Brill.
Dijkstra, J. (2011). The fate of the temples in late antique Egypt. In Lavan, L., & Mulryan, M. (Eds.), *The Archaeology of Late Antique 'Paganism'* (pp. 389–436). Leiden.
Di Salvo, M. (2017). *The Basilicas of Ethiopia: An Architectural History*. I.B. Tauris.
Durante, A. M. (2003). La città vescovile di Luna nell'alto medioevo. In Marcenaro, M. (Ed.), *Roma e la* Liguria maritima*: secoli IV – X; la capitale cristiana e una regione di confine; atti del corso e catalogo della mostra, Genova 14 febbraio – 31 agosto 2003* (pp. 203–214). Istituto Internazionale di Studi Liguri.
Duval, N. (1996). L'urbanisme de Caričin Grad. Une ville artificielle et ses bâtiments d'apparat: une spécificité locale ou une étape décisive dans la typologie des principia militaires? *Antiquité Tardive*, *4*, 325–339.

Fattovich, R. (1987). Some remarks on the origins of the Aksumite Stelae. *Annales d'Éthiopie*, *14*, 43–69.

Fiaccadori, G. (2010). 'Salama (Kasate Berhan)' [The Ethiopian name of Frumentius]. In Uhlig, S. (Ed.), *Encyclopaedia Aethiopica, Volume 4. O-X* (pp. 484–488). Harrassowitz Verlag.

Finster, B., & Schmidt, J. (1994). Die Kirche des Abraha in San'a. In Nebes, N. (Ed.), *Arabia Felix: Beiträge zur Sprache und Kultur des vorislamischen Arabien. Harrassowitz* (pp. 67–86). Wiesbaden.

Fritsch, E. (2007). Pastophoria and altars: Interaction in Ethiopian Liturgy and Church architecture. *Aethiopica*, *10*, 7–51.

Fritsch, E. (2012). Liturgie et architecture ecclésiastique Éthiopiennes. *Journal of Eastern Christian Studies*, *63*, 91–124.

Gajda, I. (2009). *Le royaume de Ḥimyar à l'époque monothéiste. L'histoire de l'Arabie du Sud ancienne de la fin du IVe siècle de l'ère chrétienne jusqu'à l'avènement de l'islam*. Académie des Inscriptions et Belles-Lettres.

Giostra, C. (2017). La diffusione del Cristianesimo lungo il Mar Rosso alla luce dell'archeologia: la città porto di Adulis e il regno di Aksum. *Rivista di Archeologia Cristiana*, *93*, 249–313.

González-Ruibal, A. (2021). The cosmopolitan borderland: Western Ethiopia c. AD 600–1800. *Antiquity*, *95*, 530–548.

Greatrex, G. (1998). *Rome and Persia at War, 502–532*. Francis Cairns Publications.

Hahn, J., Emmel, S., & Gotter, U. (2008). *From Temple to Church. Destruction and Renewal of Local Cultic Topography in Late Antiquity*. Brill.

Harrower, M., Dumitru, I. A., Perlingieri, C., Nathan, S., Zerue, K., Lamont, J. L., Bausi, A., Swerida, J. L., Bongers, J. L., Woldekiros, H. S., Poolman, L. A., Pohl, C. M., Brandt, S. A., & Peterson, E. A. (2019). Beta Samati: Discovery and excavation of an Aksumite town. *Antiquity*, *372*, 1534–1552.

Insoll, T. (2003). *The Archaeology of Islam in Sub-Saharan Africa*. Cambridge University Press.

Kaplan, S. (1982). Ezana's conversion reconsidered. *Journal of Religion in Africa*, *13*, 101–109.

Kominko, M. (2013). *The World of Kosmas. Illustrated Byzantine Codices of the Christian Topography*. Cambridge University Press.

Krautheimer, R. (1965). *Early Christian and Byzantine Architecture*. Penguin Books.

Lavan, L. (2003a). Late antique urban topography: From architecture to human space. In Lavan, L., & Bowden, W. (Eds.), *Theory and Practice in Late Antique Archaeology* (pp. 171–195). Brill.

Lavan, L. (2003b). The political topography of the late antique city: Activity spaces in practice. In Lavan, L., & Bowden, W. (Eds.), *Theory and Practice in Late Antique Archaeology* (pp. 314–337). Brill.

Lavan, L. (2021). *Public Space in the Late Antique City*. Brill.

Loiseau, J., et al. (2021). Bilet and the wider world: New insights into the archaeology of Islam in Tigray. *Antiquity*, *95*, 508–529.

Mathews, T. (1962). An early Roman chancel arrangement and its liturgical functions. *Rivista di Archeologia Cristiana*, *38*, 73–95.

Mathews, T. (1971). *The Early Churches of Costantinople: Architecture and Liturgy*, Pennsylvania University Press.

Meskell, L. M. (2009). *Cosmopolitan Archaeologies*. Duke University Press.

Mulholland, B. (2014). *The Early Byzantine Christian Church: An Archaeological Re-assessment of Forty-Seven Early Byzantine Basilical Church Excavations Primarily in Israel and Jordan, and Their Historical and Liturgical Context*. Peter Lang.

Munro-Hay, S. (1984). *The Coinage of Aksum*. Manohar.

Munro-Hay, S. (1988). The dating of Ezana and Frumentius. *Rassegna di Studi Etiopici*, *32*, 111–127.

Munro-Hay, S. (1991). *Aksum: An African Civilization of Late Antiquity*. Edinburgh University Press.

Munro-Hay, S. (1997). *Ethiopia and Alexandria. The Metropolitan Episcopacy of Ethiopia*. Zaś Pan.

Obłuski, A. (2016). Nobadian and Makurian church architecture: Qasr el-Wizz, a case study. In Łajtar, A., Obłuski, A., & Zych, I. (Eds.), *Aegyptus et Nubia Christiana: The Włodzimierz Godlewski Jubilee Volume on the Occasion of His 70th Birthday* (pp. 481–512). University of Warsaw.

Peacock, D., & Blue, L. (2007). *The Ancient Red Sea Port of Adulis, Eritrea. Results of the Eritro-British Expedition 2004–2005*. Oxbow Books.

Phillipson, D. (1994). The significance and symbolism of Aksumite Stelae. *Cambridge Archaeological Journal*, *4*(2), 189–210.

Phillipson, D. (1997). *The Monuments of Aksum*. Addis Ababa University Press.
Phillipson, D. (1998). *Ancient Ethiopia. Aksum: Its Antecedents and Successors*. British Museum Press.
Phillipson, D. (2009). *Ancient Churches of Ethiopia. Fourth – Fourteenth Century*. Yale University Press.
Phillipson, D. (2012). *Foundations of an African Civilisation Aksum & the Northern Horn 1000 BC – AD 1300*. Boydell & Brewer Ltd.
Poissonier, B. (2012). The giant Stelae of Aksum in the light of the 1999 excavations. *Palethnology, 4*, 49–86.
Robin, C. J. (2012). Arabia and Ethiopia. In Johnson, S. F. (Ed.), *The Oxford Handbook of Late Antiquity* (pp. 247–332). Oxford University Press.
Robin, C. J. (2015). Himyar, Aksum, and Arabia Deserta in Late Antiquity. The epigraphic evidence. In Fisher, G. (Ed.), *Arabs and Empires Before Islam* (pp. 127–171). Oxford University Press.
Shahîd, I. (1960). Byzantium and Kinda. *Byzantinische Zeitschrift, 53*(2), 57–73.
Sundström, R. (1907). Archaeological work at the ruins of Adulis and Gabaza. In Littmann, E. (Ed.), Preliminary report of the Princeton University expedition to Abyssinia. *Zeitschrift für Assyriologie, 20*, 151–182.
Sweetman, R. (2015). The Christianisation of the Peloponnese. The case for strategic change. *The Annual of the British School at Athens, 110*, 285–319.
Taft, R. F. (1995). Church and liturgy in Byzantium: The formation of the Byzantine church. In Akentiev, C. C. (Ed.), *Liturgy, Architecture, and Art Byzantine World. Papers of the XVIII International Byzantine Congress (Moscow, 8–15 August 1991) and Other Essays Dedicated to the Memory of Fr. John Meyendorff* (pp. 13–29). Society for Byzantine and Slavic Studies.
Turkovič, M. N. (2017). Late antique changes in the urban structure of the Dalmatian metropolis: The episcopal complex in Salona. In Blasco, D. (Ed.), *Cities, Lands and Ports in Late Antiquity and the Early Middle Ages. Archaeologies of Change* (pp. 39–52). BraDyPus.
Wüstenfeld, F. (1857–1861). *Abū l-Walīd Muhammad ibn 'Abdallāh al-Azraqī, Kitāb Ahbār Makka, ed. F. Wüstenfeld (Die Chroniken der Stadt Mekka I)*. Brill.
Zazzaro, C., Cocca, E., & Manzo, A. (2014). Towards a chronology of the Eritrean Red Sea Port of *Adulis* (1st – early 7th century AD). *Journal of African Archaeology, 12*, 43–73.

7 Making space in the Visigothic Kingdom

Church founders in sixth- and seventh-century Iberian epigraphy

Carolyn T. La Rocco

This chapter considers epigraphic accounts of church founders in Visigothic Iberia.[1] As contemporary, direct, and permanent testimonies, these inscriptions record information about religious and civil aspects of the societies in which they were created (Santos Yanguas, 2014). My analysis will focus on the sixth and seventh centuries, when the majority of the surviving Iberian evidence was produced (Santiago Fernández, 2020). This is in contrast to areas such as North Africa and Gaul, which present significantly earlier evidence (Brown, 2012).[2] This work explores how sacred spaces and new civic and community *foci* were established. As I will argue, though connected to Christian authorities, church building was open to different social groups and was as a result more idiosyncratic and variegated than typically assumed. Members of the lay elite played a defining role alongside monastic, clerical, and royal patrons. The consolidation of Christianity in the Iberian Peninsula was the result of initiatives by different social groups, often in collaboration but at times in competition with one another.

7.1 Aims and methods

Traditional scholarship assumes that the instigation of late antique Iberian church foundation was dominated by bishops (e.g. Kulikowski, 2004). Recent work argues for a broader range of social actors, with lay foundations playing a larger part in ecclesiastical building in Visigothic Iberia (e.g. López Quiroga, 2013; Utrero Agudo & Moreno Martín, 2015; Fernández, 2016; Utrero Agudo, 2017–2018; La Rocco, 2023). Here, I take this shift further and more concretely illustrate the involvement of other parties. In the sections that follow, I assess the role played by episcopal authorities in the foundation of churches, highlighting the extent to which literary and legal evidence are marked by a strong ecclesiastical bias. I then use the epigraphic material, especially church dedications and epitaphs, to show that the Iberian lay elite played a fundamental role in these initiatives. I have narrowed my study down to 25 representative church foundations from the sixth and seventh centuries, the vast majority of which being epigraphic evidence, all of which I have mapped (Figure 7.1) and collated in Table 7A, 'Inscriptions', at the end of the chapter.

7.2 Bishop Masona and the *Vitas Sanctorum Patrum Emeritensium*

As noted earlier, scholarship traditionally viewed the Christianisation of Iberia as an episcopal undertaking due largely to the nature of the surviving literary evidence, predominantly produced in ecclesiastical circles (Utrero Agudo & Moreno Martín, 2015). Most influential is the mid-seventh-century *Vitas Sanctorum Patrum Emeritensium* (*VSPE*), or the 'Lives of the Fathers of Mérida'. The *VSPE* presents biographies of important sixth- and seventh-century Méridan

DOI: 10.4324/9780429427152-9

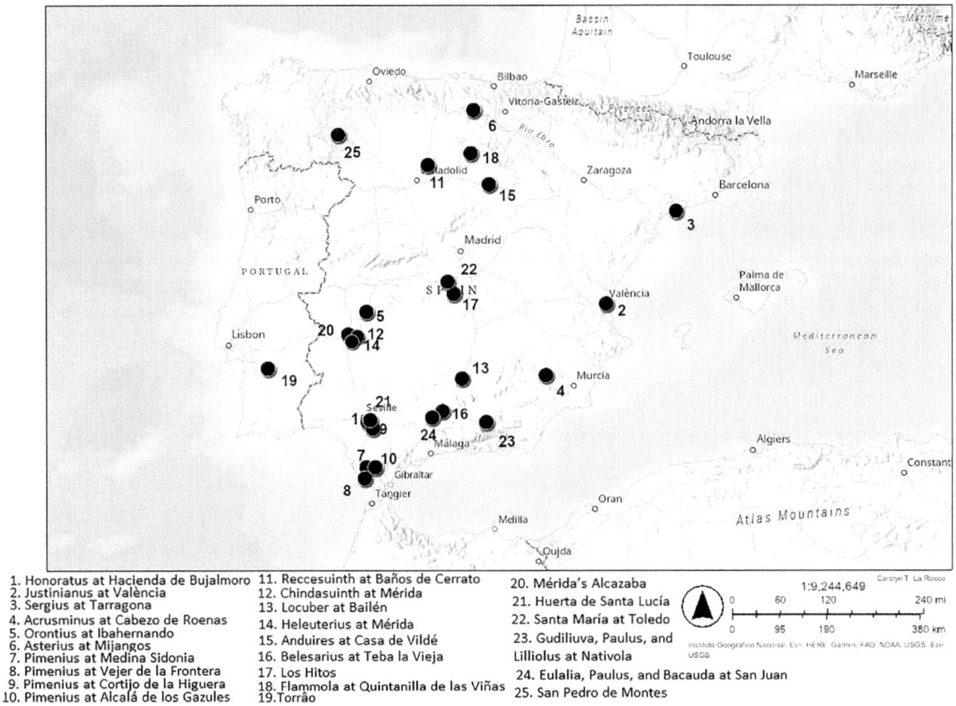

Figure 7.1 Map of church foundations analysed in the text.

church fathers. Produced by a deacon at Mérida's ecclesiastical school of Santa Eulalia, the text is a sophisticated literary composition, which prioritises political and religious concerns over accuracy (Fear, 1997; Fernández, 2016). Echoing the perspective of sixth-century Gallic and Italian accounts – most notably the Roman *Liber Pontificalis* (Brown, 2012; Machado, 2019; McKitterick, 2020) – the *VSPE* idealises its subjects, so that 'all those reading or listening may be strengthened with greater and stouter belief' (*VSPE*, *Life of Paul*, Pref. 1.3). This ecclesiastical bias appears clearly in the life of Masona, who held the bishopric of Mérida from *c.* 573 until his death, sometime between 600 and 610. In the text, Masona is depicted as a keen builder and generous patron:

> [Masona] founded and richly endowed many monasteries, built an even larger number of basilicas of wondrous appearance, and consecrated in these places many souls to God. Then he built a *xenodochium*, enriching it with a large patrimony and appointing ministers and doctors to serve travellers and the sick.
>
> (*VSPE*, *Masona* 3.3, trans. Fear, 1997)

This passage has played a central role in modern studies of late antique Iberia. Mérida, with Masona's abundant initiatives, is often portrayed as the key example of episcopal power and wealth (e.g. de Juan, 1998; Olmo Enciso, 1998; Arce, 2011; Brown, 2012; Utrero Agudo & Moreno Martín, 2015). None of Masona's building projects, however, have ever been archaeologically identified, with the possible exception of the *xenodochium* (Mateos Cruz, 1995, 2000; Collins, 2004).

Due to the picture presented by these sources, scholars have often assumed that bishops were responsible for all (or most) public building during this period, or that episcopal impetus was behind instances of construction without sufficient evidence (see Ribera i Lacomba & Rosselló Mesquida, 2014; cf. Martínez Jiménez et al., 2018; Bowes, 2018). As more recent scholarship has begun to show, however, this overly simplistic picture obscures the social, civil, and religious dynamics at play (e.g. Fernández Delgado et al., 2013). And yet there is work to be done. This is exemplified by an important study by Utrero Agudo and Moreno Martín, which aims to 'demythologize and relativize' (2015, 109) the role bishops played in sixth- and seventh-century Iberia by classifying church foundations as episcopal, aristocratic, or royal. However, influenced by the episcopal bias of our sources, the authors still misattribute many foundations to bishops, as we will see. That even a study nuancing episcopal bias is impacted by it demonstrates its entrenchment.

7.3 Bishops and secular society in the Visigothic Kingdom

The social and civil dynamics at play in Visigothic Iberia are difficult to fully unpick, but the evidence cannot be viewed in a sociopolitical vacuum (Wood, 2019, 235; cf. Utrero Agudo & Moreno Martín, 2015). Given the biases of texts like the *VSPE*, other evidence is needed to illuminate the complex power relations that informed the development of Iberia's Christian topography. The *Life of Paul*, an account of Mérida's bishop in the mid-sixth century, is illustrative of some of these social complexities. Though likely exaggerated, the text suggests that Mérida's wealth came largely from one lay aristocratic family. In his *Life*, Paul, a trained doctor, is approached by a wealthy Méridan senator (*primarii ciuitatis ex genere senatorum nobilissimi uiri*) whose equally high-status wife (*egrotasse matrona, que et ipsa inlustri stigmate progenita nobilem traebat prosapiem*) was ill after miscarrying (4.2.1). Paul refused to assist her several times (4.2.4), and her condition worsened. On her deathbed, Paul healed her:

> [T]hey fell at his feet, thanked him. . . . Then they drew up a document about their affairs to the effect that the holy man should immediately receive half of all they owned and that the other half should come under his control in his entirety and inviolate after their death . . . no richer senator than them could be found in the province of Lusitania. . . . On their death the most holy bishop Paul received their entire patrimony and while he had arrived as a stranger owning nothing, he now became more powerful than all the potentates of the town so that all the belongings of the church were considered as nothing in comparison to his possessions.
>
> (*VSPE*, *Paul* 4.2.9–4.2.18, trans. Fear, 1997)

The source thus presents a clear and linear origin for the economic and political power of a successful bishop, while also stressing the status of the pious givers (Fernández, 2016, 534). At the same time, the story can be read as a manoeuvre to maximise episcopal authority and economic opportunities, with Paul's acquiescence contingent upon financial compensation. It is a reminder of the fact that different and interlinked social groups participated in religious life in late antique Iberia – as elsewhere (Ziche, 2006). By the sixth century, many Iberian bishops were of aristocratic origin (Wood, 2019; Souvirón Bono, 2009; Osland, 2011). Masona 'spr[a]ng of what counts as noble stock on this earth' (*sanctus Masona antestis nobilis hortus in hoc seculo origine*) (*VSPE*, *Masona* 2.1; trans. Fear, 1997). Isidore, Archbishop of the See of Sevilla for three decades in the early seventh century, also came from the nobility. His brothers Fulgentius and Leandro were also bishops – Fulgentius, Bishop of Écija, and Leandro preceding

Isidore in Sevilla. The mid-sixth-century Bishop Justinian of València had three brothers, who were also bishops (see Ribera i Lacomba & Rosselló Mesquida, 2014).[3] Iberian bishops maintained these social networks through continued interaction and collaboration with their laic partners as a form of gaining or maintaining authority (Wood, 2019), remaining immersed in aristocratic society and culture.

Echoing the situation in late antique Italy (Izdebski, 2012; Machado, 2019), Iberia's royal administration was heavily dependent on local elites, the social group from which many bishops came. This was particularly true after the kingdom's conversion to Catholicism in the late sixth century (Osland, 2011; Fernández Delgado et al., 2013). The 590 CE *De fisco Barcinonensi* allowed Barcelona's bishop to collect taxes in King Reccared's name (Osland, 2011; Martínez Jiménez, 2013; Martínez Jiménez et al., 2018; *contra* Fernández, 2006). Attempting to further cement their integration into the political system, Canon 75 of the fourth ecclesiastical council of Toledo (633 CE) provisioned that, following a monarch's death, bishops and other leading aristocrats should choose his successor (Castellanos, 2020).[4] King Leovigild (r. 568–586 CE) even supposedly worked with the Arian bishop Sunna to try and remove Masona from power (*VSPE*, *Masona* 5.2). The Visigothic lay aristocracy, church hierarchy, and royal administration collaborated administratively so often that their overlapping roles are one of the kingdom's defining characteristics (Castellanos, 2020).

These close interactions and collaborations complicate attempts to determine whether church foundations can be exclusively attributed to bishops or secular donors. Repeated episcopal and lay aristocratic complaints at Iberian ecclesiastical councils indicate tensions and negotiatons between aristocrats and clerics regarding who controlled churches, and the finances associated with them (Fernández, 2016). Canon 19 of the Third Council of Toledo (in 589 CE) states that

> many (*multi*), contrary to established canons, demand that the churches they have built be consecrated, in such a way that the dowry they have given to the churches not be subject to episcopal ordination . . . but, following the ancient constitutions, everything is subject to episcopal ordination and power.[5]

Given the resources required for church construction, a good deal of these 'many' people would have been wealthy – typically, aristocrats. Iberia in fact witnessed a progressive increase of private lay aristocratic church-building (Utrero Agudo & Moreno Martín, 2015), as confirmed by, for example, Canon One of the Third Council of Zaragoza's (691 CE), which chastises bishops for improperly consecrating lay aristocratic churches.[6]

Overlapping social networks also created confusion (real or feigned) over appropriate financial conduct (de Juan, 1998; Souvirón Bono, 2009; Fernández, 2016), particularly regarding bishops' personal wealth, separate from that intended to service their churches (Utrero Agudo & Moreno Martín, 2015). The aforementioned inheritance left to Bishop Paul by the aristocratic couple he assisted is possibly an idealised representation of this (de Juan, 1998), particularly as he illegally leveraged his wealth to elect his nephew Fidelis as his successor (*VSPE*, *Paul* 4.3–4.4). Several Iberian canons address the need to establish clear boundaries between different categories of resources, indicating that episcopal misspending was taking place. As Canon Seven of the First Council of Braga in 561 establishes,

> from the church revenues, three equal portions should be made: one episcopal, one clerical, and the third for recovery [repair of buildings] or for lighting in the church: if either arch-presbyter or arch-deacon is administering, let him make an account of it to the bishop.[7]

Canon Six of the Ninth Council at Toledo (655 CE) reiterates episcopal spending limits:

> Taking account of previous sanctions regarding the portion of parochial ecclesiastical income to be conferred on the bishop, we have decided that the bishop may spend the third which belongs to him on that church or another of his choosing; this is licit and irrevocable.[8]

Conciliar references suggest episcopal abuse of these regulations and reiterate episcopal immersion in Visigothic society. Canons Three and Five of the Second Council at Braga (572 CE) concern improper episcopal remuneration. Canon Three states that 'clerical ordinations to the episcopacy must be imparted without remuneration', and Canon Five states that 'when bishops are invited to consecrate a church, they may not demand a gift from the founders'.[9] Canons One to Four and Seven from the Ninth Council of Toledo (655) suggest episcopal financial abuse; Canons Five, Eight, and Nine of the Eleventh Council of Toledo (675) address appropriate punishments.[10] Repeated proscriptions suggest that this abuse was relatively widespread, aligning with a late antique wave of episcopal misspending (Caner, 2021; Leone, 2006 for North Africa; Hillner, 2007 on Rome).

Members of the lay elite also used social capital to claim spiritual pre-eminence. The Council of Lérida (546 CE) condemns misclassification of private churches as monasteries for financial benefits, and in doing so indicates that private foundation was regularly occurring:

> If anyone from the laity wants to have their self-made basilica consecrated, classifying it as a monastery though no congregation meets there nor is an episcopal rule established there, they should not dare to do so and remain separate from diocesan law.[11]

Similarly, Canon Six from the Second Council of Braga criticises the misappropriation of donations at private churches.

If anyone builds a basilica not on account of faithful devotion but for the covetous pursuit of gain, dividing half of whatever is collected through the offerings of the congregation gathered there with the clergy because they have built a basilica on their own land – which is said to be done even now in some places – this ought to be observed in the future: that no bishop consents to such a terrible vow as daring to consecrate a hall not built under the patronage of saints, but under a tributary system.[12] As we will see, these tensions between bishops and the secular elite had a considerable impact on the foundation of churches.

Visigothic social fluidity further complicates matters. Iberian social structures and hierarchies had changed throughout the Roman period, becoming significantly different by the sixth century. The decreasing utility of *civis* as indicating a specific, Roman citizenship and thus of a certain level of distinction is noteworthy (Martínez Jiménez, 2020). This long decline was facilitated by, for example, Vespasian's extension of Latin rights throughout Hispania *c.* 74 CE, and the early third-century expansion of Roman citizenship through the *Constitutio Antoniniana*.[13] This was compounded by the late antique decline in Iberia's integration into the Roman administrative and economic system. By the sixth and seventh centuries, *civis* and related labels were catch-all terms for other forms of citizenship or identity (Martínez Jiménez, 2020), without legal weight. By that time, ethnic boundaries had also blurred, after continued interaction between 'Roman' and 'Barbarian' groups, reducing their potency as status or kinship markers (Martínez Jiménez, 2020; Castellanos, 2020). It was, moreover, no longer clear who was elite, who was not and what being 'elite' meant (cf. Curchin, 2018). Visigothic legislation instead emphasised free or unfree legal status (Fernández, 2016). The earliest extant Visigothic legal code, the 475 CE *Codex Euricianus*, and later legal texts – the *Lex Visigothorum*, *Leges Visigothorum*, and *Formulae Visigothicae* – possess much higher percentages of laws concerning free or unfree status than contemporary Roman legislation, comprising, for example, 40 per cent of the *Leges Visigothorum*'s titles (Lenski, 2022).

These more porous and flexible boundaries created opportunities for individuals of lower economic or social status to renegotiate their social standing and power. Simultaneously, the most distinguished individuals policed social and civil boundaries through practical or symbolic acts to avoid being confused with 'lower-status' individuals (Fernández, 2016). In sixth- and seventh-century Iberia, when earlier monumental building patterns had declined, and older means of asserting social distinction were no longer as effective, Christian public building took on new significance as one of these practical and symbolic acts – through it, one could demonstrate *nobilitas*, gaining or maintaining status, wealth, and power (Fernández, 2016, 513; Goddard, 2021). By founding a church, one created a new religious space and a new focal point for civic and community activity (Martínez Jiménez, 2020). This would bolster their image, like *cursus honorum* positions (Souvirón Bono, 2009; Wood, 2019) or monumental public building in earlier centuries. Thus, both traditional and new elites were incentivised to build churches, as we will see below.

7.4 The epigraphic evidence

Having explored socio-civil dynamics and tensions in Visigothic Iberia, we can consider the church founders represented in the epigraphic evidence, which is largely dedicatory and funerary in nature. Inscriptions are complex expressions of social life, and the social or ethnic origin of their commissioners is not always easily identifiable (Cooley, 2012). This is compounded by the fact that the Visigothic administration was relatively decentralised (when compared to Roman practice), making room for different groups to participate in social and civic life (Osland, 2011). As such, society was full of tensions, ambiguities, and fluidity, making the very notion of 'social identification' of limited use. One could belong to various social groups (Rapp, 2005) – being simultaneously, for example, part of a monastic community and from an elite background. Some were excluded from or misrepresented in inscriptions, moreover, while the achievements of others were exaggerated. This epigraphic ambiguity reflects the societal ambiguity experienced by these individuals and groups as they sought to create social and civil spaces for themselves by creating physical and (particularly) sacred, Christian, spaces. In such a fluid context, individuals could have been *de facto* or *de jure* aristocrats and left nearly identical archaeological footprints, as several inscriptions discussed later demonstrate. At times, we can infer status, wealth, or power based on the inscriptions' physicality – for example, material makeup, regularity and intricacy of their carving, or location (Handley, 2003; Sanchez Velasco, 2017). Though such boundaries were in reality more porous, then, I have sorted the evidence to facilitate analysis, considering bishops, non-episcopal elites, and collaborative foundations.

7.4.1 Episcopal church foundations

The epigraphic evidence for the involvement of bishops in church foundation is predominantly dedicatory in nature. These dedicatory inscriptions are often focused primarily on consecration, identifying the church, the (typically sixth or seventh century) date of consecration, and the consecrating bishop, but leaving aside other important steps in the making of Christian spaces, such as providing the land, funding the construction, or organising the works. In the sixth-century consecratory inscription from Cabezo de Roenas in Cehegín (Murcia), for example, we are told: 'In the name of God, Acrusminus, Bishop of the Bigastrense[14] Church, sanctified this basilica to Saint Vincent in the third year of his pontificate' (AEHTAM 518/ICERV 319/IHC 406). No other involvement is mentioned. Likewise, a neatly carved marble dedicatory plaque found in Ibahernando, Cáceres reads: 'In God's name. This basilica is consecrated (*sacrata est*) to Saint

Mary . . . under the Pontiff Orontius ([*Oron*]/*tio pontif*[*ice –*]) . . . April, Era 673 [635 CE]' (AEHTAM 3088/ICERV 549). Orontius was consecrator, but he is not credited with other forms of participation, presumably indicating a patron's involvement (Sastre de Diego, 2012). That only Orontius is named suggests that he was responsible for the inscription. Similarly, the dedication of Santa María of Mijangos, Burgos in 601 CE (Figure 7.2) reads: 'This place of Saint Mary was consecrated by the *pontifex* Asterius, on 6th May of the . . . 16th [year] of our glorious lord [King] Reccared. Era 639' (AEHTAM 2764/Hep 7,258). Again, no other involvement is mentioned, but Asterius alone is named. That Asterius was responsible for this inscription may also be suggested by what may be his seal in its bottom left corner (Santiago Fernández, 2009). That bishops are often commemorated in these inscriptions seemingly without financial contribution or labour makes possible social or professional incentives for involvement in church foundations (even aside from evangelical concern) clear.

The case of Bishop Pimenius of Assidonia (Medina Sidonia, Cádiz) is unique, with multiple dedicatory inscriptions recording various consecrations around Bætica. The earliest appears on a reused polytheistic altar at Medina Sidonia's Hermitage of Saints:

> The relics of martyr saints . . . i, Stephen, Julian, Felix, Justo, Pastor, Fructuosus, Augurius, Eulogius, Acisclus, Romanus, Martin, Quiricius, and Zoilus are deposited here. This basilica was dedicated (*dedicata hec basillica*) on the 16th of December in the second year of Pimenius' pontificate, Era 668 [630 CE].
>
> (AEHTAM 363/ICERV 304)

Figure 7.2 Dedicatory inscription from Santa María of Mijangos, with Asterius's seal in the bottom left.

Source: Complutense University of Madrid, Department of History and Anthropology of America, Historiographic Sciences and Techniques and Medieval History

Three other inscriptions come from the Hermitage of Saint Ambrose (Sastre de Diego, 2011) at Vejer de la Frontera (Cádiz) in 644 CE (AEHTAM 377/ICERV 305/IHC 111), Cortijo de la Higuera (Salpensa, Utrera) in 648 (AEHTAM 109/ICERV 306), and Alcalá de los Gazules (Cádiz) in 662 (AEHTAM 108/ICERV 309). Each opens with a reference to the deposition of a saintly relic – another method of sacralisation (Martínez Jiménez, 2020) – before dating Pimenius's consecration. Attaching his name to these foundations, Pimenius created sacred spaces and consolidated and maintained power in social and religious hierarchies. Three of Pimenius's dedicatory inscriptions, moreover, were carved into reused polytheistic altars – perhaps for practical or financial reasons, perhaps also a more pointed religious *damnatio memoriae* (Sastre de Diego, 2008, 2013; cf. Caseau, 2001). By placing the consecratory inscriptions of the churches on these altars, Pimenius bolstered his image and authority and indicated Christianity's 'triumph' over polytheism. Lending support to this theory is the fact that reuse of polytheistic altars for Christian dedications by bishops was unusually common in this region of the peninsula; contemporary examples are documented at Vejer de la Miel and in Egabro in Cabra, in addition to Roman aras being reused as Christian loculi or other liturgical elements, likely serving the same symbolic function in addition to being practical and cost-effective (Sastre de Diego, 2008, 469–470).

Though bishops are often the only individuals named in dedicatory inscriptions (González Fernández, 2014), they appear overwhelmingly only as consecrators (Sánchez Ramos et al., 2015). Although a crucial part of the process of creating a sacred place, the act of consecration does not indicate other involvement in church foundation, such as funding or construction (cf. Leone, 2011–2012). The only dedication memorialising a multifaceted episcopal role is the seventh-century dedication from Hacienda de Bujalmoro (Dos Hermanas, Sevilla) (AEHTAM 430 / ICERV 313 / IHC 363 / CILA 3,606):

> this sanctified and venerable temple to Christ was founded (*fundavit*) by Bishop Honoratus, the honour of whose name will remain strong for eternity, and who will be celebrated for his achievements; here also, in the *ara* in the middle of the sacred altar, he stored the remains of the three holy [martyr] brothers of Córdoba, after which he rightly dedicated the temple (*sanctorum iure dicavit*) to that trio . . . time hides in these verses, revealing its years throughout the centuries Era 66 . . . [622 CE or later].

There is, then, extremely little epigraphic evidence indicating that bishops acted outside of a consecratory role. While this may be due in small part to the focus of these inscriptions on consecration, the abundance of epigraphic evidence wherein this is their only named contribution – both when they are the only figures named and when they appear in inscriptions alongside other involved parties – suggests that this was in fact typically the extent of their contribution.

Bishops are presented differently in their epitaphs, closer to what we see in hagiographical accounts. The epitaphs of Justinianus of València (Bishop *c.* 527–548 CE) and Sergius (Bishop of Tarragona 540–*c.* 555 CE) open with church construction or refurbishment:

> Justinianus: Pious, famous teacher, cheerful, eloquent, Justinian, celibate pontifex, [*sacerdos*?], constructed new temples and restored old ones, and prepared festivals in his name for the people. He established and governed monasteries and convents, and wrote many things down for future posterity.
>
> (AEHTAM 3298/ICERV 279)

> Sergius: Skilful, magnanimous, pious, clever and knowledgeable, here in this tomb reposes Sergius, holy *pontifex*, and who restored the slipping roofs of the sacred temples, and constructed a *cenobium* not far from the city.
>
> (AEHTAM 220/ICERV 278)

This more personal description illuminates the legacies chosen by the party responsible for the inscription, likely the bishops themselves or their successors, strengthening their pedigree and authority through association. The foundation of churches was an important element of these legacies. Unsurprisingly, these epitaphs do not mention the contributions of other agents to the foundations in question, distorting our perception of the process (Handley, 2003, 3). In fact, only one dedicatory inscription and two epitaphs document episcopal involvement in church foundation outside of a specifically religious role. This does not prove that bishops did not perform other functions, as this might have been mentioned in accompanying inscriptions (and perhaps some bishops chose to remain anonymous) but suggests this was often the case. What was the 'usual' manner of church foundation, then, and who does the epigraphic evidence suggest were the people responsible for creating sacred spaces in Visigothic Iberia?

7.4.2 Non-episcopal church foundations

Contrary to what is usually assumed, various non-episcopal elites participated in the foundation of churches in sixth- and seventh-century Iberia. This is clearly indicated by the epigraphic evidence, including dedicatory inscriptions and epitaphs, which documents the active involvement of royal, clerical (non-episcopal), and lay aristocratic founders. Here, I will consider this evidence, beginning with church dedications before moving on to the funerary evidence.

Though not without interpretative challenges, there are some inscriptions indicating royal church foundation. That King Chindasuinth founded a church in Mérida may be suggested by a lacunose mid-seventh-century inscription: 'The most pious King Chindasuinth . . . in Chindasuinth's reign' (IHC 24/ICERV 366; cf. Martínez Jiménez et al., 2018, 179). Given structural similarity to King Reccesuinth's dedication of the church of San Juan Bautista in Baños de Cerrato, Palencia, the inscription would likely have introduced Chindasuinth and commemorated his role in the foundation, and then closed with the regnal year of the project's completion. In King Reccesuinth's late seventh-century dedication of San Juan Bautista, the king directly addresses John the Baptist:

> Precursor of God, martyr John the Baptist, take hold of this basilica constructed as an eternal gift to you; this [basilica] which I, the devout King Reccesuinth, lover of your name, have dedicated (*ego . . . propio de iure dicavi*), in the third year following a decade of my illustrious reign, Era 690 [652 CE].
>
> (AEHTAM 15/IHC 143/ICERV 314).[15]

In contrast with the episcopal dedications mentioned earlier, no consecration is mentioned, and Reccesuinth dedicated the basilica himself. Perhaps the church was not consecrated (del Hoyo, 2006); perhaps a bishop consecrated it, but, as a uniquely powerful social and political actor, Reccesuinth ensured he alone was commemorated in the inscription. This was not the full extent of the function this inscription served for Reccesuinth – the inscription was not just augmenting his power, but it was also an exercise of power (cf. Javier Martínez Jiménez in this volume for other, secular Visigothic royal foundations as exercises of power). In addition to appearing alone in the inscription, the dedication presents the king as the person responsible for choosing which saint would be commemorated in the church. In this sense, Reccesuinth was using his religious power to inscribe the worship of a saint into the local landscape.

The late seventh-century abbot Locuber in Bailén (Jaén) adds another layer of complexity: a dedication from 691 CE records him constructing two choruses, and a church consecration:

> In the name of God, Locuber, abbot though unworthy, made this, and constructed two choruses, and the churches of God were sanctified (*sacra/te*) 14th May, Era 729 [691 CE], the fourth year of the reign of our glorious lord Égica.
>
> (AEHTAM 429/ICERV 312/CILA 6, 213).

It is possible that Locuber himself was responsible for this sanctification. Locuber's blessing seemingly fulfils the same function as episcopal consecration, suggesting alternative routes to creating new sacred spaces and community focal points.

A single-naved, sixth- or seventh-century church with sanctuary, chorus, sacristy, and two porticoes was excavated at the monumental villa complex of Los Hitos (Arisgotas, Toledo) (Morín de Pablos et al., 2019; cf. Barroso Cabrera et al., 2015). A commemorative inscription dated stylistically to the sixth or seventh century found near the northern portico names the church's founder, though unfortunately a lacuna interrupts their name (*–do fecit*) (Morín de Pablos et al., 2019). In the seventh century, the church at Los Hitos was transformed into a funerary complex containing a prominent lay aristocrat's (*potens*) burial (Morín de Pablos et al., 2022). According to the excavators, given the church's layout and seventh-century modifications, Los Hitos likely belonged to an influential, wealthy aristocratic family connected to the royal administration in Toledo (Morín de Pablos et al., 2019, 68; Sánchez Ramos, 2020, 325–326). A monumental church in the Toledan countryside, however, also fits the bill of a monastic site, and further exploration is needed to establish Los Hitos's character (Morín de Pablos et al., 2019).

Santa María at Quintanilla de las Viñas (Burgos) presents a complicated dedication. The original apse and portions of the church's transept are preserved in a small, trilobular chapel (Caballero Zoreda, 2015). The church's interior is beautifully decorated. An elaborate, finely executed, votive inscription depicting the sun (*SOL*) flanked by two winged figures (Figure 7.3) commemorates an offering from a woman named Flammola: 'The modest Flammola offers this humble votive to God' (AEHTAM 1920/ICERV 511). The stone is prominent, placed as the right-hand capital at the entrance to the apse (Figure 7.4).

Figure 7.3 Flammola's votive stone at Santa María de Quintanilla de las Viñas.

Source: From 'Ermita de Santa María (interior)' by Ángel M. Felicísimo (Flicker).

Figure 7.4 Modified photo of Quintanilla de las Viñas's apse.

Note: (1) Flammola panel. (2) LVNA panel. (3) Panel with nimbate central figure. (4) Panel with non-nimbate central figure. (5) Christ Pantokrator. (6) Horseshoe arch.

Source: Adapted from 'Santa María. Quintanilla de las Viñas. Arco triunfal, lado de la nave, y vista del ábside' by Ignacio Cobos Rey (Wikimedia Commons). CC BY-SA 4.0

Flammola's inscribed capital matches the church's interior decoration, which was seemingly installed as part of the original construction (Mierse, 2017). The adjacent capital depicts the moon (*LVNA*) personified, forming a matching set. Five other panels also match Flammola's frieze stylistically (see Figure 7.4), with all of the figural friezes bearing similarly stylised facial features, hair, clothing, and borders. The first two each depict two angels flanking a central figure, one of whom is nimbate and cross-bearing. The third and fourth depict two Evangelists and likely decorated the transept (Caballero Zoreda, 2015). Christ Pantokrator tops the apse's archway. Given the Flammola inscription's quality, its prominent placement, and the church's decorative coherence, Flammola was quite possibly Quintanilla's founder (Lambert, 1955; Caballero Zoreda, 2015); her votive inscription may perform a similar function to the Iberian Bishop Théodulf of Orléans' 806 CE dedicatory inscription on a capital at the church of Germigny-des-Prés in Loiret, France (cf. Treffort, 2019; Croutelle, 2019).[16]

Due to the nature of twentieth-century excavation and recording practices, Quintanilla de las Viñas's dating is controversial (Utrero Agudo, 2006; Chavarría Arnau, 2017; Utrero Agudo, 2017–2018). It has been considered a Visigothic, late seventh- or early eighth-century foundation (de Orueta, 1929; Camps Cazorla, 1939–1940; Lambert, 1955; Escalona Monge, 1995; Nees, 1995; Mierse, 2017), based on contemporary epigraphic convention (Hoppe, 1984) and architectural (Camps Cazorla, 1939–1940), planimetric (Utrero Agudo, 2006), and iconographical (Paysás, 1998) similarities to San Pedro de la Nave (El Campillo, Zamora). The church's foundation has also been dated to the post-Islamic conquest, ninth- or tenth-century period (Cruz Villalón, 2002–2003; Caballero Zoreda, 2015; Utrero Agudo, 2017–2018). This is due to the stylistic difference between its altar and other Visigothic examples (Caballero Zoreda, 2015), as well as because of its horseshoe archway, usually associated with buildings of Islamic date (Montéquin, 1991, 72). Other 'Flammolas' – for example, the 'Flámula' recorded in the *Cartulario de San Pedro de Arlanza* as founding a monastery in Hortigüela, Burgos in 912 – are assumed to reference Quintanilla de las Viñas (de los Angeles Sepúlveda González, 1986).[17]

Neither chronology is secure. San Pedro de la Nave's chronology is debatable (cf. Caballero Zoreda & Arce, 1997; Rodríguez Trobajo et al., 1998; Llidó-Miravé, 2019). Horseshoe archways were also a Visigothic architectural feature (Montéquin, 1991; Nees, 1995). A two-phased, Visigothic and post-Umayyad construction lacks archaeological support (Schlimbach, 2011, 204); the evidence suggests a single phase of construction using Roman *spolia* (Utrero Agudo, 2006; Caballero Zoreda, 2015). The name Flammola, additionally, was common enough that occurrences do not automatically indicate Quintanilla (Orueta, 1929; cf. Utrero Agudo, 2006). Whether Visigothic or later, Flammola indicates continued lay church foundation into the mediaeval period.

Lay elite church founders are also documented in funerary inscriptions. The distinguished (*inlis*) woman Anduires founded a church at Casa de Vildé in Burgos de Osma (Soria) in the sixth (García Merino, 2010) or seventh (Fernández, 2016) century. Her turriform, lacunose tombstone records that Anduires and her distinguished husband (*uiro suo . . . uiro inluste*) raised 'this church' (*hanc ecles[siam]*) where she was buried (Figure 7.5):

> In the name of the Lord. Burial place of the distinguished Anduires, who built this church with her husband, the distinguished man. . .. They were married for thirty years and raised four children. The aforementioned woman peacefully and chastely . . . [passed away].
>
> (AEHTAM 3062/ICERV 505)

Anduires's epitaph opening with her church foundation emphasises its significance. The resources to build and commemorate a church suggests high social standing. The standard

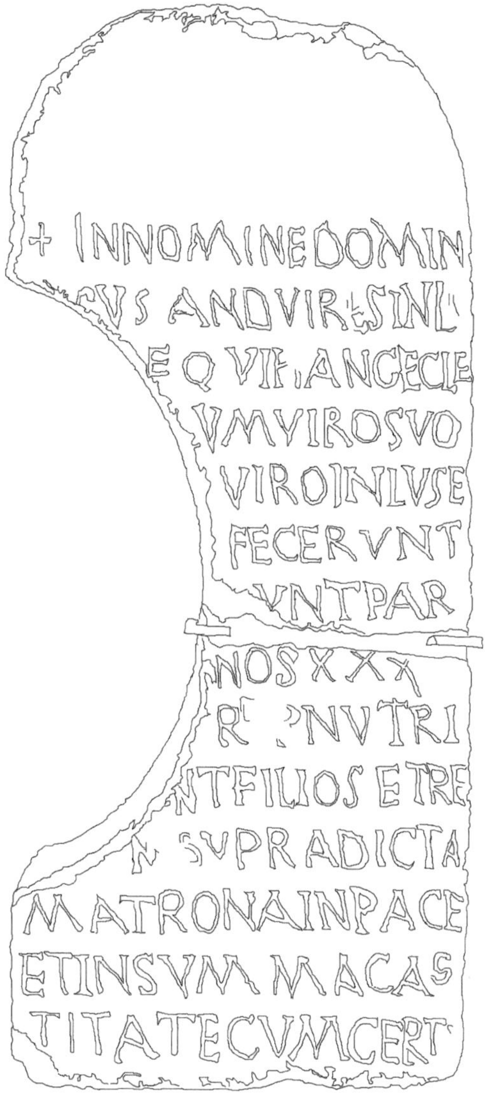

Figure 7.5 Anduires's funerary stele.

Source: Drawn by La Rocco from 'AEHTAM 3062' (Hesperia)

contemporary indications of rank *inlis* and *inluste* recall the earlier Roman '*inlustris*', that is, the title held by high-ranking Roman senators. Though almost certainly no longer being official titles at this time, the terms suggest claims to great nobility. The phrasing of the inscription on Anduires's epitaph (*hanc eclesiam*) suggests that she was buried very near the church in question. This is significant: in a sense, the entire church would have functioned as her funerary monument. Overall, the funerary inscription depicts Anduires as a powerful and wealthy female social actor and an important figure in her local society. By advertising the family's donations and status, moreover, the inscription would also have benefitted Anduires's heirs, reiterating the value of being epigraphically commemorated as a church founder (Fernández, 2016; Santiago Fernández, 2020).

Similarly, an elaborate funerary inscription dated to 662 CE found in Teba la Vieja, Córdoba, commemorates Belesarius:

> In this tomb rests the body of Belesarius, servant of Christ (*famuli (Christ)i*) . . . founder of this basilica (*condi/tori huius baselic[a]e*) . . . who lived for more or less . . . years during this century . . . who rested in peace on the . . . day of Era 700.
>
> (AEHTAM 1277/IHC 99/ICERV 157)

There is no official indication of nobility, but the reference to the foundation of a basilica foundation and the ornate funerary inscription suggest considerable disposable income; perhaps in epitaphs like Belesarius's, we are seeing new or affluent middle elites creating sacred spaces and a name for themselves in the process.[18]

Through the sampling of inscriptions presented here, we see the frequency with which church foundation by various elite, episcopal and non-episcopal groups and individuals was occurring in sixth- and seventh-century Iberia. It is clear that in addition to its impact on Iberia's physical topography, church building played an important social role, altering the 'mindscapes of towns', to borrow a phrase from Gabriele Castiglia's valuable analysis of the impact of early Christianity on the Horn of Africa in Late Antiquity in this volume. That so many members of the Iberian elite – including Visigothic monarchs, the most powerful and high-ranking members thereof – insisted upon having their names epigraphically connected to these spaces reiterates, however, that the inscriptions associated with these churches were as important as the churches themselves in securing social, civil, or spiritual authority (Isla Frez, 2007, 16).

7.4.3 Collaborative church foundations

Visigothic bishops and secular patrons seem to have frequently collaborated on church foundations. A large, neatly lettered stone plaque commemorates the late-sixth-century establishment of three basilicas in Nativola, outside Acci (Guadix, Granada) (Duval, 1991; Velázquez, 2007; Carbonell i Manils & Gimeno Pascual, 2010; Sánchez Ramos et al., 2015). Although Utrero Agudo and Moreno Martín (2015) identify this inscription as evidencing strictly episcopal church foundation, it records the distinguished (*in[lustri]*) Gudiliuva – possibly a *dux*[19] – funding the churches' construction, as well as consecration by Bishops Paulus and Lilliolus:

> In the name of Jesus Christ our God, this basilica of Saint Stephen, the protomartyr, was consecrated (*consacrata est*) in the place of Nativola by holy Paulus, *pontifex accitano*,[20] the . . . . day of the . . . year of our Lord, King Witteric, Era 645 [577 CE]. Also consecrated is the Church of Saint John the Baptist . . . . also consecrated is the Church of Saint Vincent, Valencian martyr, by holy Lilliolus, *pontifex accitano*, the 11th day of the kalends of February [22 January] of the eighth year of the reign of our most glorious lord King Reccared, Era 632 [594 CE]. These three sacred *tabernacula* for the glory of the most Holy Trinity were constructed with sacred assistance by the illustrious Gudiliuva, with his own labourers and from his own wealth . . . (*H(a)ec s(an)c(t)a tria tabernacula in gloriam trinitatis [S(an)c(t)] i(ssi)mae cohoperantib(us) s(an)c(t)is aedificata sunt ab inl(ustri) Gu(n)dilu[ua duce?] cum operarios uernolos et sumptu proprio*).
>
> (AEHTAM 360/IHC 115/ICERV 303)

The inscription documents a collaborative dynamic wherein each party provides something the other lacks: a wealthy member of the elite finances the foundation, and bishops perform the sacralisation of the space. In this, the Nativola inscription echoes the dynamic seen in earlier

Figure 7.6 One side of the San Juan altar inscription.

Source: From 'AEHTAM 1727' (Hesperia). Reprinted with permission

Roman inscriptions, such as the late antique dedication of Santo Stefano on the via Latina near Rome, which details the patron (Amnia Demetria), the consecrating pope (Leo), and even the overseeing presbyter of the building, Tigrinus.[21] That Gudiliuva could finance three churches and commemorate himself and two bishops, moreover, suggests that he was a particularly wealthy and high-status member of the late antique Iberian elite, with the inscription advertising 'his ability to perform a pious act that other (most) people, including free people, could not imitate' (Fernández, 2016, 530).

In the mid-sixth century, a woman named Eulalia and her son Paulus founded a church to St John in Zambra (Córdoba). A dedicatory inscription on a reused altar commemorates them and names Bacauda as the consecrating bishop:

> This basilica was consecrated (*consecrâta e[st]*) to Saint Mary on the 2nd day before the kalends of June, Era 698 [660 CE]. This building was dedicated (*dedicavit*) by master Bishop Bacauda. Founded for God 'the highest' by Eulalia and her son Paulus, the monk (*Paulum moñac[h]u[m]*). Holy Altar of God.
>
> (AEHTAM 1727/IHC 10/ICERV 308)

There is no status indicator like Anduire's '*inlis*' or Gudiliuva's '*inlustri*', official rank, or title. The inscription is not particularly ornate, with thin, irregular, lettering, nor is the altar heavily decorated (Figure 7.6). Perhaps this was a conscious stylistic choice to indicate humility, or perhaps financial or professional resources for a more elaborate inscription were unavailable. In any case, Eulalia and Paulus possessed sufficient wealth to found and commemorate a church – perhaps we are seeing a 'middle' or 'newly' elite laywoman and her son, a monk, creating a church and making a name for themselves, and a bishop doing the same.

The same collaborative dynamic appears in contemporary non-episcopal literature. The seventh-century *Replicatio Sermonum ad Prima Conuersione*, an autobiographical account by the ascetic Valerius of Bierzo, documents the construction of a church at San Pedro de Montes. Valerius credits the monk Saturninus (*frater Saturninus*) with the idea (*Replicatio* 9.3). Valerius opposes the endeavour, citing insufficient funding (*Replicatio* 9.7), until:

> Divine Love let it be clearly revealed to me that it was his will that this should be done. . . . As soon as the Lord's will was known . . . and Divine Providence ordained to grant it to me, many assisting workmen were hired through the munificent bounty of good Christians, through whom, with the assisting power of the Lord, it was finished. . . . It was consecrated with all care to the Lord by the man of God, the most reverend bishop Aurelio. At the same time with the power of the Lord he ordained Saturnino, the designer of this building, a priest.
>
> (9.8–9.22, trans. Aherne, 1949)

No physical evidence for the church survives, and it is possible that Valerius is employing a Christian *topos* of church-building through 'Divine Love' and 'Divine Providence', and no church was actually constructed. However, late antique hagiographies 'are *necessarily* subjected to social constraints . . . [and] must present a coherent image of the social world and of social practices' (Pachá, 2014, 256). In this sense, Valerius's account lends further weight to the impression that collaborative church foundation among various social classes and groups was not only acceptable but also seen as common or usual. *Frater* Saturninus was the initial instigator of the project, and his subsequent ordination shows that there were clerical incentives for church foundation. Valerius, an ascetic, arranged for unnamed Christian laymen to provide funding, through which labourers were hired. Finally, Bishop Aurelius 'created' the sacred space by consecrating the church. Interestingly, if this church *were* constructed, its epigraphic commemoration might only list Aurelius, like Orontius, Asterius, or Pimenius. Without Valerius, the church could be thought of as an episcopal foundation (as in, e.g. Utrero Agudo & Moreno Martín, 2015), but the reality was more complex: an intricate network of relations and mutual support leading to a single magnificent (even if imaginary) project.

7.5 Conclusion

As this study makes clear, when considered in its historical context, the surviving epigraphic evidence from sixth- and seventh-century Iberia is a vital source of information about society and the social and physical aspects of its Christianisation. The complex social, religious, and civil dynamics that are hinted at being at play by episcopal council canons and late antique hagiographies shine through in these commemorative dedicatory and funerary inscriptions. Many different agents (royal, monastic, episcopal, clerical, lay aristocratic, and familial) are documented advertising their involvement in creating these spaces, asserting themselves within local communities and social and religious hierarchies (Cooley, 2012; Sitz, 2019; Hellström & Russell, 2020) and Christianising Iberia in the process. The private building initiatives and organisation that many of these inscriptions document also reflect a break with the Roman tradition of primarily state-led building commissions (Utrero Agudo, 2017–2018, 110).

Through these inscriptions, it is also clear that Iberia echoes the late antique 'general rule' (Haensch, 2006, 53) of collaborative church foundation (Santiago Fernández, 2020), with inscriptions celebrating cooperation between lay aristocrats and bishops, and between lay aristocrats and monks; simultaneously, we see familial collaboration. This collaborative church building is not geographically isolated. It is seen in the southern peninsula via the Nativola inscription outside Acci (Granada), and further west via the dedicatory inscription from San Juan in Zambra (Córdoba). Valerius of Bierzo's account of a church foundation at San Pedro de Montes suggests that communal church foundation was common in the north-western peninsula. Communal foundation may be a chronological feature: the Nativola inscription dates to the late sixth century, the San Juan inscription to the mid-late sixth century, and Valerius's account to the seventh century.[22] While this is perhaps partially due to the later evidence's more complete preservation, it may also illuminate and reflect the mechanisms by which Christianity had effectively permeated society by this time.

We also have a clearer picture of episcopal involvement. Bishops appear almost exclusively as consecrators. It is true that other factors could also have contributed to this picture: perhaps in some cases, bishops *were* involved in other capacities, but only had their consecratory involvement documented. In some instances, accompanying inscriptions perhaps named other contributors. The fifteenth disciplinary proceeding from the Third Council of Toledo (589 CE) reminds bishops that '[i]f some of the servants of the treasury may have built churches and enriched them from their poverty, the bishop is to take this forward to be confirmed by [the King's] royal authority'.[23] Perhaps some inscriptions only commemorating episcopal consecration represent such foundations being 'taken forward' for confirmation. The epigraphic evidence, however, speaks so strongly against an episcopal role outside of consecration that it seems these extenuating circumstances would be insufficient explanation of the extent to which bishops are limited to a consecratory role. Of the 22 sixth- and seventh-century Iberian dedicatory church inscriptions known to me,[24] 11 (50 per cent) mention a bishop, and do so in a strictly consecratory capacity,[25] while eight (36 per cent) mention neither bishops nor sanctification.[26] Only three (14 per cent), interestingly, suggest some form of sanctification without mentioning episcopal involvement, suggesting that these cases were outliers or exceptions to the rule.[27] Notably, none of the 22 foundational inscriptions (0 per cent) mention episcopal involvement other than consecration. Together, this information illustrates even more clearly that bishops were overwhelmingly only consecratory participants.

Understanding the nature and extent of episcopal involvement in the foundation of churches allows us to realise the crucial role played by other elite groups in the establishment of sacred spaces across late antique Iberia. Bishops played an important role in creating sacred spaces, but others' contributions were equally necessary. The epigraphic evidence depicts a society being Christianised by different social actors, all of whom had various social, civil, or religious motivations for founding churches. Visigothic kings, members of the local lay elite, and monks were important agents in this process, involved in the commissioning, funding, and actual construction of churches.

Notes

1 I examined the Banco de Datos Archivo Epigráfico de Hispania Tardoantigua y Medieval (AEHTAM), which collects relevant archaeological and historiographical information about late antique and medieval Iberian inscriptions (as of May 2023, 5,886 entries). Accessible at http://hesperia.ucm.es/consulta_aehtam/web_aehtam/index.html. I also analysed inscriptions not yet catalogued in AEHTAM.

2 This chapter draws on my PhD at St Andrews, sponsored by the Wolfson Foundation and a Banco Santander Research Mobility Award. Translations are mine, unless otherwise indicated. The Latin text of each inscription is provided at the end of the chapter.

3 Echoing Gaul, where, for example, Honoratus of Arles, Germanus of Auxerre, or Gregory of Tours was of aristocratic background (see Van Dam, 1992, for Honoratus and Germanus; Rapp, 2005, for Gregory).

4 Though as Castellanos (2020, 61) notes, this was likely prescriptive; the only seemingly certain evidence comes from King Wamba's reign. '*[P]rincipe primatus totius gentius cum sacerdotibus successorum regni consilio communi constituent*' (ed. Tejada y Ramiro, 1859).

5 '*Multi contra canonum constituta sic ecclesias quas aedificaverint postulant consecrari, ut dotem quam ei ecclesiae contulerint censeant ad episcopi ordinationem non pertinare, quod factum et in praeterito displicet et in futurem prohibitur; sed omnia secundum constitutionem antiquam ad episcopi ordinationem et potestatem pertineant*' (ed. Tejada y Ramiro, 1859).

6 '*In sanctum et satis venerabile concilium, residentibus nuntiatum est nobis, eo quod aliqui pontifices regulam veritatis praetermittentes ecclesias quae a fidelibus pia devotione construuntur extra dies dominicos consecrent . . . . nulli penitus pontificum in quibuscumque provinciis constituto amodo liceat praeter certos dies dominicos ecclesias sanctas consecrare*' (ed. Tejada y Ramiro, 1859).

7 '*Item placuit, ut ex rebus ecclesiasticis tres aequae fiant portiones, id est una episcopi, alia clericorum, tertia in recuperationem vel in luminaria ecclesiae: de qua parte sive archipresbyter sive archidiaconus illam administrans episcopo faciat rationem*' (ed. Barlow, 1950).

8 '*Cum praeteritis sanctionibus notissimum habeatur, quae de rebus parochialium ecclesiarum pars episcopo conferatur, opportune duximus decernendum, ut si episcopus tertiam, quam de rebus eisdem sanctione paterna sibi debitam novit, aut ipsi ecclesiae cuius res esse patescit aut alteri ecclesiae cui elegerit conferre decreverit, et licitum maneat, et irrevocabile robur eius sententia ferat*' (ed. Migne, 1850).

9 '*Placuit ut de ordinatione clericorum episcopi munera nulla suscipiant*' and '*Placuit ut quotiens de aliquo fidelium ad consecrandas eclesias episcopi invitantur, non quasi ex debito munus aliquod a fundatore requirant*' (ed. Tejada y Ramiro, 1859).

10 See also Canon 4 of the Seventh Council of Toledo (646 CE).

11 '*Si autem ex laicis quisquam a se factam basilicam consecrari desiderat, nequaquam sub monasterii specie, ubi congregation non colligitur vel regula ab episcopo non constituitur, ea a diocesana lege audeat segregare*' (ed. Vives, 1963).

12 '*Placuit ut si quis basilicam non pro devotione fidei sed pro quaestu cupiditatis aedificat, ut quidquid ibidem oblatione populi colligitur medium cum clericis dividat, eo quod baselicam in terra sua ipse condiderit, quod in aliquibus locis usque modo dicitur fieri, hoc ergo de cetero observari debet, ut*

nullus episcoparum tam abominabili voto consentiat, ut baselicam quae non pro sanctorum patrocinio sed magis sub tributaria conditione est condita, audeat consecrare' (ed. Tejada y Ramiro, 1859). Several other Iberian canons also reference private church foundation; cf., for example, Canon Five of the Second Council of Braga (572 CE), Canon 19 of the Third Council of Toledo (589 CE), and Canon Two of the Ninth Council of Toledo (655 CE).

13 On Vespasian, see Pliny the Elder, *NH* 3.30; cf. Andreu Pintado (2004). On the *Constitutio Antoniniana*, see Bravo Bosch (1999); Garnsey (2004); Imrie (2018); Lavan (2019); Martínez Jiménez (2020).

14 From *Begastri*/Bigastro.

15 Given the plaque's placement far above the nave, it is illegible *in situ*; this and its stylistic aspects suggest it may be a later copy or *spolium* (Caballero Zoreda & Feijoo Martínez, 1998; del Hoyo, 2006; Martínez Jiménez et al., 2018).

16 Thank you to Fedor Schlimbach for drawing my attention to Théodulf's inscription.

17 Cf. Serrano's (1925) transcription of the twelfth-century copy of the text.

18 '*Famuli Christi*' simply identifies Belesarius as Christian, like contemporary Iberian inscriptions reading *famulus Dei* (Osland, 2019; Bar-Magen Numhauser, 2021).

19 Velázquez (2007, 262): '*inl(ustri) Gu(n)dilu(ua duce)*'; Duval (1991), 820: '*in(listri) Gudiliu*[. . . *comite?*]'.

20 From Guadix. *Inlustri* is descriptive, not rank-bearing.

21 *Inscriptiones Christianae Urbis Romae septimo saeculo antiquiores* (ICVR) n.s. VI, 15764; cf. Machado (2011).

22 Though this is perhaps partially due to the later evidence's more complete preservation.

23 '*Si qui ex servis fiscalibus fortasse ecclesias construxerint easque de sua paupertate ditaverint, hoc procuret episcopus prece sua auctoritate regia confirmari'*.

24 An additional inscription (AEHTAM 265; ICERV 354; IHC 109; CIL II 2/5, 154) names a bishop but cannot definitively be connected with the foundation of a church; another (CIL II2/7, 640) is too fragmentary to know who may have been mentioned therein.

25 See AEHTAM 1749 (IHC 146/ICERV 513); AEHTAM 1727 (IHC 10/ICERV 308); AEHTAM 360 (CIL II 2/5, 652/IHC 115/ILCV 1815/ICERV 303); AEHTAM 3088 (ICERV 549); AEHTAM 2764 (Hep 7,258); AEHTAM 518 (ICERV 319/IHC 406); AEHTAM 517 (ICERV 318/IHC 407/IHC 181/HEp 6, 675/HEp 18, 258); AEHTAM 430 (ICERV 313/IHC 363/CILA 3, 606); AEHTAM 377 (ICERV 305/IHC 111); AEHTAM 363 (ICERV 304); AEHTAM 109 (ICERV 306).

26 See AEHTAM 358 (IHC 50, 357/ICERV 301); AEHTAM 115 (ICERV 548); AEHTAM 106 (ICERV 350; IHC 361b); AEHTAM 21 (IHC 1; ICERV 311); the dedicatory inscription from Los Hitos; AEHTAM 2127 (CIL II 2/5, 769); AEHTAM 339 (HEp 15, 257/ICERV 370/IHC 397); AEHTAM 15 (IHC 143/ICERV 314).

27 See AEHTAM 263 (ICERV 335/IHC 107/CIL II2/5, 227); AEHTAM 38 (IHC 155/ICERV 302); AEHTAM 429 (ICERV 312; IHC 172; IHC 401; CILA 6, 213).

Editions and abbreviations used

AEHTAM – Archivo Epigráfico de Hispania tardoantigua y medieval online database. http://hesperia.ucm.es/consulta_aehtam/web_aehtam/index.html

Aherne, C. M. (Ed.). (1949). *Valerio of Bierzo, an Ascetic of the Late Visigothic Period* [Doctoral Dissertation, Catholic University of America].

Barlow, C. (Ed.). (1950). *Martini episcopi Bracarensis opera omnia (Papers and Monographs of the American Academy in Rome)*. Yale University Press.

Bradbury, S. (Ed.). (1996). *Severus of Minorca: Letter on the Conversion of the Jews*. Oxford University Press.

CIL – Corpus Inscriptionum Latinarum.

CILA – Corpus de inscripciones latinas de Andalucía, vols. I-IV. (1991–2002). Consejería de Cultura de Andalucía.

HEP – Hispania Epigraphica online database. https://eda-bea.es/

ICERV –Inscripciones cristianas de la España romana y visigoda, ed. J. Vives (1942). Consejo Superior de Investigaciones Científicas.

ICVR – Inscriptiones Christianae Urbis Romae septimo saeculo Antiquiores, ed. G. B. de Rossi (1861). Libraria Pontificia.

IHC – Inscriptiones Hispaniae christianiae, ed. E. Hübner (1871). Georgium Reimerum.
Loaysa y Girón, G. (Ed.). (1593). *Collectio conciliorum Hispaniae, diligentia Garsiae Loaisa elaborata, eiusque vigilijs aucta*. Petrus Madrigal.
MAN – Museo Arqueológico Nacional (Madrid).
Migne, J. P. (Ed.). (1850). *Patrologiae Cursus Completus vol 84: Sancti Isidori Hispalensis Episcopi Opera Omnia*. Petit-Montrouge.
Serrano, L. (Ed.). (1925). *Cartulario de San Pedro de Arlanza: Antiguo Monasterio Benedictino*. Madrid.
Tejada y Ramiro, J. (Ed.). (1859). *Colección de cánones y de todos los concilios de la Iglesia española: Tomo II*. Pedro Montero.
Vives, J. (Ed.). (1963). *Concilios visigóticos e hispanorromanos*. Consejo Superior de Investigaciones Científicas.
VSPE – Vitas Sanctorum Patrum Emeritensium. Lives of the Visigothic Fathers, trans. A. T. Fear (1997). Liverpool University Press.

Bibliography

Andreu Pintado, J. (2004). *Edictum, Municipium y Lex: Hispania en época Flavia (69–96 d.C.)*. BAR Publishing.
Arce, J. (2011). *Esperando a Los Árabes: Los Visigodos En Hispania (507–711)*. Marcial Pons Ediciones de Historia.
Bar-Magen Numhauser, A. (2021). *Hispanojewish Archaeology: The Jews of Hispania in Late Antiquity and the Early Middle Ages Through Their Material Remains* (Vol. 1). Brill.
Barroso Cabrera, R., J. Carrobles Santos, D. Blasco & J. Morín de Pablos. (2014). La evolución del *suburbio* y el *territorio* ercavicense desde la Tardía Antigüedad a la Época visigoda: El monasterio *servitano* y *Recópolis*. In J. López Quiroga & A. Martínez Tejera (Eds.), *In concavis petrarum habitaverunt. El fenómeno rupestre en el Mediterráneo Medieval: De la investigación a la puesta en valor* (pp. 257–294). Archaeopress.
Barroso Cabrera, R., J. Carrobles Santos, M. de Pablos & I. M. Sánchez Ramos. (2015). *Los Hitos. Arisgotas (Orgaz, Toledo). De palacio a panteón visigodo*. Ayuntamiento de Orgaz.
Barroso Cabrera, R., J. Carrobles Santos & J. Morín de Pablos. (2018–2019). 'Vrbs, praetorium, svbvrbia'. Centros de poder en la 'civitas' regia toledana y su teritoria en época visigoda. *Antigüedad y Cristianismo, 35–36*, 391–444.
Barroso Cabrera, R. & J. Morín de Pablos. (1996). La ciudad de Arcávica y la fundación del Monasterio Servitano. *Hispania Sacra*, *48*(97), 149–196.
Barroso Cabrera, R. & J. Morín de Pablos. (2007). Las ciudades de Arcávica y Recópolis y la fundación del monasterio servitano: organización territorial de un asentamiento monástico en la España visigoda. In J. López Quiroga, A. M. Martínez Tejera & J. Morín de Pablos (Eds.), *Monasterio et territoria. Elites, edilicia y territorio en el Mediterráneo medieval (siglos V-XI): Actas del III Encuentro Internacional e Interdisciplinar sobre la alta Edad Media en la Península Ibérica* (pp. 233–258). AUDEMA.
Bowes, K. (2018). Christianization of Villas. In A. Marzano & G. P. R. Métraux (Eds.), *Roman Villas in the Mediterranean Basin* (pp. 449–460). Cambridge University Press. https://doi.org/10.1017/9781316687147.025
Bravo Bosch, M. J. (1999). La *Constitutio Antoniniana*: una reflexión subjetiva. *Dereito: Revista xuridica da Universidade de Santiago de Compostela*, *8*(1), 71–79.
Brown, P. (2012). *Through the Eye of a Needle: Wealth, the Fall of Rome, and the Making of Christianity in the West, 350-550 AD*. Princeton University Press. https://doi.org/10.1515/9781400844531
Caballero Zoreda, L. (2015). Un conjunto constructivo altomedieval. Quintanilla de Las Viñas y las iglesias con cúpulas sobre pechinas de piedra toba de las provincias de Álava, La Rioja y Burgos. *Arqueología de la Arquitectura*, *12*, Article e028. https://doi.org/10.3989/arq.arqt.2015.011
Caballero Zoreda, L. & F. Arce. (1997). La iglesia de San Pedro de la Nave (Zamora): Arqueología y arquitectura. *Archivo Español de Arqueología, 70*(175–176), 221–274.
Caballero Zoreda, L. & S. Feijoo Martínez. (1998). La iglesia altomedieval de San Juan Bautista en Baños de Cerrato (Palencia). *Archivo Español De Arqueología*, *71*(177–178), 181–242.

Caballero Zoreda, L. & F. Moreno Martín. (2016). Sobre la dimensión epistemológica e histórica de una propuesta historiográfica. El modelo explicativo mozarabista. In I. Käflein, J. Staebel, Untermann, & M. Untermann (Eds.), *Im Schnittpunkt der Kulturen. Architektur und ihre Ausstattung auf der Iberischen Halbinsel* (pp. 299–330). Vervuert Verlag/Iberoamericana.

Camps Cazorla, E. (1939–1940). El visigotismo de Quintanilla de las Viñas. *Boletín del Seminario de Estudios de Arte y Arqueología*, *6*, 125–134.

Caner, D. (2021). Bishops and the politics of Lithomania in early Byzantium. In M. D. Gygax & A. Zuiderhoek (Eds.), *Benefactors and the Polis: The Public Gift in the Greek Cities From the Homeric World to Late Antiquity* (pp. 267–296). Cambridge University Press.

Carbonell i Manils, J. & H. Gimeno Pascual. (2010). A vueltas con la placa de Nativola (*CIL* II[5]/5, 652). Nuevos elementos para la reflexión. *Sylloge Epigraphica Barcinonensis*, *8*, 73–96.

Caseau, B. (2001). Πολεμειν λιθοις: La désacralisation des espaces et des objets religieux païens durant l'antiquité tardive. In M. Kaplan (Ed.), *Le sacré et son inscription dans l'espace à Byzance et en Occident: Études comparées* (pp. 61–123). Presses de l'université Paris-Sorbonne.

Castellanos, S. (2020). *The Visigothic Kingdom in Iberia: Construction and Invention*. University of Pennsylvania Press.

Cerrato Casado, E. (2018). *Arquitectura religiosa en la Bética (Hispania) durante los siglos IV-VIII d.C.* [Doctoral Dissertation, Università degli Studi di Padova].

Chavarría Arnau, A. (2017). Finding invisible Arians: An archaeological perspective on churches, baptism and religious competition in 6th century Spain. *Hortus Artium Medievalium*, *23*(2), 674–685.

Collins, R. (2004). *Visigothic Spain 409–711*. Blackwell.

Cooley, A. E. (2012). Epigraphic culture in the Roman world. In *The Cambridge Manual of Latin Epigraphy* (pp. 117–326). Cambridge University Press.

Croutelle, J. (2019). Regard sur une construction fameuse: l'église de Germigny-des-Prés à l'épreuve des sources. *Bulletin du centre d'études médiévales d'Auxerre*, *11*, 1–30.

Cruz, P. M. (1995). Identificación del Xenodochium fundado por Masona en Mérida. In *IV Reunio d'Arqueologia Cristiana Hispanica* (pp. 309–316). Institut d'Estudis Catalans: Universitat de Barcelona.

Cruz, P. M. (2000). *Avgvsta Emerita*, de capital de la *diócesis Hispaniarum* a sede temporal visigoda. *Memorias De La Real Academia De Buenas Letras De Barcelona*, *25*, 491–520.

Cruz Villalón, M. (2002–2003). Quintanilla de las Viñas y el arte cordobés. *Norba Revista de arte, XXII-XXIII*, 341–349.

Curchin, L. A. (2018). Curials and local government in Visigothic Hispania. *Antiquité Tardive*, *28*, 225–240.

de Juan, M. T. (1998). La gestión de los bienes en la iglesia hispana tardoantigua: confusión patrimonial y sus consecuencias. *POLIS. Revista de ideas y formas políticas de la Antigüedad Clásica*, *10*, 167–180.

del Hoyo, J. (2006). A propósito de la inscripción dedicatoria de San Juan de Baños. In F. Martínez & G. Pallarés (Eds.), *Temptanda viast. Nuevos estudios sobre la poesía epigráfica latina* (pp. 1–22). Servei de Publicacions de la Universitat Autonoma de Barcelona.

de los Angeles Sepúlveda González, M. (1986). Los anagramas y el programa iconográfico de Quintanilla de las Viñas: una hipótesis de interpretación. *En la España Medieval*, *9*, 1217–1248.

de Montéquin, F.-A. (1991). Arches in the architecture of Muslim Spain: Typology and evolution. *Islamic Studies*, *30*(1/2), 67–82.

de Orueta, R. (1929). Informe académico acerca de Santa María de las Viñas. *Boletín de la Comisión Provincial de Monumentos Históricos y Artísticos de Burgos*, *8*(19), 481–491.

de Santiago Fernández, J. (2009). El hábito epigráfico en la Hispania visigoda. *Jornadas Científicas sobre Documentación de la Hispania altomedieval: (siglos VI-C)*, *8*, 1–54.

de Santiago Fernández, J. (2020). Epigraphic habit and power in Visigothic Hispania. In S. Panzram & Pachá (Eds.), *The Visigothic Kingdom: The Negotiation of Power in Post-Roman Iberia* (pp. 273–292). Amsterdam University Press.

Duval, Y. (1991). "Nativola-les-trois-églises" (évêché d'Acci, 594–607) d'après Vivès [sic], ICERV, 303. *Mélanges de l'école française de Rome. Antiquité*, *103*(2), 807–820.

Escalona Monge, D. J. (1995). *Transformaciones sociales y organización del espacio en el Alfoz de Lara en la Alta Edad Media* [Doctoral Dissertation, Universidad Complutense de Madrid].

Fernández, D. (2006). What is the *De fisco Barcinonensi* about? *Antiquité Tardive*, *14*, 217–224.

Fernández, D. (2016). Property, social status, and church building in Visigothic Iberia. *Journal of Late Antiquity*, *9*(2), 512–541.

Fernández Delgado, A., J. Martínez Jiménez & C. Tejerizo García. (2013). Old and new elites in the Visigothic Kingdom (AD 550–650). In E. M. van der Wilt & J. Martínez Jiménez (Eds.), *Tough Times: The Archaeology of Crisis and Recovery. Proceedings of the Graduate Archaeology at Oxford Conferences in 2010 and 2011* (pp. 161–170). Archaeopress.

García Merino, C. (2010). El poblamiento rural en el Valle del Duero durante el Bajo Imperio. In *Actas Coloquio Internacional "Patrimonio Cultural y Territorio en el Valle del Duero"* (pp. 125–136). Zamora.

Garnsey, P. (2004). Roman citizenship and Roman law in the late empire. In S. Swain & M. Edwards (Eds.), *Approaching Late Antiquity: The Transformation from Early to Late Empire* (pp. 133–155). Oxford: Oxford University Press.

Goddard, C. J. (2021). Euergetism, Christianity and Municipal culture in Late Antiquity, from Aquileia to Gerasa (fourth to sixth centuries CE). In M. D. Gygax & A. Zuiderhoek (Eds.), *Benefactors and the Polis: The Public Gift in the Greek Cities from the Homeric World to Late Antiquity* (pp. 297–329). Cambridge University Press.

González Fernández, J. (2014). Epigrafía cristiana de la Bética. Nuevos testimonios. *Anejos de Veleia Series minor*, *32*(1), 355–367.

Haensch, R. (2006). Le financement de la construction des églises pendant l'Antiquité tardive et l'évergétisme antique. *Antiquité Tardive*, *14*, 47–58.

Handley, M. A. (2003). *Death, Society and Culture: Inscriptions and Epitaphs in Gaul and Spain, AD 300–750* (BAR International Series 1135). Oxford: Archaeopress.

Hellström, M., & Russell, A. (2020). Introduction: Imperial imagery and the role of social dynamics. In A. Russell & M. Hellström (Eds.), *The Social Dynamics of Roman Imperial Imagery* (pp. 1–24). Cambridge University Press.

Hillner, J. (2007). Families, patronage, and the titular churches of Rome, c. 300–c. 600. In K. Cooper & J. Hillner (Eds.), *Religion, Dynasty, and Patronage in Early Christian Rome, 300–900* (pp. 225–261). Cambridge University Press.

Hoppe, J.-M. (1984). *Le décor sculpté sur pierre des monuments chrétiens de l'Espagne visigothique. Représentations anthropomorphes. v. 1* [Doctoral Dissertation, Université Libre de Bruxelles].

Hoyo, J. D. (2006). A propósito de la inscripción dedicatoria de San Juan de Baños. In F. Martínez & G. Pallarés (Eds.), *Temptanda viast. Nuevos estudios sobre la poesía epigráfica latina* (pp. 1–22). SPUAB.

Imrie, A. (2018). *The Antonine Constitution: An Edict for the Caracallan Empire*. Brill.

Isla Frez, A. (2007). El lugar de habitación de las aristocracias en época visigoda, siglos VI-VIII. *Arqueología y territorio medieval*, *14*, 9–19.

Izdebski, A. (2012). Bishops in late antique Italy: Social importance vs. political power. *Phoenix*, *66*(1), 158–175.

Kulikowski, M. (2004). *Late Roman Spain and Its Cities*. The Johns Hopkins University Press.

Lambert, É. (1955). L'église visigothe de Quintanilla de Las Viñas. *Comptes rendus des séances de l'Académie des Inscriptions et Belles-Lettres*, *99*(4), 483–493.

La Rocco, C. T. (2023). *The "Christianisation" of the Iberian Peninsula: An Archaeological and Historical Consideration of Christian Space Creation by Late Roman and Visigothic Elites from the Fourth to the Eighth Century CE* [Doctoral dissertation, University of St Andrews].

Lavan, M. (2019). Constitutio Antoniniana. *Oxford Classical Dictionary*. https://doi.org/10.1093/acrefore/9780199381135.013.1794

Lenski, N. (2022). Slavery Among the Visigoths. In C. L. de Wet, M. Kahlos, & V. Vuolanto (Eds.), *Slavery in the Late Antique World, 150–700 CE* (pp. 251–280). Cambridge University Press.

Leone, A. (2006). Clero, proprietà, cristianizzazione delle campagne nel Nord Africa tardoantico: *status quaestionis*. *Antiquité Tardive*, *14*, 95–104.

Leone, A. (2011–2012). Bishops and territory: The case of late Roman and Byzantine North Africa. *Dumbarton Oaks Papers*, *65*(6), 5–23.

Llidó-Miravé, J. (2019). El papel de la epigrafía en la datación de San Pedro de la Nave. *Boletín del Archivo Epigráfico*, *4*, 26–49.

López Quiroga, J. (2013). Obispo y territorio en Hispania. La vertebración eclesiástica del territorio en el occidente de Hispania a partir de la imagen ofrecida por el Parroquial Suevo. In O. Brandt (Ed.), *Acta XV Congressus internationalis archaeologiae christianae: Toleti, 8–12.9.2008: episcopus, civitas, territorium* (pp. 463–481). Pontificio istituto di archeologia cristiana.

Machado, C. (2011). Roman Aristocrats and the Christianization of Rome. In R. Lizzi Testa & P. Brown (Eds.), *Pagans and Christians in the Roman Empire: The Breaking of Dialogue* (pp. 493–516). Lit-Verlag.

Machado, C. (2019). Aristocrats, Imperial institutions, and the topography of power. In *Urban Space and Aristocratic Power in Late Antique Rome: AD 270–535* (pp. 27–61). Oxford University Press.

Martínez Jiménez, J. (2013). *Aqueducts and Water Supply in the Towns of Post-Roman Spain (AD 400–1000) Volume I* [Doctoral Dissertation, University of Oxford].

Martínez Jiménez, J. (2020). Urban identity and citizenship in the west between the fifth and seventh centuries. *Al-Masāq*, *32*(1), 87–108.

Martínez Jiménez, J., I. Sastre de Diego & C. Tejerizo García. (2018). *The Iberian Peninsula Between 300 and 850: An Archaeological Perspective*. Amsterdam University Press.

Mazzoleni, D. (1989). Vescovi e cattedrali nella documentazione epigrafica in Occidente (Gallia, Iberia, Africa). In *Actes du XIe Congrès International d'Archéologie Chrétienne (Lyon, Vienne, Grenoble, Genève et Aoste, 21–28 Septembre 1986)* vol. 1 (pp. 794–795). École Française de Rome.

McKitterick, R. (2020). *Rome and the Invention of the Papacy: The Liber Pontificalis*. Cambridge University Press.

Mierse, W. E. (2017). Traces of a Central Asian textile on the walls of an early mediaeval Spanish Church. *AWE*, *16*, 249–270.

Morín de Pablos, J. & R. Barroso Cabrera. (2003). El monasterio Servitano: auge y caída de un cenobio visigodo. *Codex aquilarensis: Cuadernos de investigación del Monasterio de Santa María la Real*, *19*, 9–25.

Morín de Pablos, J., I. Sánchez Ramos, R. Barroso Cabrera, M. Á. Díaz Moreno, J. R. González de la Cal & A. Malalana Ureña. (2019). Los Hitos. Últimas novedades arqueológicas: Campañas 2014 a 2016. *Urbs Regia: orígenes de Europa*, *4*, 62–73.

Morín de Pablos, J., I. M. Sánchez Ramos & J. R. González de la Cal. (2020). Novedades arqueológicas del yacimiento de época visigoda de Los Hitos, Arisgotas (Orgaz, Toledo). In *Actualidad de la investigación arqueológica en España. Ciclo II (2019–2020). Conferencias impartidas en el Museo Arqueológico Nacional* (pp. 83–95). Ministerio de Cultura y Deporte.

Morín de Pablos, J., I. M. Sánchez Ramos & J. R. González de la Cal. (2022). Los Hitos (Arisgotas, Toledo). Nuevos datos para el conocimiento de un asentamiento rural de prestigio de época visigoda en la península ibérica. *Pyrenae*, *53*(1), 217–239.

Nees, L. (1995). Art and architecture. In R. McKitterick (Ed.), *The New Cambridge Medieval History II: c.700-c.900* (pp. 809–844). Cambridge University Press.

Olmo Enciso, L. (1998). Consideraciones sobre la ciudad en época visigoda. *Arqueología y Territorio Medieval*, *5*, 109–118.

Osland, D. (2011). Tribute and coinage in the Visigothic Kingdom: On the role of the bishop (Moneda y Tributo en el reino Visigodo: sobre las funciones del obispo). *ANAS*, *24*, 71–95.

Osland, D. (2019). Text and context: Patronage in late antique Mérida. *Studies in Late Antiquity*, *3*(4), 581–625.

Pachá, H. (2014). Gift and conflict: Forms of social domination in the Iberian Early Middle Ages. *Networks and Neighbours*, *2*(2), 251–277.

Paysás, J. M. (1998). Símbolos cósmicos y cristológicos en la Iglesia de Quintanilla de las Viñas. *Cuadernos de Historia de España*, *LXXV*, 17–40.

Rapp, C. (2005). *Holy Bishops in Late Antiquity: The Nature of Christian Leadership in an Age of Transition*. University of California Press.

Ribera i Lacomba, A. & M. Rosselló Mesquida. (2014). Los primeros obispos de Valencia. *Quaderns de difusió arqueològica*, *11*, 3–39.

Rico Camps, D. (2009). Arquitectura y epigrafía en la Antigüedad Tardía. Testimonios hispanos. *Pyrenae*, *40*(1), 7–53.

Rodríguez Trobajo, E., F. A. Matthias & L. Caballero Zoreda. (1998). Datación de una viga de la iglesia de San Pedro de la Nave (Zamora). *Archivo Español de Arqueología*, *71*, 283–294.

Sánchez Ramos, I. M. (2020). Looking through landscapes: Ideology and power in the Visigothic Kingdom of Toledo. In M. J. Kelly & M. Burrows (Eds.), *Urban Interactions: Communication and Competition in Late Antiquity and the Early Middle Ages* (pp. 305–341). Punctum Books. https://doi.org/10.2307/j.ctv19cwdm9.12

Sánchez Ramos, I. M., R. Barroso Cabrera, J. Morín de Pablos & I. Velázquez Soriano. (2015). Topografía eclesiástica de la Bética en la antigüedad tardía a través del corpus epigráfico. *Romvla*, *14*, 221–265.

Sanchez Velasco, J. (2017). *The Christianization of Western Baetica: Architecture, Power, and Religion in a Late Antique Landscape*. Amsterdam University Press.

Santos Yanguas, N. V. (2014). Fuentes de información: epigrafía y religiosidad. *Ilu. Revista de ciencias de las religiones. Anejos*, *25*, 163–179.

Sastre de Diego, I. (2008). Iconographic influences of Roman Aras in early Christian altars. Prevalence of formal and conceptual elements in Hispania. In M. J. Castillo Pascual, S. Knippschild, M. García Morcillo & C. Herreros González (Eds.), *Congreso Internacional 'Imagines'. La Antigüedad en las Artes escénicas y visuales: Universidad de La Rioja, Logroño. 22–24 de octubre de 2007* (pp. 425–444). Universidad de la Rioja.

Sastre de Diego, I. (2011). La reutilización de epígrafes romanos en la Hispania tardoantigua: los altares cristianos y la polémica ara de Guadix. In J. Carbonell Manils, H. Gimeno Pascual & J. L. Moralejo Álvarez (Eds.), *El monumento epigráfico en contextos secundarios: Procesos de reutilización, interpretación y falsificación* (pp. 67–87). Anmar.

Sastre de Diego, I. (2012). Aristocracia, cristianismo y epigrafía laica en la Hispania tardoantigua. *Veleia*, *29*, 17–28.

Sastre de Diego, I. (2013). *Los altares de las iglesias hispanas tardoantiguas y altomedievales: Estudio arqueológico (BAR International Series 2503)*. Archaeopress.

Schlimbach, F. (2011). Byzantinische Einflüsse auf den westgotenzeitlichen Kirchenbau in Hispanien? Bemerkungen zur Herleitung der Motive innerhalb der Baudekoration von Santa María de Quintanilla de las Viñas (Burgos) im Streit zwischen *Visigotistas* und *Mozarabistas*. In M. Altripp (Ed.), *Byzanz in Europa. Europas östliches Erbe. Akten des Kolloquiums,Byzanz in Europa vom 11. bis 15. Dezember 2007 in Greifswald* (pp. 184–243). Brepols.

Sitz, A. M. (2019). Beyond spolia: A new approach to old inscriptions in Late Antique Anatolia. *American Journal of Archaeology*, *123*(4), 643–674.

Souvirón Bono, S. (2009). Fiscalidad y control eclesiástico en la *Hispania* visigoda: supervisión de almas e Impuestos. *Baetica*, *31*, 275–289.

Stevens, S. T. (2019). Incorporating Christian communities in North Africa: Churches as bodies of communal history. In D. K. Pettegrew, W. R. Caraher & T. W. Davis (Eds.), *The Oxford Handbook of Early Christian Archaeology* (pp. 645–664). Oxford University Press.

Treffort, C. (2019). Les inscriptions de Germigny et la production épigraphique de Théodulf d'Orléans. *Bulletin du centre d'études médiévales d'Auxerre*, *11*, 1–17.

Utrero Agudo, M. Á. (2006). *Iglesias Tardoantiguas y Altomedievales en la Península Ibérica. Análisis arqueológico y sistemas de abovedamientos*. Consejo superior de investigaciones científicas.

Utrero Agudo, M. Á. (2017–2018). The Artisans behind Visigothic buildings: The materiality of identity. *Visigothic Symposium 2: Iberian Spaces, Iberian Identities*, 99–113.

Utrero Agudo, M. Á., E. Álvarez Areces, J. M. Baltuille Martín, R. Martín Talaverano, F. J. Moreno Martín, J. I. Murillo Fragero, M. Rielo Ricón & A. Villa del Castillo. (2016). San Pedro de la Mata (Sonseca, Toledo). Construir y decorar una iglesia altomedieval en piedra. *Archivo Español De Arqueología*, *89*, 45–69.

Utrero Agudo, M. Á. & F. J. Moreno Martín. (2015). Evergetism among the Bishops of Hispania between the sixth and seventh centuries: A dialogue between archaeological and documentary sources. *Journal of Early Christian Studies*, *23*(1), 97–131.

Van Dam, R. (1992). *Leadership and Community in Late Antique Gaul*. University of California Press.

Velázquez, I. (2007). *Baselicas multas miro opere construxit* (*VSPE* 5.1.1). El valor de las fuentes literarias y epigráficas sobre la edilicia religiosa en la *Hispania Visigoda*. *Hortus Artum Medievalium*, *13*(2), 261–269.

Wood, J. (2019). Building and breaking episcopal networks in late antique Hispania. In C. Cvetković & P. Gemeinhardt (Eds.), *Episcopal Networks in Late Antiquity* (pp. 227–248). De Gruyter.

Ziche, H. G. (2006). Administrer la propriété de l'Église: l'évêque comme clerc et comme entrepreneur. *Antiquité Tardive*, *14*, 69–78.

Table 7A. Inscriptions analysed in the chapter.

Inscriptions

Reference	*Location*	*Date*	*Text*	*Translation*
Episcopal				
AEHTAM 430/ ICERV 313/ IHC 363/CILA 3, 606	Hacienda de Bujalmoro (Dos Hermanas, Sevilla)	622 or later	*Fundavit s(an)c(tu)m hoc Xpi(sti) et venerabile t(e)mp(lum) / antistes Honoratus honor de nomine cuius / pollet in aeternum et factis celebratur in istis / hic aram in medio sacrans altare recondit / Tres fratres sanctos retinet quos Corboba passos / aedem deinde trium sanctorum iure dicavit / versibus aera subest annos per saecla resignans era DCLX[--]*	This sanctified and venerable temple to Christ was founded by Bishop Honoratus, the honour of whose name will remain strong for eternity, and who will be celebrated for his achievements; here also, in the *ara* in the middle of the sacred altar, he stored the remains of the three holy [martyr] brothers of Córdoba, after which he rightly dedicated the temple to that trio . . . time hides in these verses, revealing its years throughout the centuries Era 66. . .
AEHTAM 3298/ ICERV 279	--	*c.* 548	*Pius preclarus doctor alacer facundus,/ Iustinianus caelebs pontifex sacer[dos?] / noba templa construens vetustaque rest[aurans] / ornabit festa dictis predicans in populis. / Virgines instituens monacosque guvernans, / scripsit plura posteris profutura [---]. . . . . .*	Pious, famous teacher, cheerful, eloquent, Justinian, celibate pontifex, [*sacerdos*?], constructed new temples and restored old ones, and prepared festivals in his name for the people. He established and governed monasteries and convents, and wrote many things down for future posterity. . . .
AEHTAM 220/ ICERV 278	--	*c.* 555	*Sollers magnanimus pius ingenio cato / hic quiescit in tumulo Sergius pontifex s(an)c(tu)s / [qu]i sacri labentia restaura(n)s culmina templi / haud procul ab urbe construxit cenobium s(an)c(t)is / . . . .*	Skilful, magnanimous, pious, clever and knowledgeable, here in this tomb reposes Sergius, holy pontifex, and who restored the slipping roofs of the sacred temples, and constructed a *cenobium* not far from the city. . . .
AEHTAM 518/ ICERV 319/ IHC 406	Cabezo de Roenas, Cehegín (Murcia)	Sixth century	*[In] no(m) Do[mini] Acrvsminvs / Bigastrens Ecclesie eps / sacrauit anc baselicam / sci Vincentii anno III pontificaus / sui.*	In the name of God, Acrusminus, Bishop of the Bigastrense church sanctified this basilica to Saint Vincent in the third year of his pontificate
AEHTAM 3088/ ICERV 549	Dehesa Magas-quilla de los Donaires (Iba-hernando, Cáceres)	635	*In n(omine) D(omi)ni saÌ£crata est he[c ba] selica s(an)c(ta)e / Mar[iae ---] / quendis aÌ£[b] [Oron]/tio pontif[ice ---] / [---] aÌ£pri/ [les era D]CLXXXII*	In the name of God. This basilica is consecrated to Saint Mary . . . under the Pontiff Orontius . . . April, Era 673.

(*Continued*)

Table 7A. (Continued)

<table>
<tr><th colspan="5">Inscriptions</th></tr>
<tr><th>Reference</th><th>Location</th><th>Date</th><th>Text</th><th>Translation</th></tr>
<tr><td>AEHTAM 2764/ Hep 7,258</td><td>Santa María de Mijangos (Burgos)</td><td>601</td><td>Consacratus est / (lo)cus s[an]c[ta]e Mariae/ [p]ontifice Asterio / sub d[ie] pri[die] non[as] M[a]ias / XVI glo[riosi] dom[i]ni n[o]s[tri] Recca[redi]</td><td>This place of Saint Mary was consecrated by the pontifex Asterius, on 6 May of the . . . 15th [year] of our glorious lord [King] Reccared.</td></tr>
<tr><td>AEHTAM 363/ ICERV 304.</td><td>Hermitage of Saints, Medina Sidonia</td><td>630</td><td>hic sunt reliquie / [sc]rm condite, id. / [. . .] i Stefani, Iulia[ni,/ Fe]lici, Iusti, Pastor/[i, Fru]ctuosi, Auguri. / [Eu]logi, Aciscli, Rom/ [an]i, Martini, Quirici / [e]t Zoyli martirum./ dedicata hec basilica d. XVII kal. / [I] anuarias anno se/[c]undo pontifica/[t]us Pimeni, era dC/LXVIII.</td><td>The relics of martyr saints . . . i, Stephen, Julian, Felix, Justo, Pastor, Fructuosus, Augurius, Eulogius, Acisclus, Romanus, Martin, Quiricius, and Zoilus are deposited here. This basilica was dedicated ('dedicata hec basillica') the 16 December in the second year of Pimenius' pontificate, Era 668 [630 CE].</td></tr>
<tr><td>AEHTAM 377/ ICERV 305/ IHC 111</td><td>Hermitage of Saint Ambrose (Vejer de la Frontera, Cádiz)</td><td>644</td><td>Mazzoleni, 1989: 'i(n) n(o)m(ine) d(omi)ni n(o)s(tr)i [I]ησυ Χρι(sti) [recondite] / sunt reliqu[ie] / s(an)c(t)orum Vi(n)/centi Felicis / Iuliani marti/rum Dedica/tio illius b[asi]/lice sub d(ie) / XVIII k(a)l(endas) decem(bres) / anno sexto / decimo domi/ni Pimeni epi(scopi)/ aera DCLXXXII'.</td><td>In the name of our God, Jesus Christ, the relics of the Martyr Saints Vincent, Felix, and Julian are stored here. This basilica was dedicated on the 18th kalends of December (14 November) by Bishop Pimenio. Era 692.</td></tr>
<tr><td>AEHTAM 109/ ICERV 306</td><td>Cortijo de la Higuera, Salpensa, Utrera, Sevilla</td><td>25 May 648</td><td>Reliquie sanctorum Ioanni | Babtiste Eulalie Iuste Rufi|ne et felici martirum. Dedi|cata est hec basilica a Pime|nio antistite sub die VII kal(endas)| iunias era DCLXXXXV</td><td>The relics of the holy saints and happy martyrs John the Baptist, Eulalia, Justa, and Rufina [are buried here]. This basilica was dedicated by Bishop Pimenio on the seventh day before the kalends of June, Era 686.</td></tr>
<tr><td>AEHTAM 108/ ICERV 309</td><td>Alcalá de los Gazules, Cádiz</td><td>662</td><td>In nomine D(omi)ni. Hic / sunt recondite relique/ s(an)c(to)r(u)m Seruandi, Germani, / Saturnini, Iuste, Rufine / martir(um) et Ioanni Babtiste, / sub di(e) nonas Iunias / anno XXXIII domni / Pimeni pontificis / Era DCC'</td><td>In the name of God. The relics of the holy saints Seruandus, Germanus, Saturninus, Iuste, Rufinus the martyr, and John the Baptist are buried here, on the nonas of June, in the 33rd year of Pimenius' pontificate. Era 700.</td></tr>
</table>

Royal				
AEHTAM 15/ IHC 143/ ICERV 314	San Juan de Baños, Palencia	652	*Precursor D(omi)ni martir Baptista Iohannes / posside constructam in eterno munere sede(m)/ quam devotus ego rex Reccesvinthus amator/ nominis ipse tui propio de iure dicavi / tertii post dec(imu)m regni comes inclitus anno / sexcentum decies era nonagesima novem.*	Precursor of God, martyr John the Baptist, take hold of this basilica constructed as an eternal gift to you; this [basilica] which I, the devout King Reccesuinth, lover of your name, have dedicated, in the third year following a decade of my illustrious reign, Era 690
IHC 24/ICERV 366		Mid-seventh century	*Chindasvintus rex pi. . . . / . . . s in enperio / Chindasvintvs. . . .*	The most pious King Chindasuinth . . . in Chindasuinth's reign
Clerical				
AEHTAM 429/ ICERV 312/ IHC 172/IHC 401/CILA 6, 213	Castillo de Bailén (Jaén)	691	*In n(o)m(in)e D(omi)ni Locuber ac si indignnus abba fecit / et duos coros ic construxit et sacra/te sunt s(an)c(t)orum D(e)i eglesie pridie idus Ma/g[ias Era DCC]XXVIIII quarto regno gl(oriosi) d(omi)ni n(o)s(tr)i Egicani.*	In the name of God, Locuber, abbot though unworthy, made this, and constructed two choruses, and the churches of God were sanctified ('*sacra/te*') 14 May, Era 729 [691 CE], the fourth year of the reign of our glorious lord [King] Égica.
Lay Aristocratic				
AEHTAM 3062/ICERV 505/ MAN Inventory Number 57768	Casa de Vildé, Soria	Sixth and seventh centuries	*† in nomine Domin[I // loc]us Anduires inlis // [femina]e qui hanc ecle // [siam] cum viro suo // [. . .]uiro inluste // [---] fecerunt // [fue] runt par //[ticipes an]nos XXX // [quat]er nutri[vera]nt filios re // [cessit] supradicta // matrona in pace // et in summa ca[s] // titate cum cert[o. . . .] //.*	In the name of the Lord. Burial place of the distinguished Anduires, who built this church with her husband, the distinguished man. . . . They were married for thirty years and raised four children. The aforementioned woman peacefully and chastely [passed away]. . .
AEHTAM 1277/ IHC 99/ICERV 157	Teba la Vieja (Córdoba)	662	*In hunc tu/mulum rÌ£equi/escit corpus / Belesari fa/muli (Chist)i · condi/ tori huius base/ lic(a)e· qui vixit in / hoc s(ae)c(u)lo ann(o) s / plus minus < --- > / recessit in pace sub/ d(ie) < --- > / era DCC < --- >.*	In this tomb rests the body of Belesarius, servant of Christ . . . founder of this basilica . . . who lived for more or less . . . years during this century . . . who rested in peace on the . . . . Day of Era 700

(*Continued*)

Table 7A. (Continued)

Inscriptions

Reference	*Location*	*Date*	*Text*	*Translation*
AEHTAM 343; HEp 11, 577; CIL II2/13, 126	Los Hitos, Arisgotas (Orgaz, Toledo)	Sixth or seventh century	*-c.4-]++ [h?]unc aditu[m] ++ [- - t] empli// [in?] trantis Xp(ist)o carme(n) re[citate?] melodum //* *lumina cum[m]Aniv.(us) et corda s[usto] llite fle(n)tes // diluat ut [cul]pas Xp(istu)s et debita laxet //* *corpora emacula[t]a servet et penetralia me(n) tis // effuget ac tetrum c[o]lubrum r[ur]su(m) q(ue)* *retrudat // quo v(e)s(tr)is absit animis inlectio eius // luxur [ia] e leporisq(ue) {v}vile [ob?]lectatio* *turpis // garru[lit] as vana fastus rerumq(ue) cupido. // glorific[ate] po(r?)tas vos* ***imn[i]s ispō(n) te**** *parē(n?)tes. // † in d(e)i n(o)m(in) e anc ecl[es]ia(m) un+[-c.3-]do fecit//* *Modified from earlier readings (e.g. '*immisi sponte*' or '*immissi sponte*') after consultation with Antonello Vilella.	. . . this entrance . . . of the temple / Entrants, recite a poetic song to Christ / Raise up your crying eyes with your hands and hearts / That Christ may wash away your sins and release your debts / And keep your mind and body pure / Frightening and driving back that loathsome serpent / So that far from your soul may be / Delight in luxury and contemptible charm / Idle gossip and desire for riches / You (who are reading this inscription), voluntarily obedient, glorify the doors with hymns. / In God's name [---] do made this church
AEHTAM 1920/ ICERV 511	Quintanilla de las Viñas (Membrillas de Lara, Burgos)	Seventh to tenth century?	*Oc exiguum exigua off(ert) D(e)o Flammola votum*	Flammola offers this humble votive to God

Collaborative				
AEHTAM 360/ CIL II 2/5, 652/ IHC 115/ ILCV 1815/ ICERV 303	Nativola, Acci, Guadix Granada	*c.* 574	*[In nomin]e D(e)i n(o)s(tr)i H(s)u Xi consacrata est // [e]clesia s(an)c(t)i Stefani primi martyris · // in locum Natiuola a s(an) c(t)o Paulo Accitano pont(i)f(i)c(e) // die (uacat 10) an(no) (uacat 1-2) d(omi)ni n(o) s(tr)i Vuittirici reg(i)s // er(a) DCXLV item consacrata est eclesia / s(an)c(t)i Iohann(is) [[Bab]]tiste (uacat c. 12) // (uacat c. 32) // item consacrata est eclesia s(an)c(t) i Vicentii // martyris Valentin(i) a s(an)c(t) o Lilliolo Accitano pont(i)f(i)c(e) // [d(ie) XI] Kal(endas) Febr(uarias) an(no) VIII gl(oriosissimi) d(omi)ni Reccaredi reg(i) s er(a) DCXXXII. // H(a)ec s(an)c(t)a tria tabernacula in gloriam trinitatis [S(an) c(t)]i(ssi)mae // cohoperantib(us) s(an)c(t) is aedificata sunt ab inl(ustri) Gu(n)dilu[ua duce?] // cum operarios uernolos et sumptu proprio. . . .*	In the name of Jesus Christ our God, this basilica of Saint Stephen, the protomartyr, was consecrated in the place of Nativola by Saint Paul, pontifex *accitano,* the . . . day of the . . . year of our Lord, King Witteric, Era 645 (577 CE). Also consecrated is the Church of Saint John the Baptist . . . also consecrated is the Church of Saint Vincent, Valencian martyr by Saint Lilliolo, pontifex *accitano*, the 11th day of the *kalendas* of February [22 January] of the eighth year of the reign of our most glorious lord King Reccared, Era 632 [594 CE]. These three sacred *tabernacula* for the glory of the most Holy Trinity were constructed with sacred assistance by the illustrious Gudiliuva, with his own labourers and from his own wealth.
AEHTAM 1727/ IHC 10/ICERV 308	San Juan (Cabra, Córdoba)	660	*Consecrâta e(st) / baselica hâec / S(an)c(t)ae Mariae / · II · K(a)l(endas) · Iunias / (a)e(ra) DCLXLVIII // / Dedicavit / hânc âede (m) / d(o)m(inu)s / Bacauda / epÌ,, (i)scÌ,, (o)pÌ,, (u) s // Fundavit eam / Altissimus / per Eulaliam / et filium eius / Paulum monÌ, ac(h)u(m) // Ara / S(an)c(t)a / D(omi)ni*	This basilica was consecrated (*consecrata e[st]*) to St Mary on the 2nd *kalendas* of June of the era 698 [660 CE]. This building was dedicated by master Bishop Bacauda. Founded for God 'the highest' by Eulalia and her son Paulus, the monk. Holy Altar of God.

8 Holy objects on the move

Relics in Constantinople between city centre and urban periphery

Nadine Viermann

Around 400 CE, the people of Constantinople witnessed an extraordinary event, the details of which are recounted in two sermons held by the city's bishop, John Chrysostom (Joh. Chrys., *Hom. Mart.* [*PG* 63, 467–478]).[1] During night-time, people gathered inside Constantinople's 'Great Church', the Hagia Sophia,[2] to then make their way in a candle-lit procession to Drypia, a suburb approximately nine miles outside the city, accompanied by choirs who sang psalms in Latin, Greek, and Syriac. The aim of this nightly gathering was to transfer the physical remains of martyrs, who remain unnamed in Chrysostom's account, from the city centre to the Thracian hinterland. At a time when Christian religious rites were almost daily business in the new capital on the Bosporus, the outstanding character of this event becomes clear as members of the imperial family left their palace to join the crowds in honouring the relics. The Augusta Eudoxia, wife of Emperor Arcadius, walked next to the casket containing the holy remains all the way to Drypia – so Chrysostom tells us. Once the relics were deposited at their new resting place, Arcadius himself paid a visit to the shrine, accompanied by his imperial guards, and laid down his crown in homage to the martyrs.

This event highlights the crucial role that relics played in the religious and civic life of Constantinople, the capital of the East Roman Empire. Throughout the Middle Ages, one of Constantinople's most outstanding features was its vast collection of relics. This collection encompassed, for example, parts of the True Cross, the Crown of Thorns, the Rod of Moses, several objects connected to the Virgin Mary, and the bodily remains of apostles, as well as of almost countless saints and martyrs.[3] That the city would turn into a veritable repository for holy relics, however, was anything but clear in the first decades after its rededication by Emperor Constantine.

When inaugurated in 330 CE, at the site of the Greek settlement of Byzantion, Constantinople was one of several imperial residences established after the Roman emperors had gradually distanced themselves from the old capital in Italy (Grig & Kelly, 2012, 3–30). Constantinople, however, would quickly come to overshadow other residences like Trier, Milan, Ravenna, Antioch, or Nicomedia. The city on the shores of the Bosporus housed a second senate, established by Constantine's son and successor, Constantius, which became equal to the one in Rome (Grig & Kelly, 2012, 14–16). After Theodosius I had made particular efforts in furthering the city's position within the Roman Empire (Croke, 2010, 241–264), the reign of his son and successor, Arcadius, saw the beginning of a sedentary type of monarchy. From the early fifth century onwards, the East Roman emperors ceased travelling the empire and personally leading the armies and instead settled in Constantinople permanently (van Nuffelen, 2012, 183–202). When the Western emperor Romulus Augustulus was deposed in 476 and 'successor states' were established on formerly Roman territories, Constantinople's importance further increased as undisputed capital of what was left of the Roman Empire (Cameron, 2012, 20–38).[4]

DOI: 10.4324/9780429427152-10

In the early days of its existence, Constantinople was by no means purely Christian but was religiously ambivalent with a remarkable balance between monuments with Christian and pagan connotations. This becomes evident in Constantine's column, topped by a gilded statue of the emperor in the likeness of the god Sol/Helios (Bardill, 2012, 28–34; Kaldellis, 2016, 714–739), and his mausoleum, later known as the Church of the Holy Apostles (Johnson, 2009, 119–129); according to his biographer, Eusebius of Caesarea, Constantine was buried there amidst cenotaphs of the 12 apostles (Eus., *VC* 3.60). Only gradually, over the course of the fourth century, did the city gain its specifically Christian imprint that would shape its character for centuries to come (Stephens Falcasantos, 2020). However, whereas other cities in the Roman Empire could build on a glorious Christian past – with apostles having once set foot into their walls or famous local martyrs – Constantinople did not have much to offer from a Christian perspective (Ousterhout, 2012). The two local martyrs, SS Mocius and Acacius, who had died for their faith during the reign of Diocletian (Wortley, 1999), could not quite compete with the prestige that was derived, for example, from Rome's connection with SS Peter and Paul.[5] As the bond between Christianity and the Roman emperors grew stronger throughout the fourth century, measures were taken to balance out Constantinople's deficit of an indigenous Christian history. In order to enhance the city's Christian prestige and to form a specifically Christian identity, relics were brought from various places across the Mediterranean (Klein, 2006, 81–88). The first holy objects to arrive in Constantinople were the remains of saints Timothy, Luke, and Andrew, most likely in the years 356/7. On the initiative of Emperor Constantius II, they were placed in the Church of the Holy Apostles (Jerome, *Chronicle* [ed. Helm], 240–241; *Chron. Pasch.* a. 356–357 [Dindorf, 542]; Mango, 1990; Wortley, 2009, Ch. IV). After this practice had initially been established, the process of translating relics quickly picked up pace; various objects like the head of St John the Baptist or the relics of his father Zacharias were brought into the city, welcomed by the metropolitan people and deposited in one of the churches which by then characterised the urban space (Diefenbach, 1996, 43–48; Klein, 2006, 81–88). The formerly blank slate turned into a densely populated map, a Christian sacred topography structured by enshrined relics.

Throughout the Middle Ages, Constantinople remained the capital of the East Roman or Byzantine polity with a remarkable degree of resilience when facing geopolitical changes, external threats, and substantial reductions in territory (Haldon, 2016); its citizens bought into commonly shared ideas of divine protection (Pentcheva, 2002; Baynes, 1949). In this context, scholars have emphasised the importance of liturgical processions, rituals, and the collective engagement with relics to strengthen social cohesion in medieval Constantinople.[6] While acknowledging these strands of research, this chapter asserts that we should not take the presence of relics in Constantinople for granted, but instead problematise the very beginnings of the city's famed relics collection. As evidenced in the above overview, it was during Late Antiquity that the city established patterns of transferring and enshrining these holy objects. Recent research has highlighted the dynamic and formative character of Late Antiquity with regard to the beginnings of the cult of relics across the Mediterranean and its relevance to popular religion (Wiśniewski, 2019) as well as the creation of Christian sacred space (Yasin, 2009, 2015). Within this context, this chapter reads the translation of relics to Constantinople as highly staged acts that furthered a process of integration of the city's sacred topography as well as civic community.[7]

Relics, once having arrived in Constantinople, were far from static. Instead, they could be carried around the city in processions or moved in-between shrines as the opening episode of relics being transferred to Drypia suggests. It is the latter aspect, the relics' movements between different churches within and around Constantinople, that this chapter focuses on. On the basis of three case studies – the relics of the prophet Samuel, the aforementioned relics of the unnamed martyrs, and the robe of the Virgin Mary – I will explore the movement of relics within Constantinople and its hinterland and their impact on the formation of an entangled sacred topography in Late Antiquity. A careful contextualisation of each episode within the religious and political climate of late antique Constantinople enables me to showcase the impact of relics on forging civic cohesion.[8]

8.1 The relics of Samuel

On 19 May 406 CE, the physical remains of Samuel, the Old Testament prophet, were brought from Jerusalem to Constantinople. In his treatise *Contra Vigilantium*, the church father Jerome, based in Jerusalem at that time, describes how Samuel's bones made their way from the Holy Land to their destination: '[The people of all the Churches] welcomed them with as much joy as if they beheld a living prophet in the midst of them, so that there was one great swarm of people from Palestine to Chalcedon' (*Contr. Vigilantium* 5).[9] The *Chronicon Paschale*, from the seventh century, gives further information on the reception in Constantinople:

> [T]he remains of St. Samuel were conveyed to Constantinople by way of the Chalcedonian jetty . . ., with Arcadius Augustus leading the way, and Anthemius, praetorian prefect and former consul, Aemilianus, city prefect, and all the senate; these remains were laid to rest for a certain time in the most holy Great Church.
>
> (*Chron. Pasch.* a. 406 [Dindorf 569])[10]

From these texts, we can gather that the translation of Samuel's relics was a remarkable public event which caught the peoples' attention, not only in Constantinople but all along the way. The bones, wrapped in a precious container, were considered to have the same impact as the prophet himself; Samuel was believed to be present among the people, settling down in his new resting place, Constantinople. Once the procession had reached Chalcedon, none other than Emperor Arcadius, together with the highest urban magistrates and the senate, took the opportunity to welcome the revered guest and lead him into the Great Church, Hagia Sophia, in the city centre.

Yasin (2015, 133–151) describes the translation of relics along anthropological categories as *rite de passage*, a tripartite sequence of 'separation, transition and incorporation', which also applies to the relics that were brought to Constantinople. The most poignant visual expression for a relic's transition and incorporation is the famous ivory carving that is now part of the cathedral treasure of Trier (Germany). Although the ivory's date remains debated, with hypotheses ranging from the fifth into post-iconoclast centuries, it is commonly interpreted to depict the arrival in Constantinople of the relics of the first Christian martyr, St Stephen (Acts 7:54–60), arranged by the Augusta Pulcheria, sister of Theodosius II.[11] This ivory conveys the public character of the event (indicated by the crowds of attending people), the involvement of members of the imperial family, and the relic's impact on the urban space, in that it is about to be put to rest in a church built for this very purpose – in this case, if we follow the above interpretation, the newly constructed church of St Stephen inside the imperial palace (Theophanes Confessor *AM* 5920 [De Boor 86–87]; Janin, 1969, 473–474). The way in which the Trier Ivory depicts the relics being welcomed and incorporated into the city's fabric can be seen as emblematic of the practice of translating relics in general; many other holy objects would be received in a similar, highly ceremonialised manner, reminiscent of the imperial *adventus* – with the difference that it was not the emperor who was arriving in the city but the relic (Holum & Vikan, 1979). In that sense, the act of translation can be assessed as a ritual in the narrower sense, as it effected a change of status both for the relic and for the space into which the relic was incorporated: the formerly movable object became part of an architecturally defined space, elevating this very space due to its sacred presence.[12] For many churches, also in Constantinople, the relic was foundational not only metaphysically, in that the church's dedication was based on it, but also in a more concrete sense, being interred into the church floor or incorporated into the altar to become an integral part of the architecture.[13]

When Samuel's relics arrived in Constantinople in 406, however, the stage of incorporation or interment was not immediately completed. Another entry in the *Chronicon Paschale* tells us that five years later, in 411, Samuel's remains were again transferred; this time into what was intended to be their final resting place – a newly built church in the extramural quarter of Hebdomon (*Chon. Pasch.* a. 411; Dindorf 570–571), today's Bakırköy neighbourhood (Istanbul). The Hebdomon was located at the seventh mile outside the city centre, at the *Via Egnatia*, the core connection between Constantinople and the West; it housed a military camp and gathering ground where several emperors, especially in the fourth and fifth centuries, were proclaimed. Over the centuries, the quarter was furnished with a triumphal column for Theodosius II and an imperial palace, the *Iustinianae* (see Figure 8.1; Simeonov, 2016, 121–138; Taddei, 2014, 77–84; Bardill, 1999, 693–694; Demangel, 1945). Samuel's relics were not the first ones to be venerated at Hebdomon, as Theodosius I had already commissioned a church for the head of St John the Baptist in direct vicinity to the *Via Egnatia* to augment the religious portfolio of this extramural quarter (see Soz. 7.21; Theodor Anagnostes 268; *Chron. Pasch.* a. 391 [Dindorf 564]; cf. Diefenbach, 2002, 26–27).[14]

The exact location of the Church of St Samuel and its relation to the Church of St John unfortunately cannot be determined anymore. However, the evidence presented here suggests that Arcadius followed in the footsteps of his father, Theodosius I, in further developing the

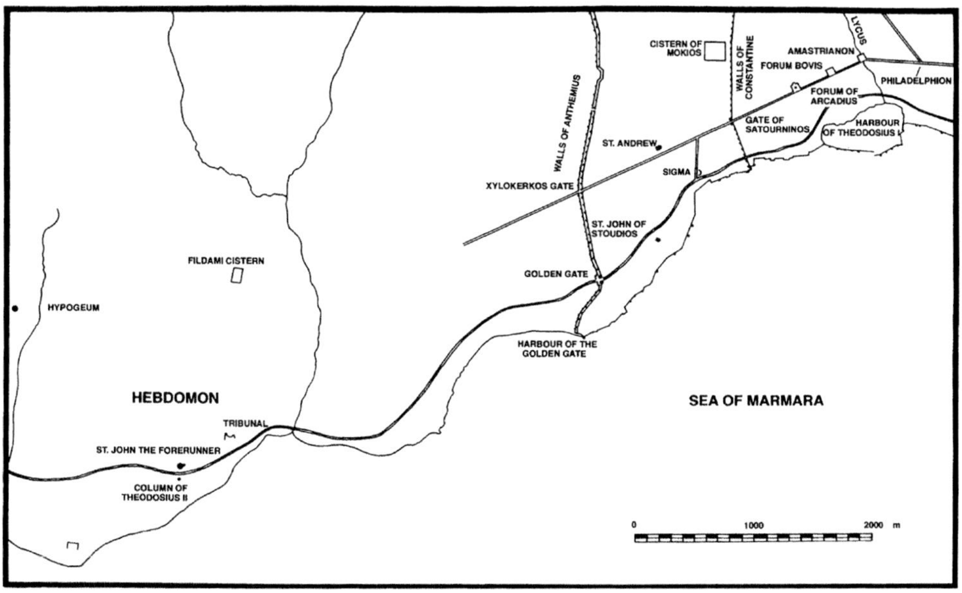

Figure 8.1 The quarter of Hebdomon and the southern end of the Theodosian Wall.

Source: From Bardill (1999, 693, Figure 17). Reprinted with permission

Hebdomon not only as military ground but also as an area with religious significance. The area offered a first impression for those approaching the city by land, a space to represent the emperors and the character of the city that travellers from the West were about to enter. By translating Samuel's remains from the city's main church into the suburban region of Hebdomon, a clear connection was drawn between urban centre and periphery by means of a holy object. Although the sources do not give any detailed information as to how Samuel's translation to Hebdomon was performed, we can assume that it was just as much of a public and lavish event as the other translations of which we are informed better. At a time, in the early fifth century, when neither the religious nor the political status of Constantinople had been entirely consolidated, the Hebdomon was established as a figurehead to represent Constantinople's ambition as Christian capital. Arcadius, being the first emperor to spend his reign in Constantinople instead of travelling the empire, had a clear interest in furthering the city's profile.

The significance of Samuel's translation, however, goes beyond bolstering the religious prestige of a suburban quarter and extending the city's sacred topography, as the relic's temporary storage in Hagia Sophia occurred during a turbulent time in the church's existence. Dedicated in 360 under Constantius II as the city's main church, Hagia Sophia was burned down when Constantinople was shaken by riots that were sparked by the deposition of Bishop John Chrysostom (Janin, 1969, 457). This controversy had its roots in competition for ecclesiastical power and influence (Liebeschuetz, 1990, 157–222). Chrysostom, who had been appointed bishop of Constantinople in 397, was a hugely popular preacher and intervened in doctrinal tensions that characterised the religious environment of the capital (Stephens Falcasantos, 2020, 110–131); he soon faced opposition within Constantinople and empire-wide. In 403, the Synod of the Oak deposed Chrysostom, based on the accusations of Origenist thinking. Although popular pressure in the capital forced Arcadius to recall him after only one day, the dispute could not be settled (Allen & Mayer, 1999, 7–10). Finally, Chrysosostom alienated the empress Eudoxia, who had once been his great admirer (Busch, 2015, 71–82). On 20 June 404, Arcadius issued a decree that again sent Chrysostom into exile, where he would die in 407 (Allen & Mayer, 1999, 10–11). The bishop's second exile was answered by a violent uprising of his supporters which left parts of the city centre, including the senate house and the Great Church, in ruins (*Chron. Pasch*. a. 404 [Dindorf 568]; Whitby & Whitby, 1989, 59; Soc. 6.18; Soz. 8.22). It took more than a decade, until 415, for the restoration and rededication of the church under Theodosius II – again in the form of a timber-roofed basilica (Janin, 1969, 457).

It is difficult to assess how severely Hagia Sophia had been destroyed in 404 – at least some parts of the building complex, like the detached *skeuophylakion*,[15] seem to have remained unharmed (Mathews, 1971, 11–17). Whatever the building's exact state, the fact that Samuel's remains were temporarily placed in a church that bore the scars of the dispute over John Chrysostom makes the act even more noteworthy. The *Chronicon Paschale* gives further hints as to the character and purpose of this act. According to this account, it was the emperor, together with urban, civic magistrates, who took centre stage in welcoming the relics into Constantinople. In contrast to other translations, there is no indication of clerics or the city's bishop playing a prominent role in the ceremony.[16] Atticus, a staunch opponent of Chrysostom who had become bishop of Constantinople in March 406 after a brief interplay of the aged Arsacius (Venables, 1911, 69), is nowhere mentioned – which is telling, given the chronicle's inclination to emphasise religious matters and personnel. If we assume that St Samuel's reception in Constantinople was performed by the emperor and secular magistrates exclusively, this might point towards an imperial policy that attempted to reconcile rather than rekindle the strife around ecclesiastical hierarchies that had left its clear traces within the urban fabric. John Chrysostom's exile had far from solved this dispute; even years later, Constantinople was still divided with regard to church political allegiances with substantial parts of the population opposing Atticus whom they considered illegitimate (Nikeph. Kall. 14.23, 27; Venables, 1911, 69). The way in which the translation of Samuel's relics was performed should be understood as an attempt to highlight common civic ground, represented by emperor, magistrates, and senate; as such, the ritual would have strengthened the devotion to the city's main church, which could be seen as a symbol of unity rather than divide, instead of offering the grand stage to the controversial bishop Atticus. The fact that the church lay partially in ruins made it even more important to emphasise its character as core of Constantinople's sacred topography.

Right before the rededication of Hagia Sophia under Theodosius II on 10 October 415, the church received another set of relics: the remains of Joseph, son of the Old Testament patriarch Jacob, and Zacharias, father of John the Baptist. This time, according to the *Chronicon Paschale*, it was the city's bishop Atticus, together with Moses, the bishop of Antaradus in Phoenicia, who carried the relics into the city, accompanied by Ursus, the city prefect, and the senate (*Chron. Pasch.* a. 415 [Dindorf 572–573]).[17] More than ten years after Chrysostom's exile, this change in protagonists – from the emperor and civic magistrates to the bishop leading the way – might point towards progress in reconciling the schism that had divided Constantinople. For the relics of Joseph and Zacharias, there is no explicit reference as to the relics' final destination, but it is highly likely that Hagia Sophia again only served as a temporary resting place (Wortley, 2009, 632–634). Evidence for the existence of a Church of St Zacharias *c.* 474 implies that the relic was transferred there to serve as an eponymous, foundational token (Janin, 1969, 133; Wortley, 2009, 634). In 406, the arrival of Samuel's relics must have emphasised the affected state of Hagia Sophia in the aftermath of the violent clashes around John Chrysostom's exile. Almost ten years later, the reconstruction works were completed and the translation of Joseph's and Zacharias's remains could be associated with the eventual recovery of the urban fabric.

Different from most other churches in Constantinople, Hagia Sophia was not founded on relics. As a dedication to God's Holy Wisdom, the second person of the Trinity, no relic could have adequately embodied its essence (Wortley, 2009, Ch. XVI).[18] During Late Antiquity, as well as later centuries, Hagia Sophia was repeatedly used as a temporary space for relics that were either stored there before being distributed to other, less famous, shrines or brought into the church to be venerated on special feast days.[19] The translation of Samuel's relics, in connection with those of Joseph and Zacharias, as well as those of the nameless martyrs further discussed later, indicate that Hagia Sophia in the city centre served as a sacred hub for collecting and dispersing relics. In the early fifth century, when the city's sacred topography was still very much in the making, the ritual of translating relics from Hagia Sophia to their final resting place in the urban periphery not only emphasised the Great Church's status within Constantinople's ecclesiastical fabric but also established a sacred network between city centre and hinterland, which was commemorated in the city's annals and kept alive in Constantinople's liturgical calendar.

8.2 The unnamed martyrs

The introduction to this chapter has already outlined the translation of unnamed martyrs' relics from Hagia Sophia to the suburban Drypia around 400 CE,[20] as evidenced in two sermons by John Chrysostom, who was himself an eyewitness of the event and at this point still on good terms with the imperial couple. The sermons give a vivid impression of the festivities that stretched over several days and attracted a huge and diverse crowd: the people of Constantinople, magistrates, and imperial guards. Even more so than the brief account on the translation of Samuel's remains in the *Chronicon Paschale*, John Chrysostom's sermon emphasises the active role of members of the imperial family. After recounting how the magistrates rubbed shoulders with the common people, Chrysostom turns towards Empress Eudoxia:

> Yet why should I speak of women or magistrates, when even she who wears the crown and is dressed in the purple did not suffer to be separated even a little from the relics along the entire journey, but followed the saints like a maid, touching the casket and the veil which covered it. Suppressing all human pride, she has allowed herself to be seen by so many people in the midst of the crowd – she whom not one of the eunuchs serving in the imperial court is allowed to look at.
>
> (*PG* 63, 469)[21]

The lofty empress, usually concealed within the imperial palace, not only took part in the procession but also ostentatiously lowered herself, walking besides the relics to showcase her humility and piety. On the following day, Arcadius joined the celebrations together with his imperial guards:

> Even the man who grants benefactions to every person in the world came to enjoy benefactions from these saints and to reap the greatest goods. For this reason, then, he put down his crown, and so did all his guards – some their spears, others their shields – and, putting aside all that fancy, all of them came with a humbled mind, as if entering from earth into heaven.
>
> (*PG 63*, 473)[22]

Like his wife, the emperor demonstrated his piety by temporarily leaving behind his elevated status – in this case by taking off his crown and letting go of his military guard.

This instance shows how the imperial family could use the translation of relics as an opportunity to connect with Constantinople's populace. As has often been pointed out, this was particularly important at a time when the emperors permanently settled in Constantinople instead of travelling the provinces (Diefenbach, 1996; cf. Pfeilschifter, 2013).[23] Arcadius, who had come to power, aged 18, after the death of his father, Theodosius I, depended on keeping a good relationship with the metropolitan people – a group that gained political weight as the city emerged as second (and later foremost) capital to the Roman Empire. The translation of relics had, one can conclude, not only a religious purpose but also a political purpose: it showcased both the emperor's piety and his *civilitas*, his care for and intimate relationship with the metropolitan people (Diefenbach, 2002, 31–39). In the context of translating relics, imperial women could also enter the spotlight and, with their ostentatious piety, contribute to the prestige of the ruling family. Eudoxia was far from a passive counterpart to her husband; her virtuous appearance in the procession to Drypia is very much in line with other active interventions in political and religious matters of the capital (Busch, 2015, 59–85). On the aforementioned Trier Ivory, it is also the empress who takes centre stage in receiving the relics in front of her newly built church.[24] Already a few years before the reception of Samuel's relics in Hagia Sophia, the procession to Drypia highlights how Arcadius and his family used the act of translating relics as an opportunity to engage with the public in their capacity as political *and* religious figureheads of the capital.

The translation of relics to Drypia should be understood as a highly significant undertaking also in spatial terms, at a time when Constantinople did not yet host the abundance of relics that it would later become famous for. Compared to the Hebdomon, the settlement of Drypia, most probably located nine miles west of the city in the midst of the peninsula (today's Bağcılar neighbourhood; Külzer, 2008, 340), is far less well explored. All we know it is that it housed a Church of St Thomas the Apostle, built in the late fourth century.[25] Unlike Hebdomon, Drypia did not have a port nor was it adjacent to one of the main access roads into Constantinople. The Hebdomon was, in a way, a flagship suburb for everyone coming from the West and an ideal space to give a preview into the city's character, its political and religious endowment; the translation of Samuel's remains, discussed earlier, further emphasised the area's prestige. In contrast to that, the translation of relics to Drypia seems to have been less directed towards a broad supra-regional public but rather towards those living in and around Constantinople. As the city gained importance as a permanent residence of the East Roman emperors, it looks as though explicit efforts were made in furthering the connection between city centre and hinterland by means of transferring Christian tokens.

The purpose of translating relics from Constantinople to Drypia, however, might only be fully grasped when read in the context of the political and military crisis that shook Constantinople and Thrace around 400 CE. In the late fourth century, the eastern part of the Roman Empire saw a line of Gothic incursions, in part triggered by the Empire's need to recruit troops (Kulikowski, 2006). At the Battle of Adrianople in 378, Gothic federations caused a severe blow to the Roman army; Emperor Valens perished on the battlefield together with a sizable number of his troops. In the aftermath, Theodosius I, having been proclaimed emperor to fix the military debacle, was left with no option but to form alliances with Gothic federations who settled on Roman ground (Liebeschuetz, 1990, 7–85; Kulikowski, 2006, 123–153; Meier, 2019, 171–190). This, however, was not the end of military clashes between Goths and Romans. After Theodosius's death in 395, Gothic contingents under the lead of Alaric ravaged Thrace and threatened to lay siege on Constantinople (Kulikowski, 2006, 165; Meier, 2019, 198). Meanwhile, the Roman army and administration itself relied heavily on Gothic generals, one of whom, Gainas, used the precarious situation in the last years of the fourth century to consolidate his position in Constantinople. In 395, Gainas instigated the assassination of the praetorian prefect, Rufinus, éminence grise behind the young emperor Arcadius's back; in 399, he pressured Arcadius into deposing his closest confidante, the *praepositus sacri cubiculi* and consul Eutropius (Liebeschuetz, 1990, 89–105). Backed by his forces that had gathered in Chalcedon, on the Asian side of the Bosporus, Gainas demanded the highest military commands and the dismissal of further imperial officials in the following year; Arcadius had no option but to concede. Subsequently, Gainas led parts of his troops into Constantinople and established a veritable military regime (Faber, 2011, 129–132; Kulikowski, 2006, 168–169). While the emperor had no capacities to oppose the warlord, the people of Constantinople concertedly turned against Gainas and massacred the city's Gothic population on 12 July 400. Beyond its political and military contexts, the conflict bore religious significance as the violence escalated between the city's Nicean populace and the Arian Goths (Liebeschuetz, 1990, 111–125). Gainas, who had already left Constantinople before the massacre, withdrew further into Thrace, was defeated by Roman troops later that year, and eventually killed by Hun forces under Uldin (Cameron & Long, 1993, 199–252).

The translation of relics from Constantinople to Drypia is not securely dated, but the details of Chrysostom's sermon suggest that it was conducted in the window between Eudoxia's proclamation as Augusta on 9 January 400 (*Chron. Pasch.* a. 400 [Dindorf 567]) and 403, when the relation between empress and bishop began to sour (Holum, 1982, 56 with n. 35).[26] Hence, a date in the immediate aftermath of the Gothic crisis is likely. With this in mind, the choicc of a rather undistinguished place in Thrace like Drypia reveals a certain rationale. The Thracian hinterland had, at that point, been shaken by enemy incursions and civil-war-like circumstances for decades. After the aforementioned massacre, Gainas's remaining Goths briefly settled outside

Constantinople, before marching further west (Cameron & Long, 1993, xii). The sources do not specify where exactly Gainas set up camp; however, one might hypothesise that it was in or around Drypia, and that the translation of relics from Constantinople was a means to make repairs for what the settlement had endured.

Beyond the disruptive effects on the Constantinopolitan hinterland, the Gainas-crisis, as outlined earlier, had challenged imperial authority over the military as well as the civic space of Constantinople. In its aftermath, Arcadius sought to assert his independence from civic and military backers and stabilise his position in the capital. The most sustained statement of imperial authority was the Column of Arcadius that commemorated the destruction of Gainas and his Goths with an elaborate narrative frieze, turning the crisis into an episode that accentuated Arcadius's triumphal capacities. Work on the monument, which was modelled on the Column of Trajan in Rome and the Column of Theodosius in Constantinople, started in 402/403 as part of the new Forum of Arcadius; it was completed in 421 when Theodosius II crowned the column with a statue of his late father who had died in 408 (Liebeschuetz, 1989). If the most visually impressive, the column was certainly not the only means for Arcadius to cope with the disruption caused by Gainas. The religious festivities in Drypia offered the emperor an opportunity to connect with the population whose intervention had been crucial in breaking the Gothic influence over the city. In the context of inner-Christian tension, the ceremony, led by members of the imperial family as well as the city's bishop, John Chrysostom, asserted Nicean faith after the Arian Goths had been expelled. Finally, the act of translating precious relics from the city centre to the urban periphery can be read as an attempt to strengthen the sacred topography of those regions that, lying outside the city walls, had suffered heaviest from recent military action. In this context, Hagia Sophia again served as sacred hub from which holy objects were dispersed. Enhancing the network of relics not only extended Constantinople's sacred topography, but it also created a transcendental line of protection in the Thracian hinterland that was most exposed to attacks. That relics had the capacity of building a protective girdle around Constantinople becomes even more obvious with regard to the next case to be discussed here.

8.3 The Virgin's robe

Whereas the first two cases discussed here have covered bodily relics, the third case deals with what can be labelled as a secondary or contact relic, the robe or veil (ἐσθής) of the Virgin Mary, stored in a Church dedicated to the Theotokos, the mother of God.[27] This church was situated in Blachernae at the shores of the Golden Horn, in the liminal area right outside the Theodosian walls (Figure 8.2; Mango, 1998). The event I am going to focus on, in which the Virgin's robe was first removed from and then returned to its Blachernae shrine, took place in the early seventh century. When Constantinople had firmly secured its status as undisputed political and religious centre of the Eastern Mediterranean, the devotion to the Virgin Mary became an ever more important aspect of the city's collective identity, centred around the belief that the city was under her special protection (Cameron, 1978, 79–108, 1979, 42–56; Mango, 2000; on the long-term development, see Pentcheva, 2002, 2006). As will become evident from the following analysis, however, it was a situation of utmost threat that brought the Virgin's cult to its full bloom.

Before delving into the early seventh century, however, a couple of words need to be said on the Marian church in Blachernae and its precious content. The initial arrival of the Virgin's robe in Constantinople and the establishment of the Blachernae shrine are covered in two conflicting traditions. According to a textual tradition reaching back, most probably, to the late fifth century, the Virgin's robe was clandestinely brought from the Holy Land to Constantinople under

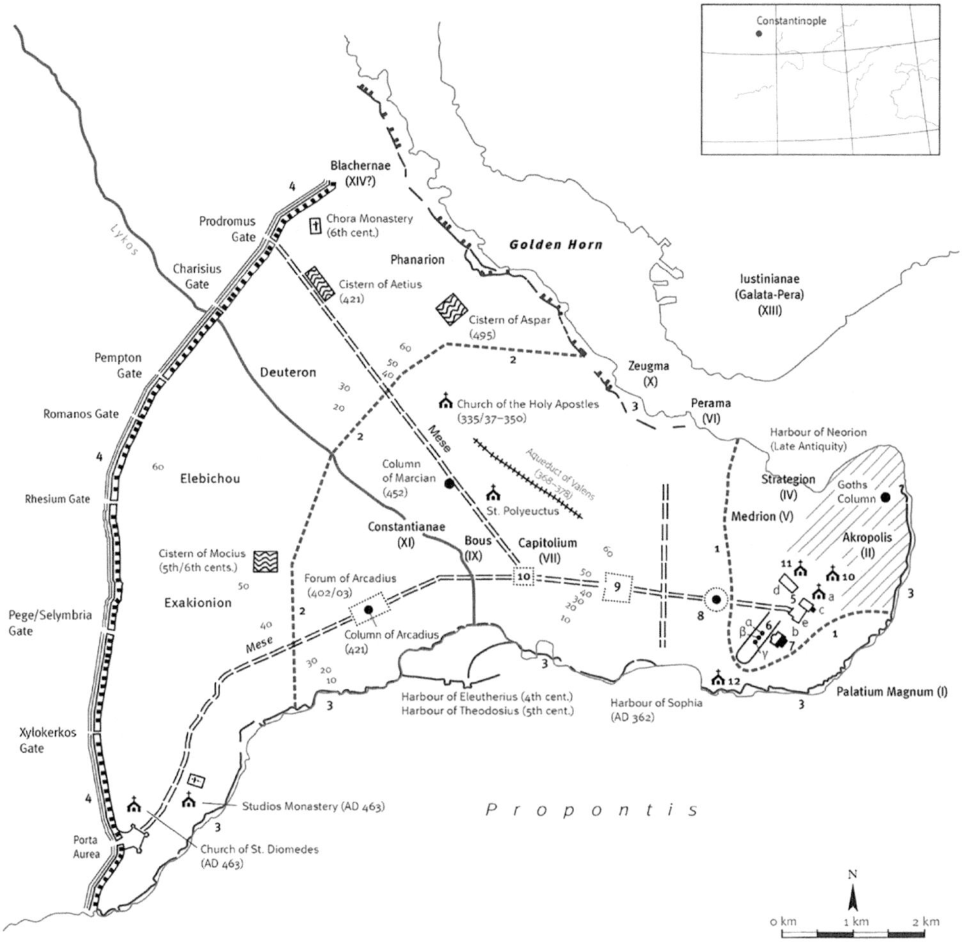

1 Land wall (conjectured course) at the time of the Emperor Septimius Severus (193–211)
2 Land wall (conjectured course) at the time of the Emperor Constantine I (306–337)
3 Sea wall at the time of the Emperor Theodosius II (408–450)
4 Land wall at the time of the Emperor Theodosius II (408–450)
5 Augusteion
 a Hagia Sophia
 b Imperial palaces
 c Senate
 d Administrative basilica
 e Baths of Zeuxippus
6 Hippodrome
 α Egyptian obelisk of Theodosius I (390)
 β Serpent Column from Delphi
 γ Walled obelisk (10th c.)
7 Palace of Lausus (5th c.)
8 Forum Constantini with column of Constantine
9 Forum Tauri (Forum of Theodosius I)
10 Hagia Eirene (4th c.)
11 Church of the Theotokos Chalkoprateia (5th c.)
12 Church of St. Sergius and St. Bacchus (527–536)

Medrion (V) Theodosian regio
Deuteron district

Acropolis of Byzantium
Church
Monastery
Cistern
Main Street

Figure 8.2 Constantinople in Late Antiquity.

Source: Courtesy of Laura Schröps

dubious circumstances by two patricians, Candidus and Galbius, during the reign of Leo I and, at first, deposited in a church dedicated to SS Peter and Marc in the Blachernae area (Mango, 1998, 67–68; Shoemaker, 2008, 56–61 following Wenger, 1955, especially 113–136).[28] After Candidus's and Galbius's death, Emperor Leo, together with his wife Verina, enclosed the relic in a precious reliquary and stored it in a church dedicated to the Theotokos (Loparev, 1895, 592–612; Wenger, 1955, 134–135, 300–303); however, it remains somewhat unclear as to whether Leo just enhanced and rededicated an already existing church or built a new one. In contrast to that, a second tradition ascribes the Blachernae Church to Empress Pulcheria, dropping Candidus and Galbius altogether. This tradition is attested in Theodor Anagnostes's ecclesiastical history, which, originally written in the early sixth century, only survives in a Middle Byzantine epitome (Theodore Anagnostes, 367). What makes this evidence dubious is that none other late antique source credits Pulcheria for sponsoring the Blachernae Church. Hence, the ascription could have been added by the epitomist at a time when a connection between Pulcheria, who was known for her sponsorship of churches, and the Blachernae Church had become popular (Mango, 1998, 65–66). Evidence of this connection is the so-called *Euthymiac History*, which can be firmly traced back to around 750 CE (Mango, 1998, 66–67; Shoemaker, 2008, 67–68; largely based on Wenger, 1955, 136–139),[29] as well as other middle and late Byzantine sources (e.g. Theophanes Confessor AM 5943 [De Boor 105]). Overall, the evidence makes it more likely to connect the Theotokos Blachernae not with Pulcheria but with Emperor Leo I and his wife Verina; the Pulcheria tradition seems to be a later fabrication (cf. Wenger, 1955, 138–139).[30] One could assume that – at a time when the Theotokos had been firmly established as divine protectress of Constantinople – a need was felt to gloss over the dubious origins of the Marian relic. To that end, a tradition might have been forged that not only emphasised the imperial connection but also ascribed the cult side to Pulcheria, who was famed for her exceptional piety and Chalcedonian orthodoxy and had remained a virgin herself (even in her late marriage to Marcian; Busch, 2015, 110–135). In contrast to that, Verina was more of a controversial figure as she had been involved in the power struggles of the later fifth century around Emperor Zenon and was sister to the usurper Basiliscus, an advocate of Miaphysitism (Twardowska, 2014).[31]

The history of the Blachernae Church is much clearer for the sixth and seventh centuries, when the Blachernae quarter received further imperial attention. According to Procopius, Justin I built a basilica on the site with two aisles and a gallery, about 50 metres long (Proc. *De aed.* 1.3.3–1.3.5). Parts of the building were restored under Justin II, as documented by two epigrams in the Palatine Anthology (*Anth. Palat.* 1.2–1.3). In addition to the basilica just mentioned, the sacred enclosure also housed a bathhouse and a martyrion with a centralised ground plan, which the sources refer to as ἡ ἁγία Σορός, containing the Virgin's robe (*De cer.* 1.27, 1.34, 2.9, 2.12; Mango, 1998, 63, 76). The bathhouse was probably supplied by the natural well that is still part of today's church, a nineteenth-century reconstruction.[32] Cyril Mango convincingly suggested that the Soros, which seems in itself to have been of considerable size, was the original church built in the fifth century to incorporate the relic (Mango, 1998, 63, 65). North of the church precinct, separated by a steep slope and terrace walls, lay an imperial palace, the oldest parts of which can be traced back to the reign of Anastasius (Müller-Wiener, 1977, 223–224). In 627, Emperor Heraclius ordered an extension of the land wall to be built, incorporating the formerly extra-mural Blachernae quarter into the massive fortifications (*Chron. Pasch.* a. 627 [Dindorf 726]; Müller-Wiener, 1977, 301–307).[33] This measure was meant to protect the area from enemy threats, especially the Avars, whose incursions had made it necessary for the Virgin's robe to be temporarily removed from its Blachernae shrine a couple of years prior. It is to this incident, the temporary removal of the Marian relic, that I now turn.

In the late sixth and early seventh centuries, the Roman Empire was shaken by the incursions of Avar tribes, who regularly crossed the Danube and pillaged the Roman province of Thrace (Pohl, 2018). In 623, Heraclius made efforts to conclude a peace treaty with the Avars, hoping to appease the situation in the west as he was preparing a counter-offensive against the Sasanian Persians under Khosrov II who had occupied the Roman provinces of Syria, Palestine, and Egypt.[34] A summit meeting was arranged between Heraclius and the Avar Khagan, which was supposed to take place in Heracleia, a city approximately 100 kilometres west of Constantinople on the coast of the Marmara Sea (*Chron. Pasch.* a. 623 [Dindorf 712]). According to the *Chronicon Paschale*, Heraclius left Constantinople with great pomp and headed west, accompanied by metropolitan dignitaries and carrying lavish gifts; but before he could reach Heracleia, where he planned to hold circus games, the imperial retinue was surprised by an Avar attack. Heraclius had to rush back into the Constantinopolitan fortification to save himself. The Avars, meanwhile, proceed to the outskirts of Constantinople with the aim of plundering the rich suburban areas (*Chron. Pasch.* a. 623 [Dindorf 713]).[35]

Facing this threat, the Virgin's robe was removed from the still unfortified quarter of Blachernae, together with other precious objects, and brought into to city for safekeeping. After having been kept in the sacristy of the Hagia Sophia for several month, the relic was returned to the Blachernae Church as the acute danger had passed (Cameron, 1979). A detailed insight into this line of events is given in a sermon, probably composed by Theodor Synkellos, a high-ranking cleric and eyewitness, which celebrates the return of the holy object to its accustomed place.[36] After recounting the story of how the relic was initially brought to Constantinople by the patricians Galbius and Candidus in the fifth century, the author turns to his own time to remember 'the miracles with which the Theotokos at Blachernae endowed the city' (Καὶ ταῦτα μὲν ἅπερ ἡ θεοτόκος ἐν Βλαχέρναις ἐθησαύρισεν τῇ πόλει μυστήρια). According to Theodor, the men who had been sent out to salvage the precious goods from Blachernae did not pay the adequate respect to the holy relic, as:

> they dared actually to lay hands on this divine casket, and to bring into the light then the wonder previously hidden from all. And inside the visible casket, which is made of gold and silver, was found a stone casket, shining with brightness, and inside this at the top was found lying the divine treasure, preserved safe in another little casket.
>
> (Cameron, 1979, 50–51; Loparev, 1895, 596–597)[37]

This passage bears crucial information on how the relic was stored within the Blachernae Church: what was visible to the eyes of the congregation was a preciously adorned casket, which seems to have simultaneously served as an altar (Wenger, 1955, 132–133). The relic itself, the piece of fabric believed to be the Virgin's veil, was concealed within an ensemble of three containers, encapsulated like a Russian doll: the visible casket made of gold and silver; one made of stone; and finally a *sorion*, a casket which seems to have been substantially smaller than the two outer layers. Even though the actual relic was, after its incorporation into the Blachernae shrine, hidden from the eyes of the congregation, its presence was clearly indicated to visitors by the precious reliquary. Theodor's description is particularly valuable as virtually no material evidence is left of late antique reliquaries from Constantinople (Klein, 2015). Archaeological evidence from other regions of the late antique Mediterranean suggests that, after the translation to their allocated resting place, relics were usually hidden away – in precious containers, which could be part of or connected to an altar, or even be incorporated into the floor (Yasin, 2015). In this context, the way the Virgin's robe was deposited in Blachernae according to the sermon seems quite typical. The practice of keeping a relic in several containers stacked into each other is corroborated by archeological findings from other parts of the empire, which preserve similar arrangements *in situ*. A particularly spectacular find is a set of three reliquaries, dating to the mid-fourth to fifth centuries, which has been discovered under the pavement of a church apse (thus, in this case, invisible to the congregation) and is today exhibited in the museum of Varna (Bulgaria).[38]

Evidence discussed in the previous sections of this chapter conveys that the translation of a relics was, in most cases, a public and highly ceremonialised act, during which the holy object was displayed to the multitude of people who took part in this event; the relic, or rather the container in which the relic was stored, was seen and in some cases even touched by the onlookers. During the act of translation, the wonder working powers of the relic could be felt. Communal affirmation played a crucial role in authenticating the object's sacred value before it was, in most cases, withdrawn from direct public access in the third stage of its rite of passage, the incorporation or interment. In the case of the Blachernae robe, however, there seems to have been, in Late Antiquity, no public memory as to a grand reception of the relics in Constantinople that could have been compared to the ones discussed earlier. The arrival of the relic, as recounted in Theodor's sermon, was, despite later imperial involvement, a clandestine act conducted by two rather obscure men, Candidus and Galbius. No public act of authentication seems to have occurred at the time of the relic's original deposition – at least it was not remembered. The incident in 623, when the Virgin's robe had to be removed from the Blachernae Church for safekeeping, presented, in a way, an opportunity to make up for this deficiency.

After having been removed from Blachernae, the innermost casket with the Virgin's robe was brought to Hagia Sophia, as Theodor's sermon tells us. Several weeks later, as the Avars had withdrawn, the patriarch Sergius ordered the relic to be restored to its accustomed resting place:

> He called together in a lofty decree all the assembly of patriarchs, the clergy and the laity, men and women, and those holding offices and ranks, and all who lived in private life. 'Here, priests and people', he said, 'Come and see the treasure till now hidden'.
>
> (Cameron, 1979, 51; Loparev, 1895, 599–600)[39]

From Hagia Sophia, the Virgin's robe was first brought into the Church of St Laurence for veneration, still concealed in its reliquary (Cameron, 1979, 52).[40] Once it had finally reached the Blachernae Church, the patriarch, in front of the masses of people that pressed into the building, opened the casket where:

> [h]e found imperial purple wrapped with myrrh and other perfumes. And opening that he found the robe of the true empress, the Mother of God, shining out with her own grace and power. . . . The divine robe, woven from perishable wool . . . had suffered no destruction as all, but was completely intact, whole and indestructible.
>
> (Cameron, 1979, 53; Loparev, 1895, 603–604)[41]

The relic's return to Blachernae was used as an opportunity to reveal the relic, usually hidden from the public eye, to the people of Constantinople. Centuries after its initial deposition, the relic was reactivated: the fact that the fabric was undefiled, like the Virgin herself, served, in front of everyone's eyes, as token for the relic's authenticity. The trajectory from Blachernae to Hagia Sophia and back again, via St Laurence, was henceforth commemorated in Constantinople's liturgical calendar, its anniversary annually celebrated (Wenger, 1955, 121). Interestingly, in 623, it was not Emperor Heraclius, but the patriarch, Sergius, who played the leading role in handling the relic, 'taking as his assistant our most pious emperor' (λαβὼν συνεργάτην τὸν βασιλέα ἡμῶν τὸν πιστότατον; Loparev, 1895, 599).[42] Sergius's prominent position in the city should become even more evident in the following years.

Only three years after the event just discussed, the Blachernae shrine stood in the spotlight again. In 626, Constantinople had to face a double siege by Avars and Persians, which almost led to the city's capture (Hurbanic, 2019). The situation was perceived as even more precarious, given that Emperor Heraclius was absent from the city (Viermann, 2020, 391–392), stationed with his army in Eastern Anatolia, where he prepared a counter-attack against the Persians who had occupied vast parts of the Roman eastern provinces (Howard-Johnston, 2021, 251–257). As Constantinople lacked its figurehead, the patriarch Sergius stepped in to galvanise resistance and offer spiritual condolence, organising processions along the walls and praying for the city's delivery (Van Dieten, 1972, 12–19). Contemporary sources agree that it was not so much the Roman weapons that eventually repelled the enemies and saved the city, but the prowess of a spiritual duo: Sergius in conjunction with the Virgin Mary herself, embodied by her robe in the Blachernae church.[43] From the liminal space around the city wall, the relic took on the function of protecting the city, offering to the people a spiritual focal point and means to cope with the emperor's absence (Viermann, 2020, 391–395, 2021, 206–228). Although the Virgin's robe was considered an important relic already before the seventh century, its (re)activation by the patriarch Sergius in connection with the Avar raids of 623 – that is, its movement within the city, storage, and veneration in Hagia Sophia; its temporary revelation; and its redeposition in the Blachernae Church – seems to have increased its popularity and the people's awareness of its existence and perceived prowess.[44] When Constantinople had to face the double siege only three years later, it was not far-fetched to revert to the idea of the Theotokos interceding in favour of the capital from her sanctuary in Blachernae. The idea of the Virgin Mary as the supernatural protectress of Constantinople, and her robe in Blachernae as the city's *palladion*, should become one of the main columns of Constantinople's communal identity throughout the Middle Ages; it could be called upon as the city had to face further onslaughts by foreign enemies (Mango, 2000; Pentcheva, 2002, 2006).

8.4 Conclusion

With the three case studies discussed here, I have argued for the crucial impact of relics on the formation of sacred topography and civic identities, and on the negotiation of religious and political hierarchies in late antique Constantinople. It has become evident that the movement of relics around the city and in between churches strengthened liminal areas around the city walls and tightened the bonds between intra- and extra-urban spaces, between city and hinterland. When considered in context, the exact way in which the relic's translation was enacted can be read as a mirror of the current political and religious climate in Constantinople: the performance could leave centre stage to the secular authorities, or it could highlight the importance of the city's patriarch; it could gloss over conflict, serve as an act of reconciliation, or bolster the city's communal identity.

The translation of relics was a ritual of unveiling: it unveiled the relics, which became visible, at least in their precious container, to the people gathered in the streets; it also unveiled members of the imperial family who usually stayed in their palace, withdrawn from the people's eyes. During the translation of relics, emperor and empress could present themselves as pious members of the community and form a connection with the public that was based on concerted commitment to the significance of a holy object: the imperial family and the crowd of people alike submitted to the relic's spiritual appeal. It was during the ritual of translation that a piece of bone or fabric was turned into a holy object; the translation served as an opportunity to authenticate and activate the relic's sacred prowess before it was incorporated into the fabric of a church. This rite of passage that welcomed the relic to its new resting place needed the performance of public affirmation that connected the congregation of believers with those prominent agents who had initiated the relic's transferral.

Even more so than the initial translation of relics to Constantinople, their intra-urban movements contributed to spatial cohesion as they usually involved trajectories connecting the city centre and the liminal space around the city walls or Constantinople's Thracian periphery. Hagia Sophia, Constantinople's main church, which itself was not based on a foundational relic, became an intermediary resting place for holy objects that would then be further distributed to shrines of lesser significance, creating a spiritual network that was commemorated and kept alive in the city's liturgical calendar. In the dynamic period of Late Antiquity, when the city's Christian sacred topography was still very much in the making, holy relics became a means to strengthen the bond between the capital and its hinterland and to extend a Christian network beyond the limits of the city walls.

Relics translated to Constantinople in Late Antiquity formed the basis of the relics collection that the city would be so famed for throughout the Middle Ages. The impact of moving relics, however, was also a very immediate one as the ritual's communal character not only authenticated the relic but also embodied a sense of communal identity centred around the belief in a holy object and its perceived prowess. The translation of relics with all the inherent qualities outlined in this chapter galvanised urban integration and a sense of Christian prestige in a city that had no glorious apostolic past. This process not only contributed to buttressing Constantinople as the new capital of the Roman Empire at a time when Christianity and Roman monarchy had irrevocably merged, but it also furthered societal integration through the enactment of commonly shared values and beliefs. Relics formed a supernatural line of defence for the capital ideologically, with people's believe in their prowess and intercession, and topographically through their dispensation around the city walls and in other liminal areas. In this capacity, translated relics contributed to the remarkable resilience of Constantinople as capital of the medieval Roman Empire.

Notes

1 The two sermons were held on two consecutive days as part of the festivities, as indicated in the text.
2 Initially, the building seems to have gone under 'the Great Church' and only later received the name 'Hagia Sophia'; both names were used throughout later centuries (Janin, 1969, 455–456).
3 On relics in Constantinople, see, for example, Hahn & Klein (2015); Wortley (2009); Klein (2006); Durand & Flusin (2004); for pilgrimage to Constantinople, see Weyl Carr (2022).
4 For the early centuries of Constantinople, see the seminal Dagron (1974).
5 For the veneration of martyrs on Rome, see Trout (2003).
6 On stationary liturgy, see Baldovin, 1986; for processions, see the recent work of V. Manolopoulou (2016, 2019); for the creation of sacred space, see Ousterhout (2012, 2019); for relics in Constantinople, see above n.3.
7 This chapter is part of a more comprehensive project in which I investigate the translation of relics to Constantinople to explore the intersection between religion, social hierarchies, and topography.
8 On the relation between city and hinterland for late antique settlements, in general, see Brogiolo et al. (2000), and Constantinople, in particular Mango & Dagron (1995).
9 *Stulti omnium Ecclesiarum populi, qui occurerunt sanctis reliquiis: et tanta laetitia, quasi praesentem viventemque prophetam cerneren, susceperunt: unde Palaestina usque Chalcedonem jungerentur populorum examina.*
10 τὰ λείψανα τοῦ ἁγίου Σαμουὴλ ἐν Κωνσταντινουπόλει διὰ τῆς Χαλκηδονησίας σκάλας, προηγουμένου Ἀρκαδίου Αὐγούστου καὶ Ἀνθημίου ἐπάρχου πραιτωρίων καὶ ἀπὸ ὑπάτων Αἰμιλιανοῦ ἐπάρχου πόλεως καὶ πάσης τῆς συγκλήτου· ἅπερ ἀπετέθη πρός τινα χρόνον ἐν τῇ ἁγιωτάτῃ μεγάλῃ ἐκκλησίᾳ.

11 See the discussions in Holum and Vikan (1979, 113–133), Spain (1977, 279–304), Wortley (1980, 381–394) and Brubaker (1999, 258–285); with an alterntive interpretation in Niewöhner (2014). The translation as arranged by Pulcheria is accounted for earliest in the ninth-century *Chronographia* of Theophanes Confessor *AM* 5920 (De Boor 86–87); for later sources, see Holum &Vikan (1979, 127).

12 On the difference between 'ceremony' and 'ritual', see Hölkeskamp (2014, 27).

13 For the incorporation of relics into church spaces, see Yasin (2015). The evidence for how exactly this incorporation looked like in Constantinople is very limited. In the Polyeuktos Church, early sixth century, the relic seems to have been stored in a crypt below the nave (Harrison, 1986, 18.); in the slightly later Church of SS Sergius and Bacchus, some recent findings suggest that the relic was incorporated into the floor (word of mouth).

14 Substructures of the church's octagonal Iustinianic reconstruction were excavated in the early twentieth century (Taddei, 2014, 79–80).

15 For the *skeuophylakion* surviving, see Mathews (1971, 11–12), based on remarks from Palladius's *Life of Chrysostom*.

16 On the Trier Ivory, it is two clerics that carry the reliquary, seated on a carriage; see above n. 11.

17 Whitby and Whitby (1989, 64) suggest that Moses's involvement means that the relics came to Constantinople from Antaradus; cf. Klein (2006, 84).

18 A later tradition states that Palestinian martyrs were laid down in Hagia Sophia for the initial dedication in 360, but Wortley argues convincingly that this is highly unlikely (see Wortley, 2009, 634–636).

19 From the seventh century on, the relic of the True Cross, having been brought to Constantinople due to the Arab conquest of Palestine, was venerated in Hagia Sophia in feast days, but usually kept in the imperial palace (see Klein, 2006, 89–90).

20 Efthymios Rizos (Cult of Saints, E02660 – http://csla.history.ox.ac.uk/record.php?recid=E02660) proposes that they may have been Italian martyrs that had been sent to John Chrysostom by Vigilius, bishop of Tridentum.

21 Καὶ τί δεῖ λέγειν γυναῖκας, ἢ ἄρχοντας, ὅπου καὶ αὐτὴ ἡ τὸ διάδημα περικειμένη καὶ τὴν πορφυρίδα περιβεβλημένη, παρὰ τὴν ὁδὸν ἅπασαν οὐδὲ μικρὸν τῶν λειψάνων ἀποσχέσθαι ἠνέσχετο, ἀλλ' ὥσπερ θεραπαινὶς παρηκολούθει τοῖς ἁγίοις, τῆς θήκης ἁπτομένη καὶ τῆς ὀθόνης τῆς ἐπικειμένης, καὶ πάντα τὸν ἀνθρώπινον καταπατοῦσα τῦφον, καὶ ἐν μέσῳ θεάτρῳ τοσούτῳ φαινομένη δήμῳ, ἣν οὐδὲ εὐνούχοις ἅπασι τοῖς ἐν ταῖς βασιλικαῖς στρεφομένοις αὐλαῖς θέμις ἰδεῖν; Transl. Efthymios Rizos.

22 καὶ ὁ πάντας εὐεργετῶν τοὺς κατὰ τὴν οἰκουμένην, ἦλθεν εὐεργεσίας ἀπολαύσων παρὰ τῶν ἁγίων τούτων, καὶ τὰ μέγιστα καρπωσόμενος ἀγαθά. Διά τοι τοῦτο καὶ αὐτὸς τὸ διάδημα, καὶ οἱ δορυφόροι πάντες, οἱ μὲν τὰς ἀσπίδας, οἱ δὲ τὰ δόρατα ἀποθέμενοι, καὶ τὴν φαντασίαν ἐκείνην ἀφέντες, μετὰ κατεσταλμένης παρεγένοντο διανοίας ἅπαντες, ὡς ἀπὸ γῆς εἰς τὸν οὐρανὸν εἰσιόντες. Transl. Efthymios Rizos.

23 For the relationship between emperor and populace, see Pfeilschifter (2013, 294–346).

24 The empress depicted on the Trier ivory is most likely Pulcheria, daughter of Anastasius and Eudoxia, and sister of Theodosius II (see Holum, 1982, 79–111; Busch, 2015, 110–135).

25 The knowledge of a church dedicated to St Thomas derives from the introduction to John Chrysostom's second homily on the translation of relics to Drypia, where it is stated that Arcadius visited this very shrine in Drypia before returning to Constantinople (cf. Janin, 1969, 251).

26 The wholeheartedly positive presentation of Eudoxia in Chrysostom's sermon, as cited above, makes it unlikely that it was held after their fall-out in 403 (for that, see Busch, 2015, 71–82). Janin (1969, 251), in contrast, assumes the translation to have happened in the last months of 398 but does not give any reason or concrete evidence for that.

27 For further Marian relics in Constantinople, like her girdle, stored in the Church of Chalkoprateia in the city centre, see Mango (2000, 17–25) and Shoemaker (2008, 53–74).

28 The most prominent example of this tradition is a sermon attributed to Theodor Syncellus dated to 624/5, which will be discussed in detail below. The text (*In depositionem pretiosae vestis*) is edited in Combefis (1648, 751–788), partially re-edited in Loparev (1895, 592–612) and partially translated in Cameron (1979). However, Cameron's translation does not cover the part on the invention and translation of the relic to Constantinople.

29 It is firmly attested in the fourth homily by Cosmas Vestitor from the eighth century, edited in Wenger (1955, 328–333). This tradition states that Pulcheria, after having married Emperor Marcian in 450, made efforts to bring Marian relics to Constantinople. Due to the lack of bodily relics of the Theotokos (her having ascended to heaven), the bishop of Jerusalem, Juvenal, conceded to send the Virgin's funeral garment to Constantinople, which had been found in the tomb in which she had been laid to rest.

30 Further evidence that speaks for Leo and Verina is listed in Mango (1998, 64–65). In contrast, Shoemaker (2008, 60–61) does not rule out the possibility that Pulcheria founded the Marian shrine.
31 Mango (1998, 74) discusses whether there might be an association between the Blachernae shrine in its early days and Miaphysitism.
32 The Blachernae church burnt down shortly before Constantinople was taken by the Ottomans. In the nineteenth century, a much smaller church was built on the spot which still stands today.
33 These fortifications were later extended under the reigns of Leo V and Manuel Komnenos.
34 On the Roman-Persian war, see recently Howard-Johnston (2021); on Heraclius's attempt to conclude a peace treaty with the Avars, see Kaegi (2003, 118–121) and Hurbanic (2011, 315–328).
35 The date of the incident is disputed as *Chron. Pasch.* (a. 623 [Dindorf 712 f.]) dates it to 623 and Theophanes Confessor (AM 6110 (De Boor 301–302)) to 619. Cameron (1979) follows the earlier date, whereas the later date is convincingly advocated for by Whitby and Whitby (1989, 203–205) and Mango (1998, 67–68, 2000, 19–20).
36 On the sermon, see above n. 28; it seems to have been composed for the anniversary of the relic's return to Constantinople; see Cameron (1979). Cameron, however, gives what seems to be the wrong date for the sermon (620 instead of 624/5); on the date, see above n. 35. Theodor Synkellos is better known for having acted as an emissary to the Avars in 626 (*Chron. Pasch.* a. 626 [Dindorf 721]), when Constantinople faced a double siege by Avars and Persians, and as the author of a sermon remembering this double siege and the Roman victory (edited in Sternbach (1900, 2–24) and translated into French by Makk (1975).
37 ἐτόλμησαν δὲ καὶ τῆς θείας ἐφάψασθαι ταύτης σοροῦ, καὶ τὸ πᾶσι τέως ἀθέατον, ἐν φανερῷ τότε καταστῆσαι μθστήριον. ἔσωθεν δὲ τῆς ὁρωμένης σοροῦ, ἥτις ἐκ χρυσοῦ καὶ ἀργυρου ἔχει τὴν ποίσιν, σορὸς εὑρέθη λίθοθ, στιλβούσης λαμπρότητα, καὶ ταύτης ἔνδον πρὸς τῷ μέρει τῷ κατὰ ἄρκτον, εὕρηται κείμενος ὁ θεῖος θησαυρὸς ἐν ἑτέρῳ σοριῷ μικρῷ φυλαττόμενος. The term *soros* is used for the chapel in Blachernae as a whole, as well as for the casket or container in which the relic was concealed.
38 In this case, the outer container, in the shape of a sarcophagus, is carved from marble; the second layer, again sarcophagus-shaped, is made of silver; the innermost element, containing the actual relic (two pieces of bone and a piece of wood), is a tiny golden box embellished with gems (Yasin, 2015, 148–149 with Figure 7.13).
39 Μεθ' ὑψηλοῦ συγκαλεῖται κηρύγματος ἅπασαν τῶν ἀρχιερέων τὴν σύνοδον, κλῆρόν τε καὶ λαὸν, ὅσος ἐν ἀνδράσι καὶ ὅσος ἐν γυναιξίν, ὅσος ἐν τέλει, καὶ ὅσος ἐν ἀξιώμασιν, καὶ ὅσος ἐν ἰδιωτικῷ βιοτεύει τῷ σχήματι. Δεῦτε, λέγων, ἴδετε, ἱερεῖς καὶ λαοὶ, δεῦτε θησαυρὸν τόν ἕως νῦν κεκρυμμένον θεάσασθαι.
40 This is most probably a church on the Golden Horn (Janin, 1969, 300–302).
41 Εὗρεν τὴν ἐσθῆτα τῆς ὄντως ἀληθινῆς βασιλίδος τῆς θεομήτρος, τὴν ἰδίαν ἀποστίλβουσαν χάριν καὶ δύναμιν. . . . ἡ δὲ ἐσθὴς ἡ θεία ἐξ ἐριων εὐφθάρτων ἐξυφασμένη, . . . διαφθορὰν τὸ σύνολον οὐχ ὑπέμεινεν, ἀλλ' ὅλη ἐστιν ἀπαθής, ὅλη συνεστῶσα καὶ ἀδιάφθορος.
42 On the well-functioning relationship between emperor and patriarch, see Viermann (2021, 184–185, 227–228).
43 The most relevant sources on the siege of 626 are Georg of Pisidia's *Bellum Avaricum*, *Chron. Pasch.* (a. 626 [Dindorf 715–726]), and another sermon by Theodor Synkellos edited in Sternbach (1900, 2–24) and translated into French by Makk (1975).
44 For the activation of relics, cf. Ousterhout (2019, 89–104, especially 98); cf. also Ousterhout's remarks on how movable the relics were in Constantinople (96).

Editions and abbreviations

Anth. Palat – *Anthologia Graeca*, 4 vols. (2nd ed.), ed. H. Beckby (1965–1968). Heimeran.

Chron. Pasch. – *Chronicon Paschale* (Corpus Scriptorum Historiae Byzantinae 16–17), ed. L. Dindorf (1832). Weber.

Cult of Saints – *The Cult of Saints in Late Antiquity*. online database. http://csla.history.ox.ac.uk

De cer. – *Constantinus Porphyrogenitus, De Ceremoniis aulae Byzantinae libri duo Graece et Latine* (2 vols), ed. J. J. Reiske (1829/1830). Weber.

Eus. *VC* – *Über das Leben des Kaisers Konstantin* (Eusebius Werke, 1.1), ed. F. Winkelmann (1975). Akademie Verlag.

Jerome, *Chronicle – Hieronymi Chronicon/Eusebius* (Eusebius Werke 7), ed. R. Helm (1956). Akademie Verlag.
Jerome, *Contr. Vigilantium – Hieronymus: Adversus Vigilantium* (CCSL 79C), ed. J.-L. Feiertag (2005). Brepols.
Joh. Chrys., *Hom. Mart.* (*PG* 63, 467–468) – John Chrysostom, *Homilia dicta postquam reliquiae martyrum*. In *S. Joannis Chrysostomi opera omnia quae exstant, Patrologia Graeca* 63 (cols. 467–472), ed. J.-P. Migne (1862). Imprimerie Catholique.
Nikeph. Kall. – *Nicephori Callisti Xanthopuli ecclesiasticae historiae libri XVIII, PG 145–147*, ed. J.-P. Migne (1865/1967). Imprimerie Catholique/Brepols.
Proc. *De aed – Procopius, De Aedificiis. In Procopii Caesariensis opera omnia, vol. 1* (pp. 5–186), ed. G. Wirth (1964). Teubner.
Soc. – *Sokrates Kirchengeschichte*, ed. G. C. Hansen (1995). Akademie Verlag.
Soz. – *Sozomenus: Kirchengeschichte*, eds. J. Bidez & G. C. Hansen (1960). Akademie Verlag.
Theodore Anagnostes – *Theodoros Anagnostes, Kirchengeschichte* (2nd ed.), ed. G. C. Hansen (1995). Akademie Verlag.
Theophanes Confessor – *Theophanis Chronographia* (2 vols), ed. C. De Boor (1883). Teubner.

Bibliography

Allen, P., & Mayer, W. (1999). *John Chrysostom*. Routledge.
Baldovin, J. F. (1986). *The Urban Character of Christian Worship: The Origins, Development and Meaning of Stational Liturgy*. Pont. Inst. Studiorum Orientalium.
Bardill, J. (1999). The Golden Gate in Constantinople: A Triumphal Arch of Theodosius I. *American Journal of Archaeology*, *103*(4), 671–696.
Bardill, J. (2012). *Divine Emperor of the Christian Golden Age*. Cambridge University Press.
Baynes, N. H. (1949). The Supernatural Defenders of Constantinople. *Analecta Bollandiana*, *67*, 165–177.
Brogiolo, G. P., Gauthier, N., & Christie, N. (2000). *Towns and Their Territories Between Late Antiquity and the Early Middle Ages*. Brill.
Brubaker, L. (1999). The Chalke Gate, the Construction of the Past, and the Trier Ivory. *Byzantine and Modern Greek Studies*, *23*, 258–285.
Busch, A. (2015). *Die Frauen der theodosianischen Dynastie. Macht und Repräsentation kaiserlicher Frauen im 5. Jahrhundert*. Franz Steiner Verlag.
Cameron, A. (1978). The Theotokos in Sixth-Century Constantinople: A City Finds Its Symbol. *The Journal of Theological Studies*, *29*, 79–108.
Cameron, A. (1979). The Virgin's Robe: An Episode in the History of Early Seventh-Century Constantinople. *Byzantion*, *49*, 42–56.
Cameron, A. (2012). *The Mediterranean World in Late Antiquity: AD 395–700*. Routledge.
Cameron, A., & Long, J. (1993). *Barbarians and Politics at the Court of Arcadius*. University of California Press.
Combefis, E. (1648). *Historia Haeresis Monothelitarum*. Antonius Bertier.
Croke, B. (2010). Reinventing Constantinople: Theodosius I's Imprint on the Imperial City. In S. McGill, C. Sogno, & E. Watts (Eds.), *From the Tetrarch to the Theodosians* (pp. 241–264). Cambridge University Press.
Dagron, G. (1974). *Naissance d'une capitale. Constantinople et ses institutions de 330 à 451*. Presses universitaires de France.
Demangel, R. (1945). *Contribution à la topographie de l'Hebdomon*. de Boccard.
Diefenbach, S. (1996). Frömmigkeit und Kaiserakzeptanz im frühen Byzanz. *Saeculum*, *47*, 35–66.
Diefenbach, S. (2002). Zwischen Liturgie und civilitas. Konstantinopel im 5. Jahrhundert und die Etablierung eines städtischen Kaisertums. In R. Warland (Ed.), *Bildlichkeit und Bildorte von Liturgie. Schauplätze in Spätantike, Byzanz und Mittelalter* (pp. 21–49). Reichert.
Durand, J., & Flusin, B. (2004). *Byzance et les Reliques du Christ*. Association des Amis du Centre d'Histoire et Civilisation de Byzance.

Faber, E. (2011). Anti-Germanism in Constantinople or the Revolt of Gainas? In D. Hernandez de la Fuente (Ed.), *New Perspectives on Late Antiquity* (pp. 124–135). Cambridge Scholars Publishing.

Grig, L., & Kelly, G. (2012). Introduction: From Rome to Constantinople. In L. Grig & G. Kelly (Eds.), *Two Romes. Rome and Constantinople in Late Antiquity* (pp. 3–30). Oxford University Press.

Hahn, C. J., & Klein, H. A. (2015). *Saints and Sacred Matter. The Cult of Relics in Byzantium and Beyond*. Harvard University Press.

Haldon, J. (2016). *The Empire That Would Not Die. The Paradox of Eastern Roman Survival, 640–740*. Harvard University Press.

Harrison, M. (1986). *Excavations at Sarachane in Istanbul, Vol I*. Princeton University Press.

Hölkeskamp, K.-J. (2014). Performative Turn Meets Spatial Turn: Prozessionen und andere Rituale in der neueren Forschung. In D. Boschung (Ed.), *Raum und Performanz. Rituale in Residenzen von der Antike bis 1815* (p. 27). Franz Steiner Verlag.

Holum, K. G. (1982). *Theodosian Empresses: Women and Imperial Dominion in Late Antiquity*. University of California Press.

Holum, K. G., & Vikan, G. (1979). The Trier Ivory, "Adventus" Ceremonial, and the Relics of St. Stephen. *Dumbarton Oaks Papers*, *33*, 113–133.

Howard-Johnston, J. (2021). *The Last Great War of Antiquity*. Oxford University Press.

Hurbanic, M. (2011). The Eastern Roman Empire and the Avar Khaganate in the Years 622–624 AD. *Acta Antiqua Academiae Scientiarum Hungaricae*, *51*, 315–328.

Hurbanic, M. (2019). *The Avar Siege of Constantinople in 626. History and Legend*. Springer.

Janin, R. (1969). *La géographie ecclésiastique de l'empire byzantin. Le siège de Constantinople et le patriarcat oecuménique, Tome III: Les églises et les monastères*. Institut français d'études byzantines.

Johnson, M. J. (2009). *The Roman Imperial Mausoleum in Late Antiquity*. Cambridge University Press.

Kaegi, W. E. (2003). *Heraclius. Emperor of Byzantium*. Cambridge University Press.

Kaldellis, A. (2016). The Forum of Constantine in Constantinople: What Do We Know About Its Original Architecture and Adornment? *Greek, Roman, and Byzantine Studies*, *56*, 714–739.

Klein, H. A. (2006). Sacred Relics and Imperial Ceremonies at the Great Palace of Constantinople. In F. A. Bauer (Ed.), *Visualisierungen von Herrschaft. Frühmittelalterliche Residenzen, Gestalt und Zeremoniell* (pp. 79–100). Ege Yayınları.

Klein, H. A. (2015). Materiality and the Sacred: Reliquaries and the Rhetoric of Enshrinement. In C. Hahn & H. A. Klein (Ed.), *Saints and Sacred Matter. The Cult of Relics in Byzantium and Beyond* (pp. 231–252). Dumbarton Oaks.

Kulikowski, M. (2006). *Rome's Gothic Wars: From the Third Century to Alaric*. Cambridge University Press.

Külzer, A. (2008). *Tabula Imperii Byzantini, Vol. 12: Ostthrakien*. OAW Verlag.

Liebeschuetz, J. H. W. G. (1989). The Gainas Crisis at Constantinople in 399. In D. H. French & C. S. Lightfoot (Eds.), *The Eastern Frontier of the Roman Empire. Proceedings of a Colloquium Held at Ankara in September 1988* (pp. 277–284). BAR.

Liebeschuetz, J. H. W. G. (1990). *Barbarians and Bishops. Army, Church, and State in the Age of Arcadius and Chrysostom*. Oxford University Press.

Loparev, C. (1895). Staroe Svidetelstvo o Polozhenii rizy Bogorodnitsy vo Vlakherniakh. *VizVrem*, *2*, 592–612.

Makk, F. (1975). *Traduction et Commentaire de l'homélie écrite probablement par Théodore le Syncelle sur le siège de Constantinople en 626*. JATE.

Mango, C. (1990). Constantine's Mausoleum and the Translation of the Relics. *Byzantinische Zeitschrift, 83*, 51–56.

Mango, C. (1998). The Origins of the Blachernae Shrine at Constantinople. In N. Cambi & E. Marin (Eds.), *Acta XIII Congressus Internationalis Archaeologicae Christianae* (Vol. 2, pp. 61–76). Instituto di Archeologia Cristiana.

Mango, C. (2000). Constantinople as Theotokoupolis. In M. Vassilaki (Ed.), *The Mother of God: Representations of the Virgin in Byzantine Art* (pp. 17–25). Skira Editore.

Mango, C., & Dagron, G. (1995). *Constantinople and Its Hinterland: Papers From the Twenty-Seventh Spring Symposium of Byzantine Studies, Oxford, April 1993*. Routledge.

Manolopoulou, V. (2016). *Processing Constantinople: Understanding the Role of Light in Creating the Sacred Character of the Landscape*. Newcastle University.

Manolopoulou, V. (2019). Processing Time and Space in Byzantine Constantinople. In C. Morris & G. Papantoniou (Eds.), *Unlocking Sacred Landscapes. Studies in Mediterranean Archaeology Series* (pp. 155–167). Aström Editions.

Mathews, T. (1971). *The Early Churches of Constantinople: Architecture and Liturgy*. Pennsylvania State University Press.

Meier, M. (2019). *Geschichte der Völkerwanderung. Europa, Asien und Afrika com 3. Bis zum 8. Jahrhundert*. C.H. Beck.

Müller-Wiener, W. (1977). *Bildlexikon zur Topographie Istanbuls: Byzantion, Konstantinupolis, Istanbul*. Verlag Ernst Wasmuth.

Niewöhner, P. (2014). Historisch-topographische Überlegungen zum Trierer Prozessionselfenbein, dem Christusbild an der Chalke, Kaiserin Irenes Triumph im Bilderstreit und der Euphemiakirche am Hippodrom. *Millennium*, *11*(1), 261–288.

Ousterhout, R. (2012). The Sanctity of Place and the Sanctity of Buildings: Jerusalem versus Constantinople. In R. Ousterhout & B. D. Wescoat (Eds.), *Architecture of the Sacred – Space, Ritual, and Experience From Classical Greece to Byzantium* (pp. 281–306). Cambridge University Press.

Ousterhout, R. (2019). Constructing and Deconstructing Sacred Space in Byzantine Constantinople. In S. Yalman & A. Hilâl Uğurlu (Eds.), *Sacred Spaces and Urban Networks* (pp. 89–104). ANAMED.

Pentcheva, B. V. (2002). The Supernatural Protector of Constantinople: The Virgin and Her Icons in the Tradition of the Avar Siege. *Byzantine and Modern Greek Studies*, *26*(1), 1–41.

Pentcheva, B. V. (2006). *Icons and Power: The Mother of God in Byzantium*. Pennsylvania State University Press.

Pfeilschifter, R. (2013). *Der Kaiser und Konstantinopel. Kommunikation und Konfliktaustrag in einer spätantiken Metropole*. De Gruyter.

Pohl, W. (2018). *The Avars. A Steppe Empire in Central Europe, 567–822*. Cornell University Press.

Shoemaker, S. (2008). The Cult of Fashion: The Earliest 'Life of the Virgin' and Constantinople's Marian Relics. *Dumbarton Oaks Papers*, *62*, 53–74.

Simeonov, G. (2016). Der Hafen und die Anlegestelle des Hebdomon. In F. Daim (Ed.), *Die byzantinischen Häfen Konstantinopels* (pp. 121–138). RGZM Verlag.

Spain, S. (1977). The Translation of Relics Ivory, Trier. *Dumbarton Oaks Papers*, *31*, 279–304.

Stephens Falcasantos, R. (2020). *Ritual, Violence, and Memory in the Making of a Christian Imperial Capital*. University of California Press.

Sternbach, L. (1900). *Analecta Avarica*. Sumptibus Academiae litterarum.

Taddei, A. (2014). Notes on the So-Called 'Palace of Ioukoundianai' at Hebdomon (Constantinople). *Hortus Artium Medievalium*, *20*(1), 77–84.

Trout, D. E. (2003). Damasus and the Invention of Early Christian Rome. *Journal of Medieval and Early Modern Studies*, *33*(3), 517–536.

Twardowska, K. (2014). Empress Verina and the Events of 475–476. *Byzantinoslavica*, *71*, 9–22.

van Dieten, J. L. (1972). *Geschichte der Patriarchen von Sergios I. bis Johannes VI. (610–715)*. Hakkert.

van Nuffelen, P. (2012). Playing the Ritual Game in Constantinople (379–475). In L. Grig & G. Kelly (Eds.), *Two Romes. Rome and Constantinople in Late Antiquity* (pp. 183–202). Oxford University Press.

Venables, E. (1911). Atticus, Archbishop of Constantinople. In H. Wace & W. C. Piercy (Eds.), *Dictionary of Christian Biography and Literature to the End of the Sixth Century* (p. 69). John Murray.

Viermann, N. (2020). Merging Supreme Commander and Holy Man. George of Pisidia's Poetic Response to Heraclius' Military Campaigns. *Jahrbuch der Österreichischen Byzantinistik*, *70*, 379–402.

Viermann, N. (2021). *Herakleios, der schwitzende Kaiser. Die oströmische Monarchie in der ausgehenden Spätantike*. De Gruyter.

Ward-Perkins, B. (2012). Old Rome and New Rome Compared: The Rise of Constantinople. In L. Grig & G. Kelly (Eds.), *Two Romes. Rome and Constantinople in Late Antiquity* (pp. 53–80). Oxford University Press.

Wenger, A. (1955). *L'assomption de la T. S. vierge dans la tradition byzantine du VIe au Xe siècle: études et documents*. Inst. Français d'Études Byzantines.
Weyl Carr, A. (2022). Pilgrimage to Constantinople. In S. Bassett (Ed.), *The Cambridge Companion to Constantinople* (pp. 310–323). Cambridge University Press.
Whitby, M., & Whitby, M. (1989). *Chronicon Paschale 284–628 AD. Translated with Notes and Introduction*. University of Liverpool Press.
Wiśniewski, R. (2019). *The Beginnings of the Cult of Relics*. Oxford University Press.
Wortley, J. (1980). The Trier Ivory Reconsidered. *Roman and Byzantine Studies*, *21*(4), 381–394.
Wortley, J. (1999). The Byzantine Component of the Relic-Hoard of Constantinople. *Greek, Roman, and Byzantine Studies*, *40*, 353–378.
Wortley, J. (2009). *Studies in the Cult of Relics in Byzantium Up to 1204*. Routledge.
Yasin, A.-M. (2009). *Saints and Church Spaces in the Late Antique Mediterranean: Architecture, Cult, and Community*. Cambridge University Press.
Yasin, A.-M. (2015). Sacred Installations: The Material Conditions of Relic Collections in Late Antique Churches. In C. Hahn & H. A. Klein (Eds.), *Saints and Sacred Matter: The Cult of Relics in Byzantium and Beyond* (pp. 133–151). Dumbarton Oaks.

9 Space and place[1]

Late antique churches and place remaking

Rebecca Sweetman

9.1 Introduction

From the early fourth century, the once familiar religious landscape of the Eastern Mediterranean began to change with the incremental construction of Christian churches. This has been suggested in part as a result of the Constantinian requests of Bishops to build churches as recorded by Eusebius (Armstrong, 1967).[2] Moreover, following centuries of shared traditions in mortuary and domestic contexts, church building represented the most evident signs of Christian place-making. Up to this point, it had been subtle: catacombs were largely hidden from view, if not general knowledge, and communal worship took place in domestic contexts and indistinguishable locations.

Once Christianity was adopted by the imperial family and became more acceptable throughout the Empire, church construction would have been a challenging endeavour; even for the many already-established dioceses, land, finance and community trust had to be secured. Consequently, while some elements of the transformation were organic, other parts had to be planned. There was significant variety in levels of strategic development and construction dates across the Empire: from very early fourth century in the Holy Land to the construction of the first church in the Peloponnese in the early fifth century. Most churches were built during the fifth and sixth centuries (Sweetman, 2015a, 2015b).

In all cases, place-making and remaking, marked by the construction of churches, was a fundamental part of both effective conversion practices and then Christian life. Literary, epigraphic and archaeological data reveal a number of ways in which this happened: from the carefully chosen location, architectural form, associated figures and stories to the use of emotion, memory and the senses. While the Imperial adoption of Christianity would have paved the way for church building, I suggest here that the context of intense displacement due to natural disasters as well as wars and invasions in the fourth century in the East increased the impetus and resources (ironically) for Christian place-making through church building. It is possible to suggest that strategically and organically, the church drew on the strong sense of displacement to make Christianity more relevant in the re-establishment of communities. One of the key values of looking at place-making and remaking in the late antique period is that it enables us to reinstate active agency into the processes of community engagement with the church, once it is constructed.

9.2 Place-making and remaking

Place-making is a rather nebulous term with variety of meaning within and between disciplines (Ellery et al., 2021). In archaeology and anthropology, the notion of place-making is often used

DOI: 10.4324/9780429427152-11

by those taking a phenomenological approach to the study of ancient societies such as Tilley (1994) and Barrett and Ko (2009) as well as in studies of prehistoric communities. For the purposes of this work, I follow Ashmore's (2014) idea of meaningful places and consider places to be significant locations in a landscape that are interconnected with people, attached to the senses, other places and communities. Places can be made and remade and do not necessarily evoke the same kinds of feelings or memories for all. Once a place is made, while different experiences and understandings are inevitable, the place encourages collective experiences and memory-making, which can be transferred to other group contexts, thereby strengthening the community (Sweetman, 2015a). As such, places are multifunctional; they provide a community focus as well as enabling community involvement.

Marchand (2015) argues that a cognitive approach that takes account of the interplay between emotion and environment highlights the extraordinary (as opposed to the everyday) in place-making. Through his study of contemporary pilgrims to the church of the Holy Sepulchre in Jerusalem, Marchand (2015) highlighted many of the cognitive processes drawn on, including memory and emotion, to create place. He wrote (2015, 79), ' "Place" (or the "sense of place" or "displacement") is a complex kind of mental representation, drawing upon information made available from a variety of diverse cognitive networks'.

Furthermore, he explained that displacement encourages the consideration of place and the construction of the idea of place as a changing process according to one's thinking. It follows that being out of one's usual realm of spaces could be considered displacement, which gave a heightened sense of occasion ensuring that memories were consolidated. Considering and remembering the space constitutes 'place-making' (Marchand, 2015, 79). And I would add that the process of getting to the place and engaging with it physically helps to involve community and consolidate the sense of belonging. As has been shown before, many of the earliest churches are located in existing gathering points (e.g. sanctuaries, wells, cemeteries), which contribute to this ordering of space (Sweetman, 2015a). Moreover, in considering the topography of the churches, we will see that place-making is further enhanced by the situation of churches along routes of communication. Place-making is pertinent in building community solidarity and particularly so in periods of displacement (Eckenwiler, 2018), but it is not automatic nor is it passive (Estey, 2019). It provides opportunities for displaced people to become part of a community but only if there is engineering behind this.

Place-making is not limited to the immediate space occupied (for the purposes of this discussion, by a church), but other spaces can be transformed by the incorporation of and association with place. For example, processions to a place can help to give religious meaning to otherwise everyday space (such as an urban or rural landscape) (Bull & Mitchell, 2015). By the middle of the fourth century, church building became a fundamental part of the Christianisation process and networks of churches across the landscapes became a visible reminder of that. This process was not limited to Christianity, as the chapters by Javier Martínez Jiménez and Esen Ogus in this volume remind us, but Christian place-making was a distinctive phenomenon of the late antique period.

9.3 Context

While the earliest churches were being constructed from the fourth century, this was also a period of some uncertainty and tension for many communities across the Empire. As discussed in detail below, from the fourth century, areas of the Eastern Empire experienced earthquakes, tsunamis, crop failure, disease, incursions, community rifts, exiles, forced displacement, in addition to uncertainty in religious identity. Not everyone experienced the same issues at the same time and Allen and Neil's (2013, 3) analysis of episcopal letters of the fifth and sixth centuries has shown that much of what damned the period were small-scale local or regional crises.

9.4 Impact of displacement

To understand the role of the church as an institution in place-making, it is necessary to have a brief insight into the psychological impact of displacement on both the displaced and the host communities. Studies of displacement in both ancient and modern contexts tend to focus on the impact of being forced to move and less on the processes of trying to build a new life (Mazumdar & Mazumdar, 2004; Sweetman, 2016). Furthermore, internally displaced people, for example a community that may be forced to leave a city for rural spaces, are often overlooked in the scholarship (Naidu, 2016). Porter and Haslam (2001) note that the impact of forced displacement varies considerably and is dependent on elements such as the location of displacement, whether there is community/family separation, pre-migration trauma and age. While the impact may vary, scholars and practitioners agree that loss of home, identity, community and culture is acute (Phillimore, 2011) and has significant mental health and well-being issues in forced migrants (Chatterjee et al., 2020). Bhugra (2004, 130) notes that forced migrants experience 'at first a sense of loss, dislocation, alienation and isolation, which will lead to processes of acculturation'. Furthermore, the decision to leave is one made often under pressure and as a result of a shocking experience (Verme & Schuettler, 2021).

While one cannot generalize about individuals, many with lived experience similarly note the overwhelming sense of insecurity and isolation post migration (Chatty, 2014; Sigona, 2014). Sigona (2014) has also stressed the importance of shared memories as a means of bearing witness to the trauma of displacement and subsequently in finding a sense of belonging. I would add here that loss of place disrupts memory-making, which has an impact on community cohesion. Hack-Polay et al. (2021) have highlighted the nature of identity crisis that forced migrants to suffer in leaving their communities and in trying to negotiate and settle on their identities in their new host societies. They note that forced migrants often experiment with different identities before settling on one (Hack-Polay et al., 2021). Furthermore, this identity crisis is also manifested in younger generations of immigrants who become distanced from their original communities and who gain understandings of cultural identity via their parents, rather than through direct experience, as Valentine et al. (2009) surmised based on studies of Somalian refugees in the United Kingdom and Denmark.

These uncertainties of identity are compounded by the difficulties of being able to interact with host communities, partly because of residential patterns but also because of ideas of nation state (Valentine et al., 2009). Consequently, the already-fragile sense of belonging and place is further eroded. It is not surprising that immigrants turn to religion to help adapt to and settle into new lives and communities (Hollenbach, 2014; Naidu, 2016).[3] At the same time, the impact of displacement on host communities is not well understood in ancient society (Cohen, 2006) and, in contemporary studies, research tends to focus on the economic effects (Verme & Schuettler, 2021). Corcoran (2002) suggests that host communities, particularly ones in economically disadvantaged areas, may begin to withdraw from their traditional civic life when migrants move in, thereby creating further wedges between the existing and new communities. The sense of place is fundamental in collective memory-making (Sennett, 2001) and is disrupted among the forcibly displaced, as well as often in host communities who commonly change the way they interact with their places with the arrival of new groups.

Displaced individuals and communities need to regain their agency, and this is done in part through finding a sense of belonging and identity. However, this is not always possible, and displacement is often not just long term but permanent. There is often a draw to return home, as can be seen in the way places devastated by natural disasters, such as earthquakes, are quickly rebuilt. Although, even in these cases, Zetter and Boano (2010) note that while homes are restored, the disruption in place is more enduring because of the lack of attention to the reconstruction of public space.

It is possible to argue here that Christianity helped to create order and a sense of place in areas that had experienced significant crises and natural disasters, as well as in communities that hosted displaced people. As such, it is possible to argue that part of the success of the church in Late Antiquity was through its place-making and therefore allegiance in among these various crises which precipitated significant forced mobility across the Eastern Empire, which we will now turn to.

9.5 Context of displacement

During the fourth century, communities around the eastern Mediterranean were exposed to a range of disasters, events and processes that made the issue of displacement central across society. In the first half of the fourth century, there was significant trouble around the Danube, which had been brewing since the second half of the third century (Halsall, 2007, 138). This was followed by the abandonment of Dacia (Halsall, 2007, 138). While Constantine quashed many of the efforts of the Goths to cross the Danube, he transplanted (or settled) communities of Sarmatians on the north bank of the Danube to help shore up the line against the Goths (Heather, 2017).

The 365 earthquake destroyed most Cretan cities and many cities in southern Greece and Cyprus (Stiros, 2001). In his *Res Gestae*, not only does Ammianus describe the tsunami that devastated cities on the northern coasts of Cyrenaica and Egypt following the earthquake (Kelly, 2004), but he also notes the significant negative impact that this would have had on food production. These events would have displaced populations and the need for urban reconstruction would have created a market for architects and craftspeople to travel too. Shortly after these earthquakes, the invasions of the Huns eastwards across the Pontic Steppes forced Goths to flee and move south into lands of the Roman Empire along the Danube from 376 sanctioned by Valens (Halsall, 2007, 21). A few decades later, Alaric led invasions through mainland Greece – sacking Athens in 395 – and through Italy – creating havoc in Rome in 410 – creating yet more displacement (Zosimus 5.5.4–5.5.8; Kaegi, 2015). The fourth- to sixth-century earthquakes caused significant geomorphological change on the coastline and the full or partial submersion of at least 21 coastal settlements and ports, including Kenchreai and Gytheion, whose ports were submerged in 375 (Avraméa, 1997, 44).

In addition to environmental and conflict displacement, political and religious exile continued to be common (Braginton, 1944; Barry, 2019). A series of fourth-century examples is recorded in Ammianus's work, including that Phronimius (commander of Julian's armies) was sent by Valens to exile in the Crimea, Constantius II exiled Gerontinus and Julian exiled Florentius (Stini, 2011). Stini (2011) notes that, while accounts of infamous exiles survive, records for the less notable and internal exiles were unlikely to have been preserved. Bishops and other clergies were also frequently banished and actually, in the fourth century, there is a change in attitude to clergy in exile (Barry, 2019). Where once writers such as Tertullian and even Cyprian of Carthage railed against those who took flight instead of accepting martyrdom, by the early fourth century, this had changed to positive echoes of Christ fleeing to the desert as it wasn't quite his time yet (Barry, 2019). Bishop Athanasius (*c.* 298–373) became something of a poster child in this regard. He was exiled five times on the orders of four different emperors and was lauded as a significant force in the spread of Christianity (Barnes, 1993). This is indicative of a wider acceptance of displacement in all forms.

This protracted displacement in Greece in the latter half of the fourth century impacted a broad swathe of the population, which would have created something of a fluid sense of belonging in the East. As noted earlier, the displaced as well as the host communities were affected by placelessness and creation of in-groups and out-groups, which altogether would have had a stress on community identity. Place-making helps to create shared experiences and, as Estey (2019) notes, it is a key way of helping to create tolerance for religious diversity at a time when Christian religious identity itself was being debated, for example with the rise of Arianism and the debates in the Councils of Nicea and Constantinople (Harland, 2009). Place-making is an obvious way to help restore communities and it is possible that while the church itself was experiencing process of self-reflection, Christian communities may have seen church building as an opportunity for supporting others, as well as proselytising. In overt and covert ways, the Christian church catered to communities experiencing displacement through the provision of charity and social services (Doerfler, 2019) as well as encouraging congregations to be charitable to strangers. Furthermore, it seems that place-making and remaking was a fundamental part of the growth of Christianity in this period – and not just tacitly.

9.6 Church building and displacement

The idea of displacement (forced or voluntary) was never far from the idea and reality of Christianity. Displacement was articulated by the movement of both the living and the dead. The earliest proselytisers, including St Paul and his companions, travelled widely along well-traversed routes to convert populations. In turn, bishops did the same, and we have records of individuals such as Titus (first century CE) moving between Ephesus, Corinth, Jerusalem, Crete, Macedonia and Epirus (*Corinthians* 2:1–5; 7:6–15). According to Paul's letter to Titus, he was eventually appointed Bishop of Crete (*Titus* 1:5). Later, his relics were moved to Venice but are now back in Crete. Barnabas is similarly peripatetic, having travelled with Paul around Asia Minor (in particular Antioch), Jerusalem and Cyprus (*Acts* 13.9–16.2) and he may have been an early Bishop of Milan too. A later Bishop of Cyprus, Anthemios, used his miraculous discovery of the buried body of Barnabas with the Gospel of St Matthew to help convince Emperor Zeno of the importance of the autocephaly of the Cypriot church in the face of a threat of a takeover by the church of Antioch (Luijendik, 2020).

Later, bishops, and indeed members of the imperial family, became associated with church building in the wake of travelling. Empress Helena travelled in search of the Holy Cross and other places of Christian importance, and in the process commissioned churches to be built at Jerusalem and possibly Paros in the early fourth century (Drijvers, 1992). In many respects, she was following the programme instigated by her son Constantine who encouraged bishops to have churches built in Rome, Trier, Jerusalem, Bethlehem, Antioch, Nicomedia and Constantinople – among other locations (Armstrong, 1967).

In the early church, bodies of the dead could not be moved. However, by the fourth century, this had changed, and relics and bodies began to be moved for the purposes of establishing churches and holy spaces. In the mid-fourth century, Constantinople was infamously a city without martyrs and, for this reason, Constantius II brought the bodies of Andrew and Luke from Achaia and Timothy from Ephesus to the church of Agia Sophia in 365–367 (Philostorgius, *Ecclesiastical History* 3.2; Woods, 1991). Bishop Ambrose famously brought the bodies of the protomartyrs of Gervasis and Protasius to his Basilica Ambrosiana in Milan in 386 (Gregory of Tours, *Glory of the Martyrs* 46). In discussing the holy revelation that led Bishop Ambrose to find the concealed bodies of the martyrs, Gregory of Tours noted that their relics (including bloodstained clothes) had made their way across vast swathes of Italy. Although the worship of relics was originally an eastern phenomenon, with the earliest reliquaries being dated to the fourth century, the interest in them spread to the west – albeit not with the same kind of intensity as in the east. By the late fourth century, they were commonly given monumental shrines or graves and became a pilgrimage locus (Thunø, 2018).

One of the main values of undertaking a pilgrimage was the creation of a tangible relationship with religious centres (Bitton-Ashkelony, 2005), and it was the same with the movement of bishops, bodies or relics. Encouraging a rhetoric of movement as part of Christianity would also have enabled a sense of shared experiences, and I would argue that the connection between movement of people (dead or alive!) and church foundation was carefully orchestrated to create a new community place.

The displacement of people in this period would have added to the impetus for change to take place, as well as new opportunities for place-making. A carefully integrated approach would have been required in order to ensure that place-making or remaking did not become displacement and alienate parts of the population, for example with the appropriation of polytheistic sanctuary space. For many bishoprics, it would have been difficult to secure the land, the financing and trust of the community to build the first church within. To do this, careful planning and consideration was needed (Sweetman, 2015b).

We will now turn to the Church and place-making with a focus on examples from Cyprus, Crete and the Peloponnese. The rationale for this is to compare different processes of place-making in areas that experienced different levels of displacement in the fourth century, with a view to understanding the role of the Church, if any, in place-making and remaking. Various Cypriot and Cretan cities succumbed to earthquake destructions during the fourth century, including Salamis in 332, Paphos in 342, Kourion and nearly all of Crete in 365. Conversely, the Peloponnese was an area that was left relatively unscathed in this period.

Intentional and organic processes of adaptation and transferal of memory from existing to new religious places have been explored in the case of the Peloponnese, Crete and Cyprus (Sweetman, 2015a, 2015b, 2017). We will see how the Church as an institution used place-making, and indeed remaking in the case of sanctuary spaces, in the context of displacement to both draw in the local community and provide a sanctuary of steadfastness in an otherwise restless world. In many cases, place-making (or remaking) draws on existing associations with the landscape or pre-existing religious spaces. As such, it is likely that the Church drew on an element of communal commonality (be that through secular or religious spaces) to consolidate the shared sense of identity. This helps to explain why the earliest churches in the Peloponnese, for example, are located outside the *temenos* walls of sanctuaries at places such as Epidavros (Figure 9.1) and Olympia (Figure 9.2) (Sweetman, 2015a). While experiencing the specifics of the place, the visitor (be they local or even a pilgrim from further afield) can then transmit the memory and experience of places to their home contexts, helping to extend the sense of community and connection with the church (Marchand, 2015, 64).

Figure 9.1 Late antique church East of *Propylon* at Epidavros.

Source: Author's image

Figure 9.2 Late antique church in workshop of Pheidias at Olympia.

Source: Author's image

9.7 Place-making and landscape

It is likely that the earliest churches drew on associations with key figures in Christian stories. For example, the earliest known churches at Jerusalem and Bethlehem were connected to Christ through the places of his burial and birth. The church of the Holy Sepulchre in Jerusalem, the site of the crucifixion of Christ, was likely venerated from the time that he died there. Hadrian had a temple of Aphrodite constructed at the site, which was later replaced by Empress Helena's church in 326 CE, outside the walls of the city (Marchand, 2015). Marchand's (2015) work on modern pilgrimage has highlighted the myriad of emotions that a visit to the church – through crowds, noises, smells and heat – can elicit. This can be paralleled with visits to the first late antique churches, many of which were constructed on the edges of towns (e.g. in Corinth and Argos), at ports (Kiato and Kenchreai) or outside sanctuary walls (Epidavros and Olympia) (Sweetman, 2015b). While the churches at existing sanctuaries might have made a connection with the existing place, other early churches would have sought this through their construction in, for example, existing cemeteries (e.g. Kodratos in the North Cemetery of Corinth: Figure 9.3) or by gathering points such as wells or highly visible locations (Sweetman, 2015a). It can be seen from the topography of the churches in places such as Crete, and arguably the Cyclades, that wayfinding was an important element in the construction of the churches in their establishment throughout this period (Sweetman, 2017). Many of the late antique churches are purposely located along routes, as can be seen in the arrangement of the churches from the south of Crete (Agia Galini) to the north (Stavromenos) through the Amari Valley (Vizari and Sybrita: Figure 9.4). In this case, the churches which may have been used in wayfinding are replaced by route-planning or 'destination-orientated transport' (Ingold, 2016, 77).

Figure 9.3 Late antique church of Kodratos, Corinth.

Source: Author's image

Figure 9.4 Map of Crete showing location of late antique churches.

Source: Author's image

While the destination may be in mind, the careful placement of churches as nodes in a network would have reinforced the Christianisation of a more substantial space. People encountered as part of the journey to a church or Christian place would have formed part of the network, and Ingold (2016, 79) likens this to a 'carrying across' of life rather than a significant change. In a similar way, the coastal churches of the Cyclades might suggest an analogous function (e.g. the church at Trypiti on Melos, Paradise Bay on Amorgos and Agia Sofia on Sifnos: Figure 9.5). The intentionality of their location is seen in examples that are primarily visible from the sea and not the land (such as Episkopi on Sikinos and Agia Theodori on Naxos; Figure 9.6). There is a significant concentration of churches along the south-east coast of Cyprus (Salamis, Amathous and Kourion), lining the sea route from Antioch to the holy land. It is tempting to suggest that this was a deliberate emphasis to not only remind the traveller of the religion but also draw attention to the space of the island (Sweetman, 2017).

The presence of very early churches in places such as Cyprus (Salamis) and Crete (Knossos) may have a rather prosaic explanation as a tangential result of the seismic destruction there. The rebuilding programmes would have attracted craftspeople and architects from far afield, creating opportunities to provide new buildings such as churches.

In some cases, reused material may have helped to underline particular experiences of displacement, particularly if they were intentionally positioned to be recognisable as belonging elsewhere. The newly founded churches marked space in an obvious way in the landscape. However, once constructed, the church building itself functioned as a means of subtle control. They may have been located beside important routes or in control of economic resources or land. The architecture too, with its foreboding exterior and selective accessibility and secrecy, would have created a level of curiosity and also a place of refuge. In both modern and ancient visits, the sense of anticipation of the visit to the church would have been raised by the imposing building and concealment of the interior (Sweetman, 2015a, 2015b). Daelemans (2020) has convincingly discussed the value of the healing dimension in church architecture and it is possible to see where the presence of the church might have contributed to a healing function of a once displaced community. Visitors would be further awed by the impressive interiors

Figure 9.5 Late antique church, Paradise Bay, Amorgos.
Source: Author's image

Figure 9.6 Mausoleum, Episkopi, Sikinos.
Source: Author's image

Figure 9.7 Lechaion Basilica.

(mosaics, wall paintings, imported marble architectural features) (e.g. Sparta Acropolis Basilica or the Lechaion Basilica: Figure 9.7) (Sweetman, 2015a). Ancient congregations would have had interrupted views of the rituals and careful arrangement of the architecture to enable visual engagements with natural and artificial light. Furthermore, with a variety of different sounds from music, bells and voices and smells from incense, congregations, as with modern visitors, would have experienced similar sensory engagements in church visits (Marchand, 2015). These sensory experiences play on the emotions of the ancient and modern visitor alike, creating senses of displacement (Marchand, 2015, 72).

In addition to the architecture and location, the stories connected to late antique churches were fundamental to their attraction. Just as the journey to the church would have involved stories and people, the early church may have echoed this by being imbued with similar associations. It is arguable that the success of place-making may have included the carefully curated stories and individuals that became associated with the churches after the initial foundation. It is possible to take this a step further and even suggest that, in many cases, these associations echoed a sense of displacement. We can see that, even with a brief survey, many of the individuals associated with the first Christian churches were by and large peripatetic, that is to say, martyrs, bishops, imperial aristocracy and even the relics too.

9.8 Place-making, people and stories

When we are fortunate, it is possible to make associations between individuals and churches through inscriptions and or literary records. For example, the earliest church at Constantia (Salamis), the church of Agios Epiphanius, was commissioned by the bishop of Cyprus, Epiphanius (d. 403), and his body was subsequently placed in the church. Epiphanius himself was a wandering bishop and known as an outsider; he was born in Palestine, lived in Egypt and worked again in Palestine before moving to Cyprus. Even while bishop he travelled widely (Kim, 2015). The bishop of Alexandria, Athanasius, also travelled to North Africa and was notoriously exiled five times from Alexandria, whereupon he spent time in Trier, Rome and the Egyptian desert, among other places. When he finally returned to Alexandria, he had his church constructed in 370, which was later converted into the Attarine mosque. Bishop Babylos of Antioch died as a martyr in Antioch in 253 during the so-called Decian persecution. However, his remains were transferred a century later to the church of Babylas (in Daphne), commissioned by Emperor Constantinius Gallus. His remains were moved again by an enraged Julian who was unable to get an answer from the nearby oracle of Apollo because of its proximity to the body of Babylas (Sozomen, *Ecclesiastical History*, 5.19.1–5.19.17).[4] The remains of the bishop were finally reinterred in another martyrium church of Babylas in Antioch by the then bishop, Meletius, towards the end of his career in the 370s. Bishop Meletius himself was Armenian and exiled a number of times. He was replaced in 381 by Bishop Flavian, who recorded his name in a mosaic in the newly constructed martyrium church of Babylus.[5] These early churches are commonly associated with figures who were displaced in life and/or death. Unfortunately, the evidence is not so rich for many other locations. At the late-fifth-century church in Tegea, an inscription in the central aisle mosaic records that Thrysos, the priest, had the church and mosaics constructed (Asimakopoulou-Atzaka, 1987, no. 21), but it is not possible to locate a clergyman Thrysos beyond this reference. Some cases allow us to identify bishops as having attended ecumenical councils but little beyond that, for example, Bishop Euphrates who likely erected the church of the Archangel Michael in Eleutherna in Crete attended the Council of Chalcedon in 451 (Themelis, 2004). In other cases, bones have been located in tombs or reliquaries in churches but, again, it is not always possible to reconstruct associations with particular individuals.

It is more common to find associations with stories of martyrs rather than the physical evidence. Many of the numerous churches in the area of ancient Corinth and its suburbs have such associations: for example, the churches of Kodratos (Figure 9.3), Kraneion and Lechaion (Figure 9.7). The martyr Kodratos is said to have grown up in the wilderness as his mother had taken him and fled into the mountains to escape persecution (Engles, 1990). As something of a hermit, he attracted a following but was eventually tortured, dragged through the city of Corinth and beheaded as part of the Decian persecutions in the mid-third century. A church in his honour was established in the fifth century where he and his followers fell and their blood allegedly turned into a miraculous healing spring (Figure 9.3). This site, with its new healing associations, was some 400 metres to the east along the same contour as the Asklepion and on the route to the Lechaion harbour. Although later in date, the Lechaion basilica was associated with the martyr Leonidas, who was a teacher originally from Troizen. Following his trial during the Decian persecutions, he was tortured along with seven virgins who were miraculously about to walk on water but were eventually weighed down. Their relics went missing until the early twentieth century, but they are now said to be in Palaia Epidavros.

Apostle Andrew was said to have been captured near the village of Leontio in the mountains to the east of the city, where his footprints are allegedly miraculous carved into the rock. The remains of a late antique church are still visible. According to Gregory of Tours, in his *Glory of the Martyrs*, the relics in the tomb of Apostle Andrew produced healing oil which was also good for crops. As was already mentioned, Andrew's body was later transferred to Constantinople by Constantius II. A similar pattern is seen in the early church in Crete where, for example, the Ten Martyrs who came from towns across the island were murdered during the Decian persecutions at Alonion, close to the modern town of Agioi Dekoi. Later, during the reign of Constantine, the bodies were reburied in Gortyna by Bishop Paul of Crete in 312.

All of these examples signify displacement; from rural to urban space, land to sea, mountains to city and, quite often, the bodies are also transferred. As noted earlier, the associated stories of peripatetic holy people and objects would have reinforced the notion of finding a place. Their fixed places come with the foundation of the church. These disassociations and reassociations between people and places would have formed part of the processes of place-making and familiarisation. Once in place, these stories and objects of displacement would furthermore help provide a link in the broken line of tradition and identity that the displaced experience with loss of place.

9.9 Conclusion

Scholarship is increasingly showing a more peaceful and incremental Christianisation process than previously considered, certainly in the Eastern provinces (Frankfurter, 2018; Sweetman, 2015a, 2015b). Contrary to the idea of victory over adversity as a means of attracting a wide population to the church, more subtle means were required to encourage conversion. As such, place-making and remaking would have been keen to this process – giving existing communities a focal point in place of their old religion (even if there was significant continuity of polytheistic festivals) and displaced communities a means of gaining a sense of belonging.

To enable the earliest church buildings to be successful in the landscape, that is to say, function as a community hub, they had to appeal to a wide group of people, some of whom had direct and indirect experience of displacement as well as in a sense seeing their gods being displaced. Locating a church in a shared placed of importance or gathering draws on an emotional attachment to the area. When a shared sense of commonality is forthcoming, this helps people who feel (or are) displaced experience ‘symbolic belonging’ (Dromgold-Sermen, 2020). In many respects, the constant movement of pilgrims, bishops, imperial and holy individuals, relics in the church provides a sense of commonality in the shared displacement. Other features in the church would have echoed the sense of displacement such as the presence of relics or stories, burials or martyria, as well as the pilgrim or visitor themselves.

The creation of place through the situation of a church would have provided structure to and opportunities for community consolidation and be a catalyst for solidarity (Eckenwiler, 2018). This would have been further articulated through movement of pilgrims, bishops, imperial elite, relics and displaced. This idea of commonality in movement created an accumulation of collective experience helping to articulate the space as meaningful. As people and relics of faith migrated, so did the faith itself.

Place-making did not serve just one purpose. It gave the new religion a public focus. It both called to mind and embraced the state of displacement while subtly emphasising the value of the new but accepted. In practical terms, it provided a means for community building in the face of internal and external displacements. And in terms of the benefits for the church, it helped to mark ownership of land (with direct economic potential or indirect through leasing of the land). Through the location, architecture and associations, the church was an effective element of Christianisation processes.

Notes

1 I wish to thank Carlos Machado for the excellent discussions and insights on drafts of this paper.
2 For example, Bishop of Jerusalem, Macarius, was ordered by Constantine to build churches at the sites of the Resurrection, the True Cross, Nativity and Ascension.
3 Naidu (2016) refers to ‘personal scaffold of religious beliefs and behaviours’.
4 See http://csla.history.ox.ac.uk/record.php?recid=E02274 (ID E02274).
5 See http://csla.history.ox.ac.uk/record.php?recid=E01753 (ID E01753).

Editions and abbreviations

Gregory of Tours, *Glory of the Martyrs – Gregory of Tours: Glory of the Martyrs*, trans. R. Van Dam (1988). Liverpool University Press.

Philostorgius, *Ecclesiastical History – Philostorgius, Ecclesiastical History*, trans. P. R. Amidon (2007). Society of Biblical Literature.

Sozomen, *Ecclesiastical History – Sozomen, Historia Ecclesiastica. In A select library of Nicene and post-Nicene fathers of Christian church. Volume II. Socrates, Sozomenus: Church Histories*, ed. and trans. P. Schaff & H. Wace (1976). W. M. B. Eerdmans Publishing Company.

Zosimus – *Zosimus. New History* (Byzantina Australiensia 2), trans. R. T. Ridley (1982). Australian Association for Byzantine Studies.

Bibliography

Allen, P., & Neil, B. (2013). *Crisis Management in Late Antiquity (410–590 CE) (Vigiliae Christianae, Supplements 121)*. Brill.

Armstrong, G. T. (1967). Constantine's Churches. *Gesta*, *6*, 1–9.

Ashmore, W. (2014). On ancient placemaking. In O. Harmansah (Ed.), *Of Rocks and Water: Towards an Archaeology of Place* (pp. 40–46). Oxbow.

Asimakopoulou-Atzaka, P. (1987). *Σύνταγμα των παλαιοχριστιανικών ψηφιδωτών δαπέδων της Ελλάδος II: Πελοπόννησος – Στερεά Ελλάδα*. Κέντρον Βυζαντινών Ερευνων Θεσσαλονίκη.

Avraméa, A. (1997). *Le Péloponnèse du IVe au VIIIe siècle: Changements et persistances*. Byzantina Sorbonensia 15. Publications de la Sorbonne.

Barnes, T. D. (1993). *Athanasius and Constantius: Theology and Politics in the Constantinian Empire*. Harvard University Press.

Barrett, J. C., & Ko, I. (2009). A Phenomenology of Landscape: A Crisis in British Landscape Archaeology? *Journal of Social Archaeology*, *9*(3), 275–294.

Barry, J. (2019). *Bishops in Flight: Exile and Displacement in Late Antiquity*. University of California Press.

Bhugra, D. (2004). Migration, Distress and Cultural Identity. *British Medical Bulletin*, *69*(1), 129–141.

Bitton-Ashkelony, B. (2005). *Encountering the Sacred. The Debate on Christian Pilgrimage in Late Antiquity*. University of California Press.

Braginton, M. (1944). Exile Under the Roman Emperors. *The Classical Journal*, *39*(7), 391–407.

Bull, M., & Mitchell, J. (2015). Introduction. In M. Bull & J. Mitchell (Eds.), *Ritual, Performance and the Senses* (pp. 1–10). Bloomsbury.

Cadge, W., & Howard Ecklund, E. (2007). Immigration and Religion. *Annual Review of Sociology*, *33*(1), 359–379.

Chatterjee, H., Clini, C., Butler, B., Al-Nammar, F., Al-Asir, R., & Katona, C. (2020). Exploring the psychosocial impact of cultural interventions with displaced people. In E. Fiddian-Qasmiyeh (Ed.), *Refuge in a Moving World. Tracing Refugee and Migrant Journeys Across Disciplines* (pp. 323–345). UCL Press.

Chatty, D. (2014). Anthropology and forced migration. In E. Fiddian-Qasmiyeh, G. Loescher, K. Long, & N. Sigona (Eds.), *The Oxford Handbook of Refugee and Forced Migration Studies* (pp. 74–86). Oxford University Press.

Cohen, S. (2006). Cicero's Roman exile. In J. Gaertner (Ed.), *Writing Exile: The Discourse of Displacement in Greco-Roman Antiquity and Beyond* (pp. 109–128). Brill.

Corcoran, M. (2002). Place Attachment and Community Sentiment in Marginalised Neighbourhoods: A European Case Study. *Canadian Journal of Urban Research*, *11*(1), 201–221.

Daelemans, B. (2020). Healing Space: The Synaesthetic Quality of Church Architecture. *Religions*, *11*(12), 635.

Doerfler, M. (2019). Gone but Not Forgotten? Retrieving the Migrant in Late Antiquity. *Journal of the American Academy of Religion*, *87*(4), 1153–1177.

Drijvers, J. W. (1992). *Helena Augusta*. Brill.

Dromgold-Sermen, M. S. (2020). Forced Migrants and Secure Belonging: A Case Study of Syrian Refugees Resettled in the United States. *Journal of Ethnic and Migration Studies*, *48*(3), 635–654.

Eckenwiler, L. (2018). Displacement and Solidarity: An Ethic of Place-Making. *Bioethics*, *32*, 562–568.

Ellery, P. J., Ellery, J., & Borkowsky, M. (2021). Toward a Theoretical Understanding of Placemaking. *International Journal of Community Well-Being*, *4*, 55–76.

Engles, D. (1990). *Roman Corinth: An Alternative Model for the Classical City*. University of Chicago Press.

Estey, K. (2019). Place-making and religion: A solidarity psychology of the commons. In V. Counted & F. Watts (Eds.), *The Psychology of Religion and Place* (pp. 261–284). Palgrave Macmillan.

Frankfurter, D. (2018). *Christianizing Egypt: Syncretism and Local Worlds in Late Antiquity*. Princeton University Press.

Hack-Polay, D., Mahmoud, A. B., Kordowicz, M., Madziva, R., & Kivunja, C. (2021). "Let Us Define Ourselves": Forced Migrants' Use of Multiple Identities as a Tactic for Social Navigation. *BMC Psychology*, *9*(1), Article 125.

Halsall, G. (2007). *Barbarian Migrations and the Roman West 376–568*. Cambridge University Press.

Harland, P. (2009). *Dynamics of Identity in the World of the Early Christians*. T. and T. Clark.

Heather, P. J. (2017). Refugees and the Roman Empire. *Journal of Refugee Studies*, *30*(2), 220–242.

Hollenbach, D. (2014). Religion and forced migration. In E. Fiddian-Qasmiyeh, G. Loescher, K. Long, & N. Sigona (Eds.), *The Oxford Handbook of Refugee and Forced Migration Studies* (pp. 447–460). Oxford University Press.

Ingold, T. (2016). *Lines: A Brief History*. Routledge.

Kaegi, W. (2015). *Byzantium and the Fall of Rome*. Princeton University Press.

Kelly, G. (2004). Ammianus and the Great Tsunami. *Journal of Roman Studies*, *94*, 141–167.

Kim, Y. R. (2015). *Epiphanius of Cyprus: Imagining an Orthodox World*. University of Michigan Press.

Lems, A. (2016). Placing Displacement: Place-Making in a World of Movement. *Ethnos*, *81*(2), 315–337.

Luijendik, A. (2020). The Gospel of Matthew in the acts of Barnabas through the lens of a book's history: Healing and burial with books. In L. Nasrallah, A. Luijendijk, & C. Bakirtzis (Eds.), *From Roman to Early Christian Cyprus* (pp. 169–194). Mohr Siebeck.

Marchand, T. H. (2015). Place-making in the "Holy of Holies": The church of the Holy Sepulcher, Jerusalem. In M. Bull & J. P. Mitchell (Eds.), *Ritual, Performance and the Senses* (pp. 63–83). Routledge.

Mazumdar, S., & Mazumdar, S. (2004). Religion and Place Attachment: A Study of Sacred Places. *Journal of Environmental Psychology*, *24*(3), 385–397.

Naidu, M. (2016). Displaced Sense: Displacement, Religion and Sense-Making. *Journal for the Study of Religion*, *29*(1), 104–126.

Phillimore, J. (2011). Refugees, Acculturation Strategies, Stress and Integration. *Journal of Social Policy*, *40*(3), 575–593.

Porter, M., & Haslam, N. (2001). Forced Displacement in Yugoslavia: A Meta-Analysis of Psychological Consequences and Their Moderators. *Journal of Traumatic Stress*, *14*, 817–834.

Sennett, R. (2001, February). New Capitalism, New Isolation: A Flexible City of Strangers. *Le Monde Diplomatique*. https://mondediplo.com/2001/02/16cities

Sigona, N. (2014). The politics of refugee voices: Representations, narratives, and memories. In E. Fiddian-Qasmiyeh, G. Loescher, K. Long, & N. Sigona (Eds.), *The Oxford Handbook of Refugee and Forced Migration Studies* (pp. 369–384). Oxford University Press.

Stini, F. (2011). *Plenum exiliis mare: Untersuchungen zum. Exil in der römischen Kaiserzeit*. Franz Steiner Verlag.

Stiros, S. (2001). The AD 365 Crete Earthquake and Possible Seismic Clustering During the Fourth to Sixth Centuries AD in the Eastern Mediterranean: A Review of Historical and Archaeological Data. *Journal of Structural Geology*, *23*(2–3), 545–562.

Sweetman, R. J. (2015a). Memory, Tradition and the Christianization of the Peloponnese. *American Journal of Archaeology*, *119*(4), 501–531.

Sweetman, R. J. (2015b). The Christianization of the Peloponnese: The Case for Strategic Change. *Annual of the British School at Athens*, *110*, 285–319.

Sweetman, R. J. (2016). Networks: Exile and tourism in the Roman Cyclades. In J. F. D. Frakes, M. Egri, & S. E. Alcock (Eds.), *Beyond Boundaries: Connecting Visual Cultures in the Provinces of Ancient Rome* (pp. 46–61). Getty Publications.

Sweetman, R. J. (2017). Networks and Church Building in the Aegean: Crete, Cyprus, Lycia and the Peloponnese. *Annual of the British School at Athens*, *112*, 207–66.

Themelis, Π. (2004). *Πρωτοβυζαντινή Ελεύθερνα: Τομέας 2*. The University of Crete Publications.

Thunø, E. (2018). Reliquaries and the cult of relics in Late Antiquity. In R. Jensen & M. Ellison (Eds.), *The Routledge Handbook of Early Christian Art* (pp. 150–168). Routledge.

Tilley, C. (1994). *A Phenomenology of Landscape: Places, Paths, and Monuments*. Berg.

Valentine, G., Sporton, D., & Nielsen, K. B. (2009). Identities and Belonging: A Study of Somali Refugee and Asylum Seekers Living in the UK and Denmark. *Environment and Planning D: Society and Space*, *27*(2), 234–250.

Verme, P., & Schuettler, K. (2021). The Impact of Forced Displacement on Host Communities: A Review of the Empirical Literature in Economics. *Journal of Development Economics*, *150*, Article 102606.

Woods, D. (1991). The Date of the Translation of the Relics of SS. Luke and Andrew to Constantinople. *Vigiliae Christianae*, *45*(3), 286–292.

Zetter, R., & Boano, C. (2010). Space and place after natural disasters and forced displacement. In G. Lizarralde, C. Johnson, & C. Davidson (Eds.), *Rebuilding After Disasters: From Emergency to Sustainability* (pp. 206–230). Routledge.

Part 3

Space and meaning

10 Entanglement of public space and honorific statue habit in late antique Asia Minor

Esen Ogus

Space matters to social life, and what we call 'social' is not only about people, but is an entity constituted by a hybrid of people, places, things, and architecture (Law, 1994). The 1980s saw a spatial turn in the social sciences and humanities, which resulted in a vast body of research underlining how spatial structuring makes a significant difference to social relations (Urry, 1985, 1997; Soja, 1989; Lefebvre, 1992; Unwin, 2000). One important aspect of the inquiry into the association between space and the social is that forms of social relations (such as class and hierarchy) and forms of spatial relations (such as geography and spatiality) are entangled and mutually constructed. While the requirements of the social shape the space, the space also shapes the social in a mutually interactive way, with neither privileged over another *a priori* (Soja, 1996, 2000, 2008; Borch, 2002; Massey, 2005). Therefore, space is not only a medium manipulated to affect social life, but is also an agent that actively shapes the functioning of that life.

A second important aspect of the interaction of space with the social is the priority of viewing the former not as a static constructor of the latter but as dynamic and made up of moving elements involving humans, objects, and environments, all constantly in motion. These elements are assembled and reassembled in changing configurations and with rearticulated meanings (Urry, 1997; Cresswell, 2002). Places are never 'finished' entities but are rather always the result of processes and practices that are constantly being performed (Thrift, 2007; Anderson & Harrison, 2010; Cresswell, 2014). As such, a sense of place is developed through the interaction between the executive power that enforces the law, such as local or regional state institutions, and individual human agency (Pred, 1984; Cresswell, 2014).

These two aspects of the space and social interaction – that of the ability of space and the social to mutually shape each other and that of the construction of the social in a mobile and dynamic spatial setting – do not only concern modern spaces and urban life but also concern ancient ones, a reality that needs to be further scrutinized (Algra, 1994; Day et al., 2019). It is my aim in this chapter to consider the entanglement of space and the social in Late Antiquity, namely how the space–social binary contributes to the appropriation and remarkable longevity of a particular material element of urban life: honorific statuary. I will explore the ways in which the ideologically and socially structured space shaped the honorific statue habit and was in turn itself shaped by that habit.

The reason why the spatial and social contexts of honorific statuary deserve inquiry is the peculiar and persistent relationship of this medium of public honours with urban dwellers in Late Antiquity. Honorific statue culture was traditionally based on a sensitive balance of reciprocal favours between the city and the honorands (Stewart, 2003, 115–156; Ma, 2013, 300–306; Ward-Perkins, 2016b, 302). When accomplished and benevolent individuals were honoured and rewarded through public decrees and statues, they were further motivated for continual acts of

DOI: 10.4324/9780429427152-13

munificence (Ma, 2012, 2013, 300–306). The remarkable resilience of the statue habit in the classical world is evident from its long life that began around the fourth century BCE and its survival well into the sixth century CE. The statue culture in Late Antiquity most strongly persisted in areas where there was an earlier Imperial habit in the first place. These areas include North Africa, Italy, Greece, and Asia Minor (Smith, 2016a, 5; North Africa: Lepelley, 1981, 1994; De Bruyn & Machado, 2016; Italy: Machado, 2006, 2016; Witschel, 2007; Chenault, 2012; Machado & Ward-Perkins, 2012; Weisweiler, 2012; Machado with Lenaghan, 2016; Greece: Gehn, 2012, 2016a, 2016b; Brown, 2016; Asia Minor: Højte, 2005; Auinger & Sokolicek, 2016; Lenaghan, 2016; Smith, 2016b). In regions where this symbolic reward mechanism never took root under the Empire, namely Britain, Gaul, and the Roman Near East, the numbers of late antique statues also remained low (Smith, 2016a, 5; Hispania, Gallia, Raetia: Witschel, 2013, 2016; Near East: Gehn & Ward-Perkins, 2016).

The medium of statuary in Late Antiquity, albeit employed in select locations, defied all hostile odds aligned against it. These included factors such as cultural and religious changes, economic constraints, modified social profiles of honorands, and the broad sceptical attitude towards statuary in a late Christian Empire of cultural, religious, and administrative transformations (Brown, 1971, 111–114; Liebeschuetz, 2001). By the time the honorific habit was gradually abandoned, late antique society, in regions mentioned earlier, where the Hellenistic and early Roman statue culture was strong, had established almost a compulsive dependency on honorific statues. This is because, after at least a 500-year-old tradition of expressing gratitude and flattery to benevolent individuals, first and foremost the emperor, there was as yet no visual and cultural medium stronger than statuary that linked society to its honorands in a unique *quid pro quo* relationship (Stewart, 2003, 83–91; Bauer & Witschel, 2007; Ma, 2013, 300–306; Ward-Perkins, 2016b, 302).

As I argue here, one of the biggest factors in the persistence of honorific statuary is its intricate entanglement with urban space in Late Antiquity. On the one hand, the space was deliberately adjusted to accommodate new statuary and ensembles; on the other hand, the space itself modified the honorific habit by posing spatial and social constraints or by providing novel display opportunities. Statue viewership in the late antique city did not merely occur on its own but was rather structured and manipulated to maximize the benefits to be received from honours bequeathed by statues. For instance, maximum visibility of certain statues was prioritized in choosing a display space, such as optimizing the height of the locale and column that served as statue base or setting up statues along major urban thoroughfares where residents crowded. Similarly, new display areas were created to enhance urban life. Space is thus not a passive context but rather an ally in achieving the potential effects of statue dedications, as it supports the statue's meaning by creating a new social venue of gathering and passing.

The socially constructed space as relevant to honorific statuary operates on two main axes: visibility and mobility. 'Visibility' in the present work pertains to spatial and social arrangements intended to ensure that statuary remains more visible and meaningful to the public, hence more effectively implementing its symbolic and political agenda. The concern here is to dissect the nature of the 'representational space', which embodied complex cultural symbolisms and enhanced the semantics of the statue that it hosted (Lefebvre, 1992, 33). I define the 'mobility' aspect, on the other hand, as the set of dynamic viewing practices shaped by spatial and cultural constructs, encompassing both the mobility of the viewers and that of the statues themselves. Mobility is a factor that became particularly important in the context of the dynamic spatiality of the late antique city, as explained later, which is characterized by intensive architectural refurbishment projects and ever-changing visual cityscape with the frequent shuffling of statue ensembles and concurrent reuse of existing statues and bases.

The interplay of the visibility and mobility factors on the statue-viewing practices needs always to be comprehended on a third axis, which is 'temporality' (Urry, 1985). Formation of space is, after all, never independent of time. Spatiality and temporality are different from each other, but space is always subject to chronographic requirements, and neither can be conceptualized as the absence of the other (Massey, 1994, 264). The dynamic spatiality of Late Antiquity characterized by frequent renovations, as explained later, cannot be thought of separately from the simultaneity of time. The passage of time in Late Antiquity operated on the cityspace in an unpredictable manner, not at all consistently and in linear progression, either in favour of or against sustaining the statue habit. In other words, if we had a time-lapse video documenting a major urban centre from the late third to the sixth centuries, we would have witnessed, on the one hand, the gradually declining numbers of new dedications, and the eventual demise of the statue habit in the sixth century (Ward-Perkins, 2016b, 295–296). On the other hand, however, we would also have noticed the reuse and refurbishment of statues, and as I explain below, statues pulled down and later set up again, and a back-and-forth controversy and anxiety on statue dedications.

This social unease could perhaps best be attested by the surviving law codes promulgated over a period of 50 years. They indicate increasingly harsh measures to control the setting up of statuary for people beneath the imperial family and of imperial statues dedicated by private individuals (Ward-Perkins, 2016b, 307). A law from 398 CE made it illegal for brass, silver, or marble statues to be dedicated to governors during their term of office without the permission of the emperor (*CJ* 1.24.1). In the code of 425, a governor was mandated to be present when imperial statues were erected, and imperial images were not to be taken from private collections (*CJ* 1.24.2). In 439, the emperors banned the setting up of any imperial statue by private individuals (*CJ* 1.24.3). Finally, a law of 444 makes it mandatory that the expense of a statue should be paid by the honorand himself (*CJ* 1.24.4; Horster, 1998; Ward-Perkins, 2016b, 307). There is no good evidence whether these laws were enforced. If they had been, there would probably have been no need for subsequent legislation to further underline the issue. Moreover, the mandates for permits and the imperative to pay for a statue out of one's pocket were probably intended more to deal a final blow to the habit than to regulate it. The increasingly prohibitive language could be interpreted as a symptom of the escalating controversy surrounding the statue medium, in which we can append the dimension of time to the interaction of space and the social. This controversy is further evident from another law of 425, 'On Imperial Images', which made clear that the adoration of imperial statues or images were condemned, since 'worship in access of human dignity is reserved for the supreme God' (*CTh* 15.4.1; Setton, 1941, 211; Ward-Perkins, 2016b, 305).

In what follows is a discussion of the cultural meaning, employment, and broader display habits of honorific statuary in Late Antiquity. Two consecutive sections discuss the aforementioned dimensions of statue-viewing in Late Antiquity in relation to urban space: visibility and mobility. The broader substrate of the discussion will be the mutual interaction of space with the social, more specifically how space shaped social constructs, and how social requirements in turn shaped the space. My main geographical area of focus will be the East Empire, namely Constantinople and Asia Minor, which housed not only vivid centres of urban life and significant late antique evidence but also a remarkably strong honorific statue tradition during the period in question, second only to Rome and Italy in the numbers of statue dedications (Smith, 2016a, 5, Figure 1.4; Ward-Perkins, 2016a, Figure 2.2). The chronology of the outstanding statue record in Asia Minor could be traced uninterruptedly from the Imperial period to the sixth century, whereas in the West, with the exception of Rome, the classical urban culture and the statue habit ended around 150 years earlier than the great eastern centres (Stewart, 2003,

162–174; Lenaghan, 2016, 99–101, Figure 8.4; Smith, 2016a, 5). The two best documented cities to provide preserved statues, statue bases, and their contexts are Ephesus and Aphrodisias in Caria. While Constantinople provides many high-quality examples of honorific statuary, the contextual information from this city relies on highly controversial ancient textual evidence. The present work, therefore, for lack of extant and fully published evidence, presents the majority of its case studies from these urban centres. In addition, sites such as Corinth in Greece, Perge in Pamphylia, Sagalassos in Pisidia, and Tripolis in Lydia are brought up for comparanda.

10.1 Late antique honorific statues

Several important shifts took place in the honorific practice in Late Antiquity, in comparison to the preceding Imperial period. These shifts include a noticeable decline in the overall honours and a significant change in the composition of honorands. In the early period (first to third centuries CE), honours for civic notables were much higher in proportion to other types of honours. In the late period (fourth to sixth centuries CE), however, the main honorands came to be the emperors, imperial family, and imperial office holders (governors) (Smith, 2016a; Ward-Perkins, 2016a).

Probably the most remarkable shift in the habit is the intensive reuse of imperial statue monuments. Reuse had always been common, even in the early imperial period, but its volume and manner of execution in Late Antiquity, when almost all bases and more than half of the statuary around the Empire were reused, makes reuse the mainstream practice of dedicating statues (Smith, 2016a, 4). In the early and mid-fourth century, statue setups almost entirely rely on the already-existing imperial material, whereas in the late-fourth and fifth centuries new statue types and production were added to the bulk of reused statuary (Smith, 2016a, 8–9, 19). Intensive reuse was not the result of spontaneous and private initiatives but was part of an administrative system, frequently carried out by local and imperial authorities. It has been documented on the data of statue bases that in Rome, the highest-ranking officials, *praefecti urbi*, were responsible for more than 40 per cent of reused bases (Machado, 2017, 334). The proportion of initiatives by local magistrates and councils around the rest of Italy is an even higher 60 per cent (Machado, 2017, 334). The data suggest that reuse was more a communal and institutionalized than a personal *ad hoc* practice dependent on contingencies.

Even though new statues in new costumes were produced in the fifth and early sixth centuries, reuse and recycling remained an ongoing, 'sustainable' practice, rather than a one-case event in the form of 'converting' early imperial statues and bases into later ones. As an example, at Ephesus, a statue base was used at least three times, once in the early imperial period and twice in Late Antiquity. The base with the imperial inscription on it was first turned around 180 degrees, then inscribed for Dulcitius, the governor (proconsul) of Asia in 361/63, and for Messalinus, also governor of Asia (fourth to fifth centuries) *(LSA*-730, 733). Similarly, the statue of the governor, Flavius Palmatus, in Aphrodisias has a portrait head from the late fifth or early sixth century, but a reused century-old statue body, dating to the late fourth or early fifth century (*LSA*-198; Smith, 2016a). Judging from this high statue turnover in the urban centres such as Ephesus and Aphrodisias in Asia Minor, where the main evidence comes from, it may not at all have been unusual for a late antique urban dweller to witness during their lifetime a different statue in the same display spot.

This intense recycling activity went hand in hand with frequent seismic activity around the Mediterranean, or natural wear and tear, which probably posed structural and financial challenges in maintenance needs and coping with ruins. As attested by archaeological evidence, the ruinous state of infrastructure and derelict buildings must have made a mark on the urban landscape for decades on end. Ephesus, for example, suffered from three concurrent earthquakes in the fourth century, in 358, 365, and 368, which resulted in the destruction of public buildings, such as the structures in the so-called State Agora, Domitian's fountain, the fountain of Laecanius Bassus, Trajan's Nymphaeum, and the Scholasticia Baths (Figure 10.1, nos. 18, 28, 29, 38, 41) (Ladstätter & Pülz, 2007). Some inscriptions attest to the 'ruinous' state of the late antique city by suggesting that the patron 'saved' the monument or the statue from ruins. For instance, in an inscription from the mid-to-late fifth century built into the monumental Gate in the Place of Palms at Aphrodisias (Figure 10.2, no. 20, formerly called the South Agora by excavators), the Gate speaks in the first person: 'Stranger, sing of Dulcitius, the governor, giver of games and founder and lover of honour and Maioumarch, who, stretching out his strong hand, raised me too, who had suffered for unnumbered years' (*ALA 2004*, no. 40; Wilson, 2016). In the same city, the statue of Flavius Eutolmius Tatianus, the fallen praetorian prefect of the East, was pulled down, but later re-erected by his relative. The statue base speaks in the first person:

> [A]ll-conquering time would have destroyed me if my child, of the third generation, who has the same name and similar achievements, had not lifted me up from the ground and set me on a monument, to be seen and admired by all, local inhabitants and strangers alike.
>
> (*ALA 2004*, no. 37; *LSA*-193)

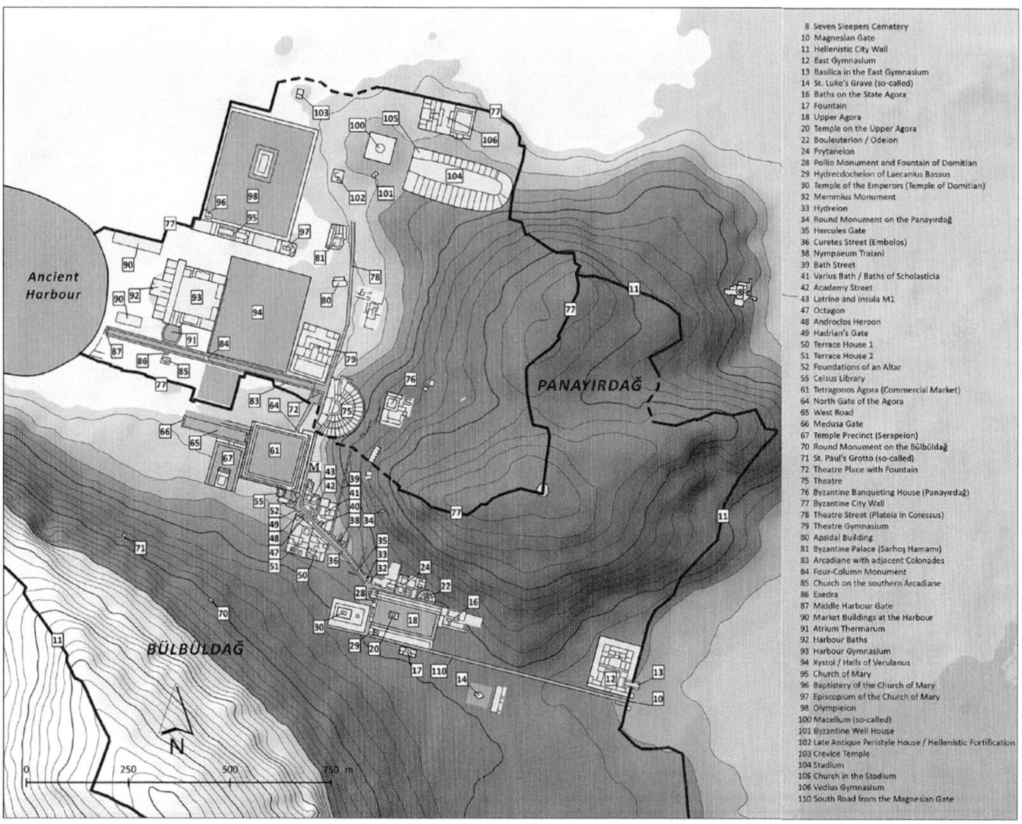

Figure 10.1 Plan of Ephesus.

Source: Courtesy of OeAI-OeAW. Copyright C. Kurtze

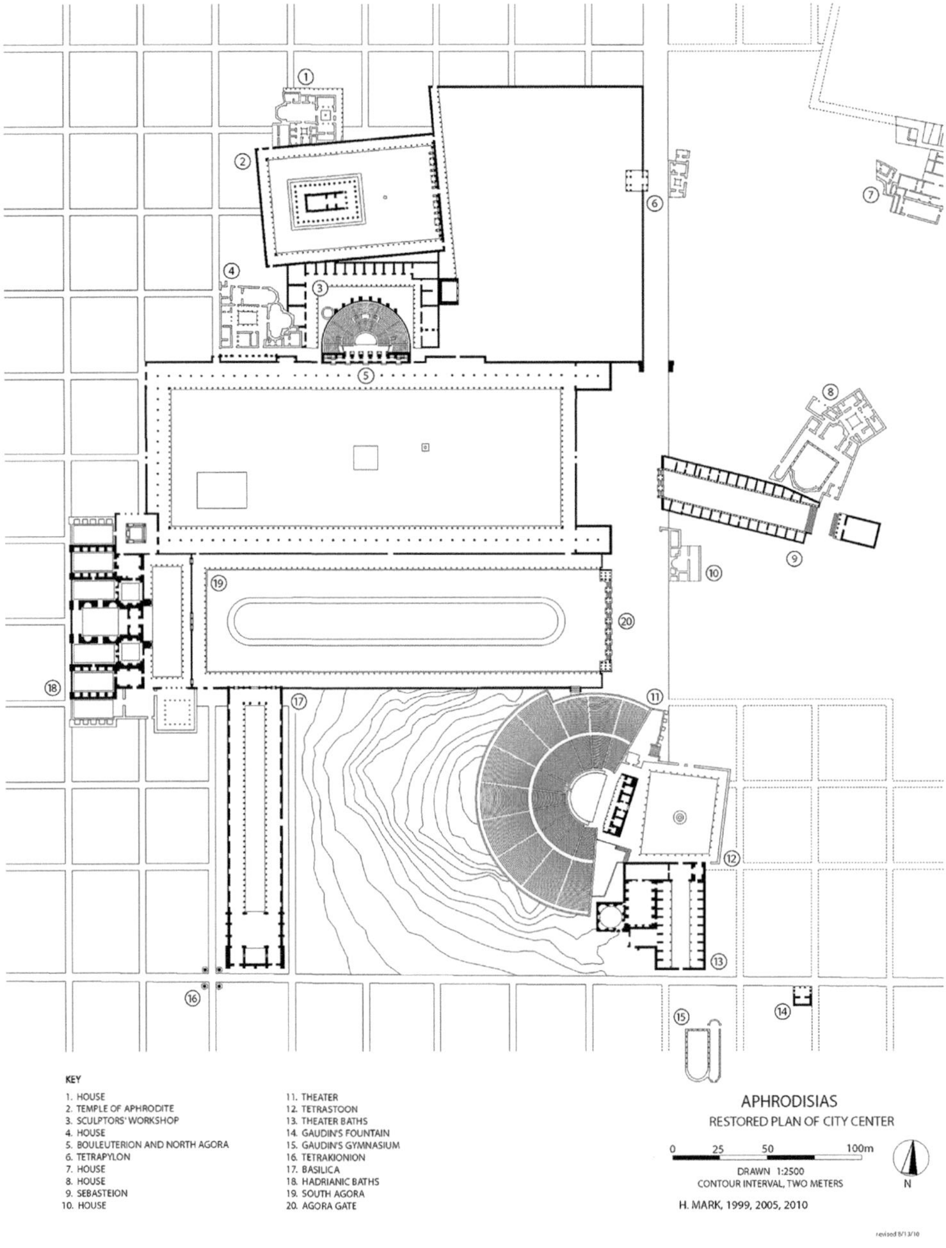

Figure 10.2 Plan of Aphrodisias.

Source: Courtesy of NYU Aphrodisias Excavations

We also have evidence for ruinous state of imperial contexts outside of Asia Minor. In Thubursicu Numidarum (Africa Proconsularis), a statue base found on the *forum novum* erected in 361 in honour of Constantine mentions that the statue and inscription had been removed from ruins (*de ruinis*) (*LSA*-1182). The inscription probably refers to the old forum, which had been abandoned in the period between Constantine and Julian, possibly because it had been destroyed by a natural disaster such as an earthquake (Lepelley, 1981, 214).

Alongside with rebuilding ruins, new construction activities also occurred, particularly during the Theodosian period, but neither consistently in all years nor ubiquitously. Even then, many new projects entirely recycled construction material from earlier buildings, which is the case also in Rome, starting from the time of the Tetrarchs and Constantine (Coates-Stephens, 2001, 231–233). An example is the fountain basin built at the end of the fourth or in the early fifth century in front of the Celsus library at Ephesus, recycling relief blocks from the so-called Parthian Monument (Miltner, 1958; Foss, 1979; Thür, 1999a; Ladstätter & Pülz, 2007; Wiplinger & Wlach, 1995). The library, whose façade had once been a display space for the statuary of Celsus and his family, now likely re-employed the same statues for the late antique restorers of the place, the proconsuls Stephanos and Philippos, who are named in an inscription above the central window of the façade of the library and on a statue base in the upper storey (*IEph* 5115).

All these replacements and recycling activities suggest that late antique urban space had a dynamic, spontaneous, and ever-changing character, perhaps even more so than the earlier imperial period, because of the intensity, frequency, and immense scale of refurbishment, exclusively relying on recycled materials (Saradi-Mendelovici, 2006; Kristensen & Stirling, 2016). Correlated to this volatile display space, statues were treated in a broad spectrum of manners, ranging from being displayed in the most prestigious place in the city to ending up as dumped waste. The imperial columns in Constantinople (more on them below), mounted with precious statues of emperors, may be thought of as examples of the former. Despite such glamorous statues dominating the cityscape, examples for disgraced statues are plenteous. At Corinth, several portrait heads, dating between third and fifth centuries, were intentionally separated from their bodies and dumped in the drains of the Peirene fountain at an unknown later date (Rothaus, 2000; Robinson, 2011, 283–284; Brown, 2016, 183, n. 42). Likewise, fifth-century tondo portraits of philosophers from Aphrodisias were shattered into pieces and discarded sometime in the sixth century in the back alley of the urban mansion in which they had once been displayed (Figure 10.2, no. 8) (Smith, 1990, 130). Considering this unstable atmosphere of statue display and its spatial context, the next section discusses how the factor of visibility played a role in the visual presentation of statues, and how this factor was manipulated by the dominant state ideology to characterize social life in the late antique city.

10.2 Visibility

Visibility appears to have been a major factor in the choice for the locations of statues, especially in large urban centres, where there were more varied display options than the small cities and where the expanse of the urban landscape posed the challenge to display the image to all residents. In Constantinople, spatial layouts of honorific statues attest to a viewing practice that mandates a visibility regimen imposed by the state. Tall imperial columns were set up in imperial fora, which were built on the visually dominating hills of Constantinople and aimed to be viewed not only by visitors to these spaces but also by all residents of the city. The topographical features of New Rome, a city laid out on hills and valleys, were exploited to serve this peculiar statue setup. Constantine I's colossal bronze statue stood on a porphyry column set up in 328 in his forum, laid out on a dominating hill of about 50 metres above sea level, on the ruins of the Hellenistic and Roman necropolis of Byzantium (Müller-Wiener, 1977; Janin, 1964). The column itself was 36 metres tall (Figure 10.3) (Ousterhout, 2014). Standing at a total height of over 80 metres above sea level, Constantine's porphyry statue would have dominated the skyline and been visible from, for instance, the main congregation space of the public, the Hippodrome, which lies at about 30 metres above sea level (Janin, 1964). Constantine's column was succeeded by that of Theodosius I (386–94 CE) in the Forum of Theodosius (Tauros), a column that

Figure 10.3 Column of Constantine, Constantinople (Istanbul).

Source: Author's image

was later reused as a base for the statue of Emperor Anastasius (506 CE). The spiral column in the Forum of Arcadius, crowned by the colossal statue of the emperor (401–421 CE), came next in the sequence (*LSA*-2458, 2459). Just like the Forum of Constantine, the Fora of Theodosius and Arcadius were also about 50 metres above sea level (Janin, 1964).

Other columnar monuments were set up in Constantinople: Eudoxia, empress and wife of Arcadius (403 CE), had a column in 'Pittakia', Theodosius II (402–450 CE) in Hebdomon, Marcian (450–52 CE) in the Forum of Marcian, and the column of Leo I (457–74 CE) in the Forum of Leo (*LSA*-27, 31, 2461, 2462). The imperial tradition of statues on tall columns continued up to Justinian and was ended by Phokas in 609 (*LSA*-2463, 2774). This last epigraphically attested statue was located in Magnaura, not in a public venue accessible by all, but in a distant imperial complex and in the middle of an arsenal. The reasoning for this rather 'withdrawn' statue display is not clear; however, frequent attacks against statues might have played a role, so would setting up imperial statues now being controversial from a Christian perspective (see Georgius Cedrenus [Bekker, 1838, 709, 9]; Leo Grammaticus [Bekker, 1842, 146, 1]; *Parastaseis*, no. 74; *Patria* 2.34; Anderson, 2016, 299–300).

The columns in Constantinople complemented a long tradition of Republican and Imperial columns in Rome, for instance those of Trajan and Marcus Aurelius (Trajan: Maffei, 1993b; Marcus Aurelius: Maffei, 1993b). Under the Tetrarchy, the Forum Romanum was restored and contained five porphyry columns that were each 12 metres high topped by statues of the Tetrarchs along with Iuppiter Optimus, and a marble column 17 metres high topped by another statue, likely of Diocletian (Weisweiler, 2012; Machado, 2019, 97–102). It has been suggested that these statues were not positioned in the centre of public squares but often at their entrance and exit points, impeding the free movement of the public (Bauer, 1996). This statue layout mandates a viewing practice implemented by the state, obligating residents to view the statues every day, rather than granting them the free choice to do so (Bauer, 1996). A similar viewing policy in Constantinople – by exploiting the convenience of topography and height – eliminates free choice and sets up dominant subjects in the skyline for maximum visibility (similar example in Coarelli, 2009).

Another issue most pertinent to the visibility of honorific statuary is the display of the honorific statues in places other than on tall columns. During the imperial period, the main display spaces for honorific statuary were the relatively static spaces in the city, such as the agora, the columnar facades of public buildings, and other closed spaces inside the porticoes. Such places, especially the agora, continued to be prominent spots for statue displays. For instance, at Sagalassos, many bases were set up in Late Antiquity in the Lower Agora (Lavan, 2013, 2020, 296–299), and at Aphrodisias, the courtyard of the Hadrianic Baths and the west portico of the Place of Palms maintained their prominence through Late Antiquity (Figure 10.2, nos. 18 and 19) (Smith, 2016b). However, from the Theodosian period onwards and over the course of the fifth century, a popular display space came to be along the colonnaded street, not inside the portico, but in front of it (Jacobs, 2012; Lavan, 2006, 2020, 296–299). Earlier in the imperial period, some statues were perhaps occasionally displayed in front of the colonnades of the porticoes; however, many surviving late antique ensembles show that, even in the newly built spaces, statues were frequently and deliberately organized and laid out to be in the space of the passer-by lining both sides of the street. Many cities in Asia Minor got caught up in the trend of laying out their streets with statues on either side. In Ephesus, the statue bases and statues mounted on them lined up both sides of Embolos, enveloping the passer-by on either side with the portrait images of the emperors and other dignitaries (Figure 10.1, no. 36; Figure 10.4). In Perge (Pamphylia), late antique statue bases lined up outside the colonnade of the West stoa on the main North-South Street (Figure 10.5).

Figure 10.4 Statue bases on Embolos, Ephesus.

Source: Reproduced from 'AW-OAI-EKU-00623'. Copyright by N. Gail

Figure 10.5 Statue bases in West portico of main North-South street at Perge.

Source: Author's image

'Lack of available space' inside the porticoes could not be the reason – at least not a ubiquitous one – for this peculiar choice. While many of the statuary from the Imperial Period remained untouched in Late Antiquity, hence their display venues have been dubbed 'frozen', the statues that were left *in situ* were the most significant and relevant for the city's past and future viability (Bauer, 1996, 2007; Witschel, 2007, 2008). At Ephesus, for instance, new statues in the Theodosian period were exclusively displayed along the streets and not in public squares, which were probably already packed with statues (Auinger & Sokolicek, 2016, 163). However, when necessary and desired, the statue inventory inside the porticoes could be cleaned out to accommodate new statues. At Aphrodisias, for instance, the late second to early third-century imperial statues of Dometeinos and his niece Tatiane inside the north portico of the North Agora were never touched throughout the city's late history, but their space was supplemented in the fourth century with the statue of Alexandros (*LSA*-152, 153), a governor, and in the late fourth or early fifth century with the statue of Oecumenius (*LSA*-150, 151) (Figure 10.2, above no. 5; Figure 10.6). These late additions likely replaced other imperial statues that were no more prominent or relevant in the late antique context. The same 'wiping clean' strategy probably also occurred in Constantinian Ephesus, where new statues of emperors and their families were set up in traditional public buildings, such as the fountain of the State Agora, Tetragonos Agora, and the Harbor Gymnasium (Figure 10.1, nos. 18, 61, 93), after cleaning these spaces or recycling the old statues there (Upper Agora: *LSA*-739, 2079; Harbour Gymnasium: *LSA*-744, 2080; Tetragonos Agora: *LSA*-713; Auinger & Sokolicek, 2016). Streets at this time played a less important role than they would in the later Theodosian period, when the statue displays were

Figure 10.6 Statue bases for Alexandros, Oecumenius, and Dometeinos (from left to right) in North portico of Agora, Aphrodisias.

Source: Author's image

exclusively along the streets (Auinger & Sokolicek, 2016, 163). Setting up statues in traditional spaces in Ephesus, and adding new statues in old ensembles in Aphrodisias, shows that opening up new spots in these spaces was a possibility. In this case, the 'pouring out' of statues from the porticoes and their lining of the colonnaded streets would best be explained as a deliberate choice that concerns viewing practices – rather than a consequence of spatial constraints – hence about the social aspect of the spatial.

Why would the colonnaded street be a prominent statue display space in Late Antiquity? While urban armatures had always been perceived as significant elements in the Roman city, they were more strongly associated with civic pride in Late Antiquity, as expressed by several contemporary authors, including Libanius, Malalas, and Procopius (Libanius, *Or.* 11.196–202; Malalas, *Chr.* 13.30; Procopius, *De Aedificiis* 3.4.18). Libanius, for instance, proudly boasts that the streets in Antioch 'are stretched out to such a length that merely to pave such a space would require a great force of men, while to go to the end from the beginning is a toil, which requires the help of horse' (Libanius, *Or.* 11.197). Moreover, around the Theodosian period, there is a widespread trend of renovating, relaying, and redecorating these urban armatures in Asia Minor (Jacobs, 2012). Streets were decorated with a combination of old and new statues, some carried from other locations. At Ephesus, for instance, the Embolos was redesigned and decorated in the late fourth century with ensembles such as 12 bronze Nike statues collected from elsewhere in the city, accompanied by the statue of Aelia Flacilla (379–386 CE) and likely also an equestrian statue of Theodosius (Figure 10.1, no. 36) (*LSA*-723; Roueché, 2002; Auinger & Aurenhammer, 2010). Cities that had existing Imperial collections of statues and bases, such as Ephesus, reused them, producing anew gradually declining numbers of statues. The small city of Tripolis in Lydia provides evidence that when no existing collections were available, new statues out of new materials could also be produced. Tripolis impressively lined up its main thoroughfare with more than 20 statues of governors newly made out of local red onyx stone, which no doubt attests to this city's wealth and enthusiasm in involvement (Duman, 2017, 4–5). Parallel to the rise of streets as centres of the civic life, some agorae lost their appeal in the course of the fifth century CE, with the exception of agorae that interrupted a colonnaded street, such as those in Constantinople or Sagalassos (Jacobs, 2012).

The emergence of streets in Late Antiquity as display spots for honorific statuary is also related to their function as a stage for traditional and widespread processional events, including imperial and ecclesiastical processions (see chapters by Rebecca Sweetman and Nadine Viermann, in this volume). Processions in early imperial streets were a daily occurrence (Rogers, 1991). In these events, celebrants carried statues of ritual objects, stopping at certain points for specific acts of ritual before heading to a temple or a sacred precinct (Rogers, 1991). A well-documented procession in Asia Minor is the one that took place in Ephesus, founded by C. Vibius Salutaris in 104 CE, which involved the procession of ephebes with 31 gold and silver statues dedicated by Salutaris along a circular route from the Temple of Artemis every two weeks during the year (Rogers, 1991). The vivid Roman tradition of processions continued in Late Antiquity, but with an enhanced liturgical flavour and enthusiasm about movement along the main thoroughfares. In Constantinople, the traditional imperial ceremony surrounding births, marriages, anniversaries, triumphs, and feasts gained a liturgical dimension in the era of Theodosius, and candlelit processions along the main street of Mese continued to grow under Arcadius and later emperors (Croke, 2010; Nadine Viermann, this volume). Statues in Constantinople were arranged along these processional routes, which united the imperial fora and precipitated the viewing of the imperial columns in these fora. The Christian processions of Constantinople were probably embraced by other cities in Asia Minor, for instance by Ephesus, which already had a native processional culture as attested by epigraphic and archaeological evidence (Rogers, 1991; Roueché, 1993, 1999). The placement of the honorific statues along these thoroughfares allowed the city's bureaucratic and cultural history to be on display. These statues became essential elements of street architecture, just like architraves, decorated friezes, and column capitals. The statues were made and set up to be viewed, revered, and paid attention, and the best space to meet these requirements was in front of the street colonnades, rather than inside.

The insistence on the visibility of statues could perhaps be interpreted as a way to structure human agency through spatial arrangements (Cresswell, 2014). The statue setups were arranged so that a passer-by has no choice whether to view them or not. These setups are directly related to the idea of control of the public, especially when we imagine the statues lining both sides of the street, narrowing the street with their bodies and bases, and their gaze hitting the walking crowd and implementing a psychological control mechanism, similarly to a panopticon. Michel Foucault's frequently cited work on space and visibility draws attention to the panopticon, a Victorian-era architectural device, which was a central tower with an invisible guard, around which various cells were arrayed (Foucault, 1979; Emmison et al., 2012). People in these cells cannot tell whether they are being watched, and therefore they constantly monitor their behaviour to conform institutional rules. Shaping the display space for a socially appropriate statue display in Late Antiquity could perhaps be interpreted as one of the symptoms of rising authoritarianism and bureaucratic infrastructure legally redesigned to tighten the grip of the centralized government (Jones, 1964, 321–365, 401–410; Liebeschuetz, 1972, 167–191; Liebeschuetz, 2001, 104–136). As mentioned, most of the honorific statues in this period, differently from the imperial period, were of emperors, the imperial family, and statue officials such as governors, rather than of local aristocracy (Smith, 2016a, 4–5, Figure 1.3). These honorific statues of emperors and state officials, as symbols of authority, would most successfully fulfill their purpose by being the most visible and controlling to the average public viewer. It is perhaps no wonder that, as has often been noted in stylistic studies, the eyes of the honorific statues in Late Antiquity are considerably larger than natural and those of the earlier period Imperial statues, a phenomenon independent of geography, but most prominent in the fifth and sixth centuries (L'Orange, 1965, 1973; Smith, 1999, 185, 2016a, 24; Marsengill, 2013). While, of course, the meaning of these eyes might vary depending on the context and to whom they belong, in general, they add to the statue the quality of awareness and ability to see, whether visually or metaphorically. It is possible to interpret these 'all-seeing' eyes, in addition to a symptom of the style of the era, as a symbolic media to direct their 'animated' gaze on the crowds of people and control their social behaviour.

The visibility issue concerning the statues, therefore, is mutual; on the one hand, it involves the crowd visually or physically directed into spaces where they would necessarily have to view the statues and, on the other hand, concerns the symbolic surveillance of the crowd by the presence and gaze of the statues surrounding them on either side. Available topographical space, either the hilltops, as in Constantinople, or the thoroughways that linked centres and major monuments, such as the Embolos at Ephesus, shaped the route of social processions. Mese in Constantinople was even built and decorated over time to connect the hills on which the imperial fora and statue displays were located, hence was a space of 'process' rather than one of static geography (Bauer, 1996, 146–147; 2001, 32). While human agency was directed through predetermined routes, spatial constraints and allowances configured the social. On the other hand, the intense visual conversation of the crowd with honorific statues shows that the visibility of statues was perceived to be a significant element of social acts, namely civic and liturgical processions. Just as there would be no point in leaving a street unpaved and undecorated with fancy architectural elements, there would also be no point in a procession passing through an empty street. In other words, the social requirement of 'engagement' with the retinue of city's past rulers defined the spatial context of statues in the late antique urban life of Asia Minor. It is another issue how mobile viewing practices shaped space, on the one hand, and are determined by spatial construct, on the other. In what follows, I describe mobility as a second spatial dimension in urban Late Antiquity.

10.3 Mobility

Mobile viewing practices in Late Antiquity are multifaceted, involving both the mobility of the viewers and the mobility of the portrait statues. In the former case, statue setups along the streets were meant to be viewed and venerated by a moving crowd. The mobility of statues, on the other hand, occurred for a number of reasons, including changing the display spot of statues for reuse purposes or by moving the statues themselves on special urban processions.

10.3.1 Moving viewers

The importance of civic and religious processions, hence the mobility of the beholder, has already been mentioned earlier. In the late antique city, various civic and liturgical processions that took place after the Roman processions – although with enhanced detail and numbers – became not an option to enliven daily lives but an unavoidable aspect of urban lifestyle (Lavan, 2020, 237–240; Nadine Viermann, this volume).

In Constantinople, processions were attested well into the Byzantine period (Baldovin, 1987). Some of these processions even involved the veneration (*prosyknesis*) of imperial honorific statues, a practice that continued into the fourth century and beyond (Setton, 1941, 202–203). For instance, a chapel was built at the base of the porphyry column of Constantine the Great, and the column continued to act as a significant stop in urban liturgical processions in Constantinople (Mango, 1993). According to the tenth-century *Typikon*, 46 out of the 68 processions had an intermediate service at the Oratory of Constantine in the base of his column (Baldovin, 1987). While intermediate stops occurred at the imperial fora adorned with statues, we have evidence that statues were also lined up along the stoas connecting to the Mese, for instance at 'Peripatos', which seems to be a stoa to the south of Mese (*LSA*-350, 352, 353, 408). The processional crowd along the main route of Mese at the time of Theodosius, included even the emperor himself alongside clergy and courtiers (Baldovin, 1987, 181–204; Bauer, 1996, 381–387, 2001, 2008; Croke, 2010).

This processional culture, hence the mobility of viewers, is attested to not only by written sources but also by archaeological evidence, namely the habit of inscribing acclamations in various urban spots starting from the late third century CE. These acclamations could be in praise of a faction, an emperor, or a significant individual. It is quite likely that such acclamations took place on special occasions such as when the portrait of an emperor was received or when new legislation was announced (Roueché, 1993). Perhaps the acclaiming groups even physically stood at those locations to make their acclamations, although that does not have to be the case; these texts might represent a record of the acclamations made at a public assembly, for instance at the theatre (Roueché, 1999). Regardless, these acclamations are not graffiti but are well-kept records of actual events.

The practice is well attested in the cities of Asia Minor, particularly Ephesus and Aphrodisias. In Ephesus, acclamations were inscribed on architectural elements in various urban contexts, especially along the streets. They are the densest in the Marble Street (Figure 10.1, labelled M) and the southern part of the Embolos (Figure 10.1, no. 36), where it joins the Marble Street around the Gate of Hadrian area (Figure 10.1, no. 49), while no acclamations were found in Arcadiane (Figure 10.1, no. 83) (Roueché, 1999, 162). On the lower end of the Embolos, at its junction with the Marble Street (Figure 10.1, no. 49), two of the column capitals, reused from the Prytaneion, were inscribed with monograms, one to the Greens, and the other to the Greens and Christian emperors (Roueché, 1999). On the west side of the Marble Street lies the 'Neronian Basilica', whose importance in Late Antiquity is suggested by the survival of several copies of sixth-century legislation in and around it (not labelled on the plan, east side of Tetragonos Agora, Figure 10.1, no. 61) (Roueché, 1999, 163). On the wall of the doorway leading from the Basilica to the Marble Street, a substantial inscription acclaimed Emperor Heraclius and his son together with the Greens. On the southern end of the Marble Street, the gate of Hadrian (Figure 10.1, no. 49) also bears an acclamation for Christian emperors and the Greens, dating from the fifth to seventh centuries. These texts strongly suggest that the Marble Street was the focal area for civic ceremonies, at least by the sixth and early seventh centuries. There is no evidence that these texts are contemporary with each other; rather, it seems that they accumulated over time.

It is unlikely to be a coincidence that the majority of the late statues were excavated in the Embolos and the Marble Street, which, alongside the Theatre Street, were hotspots of late antique statue displays (Auinger & Sokolicek, 2016). There are two coexisting dynamics here: on the one hand, the statuary were set up and the acclamations were inscribed in the busiest urban areas to ensure their visibility; on the other hand, the inscribed acclamations attest to a mobile crowd, who were structured to view the statues while on the move. Hence, the statues were set up visibly, and at the same time to visually converse with the acclamation-utterers, inscribers, and their readers.

The connections of acclamations with factions, the organizers of performers and partisans known as the Greens and Blues, might also be meaningful. As mentioned, the procession founded by Salutaris in 102 CE involved the ephebes of the city and was probably intended for the young population of the city to learn about the city's social hierarchy, history, and monumental past through active participation in rituals (Rogers, 1991). While *ephebeia* in Late Antiquity was a thing of the past, young supporters of factions, by taking part in processions, may have continued the social role historically assigned to their generation. The role and involvement of factions in public ceremonial in Late Antiquity is already well attested (Roueché, 1999). It is likely that acclaiming groups, especially members of factions, stood in those particular spots to make their acclamations, a situation very likely described for tenth-century Constantinople in the *Book of Ceremonies* (Roueché, 1999; Moffatt & Maxeme, 2017). If this is true, then the main audience of the statues was these young supporters of factions, who would most benefit from learning about and admiring past state officials, hence administrative history, and being reined in by the symbolic authority conveyed by these figures.

Acclamations disclose other aspects about mobile viewing at Aphrodisias, where a significant group of inscriptions was painted on columns in Tetrastoon, a public plaza to the east of the Theatre (Figure 10.2, no.12). There are several layers of texts here, one on top of each other, probably written as acclamations for officials – mostly governors – who frequently succeeded one another (Roueché, 1999). Meaningfully, we know from an inscription that Tetrastoon was renovated by a governor named Antonius Tatianus in the 360s and, from archaeological evidence, that it was a public plaza specifically used to display statues of emperors and governors in front of its west stoa (*ALA 2004*, n. 20; Smith, 2001, 2016b, 155–156). Well-studied statues, among the set of seven displayed here include statues of Emperors Valens and Julian (later reused for Theodosius) (*LSA*-196, 750) and Governor Flavius Palmatus (*LSA*-198). The acclamations may have taken place in front of the colonnades or in the nearby theatre.

On the Tetrapylon at Aphrodisias (Figure 10.2, no. 6), the gateway that crossed the main North-South Avenue of the city with the road to the Temple/Church, there is a monogram of the Green faction on the façade that faces the street (Roueché, 1989, 1999). It is quite likely that the main North-South Street of Aphrodisias that connected the Tetrapylon in the North with the Theatre in the South (Figure 10.2, between no. 6 and no. 11) also witnessed some processions. The Tetrapylon was, from the late fifth century onwards, the main gate leading to the cathedral church converted from the Temple. While no *in-situ* statue bases were found in front of the Street colonnades, this Street was heavily restored and built-over in many different occupation levels, so such evidence may easily have been lost. In fact, several late antique statues were discovered in excavations that had been reused as building materials in Late Antiquity and in the Byzantine period in walls and other medieval structures (Öğüş, 2016; Sokolicek, 2016). These statues were most likely once displayed along the Street or in the immediate vicinity, rather than carried off from a distant site to be recycled.

Another set of acclamations from Aphrodisias demonstrably links a local spatial setup with the mobile viewing of an honorific statue. In the west stoa of the Place of Palms (Figure 10.2, no. 19), a statue base roughly dating to the sixth century CE was dedicated to Albinus, a governor at Aphrodisias: 'To Good Fortune. The city, repaying his (good) works, set up (this statue of) Albinus, lover of his country, enjoying untold gold [?which he has given . . .]' (*ALA 2004*, n. 82). Inside the same portico, each of the 20 surviving columns of the roofed portico has an inscription on it, all interpreted to be acclamations for Albinus (Figure 10.7) (Roueché, 1984). Among these acclamations are: 'Up with the builder of the stoa!' (Figure 10.8), 'God is one, for the whole world!', 'Many years for the emperors!', 'Many years for the Senate!', 'Your buildings are an eternal reminder, Albinus, you who love to build!' (Figure 10.9), and a message for the rivals: '[E]nvy does not vanquish fortune'. These inscriptions were carved on the smooth bottom part of the columns, right under where the fluting starts at a height of 1.73 metres and

Figure 10.7 Columns inscribed with acclamations for Albinus. Arrow detail in Figure 10.8. West portico of the Place of Palms, Aphrodisias.

Source: Author's image

therefore slightly above eye level for an average person. They can be read as one walks along the portico. If these acclamations were uttered at the time of the dedication of the statue and inscribed at the spot to re-enact the occasion, then the moment of the dedication was both a visual and an auditory feast of evoking Albinus. It does not have to be the case that the acclamations took place around the statue; they could easily have been uttered somewhere else such as the Theatre or Council House. Even so, officials responsible for the spatial setting of the Albinus statue no doubt meant for the audience to walk through the space while reading the acclamations for the honorand. The commemoration of Albinus in this mobile way served to evoke him with multisensory referents, which elided and even annulled the distance between the inanimate object and the living prototype (Steiner, 2002, 19).

As this prominent urban processional culture demonstrates, honorific statues were in many cases meant to be viewed by mobile crowds, which also made appropriate acclamations for the honorands. In a few cases, such as the statue of Albinus at Aphrodisias, these acclamations and statue displays were combined in an ensemble to re-enact the emotion of the moving crowd at the time of the dedication or during the later processions. Therefore, moving through space in a crowd becomes a prominent aspect of the social experience, and moving by oneself triggered memories of the acclaiming crowd dedicating or celebrating a statue.

Figure 10.8 Inscription on column marked with arrow in Figure 10.7 ('Up with the builder of the stoa!'). West portico of the Place of Palms, Aphrodisias.

Source: Author's image

10.3.2 Moving statues

A second aspect of constructing social relations is by moving statues in the available space. Statues, like many other cultural objects, mobilize space and are involved in the reconstitution of belonging and memory (Urry, 1997, 487). They were often permanently moved and replaced, a habit that went hand in hand with reuse. A number of inscriptions attest to the habit by mentioning that statues were kept in 'squalid and obscure places' and were taken out to be reused (Ostia: *LSA*-329; Beneventum: *LSA*-1740; and Liternum: *LSA*-1921). In the archaeological record, we also find numerous statue deposits of what seem to be deliberately buried statuary. For instance, four headless statue bodies – one belonging to Hadrian, and three of second-century elites – found carefully and orthogonally buried in Amastris (Asia Minor) in the fourth century could well belong to a statue deposit that aimed to recycle them later (Ateşoğulları & Şimşek, 1995; Aydın et al., 2015).

There is further evidence that statues were frequently moved from different cities to Constantinople. It has so far been quite well attested, for example, that statues of mythological figures from various cities of the Empire were used to embellish Constantinople. The mythological statues in the spina of the Hippodrome or the Zeuxippus Baths are examples (Bassett, 2004). Honorific statues were also frequently carried to other locations and appropriated into new contexts and spaces. Emperor Theodosius reused statues of Hadrian and his predecessor Trajan,

Figure 10.9 Inscription on column ('Your buildings are an eternal reminder, Albinus, you who love to build!').

Source: Author's image

which Theodosius identified with, in a new setting in the Milion (*Parastaseis*, no. 18; Georgius Cedrenus [Bekker, 1838, 564.3–564.4]; Croke, 2010, 260). A similar undertaking is attested by the eighth-century text *Parastaseis*, which records that the statues of Marcian and his wife Pulcheria (450–57 CE) were transferred (*metathesas*) by Marcian's successor, Leo I (457–74 CE), in the Theodosian porticoes (*Parastaseis*, no. 45). The verb *metathesas* seems to indicate that they were carried from another city.

Statues were also temporarily moved across urban space for ceremonial purposes, such as for *adventus*, which was an important processional ceremony of welcoming emperors. The right of cities to host an *adventus* would be guaranteed by law, and participation could be obligatory (*Dig*. 1.16.7; Libanius *Or*. 27.42; *Or*. 56.1–2, 56.6, 56.9–12; Lavan, 2020, 154). *Adventus* was offered not only to actual emperors but also to their portrait images, which symbolically substituted for their imperial subjects and were officially welcomed in a particular city (MacCormack, 1972). We first hear of the *adventus* of images in the third century at Termessos Minor in Lycia, and later in the fourth century, when imperial portraits of Constantine and Maxentius were sent out in order to gain recognition of their rule (Third century: Robert, 1940, no. 113b [*IGGR* III 481 = *ILS* 8870]; Constantine: Lactantius, *de mortibus persecutorum* 25; Maxentius: Zosimus 2.12.1; MacCormack, 1990, 33–39; Lavan, 2020, 156, n. 23). Anthemius also sent his images to Constantinople after his accession in 467, with the same objective (Const. Porph., *De Cer.* 1.87).

Ecclesiastical *adventus* and the procession of relics gradually became a new phenomenon in Late Antiquity (Rebecca Sweetman, this volume; Nadine Viermann, this volume). The first recorded example is in *c*. 381, when the body of bishop Meletius was sent home to Antioch from the Eastern capital. Sozomen informs us that by the command of the emperor, these relics were received within the walls in every city and were honoured with singing of psalms, until they were transferred to Antioch (Soz., *HE* 7.10; Lavan, 2020, 156, n. 25). The arrival of the body of Symeon the Stylite into Antioch in 459 was in many ways similar to the arrival of an imperial image. The whole *adventus* ceremony for Symeon was considered beyond description by his biographer, being given an escort of troops as well as clergy (Lavan, 2020, 156–157). There is also a ceremonial emphasis placed upon the arrival of relics especially in the era of Theodosius (Croke, 2010; Brown, 2014). The emperor brought Paul's skull, the remains of the African martyrs Terentius and Africanus, and the head of John the Baptist to Constantinople in processions (Paul: Socr., *HE* 5.9.1–2; Terentius and Africanus: Theod. Lector, *HE* 2.62; John the Baptist: Sozomen, *HE* 7.21.5; Croke, 2010, 255).

Other urban processions involved carrying the image of the emperor as if he were attending a ceremony. For instance, in Constantinople, the annual ceremony on 11 May, the foundation day of the city, involved a gilded wooden effigy of Constantine the Great, its founder, which, according to John Malalas, 'left' the forum of Constantine, 'entered' a triumphal chariot, and was escorted to the Hippodrome, where the effigy 'arrived' before the imperial box and received the homage of the reigning emperor (Malalas, *Chr*. 13.8; Bauer, 2001). The effigies of emperors used in processions were probably made of wood, for the obvious ease to carry them (Öğüş, 2022, 242–243). Other than Malalas, as mentioned earlier, who refers to the wooden effigy of Constantine, Julian also mentions wood as a statue material for gods and emperors, although he does not specify how widespread it was nor whether these wooden statues were on permanent display or only processional (Julian, *Ep*. 294C). It is difficult to imagine that a weighty bronze or marble statue was carried around occasionally and brought back to its original locale. For one thing, lifting off such a statue from a base and later setting it up again would take workforce, time, and machinery such as crane. However, besides scanty evidence and common sense on the impracticality of moving bronze and marble statues, the material of the processional statue must remain somewhat hypothetical for lack of better evidence.

Not every case of moving statues needs to be related to recycling or veneration of said statues. Some statues – or their parts – were moved through the streets by angry mobs that intended to punish the prototype. A well-known example is the Riot of Statues in 387, when the statues of Theodosius and his family were pulled down by tying ropes around their necks and chopped to pieces; their body parts were subsequently dragged through the streets (Libanius, *Or*. 22.7–9; Theodoret, *HE* 5.19–5.20; Soz., *HE* 7.23; Downey, 1961; van de Paverd, 1991; Stewart, 1999; Anderson, 2016). The same treatment was given to statues of Anastasius on multiple occasions, which were hurled to the ground and subjected to abuse as if they were real persons (Anderson, 2016). Such 'processional' but vicious events constitute a form of mobility for the statue, albeit one that was meant to enhance the visibility of its humiliation, rather than its respectability.

The mobility of the crowd and the portrait images, as shown in the examples above, attest to a viewing practice that is not necessarily connected to a fixed spatial construct. Indeed, as the statues lining both sides of the streets and those that were moved in processions suggest, possibilities of statue display seem to have been deliberately kept flexible. Tolerance to a variety of spatial categories and fluidity to experiment with available spatial forms in Late Antiquity, on the one hand, facilitated by the volatility of social and spatial conditions and, on the other hand, intentionally manipulated to impress the viewer, probably played a role in the resilience of the habit for another three centuries beyond its heyday.

10.4 Conclusion

The visibility and mobility dimensions of statue-viewing discussed in this chapter directs us into the main conclusion that the late antique spatio-social construct was a much more fluid, dynamic and flexible one compared to the Imperial period. Urban dwellers in the late Empire were familiar with *ad hoc* solutions about displaying statues, as well as continuously modified statue ensembles, recycled bases and statue parts, and overall a 'thinking outside the box' mentality when it came to maintaining the statue habit.

A number of issues raised in this chapter need reiterating. First, the spatial substrate of statue-viewing in Late Antiquity was never static but was a dynamic phenomenon that changed forms spontaneously through time. New spaces such as the Tetrastoon at Aphrodisias were built, and old ones such as the State Agora at Ephesus fell into ruins. Urban space to be employed for statuary was always a volatile one, modified to different degrees in different cities and centuries. Second, visibility was a major factor in the choice of statue locations. Special importance was attributed to the streets, even more so to street processions and mobility. There was continual concern about making the statues more visible by bringing the statues forth into the mobile viewer's space. Late antique statues demanded attention and were deliberately made integral elements of active city life. Third, the spatial construct was a factor that engaged in a dynamic interplay with social relations, not an *a priori* static element that was available for manipulation. Much like space was adjusted to meet new social formations, such as adding stops for liturgical processions, social requirements were also often adjusted based on spatial constraints, such as determining the route of processions based on where the main monuments were situated.

In conclusion, the dynamism and volatility of the spatial construct and the entanglement of the spatial and social – whether deliberately implemented or not – strike one as important factors that helped sustain the statue habit until the sixth century. If the critical questions of where and how to display had remained unresolved or had not been met with the flexibility that we witness, the habit would potentially have met its demise much sooner.

Editions and abbreviations

ALA 2004 – *Aphrodisias in Late Antiquity: The Late Roman and Byzantine inscriptions* (revised second edition), ed. C. Rouché (2004). http://insaph.kcl.ac.uk/ala2004.

CJ – *Codex Iustinianus*. In *Corpus Iuris Civilis*, vol. 2, ed. P. Krüger (1887). Weidmann.

Const. Porph., *De Cer.* – *Constantine Porphyrogennetos. De Ceremoniis Aulae Byzantinae*, ed. A. Moffatt & T. Maxeme. (2017). Brill.

CTh – *Codex Theodosianus*. In *Theodosiani Libri XVI cum Constitutionibus Sirmondianis*, ed. T. Mommsen & P. Meyer (1905). Weidmann.

Dig. – *The Digest of Justinian*, vol. I, ed. and trans. A. Watson (1985). University of Pennsylvania Press.

Georgius Cedrenus – *George Cedrenus. Ioannis Scylitzae ope, vol. 1* (Corpus Scriptorum Historiae Byzantinae), ed. I. Bekker (1838). Weber.

IEph – *Inschriften griechischer Städte aus Kleinasien. Bd 17,2. Die Inschriften von Ephesos. Teil 7/2*: Nr: 3501–5115, ed. R. Meriç, R. Merkelbach, J. Nollé, & S. Şahin (1981). Habelt.

IGGR – *Inscriptiones Graecae ad res Romanas pertinentes*

ILS – *Inscriptiones Latinae Selectae*

Julian, *Ep.* – Julian, *Orations 6–8. Letters to Themistius, To the Senate and People of Athens, to a Priest. The Caesars. Misopogon* (Loeb Classical Library), trans. W. C. Wright (1913). Harvard University Press.

Lactantius, *De mortibus persecutorum* – *Lactance, De la mort des persécuteurs* (Sources Chrétiennes 337), ed. and trans. J. Moreau (1954). Cerf.

Leo Grammaticus – *Leonis Grammatici Chronographia* (Corpus Scriptorum Historiae Byzantinae), ed. I. Bekker (1842). Weber.

Libanius, *or.* 11 – Oration in Praise of Antioch (Oration XI), ed. and trans. G. Downey (1959). *Proceedings of the American Philosophical Society, 103*(5), 652–686.

Libanius, *or.* 22 – *Selected Works, Volume 2: Selected Orations* (Loeb Classical Library), ed. and trans. A. F. Norman (1977). Harvard University Press.

Libanius, *or.* 27 – *Libanii Opera. Vol. 3. Orationes 26–50*, ed. R. Foerster (1906). Teubner.

Libanius, *or.* 56 – *Libanii Opera. Vol. 4. Orationes 51–64*, ed. R. Foerster (1906). Teubner.

LSA – *Last Statues of Antiquity*. http://laststatues.classics.ox.ac.uk/

Malalas, *Chr.* – *The chronicle of John Malalas* (Byzantina Australiensia 4), trans. E. Jeffreys, M. Jeffreys & R. Scott (1986). Australian Association for Byzantine Studies.

Parastaseis – *Constantinople in the early eighth century. The Parastaseis Syntomoi Chronikai*, trans. A. Cameron & J. Herrin (1984). Brill.

Patria – *Accounts of Medieval Constantinople. The Patria*, trans. A. Berger (2013). Dumbarton Oaks Medieval Library.

Procopius, *De Aedificiis* – *Procopius. Buildings* (Loeb Classical Library), trans. H. B. Dewing (1940). Harvard University Press.

Socr, *HE* – Socrates, *Historia Ecclesiastica*. In *A select library of Nicene and post-Nicene fathers of Christian church. Volume II. Socrates, Sozomenus: Church Histories*, ed. and trans. P. Schaff & H. Wace (1976). W. M. B. Eerdmans Publishing Company.

Soz, *HE* – Sozomen, *Historia Ecclesiastica*. In *A select library of Nicene and post-Nicene fathers of Christian church. Volume II. Socrates, Sozomenus: Church Histories*, ed. and trans. P. Schaff & H. Wace (1976). W. M. B. Eerdmans Publishing Company.

Theod. Lector, *HE* – Theodorus Lector, *Historia Ecclesiastica* (*PG* 86).

Theodoret, *HE* – *Ecclesiastical History. Theodoret of Cyprus*, ed. B. Jackson & D. P. Curtin (2017). Dalcassian Publishing Co.

Zos. – *Zosimus. New History* (Byzantina Australiensia 2), trans. R. T. Ridley. Australian Association for Byzantine Studies.

Bibliography

Algra, K. (Ed.). (1994). *Concepts of space in Greek thought*. Brill.

Anderson, B. (2016). The disappearing imperial statue: Toward a social approach. In T. M. Kristensen & L. Stirling (Eds.), *The afterlife of Greek and Roman sculpture: Late antique responses and practices* (pp. 290–309). University of Michigan Press.

Anderson, B., & Harrison, P. (2010). *Taking-place: Non-representational theories and geography*. Ashgate.

Ateşoğulları, S., & Şimşek, Ö. (1995). Amasra ilçesi Küçüksanayi sitesi inşaasında bulunan Roma dönemi heykellerini kurtarma kazısı. In *5. Müze kurtarma kazıları semineri, Didim 25–28 Nisan 1994* (pp. 93–111). Ankara Üniversitesi Basımevi.

Auinger, J., & Aurenhammer, M. (2010). Ephesische Skulptur am Ende der Antike. In F. Daim & J. Drauschke (Eds.), *Byzanz: das Römerreich im Mittelalter* (pp. 663–696). Verlag des Römisch-Germanischen Zentralmuseums.

Auinger, J., & Sokolicek, A. (2016). Ephesus. In R. R. R. Smith & B. Ward-Perkins (Eds.), *The last statues of antiquity* (pp. 160–173). Oxford University Press.

Aydın, B., Buccino, L., & Summerer, L. (2015). Honored, beheaded and buried: A new deposit of statues from Amastris. In K. Winther-Jacobsen & L. Summerer (Eds.), *Landscape dynamics and settlement patterns in Northern Anatolia during the Roman and Byzantine period* (pp. 219–238). Franz Steiner Verlag.

Baldovin, J. F. (1987). *The urban character of Christian worship. The origins, development, and meaning of stational liturgy*. Orientalia Christiana Analecta 228. Institutum Studiorum Orientalium.

Bassett, S. (2004). *The urban image of late antique Constantinople*. Cambridge University Press.

Bauer, F. A. (1996). *Stadt, Platz und Denkmal in der Spätantike: Untersuchungen zur Ausstattung öffentlichen Raums in den spätantiken Städten Rom, Konstantinopel und Ephesos*. Philipp von Zabern.

Bauer, F. A. (2001). Urban space and ritual. Constantinople in Late Antiquity. *Acta ad archaeologiam et artium historiam pertinentia, 15*, 27–61.

Bauer, F. A. (2007). Virtuelle Statuensammlungen. In F. A. Bauer & C. Witschel (Eds.), *Statuen in der Spätantike* (pp. 79–109). Reichert.

Bauer, F. A. (2008). Die Stadt als religiöser Raum in der Spätantike. *Archiv für Religionsgeschichte*, *10*, 179–208.

Bauer, F. A., & Witschel, C. (Eds.). (2007). *Statuen in der Spätantike*. Reichert.

Borch, C. (2002). Interview with Edward W. Soja: Thirdspace, postmetropolis and social theory. *Distinktion: Scandinavian Journal of Social Theory*, *4*, 113–120.

Brown, A. (2016). Corinth. In R. R. R. Smith & B. Ward-Perkins (Eds.), *The last statues of antiquity* (pp. 174–189). Oxford University Press.

Brown, P. (1971). *The world of Late Antiquity. AD 150–750*. W.W. Norton & Company.

Brown, P. (2014). *The cult of the saints. Its rise and function in Latin Christianity, enlarged edition*. University of Chicago Press.

Chenault, R. (2012). Statues of senators in the Forum of Trajan and the Roman Forum in Late Antiquity. *Journal of Roman Studies*, *102*, 102–132.

Coarelli, F. (2009). L'edilizia pubblica a Roma in etá tetrarchica. In W. V. Harris (Ed.), *The transformations of Vrbs Roma in Late Antiquity* (pp. 23–33). JRA Supplementary Series 33.

Coates-Stephens, R. (2001). *Muri dei bassi secoli* in Rome: Observations on the re-use of statuary in walls found on the Esquiline and Caelian after 1870. *Journal of Roman Archaeology*, *14*, 217–238.

Cresswell, T. (2002). [Review of *Cultural Geography: A Critical Introduction*, by D. Mitchell]. *Economic Geography*, *78*(2), 247–248.

Cresswell, T. (2014). *Place: An introduction*. John Wiley & Sons.

Croke, B. (2010). Reinventing Constantinople: Theodosius I's imprint on the imperial city. In S. McGill, C. Sogno, & E. Watts (Eds.), *From the Tetrarchs to the Theodosians. Later Roman history and culture 284–450 CE* (Volume 34, pp. 241–264). Yale Classical Studies, Cambridge University Press.

Day, J., Hakola, R., Kahlos, M., & Tervahauta, U. (Eds.). (2019). *Spaces in Late Antiquity. Cultural, theological, and archaeological perspectives*. Routledge.

De Bruyn, G., & Machado, C. (2016). North Africa. In R. R. R. Smith & B. Ward-Perkins (Eds.), *The last statues of antiquity* (pp. 56–68). Oxford University Press.

Downey, G. (1961). *A history of Antioch in Syria from Seleucus to the Arab conquest*. Princeton University Press.

Duman, B. (Ed.). (2017). *Tripolis araştırmaları* (Tripolis ad Maeandrum I). Ege Yayınları.

Emmison, M., Smith, P., & Mayall, M. (2012). *Researching the visual*. SAGE.

Foss, C. (1979). *Ephesus after antiquity*. Cambridge University Press.

Foucault, M. (1979). *Discipline and punish*. Penguin.

Gehn, U. (2012). *Ehrenstatuen in der Spätantike: Chlamydati und Togati*. Reichert.

Gehn, U. (2016a). Greek mainland and islands. In R. R. R. Smith & B. Ward-Perkins (Eds.), *The last statues of antiquity* (pp. 87–97). Oxford University Press.

Gehn, U. (2016b). Athens. In R. R. R. Smith & B. Ward-Perkins (Eds.), *The last statues of antiquity* (pp. 190–199). Oxford University Press.

Gehn, U., & Ward-Perkins, B. (2016). Egypt, the near East, and Cyprus. In R. R. R. Smith & B. Ward-Perkins (Eds.), *The last statues of antiquity* (pp. 109–120). Oxford University Press.

Højte, J. M. (2005). *Roman imperial statue bases from Augustus to Commodus*. Aarhus University Press.

Horster, M. (1998). Ehrungen spätantiker Staathalter. *Antiquite Tardive*, *6*, 37–59.

Ingold, T. (1993). The temporality of the landscape. *World Archaeology*, *25*(2), 152–174.

Jacobs, I. (2012). The creation of the late antique city: Constantinople and Asia Minor during the 'Theodosian renaissance'. *Byzantion*, *82*, 113–164.

Jacobs, I. (Ed.). (2014). *Production and prosperity in the Theodosian period: Proceedings of ROCT 2010* (Interdisciplinary Studies in Ancient Culture and Religion; No. 14). Peeters.

Janin, R. (1964). *Constantinople byzantine*: *dévelopment urbain et répertoire topographique*. Institut Français d'Etudes Byzantines.

Jones, A. H. M. (1964). *The Later Roman Empire. 284–602* (Vol. I). The Johns Hopkins University Press.

Kristensen, T. M., & Stirling, L. (Eds.). (2016). *The afterlife of Greek and Roman sculpture: Late antique responses and practices*. University of Michigan Press.

Ladstätter, S., & Pülz, A. (2007). Ephesus in the Late Roman and Early Byzantine period: Changes in its urban character from the third to the seventh century AD. In A. G. Poulter (Ed.), *The transition to Late Antiquity. On the Danube and beyond* (pp. 391–433). Proceedings of the British Academy 141. Oxford University Press.

Laurence, R., & Newsome, D. J. (2011). *Rome, Ostia, Pompeii: Movement and space*. Oxford University Press.

Lavan, L. (2006). Fora and agorai in Mediterranean cities during the 4th and 5th c AD. In W. Bowden, A. Gutteridge, & C. Machado (Eds.), *Social and political life in Late Antiquity* (pp. 195–249). Late antique archaeology 3.1. Brill.

Lavan, L. (2013). The agorai of Sagalassos in Late Antiquity. An interpretive study. In L. Lavan & M. Mulryan (Eds.), *Field methods and post-excavation techniques in Late Antique archaeology* (pp. 289–253). Brill.

Lavan, L. (2020). *Public space in the late antique city. Streets, processions, fora, agora, macella, shops* (Vol. I). Brill.

Law, J. (1994). *Organizing modernity*. Blackwell.

Lefebvre, H. (1992). *The production of space* (Trans. D. Nicholson-Smith). Wiley Blackwell.

Lenaghan, J. (2016). Asia minor. In R. R. R. Smith & B. Ward-Perkins (Eds.), *The last statues of antiquity* (pp. 98–108). Oxford University Press.

Lepelley, C. (1981). *Les cites de l'Afrique Romaine au Bas-Empire* (Vol. II). Études augustiennes.

Lepelley, C. (1994). Le musée des statues divines: la volonté de sauvegarder le patrimonie artistique païen à l'époque théodosienne. *Cahiers archéologiques*, *42*, 5–15.

Liebeschuetz, J. H. W. G. (1972). *Antioch. City and imperial administration in the later Roman Empire*. Oxford University Press.

Liebeschuetz, J. H. W. G. (2001). *The decline and fall of the Roman city*. Oxford University Press.

L'Orange, H. (1965). *Art forms and civic life in the Late Roman Empire*. Princeton University Press.

L'Orange, H. (1973). The antique origin of Medieval portraiture. In H. P. L'Orange (Ed.), *Likeness and icon: Selected studies in Classical and early Medieval art* (pp. 91–102). Odense University Press.

Ma, J. (2012). Honorific statues and Hellenistic history. In C. Smith & L. M. Yarrow (Eds.), *Imperialism, cultural politics, and Polybius* (pp. 230–251). Oxford University Press.

Ma, J. (2013). *Statues and cities. Honorific portraits and civic identity in the Hellenistic world*. Oxford University Press.

MacCormack, S. (1972). Change and continuity in Late Antiquity: The ceremony of "adventus". *Historia: Zeitschrift für Alte Geschichte*, *21*, 721–752.

MacCormack, S. (1990). *Art and ceremony in Late Antiquity*. University of California Press.

Machado, C. (2006). Building the past: Monuments and memory in the Forum Romanum. In W. Bowden, A. Gutteridge, & C. Machado (Eds.), *Social and political life in Late Antiquity* (pp. 157–192). Brill.

Machado, C. (2016). Italy. In R. R. R. Smith & B. Ward-Perkins (Eds.), *The last statues of antiquity* (pp. 43–55). Oxford University Press.

Machado, C. (2017). Dedicated to eternity? The reuse of statue bases in late antique Italy. In K. Bolle, C. Machado & C. Witschel (Eds.), *The epigraphic cultures of Late Antiquity* (pp. 323–361). Franz Steiner Verlag.

Machado, C. (2019). *Urban space and aristocratic power in late antique Rome. AD 270–535*. Oxford University Press.

Machado, C. with Lenaghan, J. (2016). Rome. In R. R. R. Smith & B. Ward-Perkins (Eds.), *The last statues of antiquity* (pp. 121–135). Oxford University Press.

Machado, C., & Ward-Perkins, B. (2012). A population of statues. In A. Bravi (Ed.), *Aurea Umbria: una regione dell'impero nell'era Constantinia (III–VI secolo d.C.)* (pp. 37–45). Betagamma.

Maffei, S. (1993a). Columna Marci Aurelii Antonii. In E. M. Steinby (Ed.), *Lexicon topographicum urbis Romae* (Vol. I, pp. 302–305). Edizioni Quasar.

Maffei, S. (1993b). Forum Traiani: Columna. In E. M. Steinby (Ed.), *Lexicon topographicum urbis Romae* (Vol. II, pp. 356–359). Edizioni Quasar.

Mango, C. (Ed.). (1993). Constantine's porphyry column and the chapel of St. Constantine. In *Studies in Constantinople* (pp. 103–110). Routledge.

Mango, M. M. (2001). The porticoed street at Constantinople. In N. Necipoğlu (Ed.), *Byzantine Constantinople. Monuments, topography and everyday life* (pp. 29–51). Brill.

Marsengill, K. (2013). *Portraits and icons. Between reality and spirituality in Byzantine art*. Brepols.

Massey, D. (1994). *Space, place, and gender*. University of Minnesota Press.

Massey, D. (2005). *For space*. SAGE.

McGill, S., Sogno, C., & Watts, E. (Eds.). (2010). From the Tetrarchs to the Theodosians. In *Later Roman history and culture 284–450 CE* (Vol. 34, pp. 241–264). Yale Classical Studies, Cambridge University Press.

Miltner, F. (1958). *Ephesos. Stadt der Artemis und des Johannes*. Deuticke.

Müller-Wiener, W. (1977). *Bildlexicon zur Topographie Istanbuls. Byzantion – Konstantinupolis – Istanbul bis zum Beginn des 17*. Jahrhunderts. Unter Mitarbeit von Renate und Wolf Schiele mit einem Beitrag von Nezih Fıratlı. Verlag Ernst Wasmuth.

Öğüş, E. (2016). Excavations on the Tetrapylon Street, 2010–2011. In R. R. R. Smith, J. Lenaghan, A. Sokolicek, & K. Welch (Eds.), *Aphrodisias papers 5: Excavation and research at Aphrodisias, 2006–2012* (pp. 48–57). Journal of Roman Archaeology Supplementary Series 103. JRA.

Öğüş, E. (2022). "A statue of bronze, by which times of old used to honor men of rare example": Materials of honorific statues in Late Antiquity. *Byzantinische Zeitschrift*, *115*(1), 211–246.

Ousterhout, R. (2014). The life and afterlife of Constantine's column. *Journal of Roman Archaeology*, *27*, 304–326.

Pred, A. R. (1984). Place as historically contingent process: Structuration and the time-geography of becoming places. *Annals of the Association of American Geographers*, *74*, 279–297.

Robert, L. (1940). *Hellenica. Recueil d'épigraphie, de numismatique, et d'antiquités grecques* (Vol. 1). A. Bontemps.

Robinson, B. A. (2011). *Histories of Peirene. A Corinthian fountain in three millenia*. The American School of Classical Studies at Athens.

Rogers, G. M. (1991). *The sacred identity of Ephesus. Foundation myths of a Roman city*. Routledge.

Rothaus, R. M. (2000). *Corinth, the first city of Greece: An urban history of late antique cult and religion*. Brill.

Roueché, C. (1984). Acclamations in the later Roman empire: New evidence from Aphrodisias. *Journal of Roman Studies*, *74*, 181–199.

Roueché, C. (1989). *Aphrodisias in Late Antiquity*. Society for the Promotion of Roman Studies.

Roueché, C. (1993). *Performers and partisans at Aphrodisias in the Roman and Late Roman periods*. Journal of Roman Studies Monographs No. 6. Society for the Promotion of Roman Studies.

Roueché, C. (1999). Looking for late antique ceremonial: Ephesos and Aphrodisias. In H. Friesinger & F. Krinzinger (Eds.), *100 Jahre Österreichische Forschungen in Ephesos. Akten des Symposions. Wien 1995* (pp. 161–168). Verlag der Österreichischen Akademie der Wissenschaften.

Roueché, C. (2002). The image of victory: New evidence from Ephesus. In V. Déroche (Ed.), *Mélanges Gilbert Dagron. Travaux et Mémoires 14* (pp. 527–546). Association des Amis du Centre d'Histoire et Civilisation de Byzance.

Saradi-Mendelovici, H. (2006). *The Byzantine city in the sixth century: Literary images and historical reality*. Monographs of Messenian Archaeological Studies.

Setton, K. M. (1941). *Christian attitude towards the emperor in the fourth century especially as shown in addresses to the emperor*. Columbia University Press.

Smith, R. R. R. (1990). Late Roman philosopher portraits from Aphrodisias. *Journal of Roman Studies*, *80*, 127–155.

Smith, R. R. R. (1999). Late antique portraits in a public context: Honorific statuary at Aphrodisias in Caria, AD 300–600. *Journal of Roman Studies*, *89*, 155–189.

Smith, R. R. R. (2001). A portrait monument for Julian and Theodosius at Aphrodisias. In C. Reusser (Ed.), *Griechenland in der Kaiserzeit: Kolloquium zum sechzigsten Geburtstag von Prof. Dietrich Willers* (pp. 125–136). Institut für Klassische Archäologie der Universität Bern.
Smith, R. R. R. (2016a). Statue practice and the Late Roman Empire: Numbers, costumes, and style. In R. R. R. Smith & B. Ward-Perkins (Eds.), *The last statues of antiquity* (pp. 1–27). Oxford University Press.
Smith, R. R. R. (2016b). Aphrodisias. In R. R. R. Smith & B. Ward-Perkins (Eds.), *The last statues of antiquity* (pp. 145–159). Oxford University Press.
Soja, E. W. (1989). *Postmodern geographies. The reassertion of space in critical social theory*. Verso.
Soja, E. W. (1996). *Thirdspace: Journeys to Los Angeles and other real-and-imagined places*. Blackwell Publishers.
Soja, E. W. (2000). *Postmetropolis: Critical studies of cities and regions*. Wiley-Blackwell.
Soja, E. W. (2008). Thirdspace: Toward a new consciousness of space and spatiality. In K. Ikas & G. Wagner (Eds.), *Communicating in the third space* (pp. 63–75). Routledge.
Sokolicek, A. (2016). Excavations on the Tetrapylon Street, 2012–2014. In R. R. R. Smith, J. Lenaghan, A. Sokolicek, & K. Welch (Eds.), *Aphrodisias papers 5: Excavation and research at Aphrodisias, 2006–2012* (pp. 58–75). Journal of Roman Archaeology Supplementary Series 103. JRA.
Steiner, D. (2002). *Images in mind: Statues in Archaic Greek literature and thought*. Princeton University Press.
Stewart, P. (1999). The destruction of statues in Late Antiquity. In R. Miles (Ed.), *Constructing identities in Late Antiquity* (pp. 159–189). Routledge.
Stewart, P. (2003). *Statues in Roman society. Representation and response*. Oxford University Press.
Stichel, R. H. W. (1982). *Die römische Kaiserstatue am Ende der Antike*. G. Bretschneider Editore.
Thrift, N. (1983). On the determination of social action in time and space. *Environment and Planning D: Society and Space*, *1*, 23–57.
Thrift, N. (2007). *Non-representational theory: Space, politics, affect*. Routledge.
Thür, H. (1999a). Der Embolos: Tradition und Innovation anhand seines Erscheinungsbildes. In H. Friesinger & F. Krinzinger (Eds.), *100 Jahre Österreichische Forschungen in Ephesos. Akten des Symposions. Wien 1995* (pp. 421–428). Verlag der Österreichischen Akademie der Wissenschaften.
Thür, H. (1999b). Die spätantike Bauphase der Kuretenstrasse. In R. Pillinger, O. Kresten, F. Krinzinger, & E. Russo (Eds.), *Efeso paleocristiana e bizantina* (pp. 104–120). Verlag der Österreichischen Akademie der Wissenschaften.
Unwin, T. (2000). A waste of space? Towards a critique of the social production of space. *Transactions of the Institute of British Geographers*, *25*, 11–29.
Urry, J. (1985). Social relations, space and time. In D. Gregory & J. Urry (Eds.), *Social relations and spatial structures. Critical human geography* (pp. 20–48). Palgrave.
Urry, J. (1997). *Mobilities*. Polity.
van de Paverd, F. (1991). *St. John Chrysostom, the homilies on the statues: An introduction*. Pontificium Institutum Orientalium Studiorum.
Ward-Perkins, B. (2016a). Statues at the end of antiquity. The evidence of the inscribed bases. In R. R. R. Smith & B. Ward-Perkins (Eds.), *The last statues of antiquity* (pp. 28–40). Oxford University Press.
Ward-Perkins, B. (2016b). The end of the statue habit, AD 284–620. In R. R. R. Smith & B. Ward-Perkins (Eds.), *The last statues of antiquity* (pp. 295–308). Oxford University Press.
Weisweiler, J. (2012). From equality to asymmetry: Honorific statues, imperial power and senatorial identity in late-antique Rome. *Journal of Roman Archaeology*, *25*, 319–350.
Wilson, A. (2016). Water, nymphs, and a palm grove: Monumental water display at Aphrodisias. In R. R. R. Smith, J. Lenaghan, A. Sokolicek, & K. Welch (Eds.), *Aphrodisias papers 5: Excavation and research at Aphrodisias, 2006–2012* (pp. 201–136). Journal of Roman Archaeology Supplementary Series 103. JRA.
Wiplinger, G., & Wlach, G. (1995). *Ephesos. 100 Jahre österreichische Forschungen*. Böhlau.
Witschel, C. (2007). Statuen auf spätantiken Platzanlagen in Italien und Africa. In F. A. Bauer & C. Witschel (Eds.), *Statuen in der Spätantike* (pp. 113–169). Reichert.

Witschel, C. (2008). Sterbende Städte? Betrachtungen zum römischen Städtewesen in der Spätantike. In A. Lampen & A. Owzar (Eds.), *Schrumpfende Städte. Ein Phänomen zwischen Antike und Moderne* (pp. 17–78). Böhlau Verlag.

Witschel, C. (2013). Die spätantiken Städte Galliens: Transformationen von Stadtbildern als Ausdruck einer gewandelten Identität? In G. M. Müller & S. Diefenbach (Eds.), *Gallien in Spätantike und Frühmittelalter: Kulturgeschichte einer Region (5.-7. Jh.n.Chr.)* (pp. 153–200). De Gruyter.

Witschel, C. (2016). Hispania, Gallia, and Raetia. In R. R. R. Smith & B. Ward-Perkins (Eds.), *The last statues of antiquity* (pp. 69–79). Oxford University Press.

11 *Civitatem condidit.* City building and community formation in the new Visigothic urban foundations

Javier Martínez Jiménez

One of the main distinguishing (and surprising) characteristics of the Visigothic period of state formation (roughly 560–630 CE) is that the Visigoths were keen city builders (Figure 11.1). Of course, they founded no more than a handful of settlements that can be considered ‘urban’ from different perspectives, but that handful is still more than the cities created by the combination of all their immediate neighbours, the Franks, the Sueves, the Ostogoths, and the Vandals. Furthermore, the Visigoths not only founded new cities but also promoted old secondary settlements to city (*civitas*) status.

The importance of newly founded cities in the Visigothic period is manifold. First, these foundations highlight the importance of cities and civic communities in the conceptualisation (though not necessarily in the running) of the Visigothic Kingdom (Poveda Arias, 2023). Second, they underline the capabilities of the Visigothic monarchy, especially during the period of state formation (Martin, 2003). Third, and from an archaeological point of view, these new foundations allow us to understand better the elements that constituted the ideal city in the post-Roman West. They also offer a glimpse at construction techniques and architectural knowledge linked to royal constructions, which feed into the long-running debate about the real chronology of a number of buildings traditionally considered Visigothic in date.[1] Lastly, the study of these new cities offers a unique view into the daily life of the communities that settled there.

In this chapter, I will cover these topics, focusing on the planning, design, and construction of the new cities and on the key issue of how these places were settled, by whom, and how it was to live in a newly built (or partially built) late antique city. The main examples I will focus on will be Reccopolis and Eio/El Tolmo, the only true foundations known through archaeology. I will also include in the discussion the site of València la Vella, where recent excavations are revealing its seemingly urban nature. I will also deal in passing with the other cities built in this period only known from the written sources: Victoriacum and Ologicus.

11.1 Visigothic new foundations

The first and most famous example of a Visigothic new foundation is Reccopolis, built by Liuvigild in 578 in the province of Celtiberia (Figures 11.2 and 11.3). We know this through the *Chronicle* (578.4) of John of Biclar, a contemporary of the events, who narrates:

> With the tyrants completely destroyed, and the invaders of Spain defeated, King Liuvigild settled his peace with his own people. He founded a civic community (*civitatem*) named after his son in Celtiberia, which he named Reccopolis, which he endowed with wonderful works (*miro opere*) within the walls (*in moenibus*) and in the suburbs (*suburbanis*), and he established privileges (*privilegia*) for the people of the new city (*urbis*).[2]

DOI: 10.4324/9780429427152-14

Figure 11.1 Map of the Iberian Peninsula, locating the sites mentioned in the text.

Note: Black circles represent pre-existing cities; grey ones represent towns promoted to episcopal rank; white ones indicate new foundations; triangles indicate supposed Visigothic foundations; and diamonds indicate other locations discussed in the text.

Figure 11.2 View of the site of Reccopolis.

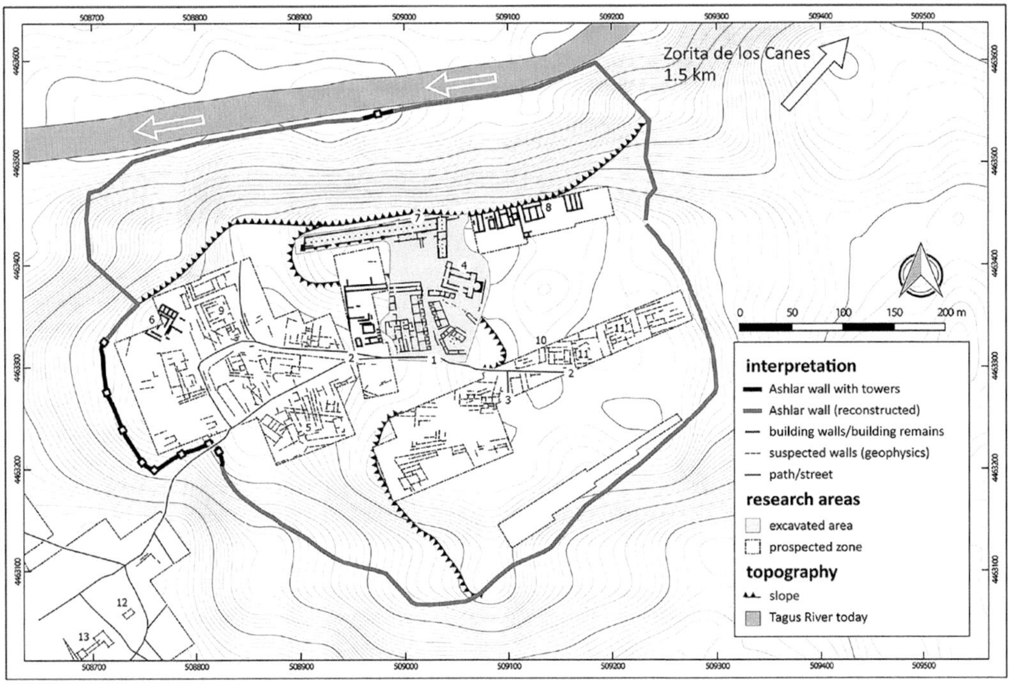

Figure 11.3 Plan of Reccopolis, including the excavations and the areas identified through geophysical survey.

Source: Adapted from Henning et al. (2019), Figure 3. Produced with permission

This is the most detailed literary description we have for the foundation of a city in the Visigothic period. We know that it was a city with walls, suburbs, and civic status, as shown by both the term *civitas* and the *privilegia* granted to the inhabitants. The city has long been identified with the remains at the Oliva Hill near Zorita de los Canes in Guadalajara (Olmo Enciso, 2008). The site is built above the Tagus River (Figure 11.2), and excavations have revealed its wall and gates, a large civic ('palace') complex, a church, and a commercial street flanked by houses (Olmo Enciso, 2020). Furthermore, geophysical surveys have shown that the walled enclosure was densely occupied with buildings and structures (Henning et al., 2019), and walking surveys have identified wall remains and pottery scatters that can be linked to suburban structures (Olmo Enciso et al., 2018) and even an aqueduct (Martínez Jiménez, 2015; Martínez Jiménez & Checa Herraiz, 2023).

Site finds confirm the chronology given by the sources, and that the city was an active urban centre into the ninth century. However, there are noticeable changes in the urban fabric from the second half of the seventh century, showing an evolution of the layout and architecture as can be expected from any living city. Besides this, the urban nature of Reccopolis is further underlined because the site has the wealthiest assemblage of imported goods of the central Iberian Peninsula, and it was a centre of glass-blowing, goldsmithing, and a royal mint (Olmo Enciso et al., 2008).

The next Visigothic foundation is Victoriacum. Its construction is dated to 581, and it is located somewhere in the region of the Basques, even if its precise location is unknown. It was founded to commemorate Liuvigild's conquests in that area, as the name would suggest. Again, this is narrated by John of Biclar (581.3):

> King Liuvigild occupied a part of Vasconia and founded (*condidit*) a civic community (*civitatem*) which is called Victoriaco.[3]

Very little is known about this foundation, as this short notice is the only piece of evidence we have. The identification of this *civitas* with any known settlement is problematic: the traditional link between Victoriacum and Vitoria-Gasteiz (due to name similarities) lacks archaeological backing (Azkarate Garai-Olaun, 2004), and the one Roman city of the area (Iruña-Veleia), which might have been the place of a re-foundation, equally fails to provide any archaeology datable beyond the fifth century (Filloy Nieva & Gil Zubillaga, 2007).

Also in the Basque region, we find Ologicus, established by Swinthila in the early 620s. Isidore of Seville (*Hist. Goth.*, 63) explains that:

> [Swinthila] had, at the beginning of his reign, a campaign against the raids of the Basques that infested the province of Tarraconensis, in which, indeed, the mountain peoples were struck with such terror at his arrival that they immediately gave him hostages and offered him their necks on their knees after abandoning their weapons and putting their hands up in prayer, almost as if they knew their rightful place. They founded Ologicus, a civic community of the Goths (*civitatem Gothorum*) with their resources (*stipendiis*) and labour, promising to obey his authority and carry out whatever they were ordered.[4]

Just as with Victoriacum, other than its broad location in the territory of the Basques, we do not really know much about Ologicus. Traditionally, this foundation has been linked to Olite in Navarre based solely on phonetic similarities (Ologicus → Olite), but the transition from one toponym to another is difficult to believe. The fact that Isidore insists that it is a '*civitas* of the Goths' has led some researchers to propose that Ologicus was nothing but a re-foundation of Victoriacum after it was lost to the Basques or the Franks (Barroso Cabrera et al., 2013).[5] However, there is no material evidence to support this.

Later sources mention two very unlikely foundations: Lugo de Llanera by Liuvigild and Bayyara by Reccared. The former appears in the twelfth century rewriting of Isidore's *History of the Goths* by Bishop Pelayo of Oviedo, who adds that '[t]his king built a city in the Asturias which he named Lugo, which is to say Luceo'.[6] The obvious intention to link this Lugo (the home of the semi-mythic founder of the Asturian kingdom, Pelayo) with the Visigothic monarchy (the ideological basis for the Leonese monarchy from the tenth century) makes this foundation doubtful.[7] The latter is mentioned in a fifteenth-century Arabic text, the *Kitab al-Rawd*, by Maghribi author Al-Himyari (Lope López de Rego, 2008). The text mentions that Bayyara was founded by 'Reccared, son of Liuvigild, King of the Goths, who unified the schismatic Christian sects of the land . . . and created eighty sees with that many bishops'. This random mention of a Visigothic foundation with a seemingly unrelated paragraph on the deeds of Reccared make this foundation equally dubious.

Other archaeological sites that are not mentioned in the sources, however, have been interpreted by their excavators as foundations of the Visigothic period with claims of urban status. The first one is the site of El Tolmo de Minateda (Figure 11.4), located in the modern province of Albacete, on the road that connected Toledo (the royal capital) to Cartagena (the Byzantine stronghold). This site had been the Ibero-Roman *oppidum* of Ilunum, which seems to have been

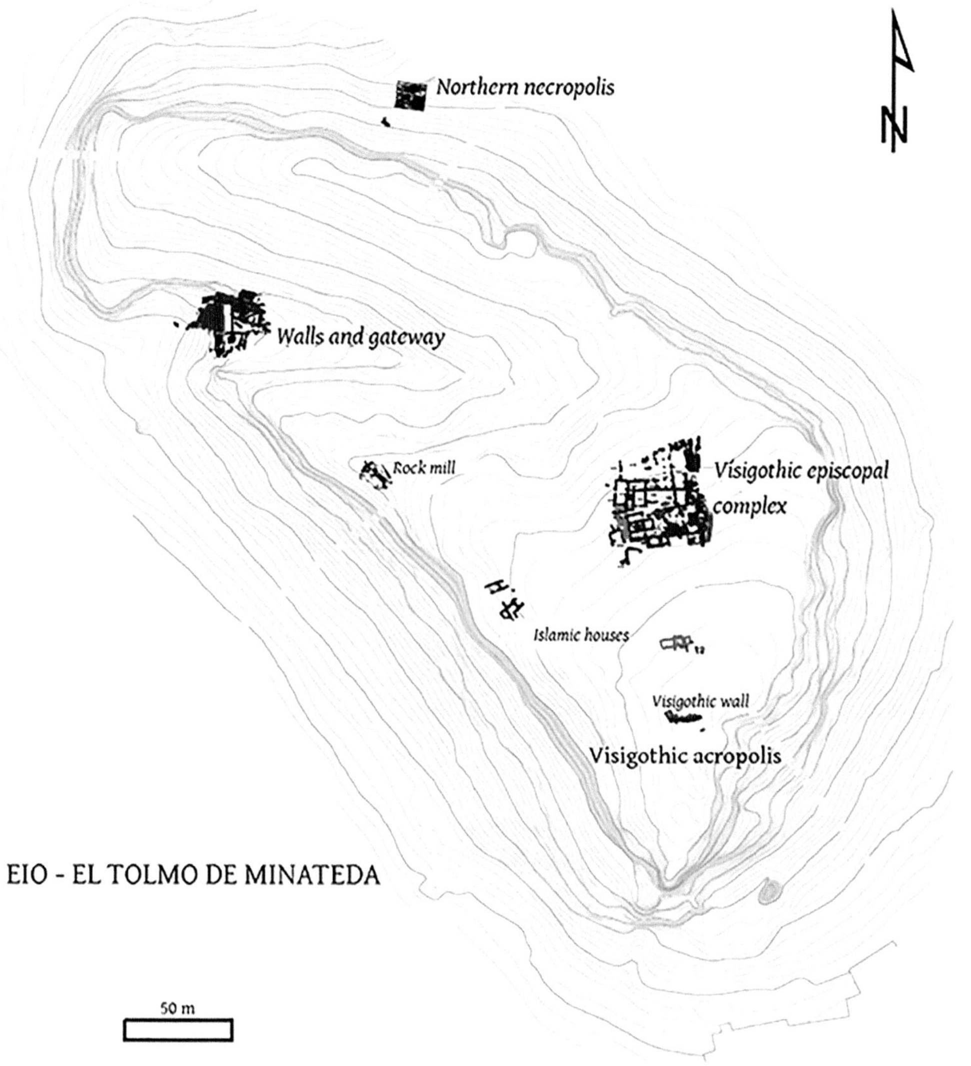

Figure 11.4 Plan of El Tolmo de Minateda, identified with the Visigothic new foundation of Eio(ta).

Source: Adapted by the author. Courtesy of Sonia Gutiérrez

abandoned before the third century (Abad Casal et al., 2008). The site was completely reconstructed in the late sixth century, when it was re-fortified, and when an episcopal complex and a citadel were built, levelling down to the bedrock the previous settlement (Abad Casal et al., 2008; Gutiérrez Lloret & Sarabia Bautista, 2013). The site continued in use after the Umayyad conquest of 711 CE but was abandoned in the early ninth century (Gutiérrez Lloret & Cañavate Castejón, 2010). Because of its strategic location between Toledo and Cartagena and the chronological sequence suggested by the finds (and because of the evolution of the toponym), it is possible to link this site with the Ecclesia Elotana or Eiotana, which is mentioned in the acts of various seventh-century episcopal councils called by the Visigothic kings in Toledo, including the synod call by King Gundemar in 610. Eio(ta)/Elo(ta) was a newly founded bishopric, created to strengthen the claims of the Visigoths over the frontier with the Imperial troops and it was one of the seats of power of Duke Theudimer during the early eighth century, the same site (Madinat Iyyuh) that the Emirs evacuated in the ninth century in order to establish the city of Murcia (Gutiérrez Lloret et al., 2005).

The second site is València la Vella, a fortified enclosure east of the city of Valencia that is believed to have acted as a central place in the late Visigothic period (Ribera Lacomba et al., 2020). The site was built in the late sixth century, although it was abandoned at some point in the seventh. There is evidence from the most recent geophysical surveys for streets that arrange the dwelling areas into different terraces and a large open space inside this fortified enclosure (Ribera Lacomba et al., 2020). The current interpretation is that this fortified space was a Visigothic urban counterpart to a Byzantine-controlled Valencia (Macias Solé et al., 2021). Excavations are still ongoing.

These sites are our best-known examples of new urban foundations, but the list of new civic sites in the Visigothic period does not end there. A handful of sites in the south-east were established in the Visigothic period as part of the reorganisation of this area after the conquest of the imperial-held territories. Sites like Begastri (Cehegín), Mula, Eliocroca (Lorca), and Auraiola (Orihuela) could be interpreted as new seats of Visigothic power, even if the archaeological interpretation of the remains is still debatable (Vizcaíno Sánchez, 2009, 199–239). But of those only one seems to have had civic status because of its episcopal nature: Begastri, like Eio, was the re-occupation of an old Roman site turned into frontier bishopric and, as such, promoted to city status. A case could be made for the civic nature of Auraiola, a hilltop fortified enclosure that became the seat of Duke Theudimer. The site is mentioned as a provincial capital in the seventh-century *Ravenna Cosmography*, but by the early eighth century, and considering the de-urbanisation of Visigothic administration (Martínez Jiménez & Tejerizo García, 2015), it is unclear if this was an urban site with any civic status. Considering the mixed nature of the evidence, the question remains as to whether these foundations can be considered cities or not, because this status depends on the criteria used.

11.2 What makes a Visigothic city?

Trying to define a city in antiquity is complex. Even nowadays it is difficult to pin down a definition for a 'city' that is accurate for all instances. Archaeologists and historians have used a range of definitions that include size, function, monumentality, and economic complexity, among others (Martínez Jiménez & Tejerizo García, 2015). For the purposes of this chapter, this problem derives from the terminology, because 'city' merges two different concepts that in antiquity were understood separately: the cityscape (*urbs*) and the city-community (*civitas*). This distinction is clearly addressed by Isidore of Seville in his *Etymologies* (15.2.1), when he describes:

> While the city (*urbs*) is itself the walls (*moenia*), it is the inhabitants, however, and not the stones that are called the civic community (*civitas*).[8]

This civic understanding of the city is equally visible in his definition of citizens (*Etym.* 9.4.2–9.4.6):

> Citizens (*cives*) are called such because they live together so they may live more honourably and safer together . . . the people (*populus*) is the totality of the citizens, including the elders of the community (*senioribus civitatis*).[9]

In these understandings of Isidore of Seville (d. 636), whose thoughts reflect the views of his day, rather than being a collection of antiquarian knowledge (cf. Curchin, 2014, 287), cities were defined by their civic communities. A similar view can be inferred from Isidore's older contemporary Gregory of Tours (d. 594) and his description of Dijon, which had all the monumental elements of a city but lacked civic status (Greg. Tur., *Hist.*, 4.19). Furthermore, despite the transformations of the Late Empire and the increasing imposition of centrally appointed figures, civic government and municipal councils did not cease to exist (Curchin, 2018; Fernández, 2020; cf. Barbier, 2021). This civic understanding of what a city was falls in line with late-Roman definitions that required *civitates* to be the seat of bishops, equating bishoprics with *civitas* territories, especially considering the consolidated role of the bishop in the later Visigothic period (Gusmão, 2016–17; Poveda Arias, 2019).

This distinction between city-*urbs* and city-*civitas* is significant enough to lead us to question the nature of the Visigothic foundations mentioned in the sources. Both John and Isidore use both *urbs* and *civitas* when they mention the foundations of Reccopolis, Victoriacum, and Ologicus. It is beyond doubt that Isidore understood the difference between the two concepts, but the way in which John underlines the foundation of the *civitas* of Reccopolis and the privileges of the inhabitants (the *cives*) of the *urbs* indicate that he was also aware of the distinction. Using *civitas* when describing these foundations (regardless of how ephemeral) could be a declaration of intent, that these establishments were constituted by a body of citizens (Martínez Jiménez, 2020b). In the case of Reccopolis (Martínez Jiménez, 2020a), it is likely that there was a body of citizens who enjoyed *privilegia*, but for the foundations in the Basque territories we can only guess that this was the original intention. It seems unlikely that either John or Isidore used *civitas* for either Ologicus or Victoriacum if they had been conceived simply as fortified outposts; regardless of whether they were established on Veleia or elsewhere (Varea?), the wealth of terms to define such fortified sites without a civic component like *castra* and *castella* (both of which our authors use, sometimes opposing *civitates* and *urbes* against *castella* – cf. Bicl., *Chron.*, 579.3) would further support that.

By right of being episcopal sees, Eio and Begastri were also *civitates*. Their reduced size (5 and 0.8 hectares, respectively – compared to Reccopolis's 21 hectares) and limited monumentality may undermine their urban nature but not their civic purpose: to establish a political community that would represent the interests of the Visigothic monarchy in the frontier. The Visigoths established a number of new bishoprics across the Iberian Peninsula during the seventh century (Poveda Arias, 2019) and in Septimania in the previous century (Schneider, 2008, 74–75), usually promoting pre-existing towns, thus granting them civic independence from their former city (Figure 11.1). The best-known examples would be Segovia, Auca, Amaia, Ávila, and Salamanca in the territory north of Toledo, Carcassone, Elne, Maguelone, and Agde in Septimania, and Egabrum (Cabra) and Italica in the Guadalquivir valley. The Sueves had already effected a similar reorganisation of their ecclesiastical administration, and places like Orense, which had been secondary settlements, were promoted to bishoprics that even acted as mints in the Visigothic period (Díaz Martínez, 2011).

Taking a step back, and looking at the archaeology, it is difficult to deny that Reccopolis was an urban centre (Figure 11.3).[10] In the case of Eio and València la Vella, the presence of fortifications and large public buildings (an episcopal complex in the former, and a yet-undefined structure in the latter), together with spaces clearly meant for housing, would also underline their urban nature – even if the civic status of València la Vella is currently a working hypothesis.

For the other sites, the evidence is unclear. Some of these new bishoprics were built on old Roman cities that had ceased to act as urban centres in the late Roman period (Brassous, 2019; Houten, 2018). It is possible that the Roman phases of Salamanca, Segovia, and Ávila were removed in order to build a Visigothic urban centre, although this seems unlikely. The almost complete lack of any archaeological remains for these cities, moreover, does not help clarify this situation (Tejerizo García, 2017, 77–80). Considering the lack of any substantial archaeological remains obtained from these cities that allow us to talk about a fully developed urbanism, it is possible that these bishoprics from the Meseta, like Orense or Egara, acted as focal points for the inhabitants of the territory, fulfilling a civic role without a clearly defined urban nucleus (Oller Guzmán, 2014). Maguelone in Septimania certainly fits this model (Raynaud, 2017).

In the south, however, John of Biclar (584.1) tells us that 'Liuvigild rebuilt the walls of the old city (*antiquae civitatis*) of Italica',[11] and a reduced fortified enclosure of late antique date has been identified by geophysical survey (Ruiz Prieto, 2013). It is clear that in Italica there was an existing civic community, but it does not appear as if the promotion and fortification were considered a foundation, as it was not stated as such. This may, in turn, underline the civic nature of other *civitates conditae*. Of course, there is always an element of propaganda in these literary descriptions of city foundations, but Reccopolis, València la Vella, Eio and, to a degree, the large suburban expansion of Toledo (Olmo Enciso, 2009; Moreno Martín, 2018) demonstrate the building capabilities of the Visigothic monarchy. Victoriacum and Ologicus might have been short-lived or not-monumental outposts in the northern frontier, but they were conceived as civic communities.

11.3 Building and designing Visigothic cities

In the late antique West, most urban foundations were, in fact, promotions of pre-existing secondary nuclei, renamed pre-existing cities, or a combination of both: Cularo became Gratianopolis; Hadrumetum became Hunericopolis, but only after being Honoriopolis; Curia Raetorum may have been the settlement renamed as Theodoricopolis. In the East, there are a number of secondary sites that, in a similar fashion, were promoted and 'founded' as cities (Rizos, 2017). There were, however, true *ex novo* foundations like Justiniana Prima (Ivanišević, 2017).

In the cases of Reccopolis and València la Vella, there is enough archaeological evidence to confirm that they were *ex novo* foundations without a pre-existing settlement. In the cases of Eio and of Begastri, these were re-foundations of pre-existing (but long abandoned) cities, which involved the levelling and destruction of pre-existing Roman remains. The impressive city-building efforts of the Visigothic monarchy make these late antique foundations stand out.

Whether *ex novo* foundations or thorough reconstructions, the foundation of these cities must have required an incredible amount of effort, skilled and unskilled workers, planning, funds, and materials – the kind of effort that only the state could muster. There is little we know with certainty about the practicalities of building a new city in this period, but there is sufficient evidence to allow a few educated guesses. Even with approximations to the construction costs of Roman buildings (DeLaine, 1997), it is impossible to calculate how much building these cities cost. It is safe to assume that the costs were high. Liuvigild's monetary reform and the minting of high-quality gold coin in this period of state formation could be linked to the large sums that were needed to pay for these construction projects (Kurt, 2020, 183), but it seems unlikely that all the required wealth came solely from taxation (Ziche, 2006). The monarchy obtained most of its income from rents from royal land (Castellanos García, 2003), but the successful military campaigns of Liuvigild (Bicl., *Chron.*, 574.2, 575.2, 576.3, 585.2), Reccared (*Hist. Goth.*, 54; Greg. Tur., *Hist.*, 9.31), Sisebut (*Hist. Goth.*, 61), and Swinthila (*Hist. Goth.*, 63) certainly secured large amounts of extra wealth in the form of loot. Confiscations of property also seem important (*Hist. Goth.*, 51: *Fiscum quoque primus iste [Liuvigild] locupletavit, primusque aerarium de rapinis civium hostiumque manubiis auxit*). Considering Isidore's phrasing on the construction of Ologicus (*stipendiis suis et laboribus conderent*), it may be the case that prisoners of war and not just loot were used to fund and build these new foundations.

In terms of resources, construction in this period is generally characterised by extensive recycling and reuse – especially in urban contexts where such materials were widely available (Utrero Agudo, 2017). In Eio (Gutiérrez Lloret & Cánovas, 2009) and Begastri (Zapata Parra, 2018–19) the remains of the earlier Roman structures (including walls, temples, houses, grave markers and aqueducts) were thoroughly reused as building material in the Visigothic phases, but in the *ex novo* constructions, this is not the case.

In Reccopolis, most of the rubble used in the mortared construction seems to have been newly quarried out of the friable local sandstone, and such quarries have been located near the aqueduct (Martínez Jiménez, 2015). The same can be said for València la Vella, where the presence of reused material is almost negligible (Macias Solé & Ribera Lacomba, 2020). However, most of the sculpted pieces from Reccopolis are new and carved in local limestone. The only pieces of reused sculpture are those made of marble, which had to be brought from nearby Roman sites, either a villa or the near city of Ercavica (Balmaseda Muncharaz, 2008) (Figure 11.5).

The widespread use of mortar in Reccopolis and València la Vella underline the need for lime, sand, and water in large quantities. Sand may be easy to obtain, even if not all sands are good for the same purposes, but limestone of good quality is more difficult to come across, and it would then need to be burnt and slaked (Dix, 1982). This would not have necessarily

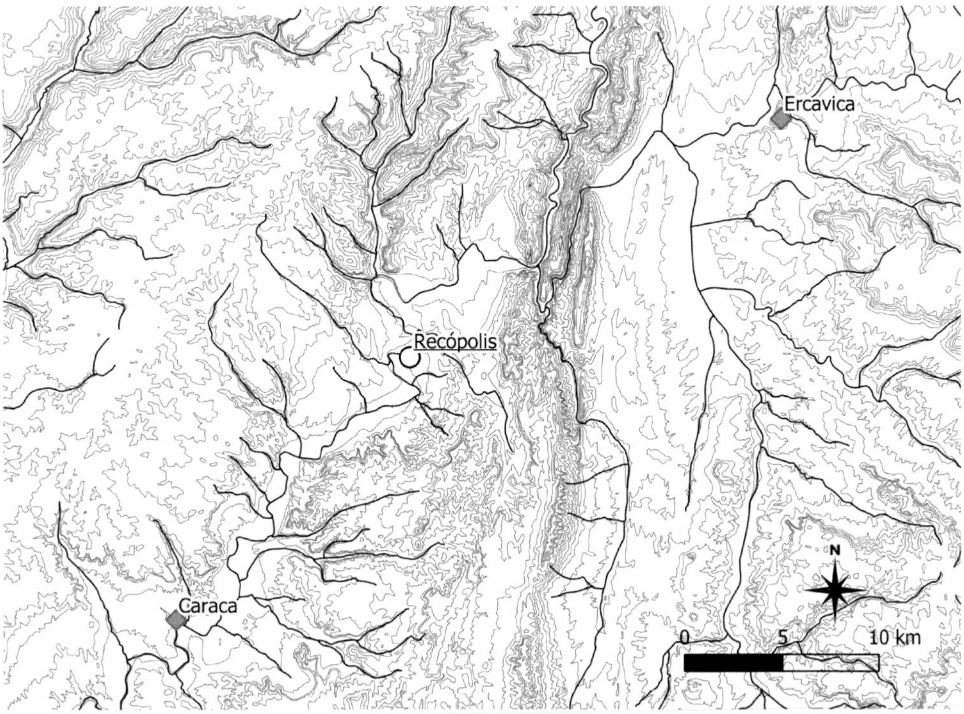

Figure 11.5 Reccopolis in the Tagus valley, roughly equidistant from the old Roman towns of Ercavica and Caraca.

happened on site, but it was yet another *chaîne opératoire* that needed to be taken into account (Martínez Jiménez, 2022b). Timber would have also been needed in bulk, with both structural (beams) and constructive (scaffolding) purposes. In Reccopolis, there is also evidence for the extensive use of roof tiles, most of which appear to have been new, adding yet another resource that had to be fabricated either *in situ* or elsewhere and then transported to site. The necessary workforce, conceivably, might have been a mixture of construction specialists and vernacular builders (Lavan, 2008, 29–30) with unskilled labourers – perhaps prisoners (as we might infer from Isidore's description of the construction of Ologicus). These workmen were most certainly local to Iberia, but design of the Reccopolis aqueduct (Figure 11.6) hints at the presence of engineers brought from outside Spain (Martínez Jiménez & González Gutiérrez, 2017). The Byzantine territories, after all, still continued a tradition of trained architects involved in complex construction projects (Cuomo, 2007, 131–45), and it was not uncommon for Roman and Byzantine emperors to send architects to foreign rulers as part of diplomatic exchanges (John Eph., 6.24; Menand., 25.1; Al-Isfahani 1.4,36; Al-Tabari, 4.690,84).

Regardless of the origins of the architects involved, we can assume that these foundations were built with a master plan in mind. This is clear in Eio, where the layout of the buildings was carved onto the bedrock after the Roman remains were levelled (Abad Casal et al., 2008), but the layout of Reccopolis may also show the existence of a preconceived city design

Figure 11.6 The Reccopolis aqueduct.

(Figure 11.3). In Late Antiquity, it is clear in both the East and the West that the grid plan did no longer represent the ideals of a city (Rizos, 2022), but the lack of grids does not imply that these new foundations were not designed with a clear urban idea in mind (Smith, 2007). Rather than a grid, these new urban foundations prioritised axes that physically and conceptually linked key focal points of the city, the hubs of political representation, and social interaction that are key in the understanding of urban communities (Lalli, 1992). These lineal axes were the main streets along which urban processions took place (Dey, 2014, 156–7); for the Visigothic period, we know of such religious (Greg. Tur., *Hist.*, 5.4), civil (*VPE*, 5.11.2), and royal (Bicl., *Chron.*, 590.3; Jul. Tol., *HWR*, 30) processions – an emphasis that, as Esen Ogus shows in this volume, is also seen in eastern cities. While in pre-existing cities these connections had to be fitted into existing street networks, in new foundations it was possible to develop these first and build the city around them (Martínez Jiménez, 2022a). The axial development of Reccopolis with a main, commercial street that links the city gate with the church and palace complex has been interpreted as a schematic imitation of Constantinople (Olmo Enciso, 2008), but it is most likely a reflection of current urban ideals, an architect's master plan, since this linear development is not unique to Constantinople. In Eio, even if the street network is not clear, the main focal points of the settlement (gate, episcopal complex, and acropolis) are all aligned.

The connection between axes and processions highlights how these streets (possibly Isidore's *plateae*; *Etym.*, 15.2.23) rather than the outdated forum may have been one of the main performative spaces for urban communities. The existence of axial avenues hint towards their *civic* nature, designed to hold processions that would bring the inhabitants together in shared rituals. The *urban* nature, on the other hand, can be seen in the monuments. The range of public monuments that were believed to belong to the cities of this period was far more limited than that of the Roman period (Diarte Blasco, 2018), but three key elements existed both in the Visigothic phases of Roman cities and in the new foundations. First are walls (Macias Solé et al., 2020), which in this period of civil wars and foreign invasions were as necessary as they were rich in symbolic value (Dey, 2010). Second are churches, from cathedrals and episcopal complexes to suburban martyrial shrines. These were focal points of religious cult and monuments to the Christian identity of the urban community (Esmonde Cleary, 2013, 151–180; Nadine Viermann in this volume). Last would be those buildings that can be linked to the civil administration: no longer basilicas, curias, or forums but rather palaces and other types of halls of representation (Martínez Jiménez et al., 2018, 170–173).

The issue remains about the housing arrangements for the new settlers. If the stratigraphic sequence of Reccopolis is any indicator, public structures were built first and private dwellings second (Olmo Enciso et al., 2008). This would fit with what we know from new foundations in the East like Dara or Justiniana Prima, where city walls seem to have been built first, then public buildings, and then houses (Rizos, 2022). Too little is known about housing in Reccopolis as a whole, but the few houses that have been excavated appear to have similar layouts and dimensions to each other (Henning et al., 2019), suggesting that these were plots assigned and distributed. The practical aspects of populating these settlements, however, will be the focus of the next section.

11.4 Experiencing a new city

Since, as Isidore tells us, it is people and not stones that make an urban community, the settlement in the new foundations should be seen as the most important aspect of city founding. The settlers who moved to Dara or Justiniana Prima were a mix of resettled soldiers and locals from nearby secondary settlements (Rizos, 2022), so we could expect a similar pattern for the Visigothic foundations. Considering the settlement of soldiers or veterans, in the Visigothic period the army was neither professional nor regular, although garrisons in cities and forts existed (Parra Romo, 2018). If the role of some of the new cities was to assert Visigothic control over frontier areas, it may be the case that the settlers were either soldiers rewarded for their service in royal campaigns or members of the royal retinue, similarly paid with land rather than with booty. The aforementioned foundations in the Basque region and at the Byzantine frontier could be seen in this light, especially considering the strong fortifications of València la Vella, Begastri, and Eio and the literary descriptions we have for Victoriacum and Ologicus.

Most of the rural settlements that surrounded Eio remained inhabited after the re-foundation (Sarabia Bautista, 2014) – although its reduced size does not completely rule out this possibility. Similarly, Isidore's use of *civitas Gothorum* in Ologicus most likely referred to the understanding of Goth as a job description, someone involved in the royal service, rather than as an indicator of Germanic descent (Buchberger, 2017; Koch, 2014). Reccopolis, far inland and away from any conceivable direct route between the Byzantine territories and Toledo, was most likely not conceived as a garrison city (Ripoll López, 2000), and while this does not exclude that some settlers were members of the royal retinue, it is likely that, as it happens in Dara or Justiniana Secunda, inhabitants of nearby villages or the remaining town dwellers of Arcavica were invited, coerced, or relocated to the new city (Rizos, 2022). The inclusion of specialised traders like glassblowers and goldsmiths in Reccopolis's original design (Castro Priego & Gómez de la Torre-Verdejo, 2008) may also point towards the relocation of specialised workmen as part of the foundational effort.

Regarding the number of inhabitants, all estimations are highly controversial and hardly accurate (Van Limbergen & Vermeulen, 2021). In terms of area, the walls of Reccopolis enclose an area approximately 20 hectares in extension.[12] Those of Eio enclose hardly 5 hectares. Taking conservative population densities from the Roman period that account for town houses rather than apartment blocks, like the 150 per hectare calculated for Pompeii (Wallace-Hadrill, 1994), we obtain that the population of Reccopolis would be around 3,000 and not even 1,000 for Eio. The presence of garrisoned troops in Eio and the existence of suburbs in Reccopolis may push the numbers slightly higher, and populations may have grown as the city developed and time passed but, if anything, these calculations give us an order of magnitude and show that we are dealing with relatively small communities of a few thousands.

From the very beginning, the founding community would have been socially differentiated, with privileged and underprivileged settlers, and we should shed any egalitarian or democratic ideas that can be linked to this type of new settlements (cf. Blanton & Fargher, 2011). This differentiation is inferred from the presence of elite areas linked to the lay and the ecclesiastical hierarchies, an upper and lower town, and suburban dwellings. From what we know from other colonial foundations like Phoenician Carthage (Fumadó Ortega, 2013), Roman Cosa (Moreno Alcaide, 2013), and the *Laws of Indies* that regulated Spanish colonial foundations in the Americas (4.7.7), urban plots allocated to new settlers did not have to be the same size, distributed evenly, or even have the same value. Some were larger or better positioned in relation to the cities' main locations, which, as it happens today, would have affected their worth and desirability. In Reccopolis, we know from John of Biclar's chronicle that there were suburbs from the very beginning and from the geophysical surveys that those domestic structures of the upper town were probably larger.

On the other hand, this spatial and social differentiation would have been also a civic one, which would have dictated the nature of social interactions in these new foundations. The free, male settlers would have formed the civic body of the new city (cf. the Roman model Bispham,

2006) – the *cives* who 'live together so they may live more honourably and safely'. As local citizens, they were the ones who would have been entitled to founders' benefits, the *privilegia*, which John of Biclar mentions, were granted to the inhabitants of Reccopolis. Other than land and membership to the civic body, privileges may have involved some sort of tax exemptions, like the ones we know from Assyrian royal establishments (Kuhrt, 1995, 539–540) and ancient Roman colonies (Diod., 15.27.4)[13] to Spanish foundations in the Americas (*Laws of Indies*, 4.3.20).[14] It is likely that the *cives* of the new foundations also enjoyed a better social status in their new homes. Improved local civic rank is something seen in Hellenistic royal foundations, where the settlers were beneficiaries of special exemptions and honours like *philanthropia* and *teimia* (Cohen, 1995, 26–30), or in the automatic granting of the rank of gentleman to the colonists in the Spanish Indies (*Laws of Indies* 4.6.6).[15] Even in Dara, local serfs were freed in order to populate the city.

Despite this, sociological and anthropological models suggest that in circumstances where a new group is forming in unfamiliar surroundings, a sense of belonging and community develops quite quickly when active participation is involved (Oishi, 2010, 12; Twigger-Ross et al., 2003). In the absence of a shared past history, local bonds of solidarity and community commitment become an essential form of cooperation to legitimise the community in newly founded cities (Lalli, 1992). The same could be said about the establishment of new social rites that could add mnemonic value to the built environment, which we can infer from the layout and the archaeology of Reccopolis and Eio.

In Reccopolis, we can guess that there was some form of inauguration ceremony, the kind that was linked to late antique *adventus* ceremonies (Dey, 2014, 148; Olmo Enciso, 2020). This is suggested by the axial development (Figure 11.3) that links the gate to Toledo with the palace complex, by the presence of a votive hoard of coins (which includes Visigothic, Frankish, Vandal, Roman, and Byzantine coins) in the basilica (Cabré Aguiló, 1946), and by the fact that Reccopolis's foundation, as given by John of Biclar, coincides with Liuvigild's ten-year jubilee or *decennalia* (578). This supposed processional route goes through the commercial street where the glass and gold traders were located, which was also where the public water cistern had been built (Martínez Jiménez and Checa Herraiz, 2023). This alone would have turned this street into an economic and socialising hub (especially since the water point was located at the crossroads of the main two axes). The basilica, as the only religious building identified so far (Caballero Zoreda & Bueno Rocha, 1989), would also have acted as a space for community gathering and definition, especially considering that it has a baptistery, clearly evidencing the building's use for rites of passage. Another rite of passage that brings the community together is burial (Williams & Sayer, 2009), and while at Reccopolis no burials have been found associated to the Visigothic phase of the basilica, the episcopal complex of Eio has an *ad sanctos* necropolis (Figure 11.7), with graves cut into the bare bedrock both inside the church and out in the street (Abad Casal et al., 2000).

While these rites could bring the urban community together, there were other elements that can be identified in the archaeological record that give us an insight into the living experience of a new city. When it comes to the water supply, for example, we find that in Reccopolis and Eio it was still offered (and, perhaps, expected) as a public service. In Reccopolis, the aqueduct was most likely connected to the cross-roads fountain, and in Eio the lack of wells might have made large cisterns (like those known from the acropolis and the episcopal complex) a public necessity (Martínez Jiménez, 2017). A couple of the water conduits that surround València la Vella (which is, itself, built high above a river) have been preliminary dated, while waiting for archaeological confirmation, to the Visigothic period and linked to the construction of the

Figure 11.7 Burials in and around the apse of the basilica of El Tolmo/Eio.

fortified settlement (Ribera Lacomba et al., 2020). This apparent expectation of a public water supply seems, at first, anachronic, especially since the aqueduct supply of most towns and cities had declined sharply during the fourth century (Martínez Jiménez, 2019), and only the main cities managed to preserve their aqueduct supply into the Visigothic period. Even so, in places such as Lisbon, we find a new fountain built during the fifth century (Mascarenhas, et al., 2012), and in Mérida we find an attempted reconstruction of an aqueduct pillar (Martínez Jiménez, 2019, 198), which show that the idea of public water supply was not a forgotten concept in this period and that it reflected both the expectations of the community and the ideals of late antique urbanism – something that is also visible in eastern foundations like Justiniana Prima (Ivanišević, 2012).

Small change is also present at Reccopolis (Castro Priego, 2014) and Eio (Doménech Belda, 2014). So far, finds have been limited to reused (or recovered) small Roman coppers, the kind that are becoming increasingly visible in Visigothic urban contexts rather than the new sixth-century coins from the south (Pliego Vázquez, 2020). But their presence already points towards urban markets and craftspeople (glassblowers, goldsmiths) who relied on coins for their daily interactions – especially in Reccopolis, where the lack of a previous settlement on site implies that this coinage had to be brought from elsewhere. Of course, the presence of gold tremisses in both sites (Reccopolis was, moreover, a city that minted coins) would suggest that agricultural output was collected as rent and the worth turned into gold coinage (Kurt, 2020). That Reccopolis and Eio were active economic hubs and tax-collecting centres, in turn, implies the presence of bureaucrats and a civic administration, probably selected from the local citizen body.

Lastly, two closely related issues that defined the localised urban experience of these new foundations need to be addressed. The first one is the pottery assemblage and the second is the rural environment. The ceramic wares of Reccopolis (Olmo Enciso & Castro Priego, 2008) and Eio (Amorós Ruiz et al., 2012) are overwhelmingly local in their typologies and petrological composition. These may have been urban or even domestic productions made for internal consumption, following patterns evident in many other cities of the Visigothic period like Mérida (Alba Calzado & Feijoo Martínez, 2003) and Toledo (Aranda González, 2013), which reflect overall trends in Iberia (Martínez Jiménez et al., 2018, 230–233). In Reccopolis, however, we find an unexpectedly high number of imports, mostly African Red Slips and amphorae, in proportions that do not match other inland cities, like Mérida, Córdoba, Eio, or Toledo (Bonifay & Bernal Casasola, 2008). Royal patronage and connections are most likely responsible for this increased connectivity, but in either case the localised nature of the common wares would have accentuated the individuality of each urban community and the material experience of the daily life. Moreover, in Reccopolis, the patent importance of locally produced glasswares may have made the urban experience even more unique and distinctive, since the Reccopolitan production could have supplied the local population with an alternative to red slip tableware (Castro Priego & Gómez de la Torre, 2008).

By the sixth century, peasant villages formed by a handful of family units with common areas that maximised their agricultural output to ensure both urban and their own supply came to substitute villa productions (Tejerizo García, 2020). Such rural settlements of Visigothic date have been identified around Eio (Sarabia Bautista, 2014) and Reccopolis (Olmo Enciso et al., 2019; Martínez Jiménez, 2018). According to field surveys, around Reccopolis most of the rural settlements identified so far seem to be individual farmsteads rather than multi-family villages (Olmo Enciso et al., 2019), so it is likely that fields were also tilled by city dwellers. Moreover, the large *horreum*-like layout of the north building of the 'palace' complex could suggest that agricultural surplus (perhaps rents extracted from royal land and dependant peasants – Castellanos García, 2003) was stored there. Field surveys show that Eio, in contrast, has larger rural settlements, a combination of villages, farmsteads, and villa re-occupations; Eio perhaps relied more on similar agricultural rents collected for the urban clergy and the royal administration than on city-dwelling farmers (Sarabia Bautista, 2014).

The distribution of rural plots has always been as important as that of urban ones for the new settlers in new, planned foundations. This can be seen in later Greek and Roman colonies (Ariño Gil & Gurt Esparraguera, 1994; Dougherty, 1993, 15–24) and in the Spanish *Laws of Indies* that instructed settlers not to build their city before fields had been tilled (4.7.15).[16] The reason behind the colonists' tax exemptions mentioned earlier was – at least partly – to allow new settlers to achieve a level of self-sufficiency during the first difficult years. Since garrisoned soldiers were (at least partially) paid in kind, and considering the charitable handouts of the church of Mérida to enfranchised locals (*VPE*, 5.3.7),[17] it is not impossible that the settlers were entitled to some form of foundational *annona* – perhaps explaining the unusually high number of imported wares. In this sense, Eio probably functioned more as an administrative (and military and religious) centre for a larger rural population than Reccopolis, where the townsfolk would have been more numerous and city life more developed.

11.5 Conclusion

While it is possible to make broad-brush descriptions of what life in Visigothic cites was like (Orlandis, 2006), local particularities (monumentality and lack thereof, existing infrastructure, continuity of local government, nature of the rural hinterland, civic status, trading connections, etc.) immediately underline how thin this homogenising veneer would have been. Mérida, Valencia, Córdoba, or Braga are different enough in this period to highlight that generalising paradigms always need to be presented with caution. However, when it comes to the new foundations of the Visigothic monarchy, it is possible to reach more concrete conclusions. The economic cycle, the development of a sense of locality and of community, the expected monuments and infrastructure, and even the potential local administration that characterise these new foundations were necessarily similar since they derived from the same late antique urban concept. Because half of these new cities lacked a pre-existing Roman armature that could constrain these ideas or dictate the lines developments had to follow, they can provide us with a raw approximation to a post-Roman city concept (both as an *urbs* and as a *civitas*) that goes beyond Isidore's definitions. We should also note that these sixth-century urban ideals were not part of a late antique shared, master plan: the urban experience of Reccopolis could be conceptually similar to those of València la Vella or Eio, but the practical logistics of founding cities in the Byzantine or the Basque frontier were not the same as those of the Visigothic core lands of Celtiberia.

Reccopolis and Eio continued to be inhabited into the ninth century, and they fulfilled a central role in the territorial administration until they were substituted by new, purpose-built Umayyad settlements (Zorita for Reccopolis and Murcia for Eio). Despite the evolution of the original layouts and the modifications to street width and public areas, the form of these late antique cities remained the same after the foundational effort. Preliminary assessments of the occupation of València la Vella suggest that it was short-lived, and little can be said about sites like Ologicus or Victoriacum, but these do not make the Visigothic city concept less viable than preceding Roman ones – although the lived experience of the new foundations was certainly unique to this period.

Notes

1 On the ongoing debate about the chronology of the so-called Visigothic churches, see Sánchez-Pardo et al. (2017).

2 'Liuvigildus rex extinctis undique tyrannis, et pervasoribus Hispaniae superatis sortitus requiem propriam cum plebe resedit civitatem in Celtiberia ex nomine filii condidit, quae Rec[c]opolis nuncupatur: quam miro opere et in moenibus et suburbanis adornans privilegia populo novae Urbis instituit'.

3 'Liuvigildus rex partem Vasconiae occupat et civitatem, quae Victoriaco nuncupatur, condidit'.

4 'Habuit quoque et initio regni expeditionem contra incursus Vasconum Tarraconensem provinciam infestantium, ubi adeo montivagi populi terrore adventus eius perculsi sunt, ut confestim, quasi debita iura noscentes, remissis telis et expeditis ad precem manibus, supplices ei colla submitterent, obsides darent, Ologicus civitatem Gothorum stipendiis suis et laboribus conderent, pollicentes eius regno ditionique parere, et quidquid imperaretur efficere'.

5 The same authors insist that, due to the lack of evidence elsewhere for a fortified position of late Antique date, Victoriacum/Ologicus was established on the Roman site of Veleia, although they vaguely allude to addressing the lack of datable material there for the Visigothic period.

6 *Historia Gothorum, additamenta Pelagiana*: 'Rex iste in Asturiis civitatem aedificavit, quam Lugo id est Luceo vocavit'.

7 Cf. the Goths and their kingdom in the Middle Ages in Bonch Reeves (forthcoming).

8 'Nam urbs ipsa moenia sunt, civitas autem non saxa sed habitatores vocantur'.

9 'Cives vocati, quod in unum coeuntes vivant, ut vita communis et ornatior fiat et tutior . . . populus universi cives sunt, connumeratis senioribus civitatis'.

10 And yet there are still attempts to contest the identification of the site (Arce Martínez, 2010).

11 'Liuvigildus muros Italicae antiquae civitatis restaurat'.

12 This is less than the 30 hectares usually given as an indicator of size (Olmo Enciso, 2008) and based on GIS calculations derived from the most recent published data (Henning et al., 2019).

13 '[T]he Romans sent out 500 colonists (ἀποίκους) with tax immunity (ἀτελείᾳ) to Sardinia' (Sherman transl.).

14 'That the settlers (*pobladores*) shall pay no sale taxes (*alcabala*) for twenty years'.

15 'That the main settlers and their sons and legitimate descendants shall be gentlemen (*hijosdalgo*) in the Indies'.
16 'That [only] after the settlers have sown [the fields], may they begin to build (*edificar*)'.
17 'Whenever one of the citizens (*de civibus urbis*) or a peasant from the rural districts came out of need to the episcopal complex to ask the dispensers for a measure of wine or oil or honey [the bishop gave more than they expected]'.

Editions and abbreviations

Al-Isfahani – Hamza al-Isfahani, *Annals*, trans. U. Daudpota (1932). K.R. Cama Oriental Institute Publication.

Al-Tabari – al-Tabari, *The History: vol. IV: The Ancient Kingdoms*, trans. M. Perlman (1987). State University of New York Press.

Bicl., *Chron.* – John of Biclar, *Chronicle*. In *Chronica Minora. Saec. IV, V, VI, VII* (*MGH AA* 11), ed. T. Mommsen (1894). Weidmann.

Diod. – Diodorus Siculus, *Library of History* (vol. VII) (Loeb Classical Library), ed. and trans. C. Sherman (1952). Harvard University Press.

Greg. Tur., *Hist.* – Gregory of Tours, *Histories*. In *Gregorii episcopi Turonensis Libri Historiarum X* (*MGH SS rer Merov.* 1), ed. B. Krusch and W. Levison (1951). Bibliopolii Hahniani.

Historia Gothorum, additamenta Pelagiana. In *Chronica Minora. Saec. IV, V, VI, VII* (*MGH AA* 11), ed. T. Mommsen (1894). Weidmann.

Isidore, *Etym.* – Isidore of Seville, *Etymologies*, ed. W. Lindsay (1911). Oxford University Press.

Isidore, *Hist. Goth.* – Isidore of Seville, *History of the Goths*. In *Chronica Minora. Saec. IV, V, VI, VII* (*MGH AA* 11), ed. T. Mommsen (1894). Weidmann.

John Eph. – John of Ephesus, *Historia Ecclesiastica* (*CCO* 105), ed. E. Brooks (1935). Peeters.

Jul. Tol., *HWR* – Julian of Toledo, *History of King Wamba*. In *Passiones vitaeque sanctorum aevi merovingici* (*MGH SS rer. Merov.* 5), ed. W. Levison (1910). Bibliopolii Hahniani.

Laws of Indies – Recopilación de las leyes de Indias. Volumes I-IV; facsimile of 1681 (1973). Ediciones Cultura Hispánica.

Menand. – Menander Protector, *Historiae Fragmenta*, transl. R. Blockley (1985). Francis Cairns Publications.

VSPE – Vitas Sanctorum Patrum Emeritensium (*CCSL* 116), ed. A. Maya Sánchez, (1992). Brepols.

Bibliography

Abad Casal, L., Gutiérrez Lloret, S., & Gamo Parras, B. (2000). La basilica y el baptisterio de El Tolmo de Minateda (Hellín, Albacete). *Archivo Español de Arqueologia*, *73*, 193–221.

Abad Casal, L., Guitiérrez Lloret, S., Gamo Parras, B., & Cánovas Guillén, P. (2008). Una ciudad en el camino: pasado y futuro de El Tolmo de Minateda (Hellín, Albacete). In L. Olmo Enciso (Ed.), *Recópolis y la ciudad en la época visigoda* (pp. 323–36). Museo Arqueológico Regional, Alcalá de Henares.

Alba Calzado, M., & Feijoo Martínez, S. (2003). Pautas evolutivas de la cerámica común de Mérida en épocas visigoda y emiral. In L. Caballero, P. Mateos Cruz, & M. Retuerce (Eds.), *Cerámicas tardorromanas y altomedievales en la Península Ibérica* (pp. 483–504). CSIC.

Amorós Ruiz, V., Cañavate Castejón, V., Gutiérrez Lloret, S., & Sarabia Bautista, J. (2012). Cerámica altomedieval en el tolmo de minateda (Hellín, Albacete, España). In S. Gelichi (Ed.), *Atti del IX Congresso Internazionale sulla ceramica medievale nel Mediterraneo* (pp. 246–57). Ca Foscary University.

Aranda González, R. (2013). Una aportación al conocimiento de las producciones cerámicas de época visigoda: el conjunto cerámico de la parcela R3 de la Vega Baja (Toledo). *Espacio, tiempo y forma. Serie I, Prehistoria y arqueología*, *6*, 381–450.

Arce Martínez, J. (2010). Recópolis, la ciudad fantasma. In A. Domínguez & G. M. Rodríguez (Eds.), *Doctrina a magistro discipulis tradita. Estudios en homenaje al profesor Dr. Luis García Iglesias* (pp. 373–394). Universidad Autónoma de Madrid.

Ariño Gil, E., & Gurt Esparraguera, J. M. (1994). Catastros romanos en el entorno de Augusta Emerita. In J.-G. Gorges & M. Salinas (Eds.), *Les campagnes de la Lusitanie romaine. Occupation du sol et habitats* (pp. 45–66). Casa de Velázquez.

Azkarate Garai-Olaun, A. (2004). El País Vasco en los siglos inmediatos a la desaparición del Imperio Romano. In P. Barruso Barés & J. A. Lema (Eds.), *Historia del País Vasco. Edad Media (siglos V-XV)* (pp. 23–50). University of the Basque Country.

Balmaseda Muncharaz, L. (2008). La escultura en Recópolis. In L. Olmo Enciso (Ed.), *Recópolis y la ciudad en época visigoda.* (pp. 141–157). Museo Arqueológico Regional, Alcalá de Henares.

Barbier, J. (2021). 'Te vir laudabilis illum defensore necnon et vos honerati que curas puplicas agitis adsidue . . . ': Regards sur la vitalité civique des cités post-romaines de Gaule. In X. Dupuis, V. Fauvinet-Ranson, C. Godard, & H. Inglebert (Eds.), *L'Automne de l'Afrique Romaine. Hommage à Claude Lepelley*. Hermann.

Barroso Cabrera, R., Carrobles Santos, J., & Morín de Pablos, J. (2013). ¿Váscones o Wascónes? Acerca del Ducado de Cantabria y la fundación de ciudades en el norte peninsular en época visigoda. *e-Spania, 3*. https://doi.org/10.4000/e-spania.22944

Bispham, E. (2006). Coloniam deducere: How Roman was Roman colonization in the middle Republic? In G. Bradley & J.-P. Wilson (Eds.), *Greek and Roman Colonization. Origins, Ideologies, and Interactions* (pp. 73–160). The Classical Press of Wales.

Blanton, R., & Fargher, L. (2011). The collective logic of pre-modern cities. *World Archaeology*, *5.1*, 502–522.

Bonch Reeves, K. (forthcoming). The Goths and their kingdom in the Middle Ages. In J. Wood, M. Lester, & J. Martínez Jiménez (Eds.), *A Companion to the Visigothic Kingdom*. Brill.

Bonifay, M., & Bernal Casasola, D. (2008). Recópolis, paradigma de las importaciones africanas en el visigothorum regnum. In L. Olmo Enciso (Ed.), *Recópolis y la ciudad en la época visigoda*. (pp. 99–115). Museo Arqueológico Regional, Alcalá de Henares.

Brassous, L. (2019). Changes in the city network in Roman Hispania. In J. Andreu & A. Blanco-Pérez (Eds.), *Signs of Weakness and Crisis in the Western Cities of the Roman Empire (c. II-III AD)* (pp. 191–206). Franz Steiner Verlag.

Buchberger, E. (2017). *Shifting Ethnic Identities in Spain and Gaul, 500–700: From Romans to Goths and Franks*. Amsterdam University Press.

Caballero Zoreda, L., & Bueno Rocha, L. (1989). De nuevo a propósito de la basílica de Recópolis. *Archivo Español de Arqueología*, *62*, 283–291.

Cabré Aguiló, J. (1946). El tesorillo de trientes da las excavaciones del plan nacional de 1944–45 en Zorita de los Canes (Guadalajara). *Informes y Memorias*, *10*, 43–44.

Castellanos García, S. (2003). The political nature of taxation in Visigothic Spain. *Early Medieval Europe*, *12.3*, 201–228.

Castro Priego, M. (2014). Recópolis y sus contextos numismáticos de época visigoda en el centro de la Península Ibérica. *Revue numismatique VI*, *171*, 463–495.

Castro Priego, M., & Gómez de la Torre-Verdejo, A. (2008). La actividad artesanal en Recópolis: la producción de vidrio. In L. Olmo Enciso (Ed.), *Recópolis y la ciudad en la época visigoda* (pp. 117–128). Museo Arqueológico Regional, Alcalá de Henares.

Cohen, G. (1995). *The Hellenistic Settlements in Europe, the Islands, and Asia Minor*. University of California Press.

Cuomo, S. (2007). *Technology and Culture in Greek and Roman Antiquity*. Cambridge University Press.

Curchin, L. (2014). The role of civic leaders in late antique Hispania. *Studia Historica. Historia Antigua*, *32*, 281–304.

Curchin, L. (2018). Curials and local government in Visigothic Hispania. *Antiquité Tardive*, *26*, 225–240.

DeLaine, J. (1997). *The Baths of Caracalla: A Study in the Design, Construction and Economics of Large-Scale Building Projects in Imperial Rome*. Journal of Roman Archaeology.

Dey, H. (2010). Art, ceremony, and city walls: The aesthetics of imperial resurgence in the late Roman West. *Journal of Late Antiquity*, *3.1*, 3–37.

Dey, H. (2014). *The Afterlife of the Roman City. Architecture and Ceremony in Late Antiquity and the Early Middle Ages*. Cambridge University Press.

Diarte Blasco, P. (2018). *Late Antique and Early Medieval Hispania: Landscapes without Strategy?* Oxbow.

Díaz Martínez, P. (2011). *El Reino Suevo (411–585)*. Akal.

Dix, B. (1982). The manufacture of lime and its uses in the Western Roman provinces. *Oxford Journal of Archaeology*, *1.3*, 331–346.

Doménech Belda, C. (2014). Moneda y espacios de poder en el reino visigodo. Los tremises de El Tolmo de Minateda. *Arqueología y Territorio Medieval*, *21*, 9–37.

Dougherty, C. (1993). *The Poetics of Colonization. From City to Text in Archaic Greece*. Cambridge University Press.

Esmonde Cleary, S. (2013). *The Roman West, AD 200–500. An Archaeological Study*. Cambridge University Press.

Fernández, D. (2020). Transformaciones institucionales y liderazgo cívico en la Hispania post-imperial. In A. Carneiro, N. Christie, & P. Diarte (Eds.), *Urban Transformations in the Late Antique West: Materials, Agents, and Models* (pp. 259–279). Coimbra University Press.

Fernández Delgado, A., Martínez Jiménez, J., & Tejerizo García, C. (2013). Old and new elites in the Visigothic kingdom (550–650 AD). In E. Van der Wilt & J. Martínez Jiménez (Eds.), *Tough Times: The Archaeology of Crisis and Recovery. Proceedings of the GAO Annual Conferences 2010 and 2011* (pp. 161–170). Archaepress.

Filloy Nieva, I., & Gil Zubillaga, E. (2007). Vida cotidiana al abrigo de las murallas. Novedades de la investigación sobre el recinto amurallado tardorromano de Veleia (Iruña de Oca, Álava, País Vasco). In A. Rodríguez Colmenero & I. Rodá de Llanza (Eds.), *Murallas de ciudades romanas en el occidente del imperio. Lucus Augusti como paradigma* (pp. 467–82). Diputación Provincial, Lugo.

Finley, M. (1973). *The Ancient Economy*. University of California Press.

Fumadó Ortega, I. (2013). ¿Quién parte y reparte? Análisis de la disposición urbana de la Cartago fenicia. *Archivo Español de Arqueología*, *86*, 7–21.

Gusmão, F. (2016–17). The role of the bishop according to the Liber Iudiciorum (Lex Visigothorum). *Visigothic Symposia*, *1*, 81–94.

Gutiérrez Lloret, S., Abad Casal, L., & Gamo Parras, B. (2005). Eio, Iyyuh y el Tolmo de Minateda (Hellín, Albacete): de sede episcopal a madîna islámica. In J. M. Gurt Esparraguera & A. Ribera Lacomba (Eds.), *VI reunió d'arqueologia cristiana hispànica* (pp. 345–370). Institut d'Estudis Catalans.

Gutiérrez Lloret, S., & Cañavate Castejón, V. (2010). Casas y cosas: espacios y funcionalidad en las viviendas emirales del Tolmo de Minateda (Hellín, Albacete). *Cuadernos de Madīnat al-Zahrā*, *7*, 123–148.

Gutiérrez Lloret, S., & Cánovas, P. (2009). Construyendo el siglo VII: arquitecturas y sistemas constructivos en El Tolmo de Minateda. In L. Caballero, P. Mateos, & M. Á. Utrero (Eds.), *El siglo VII frente al siglo VII. Arquitectura* (pp. 91–132). CSIC.

Gutiérrez Lloret, S., & Sarabia Bautista, J. (2013). The episcopal complex of Eio-El Tolmo de Minateda (Hellín, Albacete, Spain). Architecture and spatial organization. 7th to 8th centuries AD. *Hortus Artium Medeivalium*, *19*, 267–300.

Henning, J., McCormick, M., Olmo Enciso, L., Rassman, K., & Fikrit, E. (2019). Reccopolis revealed: First geomagnetic mapping of the early medieval Visigothic royal town. *Antiquity, 93*(369), 735–751.

Houten, P. (2018). *Civitates Hispaniae. Urbanisation on the Iberian Peninsula During the High Empire*. University of Leiden.

Ivanišević, V. (2012). Akveduct Caričinog Grada – Justinujane Prime. *Saopštenja*, *44*, 13–31.

Ivanišević, V. (2017). Main patterns of urbanism in Caričin Grad (Justiniana Prima). In E. Rizos (Ed.), *New Cities in Late Antiquity. Documents and Archaeology* (pp. 221–32). Brepols.

Koch, M. (2014). Arianism and ethnic identity in sixth-century Spain. In G. Berndt & R. Steinacher (Eds.), *Arianinsm: Roman Heresy and Barbarian Creed* (pp. 257–270). Ashgate.

Kuhrt, A. (1995). *The Ancient Near East. c. 3000–330 BC*. Routledge.

Kurt, A. (2020). *Minting, State, and Economy in the Visigothic Kingdom: From Settlement in Aquitaine Through the First Decade of the Muslim Conquest of Spain*. Amsterdam University Press.

Lalli, M. (1992). Urban-related identity: Theory, measurement, and empirical findings. *Journal of Environmental Psychology*, *12*, 285–303.

Lavan, L. (2008). Explaining technological change: Innovation, stagnation, recession and replacement. In L. Lavan, E. Zanini, & A. Sarantis (Eds.), *Technology in Transition. AD 300–650* (pp. xv–xl). Brill.

Lope López de Rego, J. L. (2008). "Baiyara" fundación visigoda y nombre de Villa del Río en época árabe. *Crónica de Córdoba y sus pueblos*, *14*, 71–78.

Macias Solé, J. M., & Ribera Lacomba, A. (2020). Los spolia en la formación del nuevo paisaje urbano en torno a Valentia y Edeta durante los siglos VI al VIII. In P. Mateos Cruz & C. Morán Sánchez (Eds.), *Exemplum et Spolia. La reutilización arquitectónica en la transformación del paisaje urbano de las ciudades históricas* (pp. 381–391). CSIC.

Macias Solé, J. M., Ribera Lacomba, A., & Rosselló Mesquida, M. (Eds.). (2020). *Recintos fortificados en epoca visigoda: historia, arquitectura y tecnica constructiva*. ICAC.

Macias Solé, J. M., Ribera Lacomba, A., & Rosselló Mesquida, M. (2021). València la Vella. Arqueología de una ciudad sin nombre. *Urbs Regia*, *6*, 57–67.

Martin, C. (2003). *La géographie du pouvoir dans l'Espagne visigothique*. Septentrion.

Martínez Jiménez, J. (2015). A preliminary study on the aqueduct of Reccopolis. *Oxford Journal of Archaeology*, *34.3*, 301–320.

Martínez Jiménez, J. (2017). Water supply in the Visigothic urban foundations of Eio (El Tolmo de Minateda) and Reccopolis. In E. Rizos (Ed.), *New Cities in Late Antiquity. Documents and Archaeology* (pp. 233–245). Brepols.

Martínez Jiménez, J. (2018). The rural hinterland of the Visigothic capitals of Toledo and Reccopolis, between the years 400–800 CE. In A. Delattre, M. Legendre, & Sijpesteijn (Eds.), *Authority and Control in the Countryside. Continuity and Change in the Mediterranean 6th-10th Century CE* (pp. 97–127). Brill.

Martínez Jiménez, J. (2019). *Aqueducts and Urbanism in Post-Roman Hispania*. Gorgias Press.

Martínez Jiménez, J. (2020a). Reccopolitani and other town dwellers in the southern Meseta during the Visigothic period of state formation. In M. Kelly & M. Burrows (Eds.), *Urban Interactions: Communication and Competition in Late Antiquity and the Early Middle Ages* (pp. 109–147). Gracchi Books.

Martínez Jiménez, J. (2020b). Urban identity and citizenship in the west between the fifth and seventh centuries. *al-Masaq*, *32.1*, 81–108.

Martínez Jiménez, J. (2022a). Foundational grids and urban communities in the Iberian Peninsula in Antiquity and the Middle Ages. In A. Wallace-Hadrill & S. Greaves (Eds.), *Rome and the Colonial City* (pp. 239–263). Oxbow.

Martínez Jiménez, J. (2022b). Water consumption in Roman lime mortar construction: A calculating method. *Arqueología de la Arquitectura*, *19*, e131.

Martínez Jiménez, J., & Checa Herraiz, J. (2023). The water cycle in Reccopolis. *Journal of Roman Archaeology*, *36.1*, 157–185.

Martínez Jiménez, J., & González Gutiérrez, P. (2017). Knowledge and specialised trades in the late antique West: Medicine vs. Engineering. *Journal for Late Antique Religion and Culture*, *11*, 38–58.

Martínez Jiménez, J., Sastre de Diego, I., & Tejerizo García, C. (2018). *The Iberian Peninsula 300–850: An Archaeological Perspective*. Amsterdam University Press.

Martínez Jiménez, J., & Tejerizo García, C. (2015). Central places in the post-Roman Mediterranean: Regional models for the Iberian Peninsula. *Journal of Mediterranean Archaeology*, *28.1*, 81–104.

Martínez Lemoine, R. (2003). The classical model of the Spanish-American colonial city. *The Journal of Architecture*, *8.3*, 355–368.

Mascarenhas, J. M., Bilous, F., & Neves, N. S. (2012). O aqueduto romano de Olisipo: viabilidade ou utopia? Ensaio de traçado apoiado em modelação geográfica. *Revista Portuguesa de História*, *43*, 239–264.

Moreno Alcaide, M. (2013). Condicionantes para la fundación de la colonia latina de Cosa. Aproximación desde los factores físicos, geopolíticos y religiosos para la elección del emplazamiento. *Cuadernos de Prehistoria y Arqueología de la Universidad de Granada*, *23*, 323–346.

Moreno Martín, F. (2018). Circulación de modelos y circularidad de argumentación para el conocimiento de la topografía cristiana entre la Antigüedad Tardía y la Alta Edad Media. In S. Panzram & L. Callegarin (Eds.), *Entre civitas y madina. El mundo de las ciudades en la Península Ibérica y en el norte de África (siglos IV-IX)* (pp. 153–172). Casa de Velázquez.

Oishi, S. (2010). The psychology of residential mobility: Implications for the self, social relationships and well-being. *Perspectives on Psychological Science*, *5.1*, 5–21.

Oller Guzmán, J. (2014). La *civitas sine urbe* y su funcionalidad de vertebración en el territorio provincial hispano: los casos de Egara y de Caldes de Montbui. *Pyrenae*, *45.1*, 89–110.

Olmo Enciso, L. (2008). Recópolis: una ciudad en época de transformaciones. In L. Olmo Enciso (Ed.), *Recópolis y la ciudad en la época visigoda* (pp. 40–63). Museo Arqueológico Regional, Alcalá de Henares.

Olmo Enciso, L. (2009). La Vega Baja en época visigoda: una investigación arqueológica en construcción. In A. García (Ed.), *La Vega Baja de Toledo* (pp. 69–94). Toletum Visigodo.

Olmo Enciso, L. (2020). Reccopolis: The representation of power in a complex landscape. In S. Panzram & P. Pachá (Eds.), *The Visigothic Kingdom of Toledo: The Negotiation of Power in post-Roman Iberia* (pp. 215–233). Amsterdam University Press.

Olmo Enciso, L., & Castro Priego, M. (2008). La cerámica de época visigoda de Recópolis: apuntes tipológicos desde un análisis estratigráfico. In L. Olmo Enciso (Ed.), *Recópolis y la ciudad en la época visigoda* (pp. 89–96). Museo Arqueológico Regional, Alcalá de Henares.

Olmo Enciso, L., Castro Priego, M., & Diarte Blasco, P. (2018). Beside and within the walls of Reccopolis: Social dynamics and landscape transformations of a new Visigothic urban foundation. *Groma*, *3*. https://doi.org/10.12977/groma14

Olmo Enciso, L., Castro Priego, M., & Diarte Blasco, P. (2019). Transformación social y agrosistema en el interior peninsular durante la Alta Edad Media (s. VI-VIII d. C.): nuevas evidencias desde Recópolis (Zorita de los Canes, Guadalajara). *Lucentum*, *38*, 355–377.

Olmo Enciso, L., Castro Priego, M., Gómez de la Torre-Verdejo, A., & Sanz Paratcha, Á. (2008). Recópolis y su justificación científica: la secuencia estratigráfica. In L. Olmo Enciso (Ed.), *Recópolis y la ciudad en la época visigoda* (pp. 64–75). Museo Arqueológico Regional, Alcalá de Henares.

Orlandis, J. (2006 [1981]). *La vida en España en tiempos de los godos*. Rialp.

Parra Romo, Á. (2018). El ejército visigodo en campaña, Wamba y la secesión de la Narbonense. *Studia Historica*, *36*, 221–251.

Pliego Vázquez, R. (2020). Rethinking the minimi of the Iberian Peninsula and Balearic Islands in late Antiquity. *Journal of Medieval Iberian Studies*, *12.2*, 125–154.

Poveda Arias, P. (2019). La diócesis episcopal en la Hispania Visigoda: concepción, construcción y disputas por su territorio. *Hispania Sacra*, *71*(143), 9–24.

Poveda Arias, P. (2023). Coexisting leaderships in the Visigothic cities: A 'coopetitive' model. In D. Castro & F. Ruchesi (Eds.), *Leadership, Social Cohesion, and Identity in Late Antique Spain and Gaul (500–700)* (pp. 159–183). Amsterdam University Press.

Raynaud, C. (2017). Villeneuve-les-Maguelone (Hérault). Siège épiscopal de Maguelone. *Archéologie médiévale*, *47*, 177–178. https://doi.org/10.4000/archeomed.6865

Ribera Lacomba, A., & Rosselló Mesquida, M. (2013). La ocupación tardoantigua del circo romano de Valentia. In O. Brandt (Ed.), *Acta XV congressus internationalis archaeologiae christianae Toleti, 8–12.9.2008: episcopus, civitas, territorium* (pp. 47–62). Pontificio Istituto di Archeologia Christiana.

Ribera Lacomba, A., Rosselló Mesquida, M., & Macias Solé, J. M. (2020). Historia y arqueología de dos ciudades en los siglos VI-VIII d. C. Valentia y València la Vella. *Antigüedad y Cristianismo*, *37*, 63–106.

Ripoll López, G. (2000). Sedes regiae en la Hispania de la Antigüedad Tardía. In G. Ripoll López & J. M. Gurt Esparraguera (Eds.), *Sedes Regiae (ann. 400–800)* (pp. 371–404). Real Academia de Buenas Letras de Barcelona.

Rizos, E. (Ed.). (2017). *New Cities in Late Antiquity. Documents and Archaeology*. Brepols.

Rizos, E. (2022). No colonies and no grids: New cities in the Roman East and the decline of the colonial urban paradigm from Augustus to Justinian. In A. Wallace-Hadrill & S. Greaves (Eds.), *Rome and the Colonial City* (pp. 209–238). Oxbow.

Ruiz Bueno, M. (2020). Transformaciones en la topografía del poder de Córdoba entre los siglos III y VII. *Al-Mulk*, *18*, 93–122.

Ruiz Prieto, E. (2013). Itálica tardoantigua: reflexiones y asignaturas pendientes. *Ligustinus*, *1*, 81–117.

Sánchez-Pardo, J. C., Blanco-Rotea, R., & Sanjurjo-Sánchez, J. (2017). The Church of Santa Comba de Bande and early medieval Iberian architecture: New chronological results. *Antiquity*, *91.358*, 1011–1026.

Sarabia Bautista, J. (2014). La transformación del paisaje rural tras la fundación del obispado de Eio-El Tolmo de Minateda (Hellín, Albacete, España): siglos V al IX d.C. *Hortus Artium Medievalium*, *20.1*, 216–231.

Sastre de Diego, I., & Alba Calzado, M. (2020). Bajo la protección de la mártir: Mérida en la antigüedad tardía (siglos IV-VII). In A. Carneiro, N. Christie, & P. Diarte (Eds.), *Urban Transformations in the Late Antique West: Materials, Agents, and Models* (pp. 203–230). Coimbra University Press.

Schneider, L. (2008). Aux marges méditerranées de la Gaule mérovingienne. Les cadres politiques et ecclésiastiques de l'ancienne Narbonnaise Ire entre Antiquité et Moyen Âge (V[e]-IXe siècle). In F. Mazel (Ed.), *L'espace du diocèse. Genèse d'un territoire dans l'occident médiéval (Ve-XIIIe siècle)* (pp. 69–95). Presses universitaires de Rennes.

Smith, M. (2007). Form and meaning in the earliest cities: A new approach to ancient urban planning. *Journal of Planning History*, *6.1*, 3–47.

Tejerizo García, C. (2017). *Arqueología de las sociedades campesinas en la cuenca del Duero durante la Primera Alta Edad Media*. University of the Basque Country.

Tejerizo García, C. (2020). Del modo de producción germánico al modo de producción campesino. *Nuestra Historia*, *9*, 17–36.

Twigger-Ross, C., Bonaiuto, M., & Breakwell, G. (2003). Identity theories and environmental psychology. In M. Bonnes & T. Lee (Eds.), *Psychological Theories for Environmental Issues* (pp. 203–234). Ashgate.

Utrero Agudo, M[a]. Á. (2017). The artisans behind Visigothic buildings: The materiality of identity. *Visigothic Symposia*, *2*, 99–103.

Van Limbergen, D., & Vermeulen, F. (2021). A method for estimating Roman population sizes from urban survey contexts: An application in Central Adriatic Italy. In K. Verboven (Ed.), *Complexity Economics: Building a New Approach to Ancient Economic History* (pp. 203–250). Palgrave Macmillan.

Vizcaíno Sánchez, J. (2009). *La presencia bizantina en Hispania (siglos VII-VIII). La documentación arqueológica* (2nd ed.). University of Murcia.

Wallace-Hadrill, A. (1994). *Houses and Society in Pompeii and Herculaneum*. Princeton University Press.

Williams, H., & Sayer, D. (2009). Halls of mirrors: Death and identity in medieval archaeology. In D. Sayer & H. Williams (Eds.), *Mortuary Practices and Social Identities in the Middle Ages: Essays in Burial Archaeology in Honour of Heinrich Härke* (pp. 1–23). University of Exeter.

Zapata Parra, J. A. (2018–19). Las murallas de Begastri. Análisis histórico y arqueológico. *Antigüedad y Cristianismo, 35–36*, 115–148.

Ziche, H. (2006). Making late Roman taxpayers pay: Imperial government strategies and practices. In H. Drake (Ed.), *Violence in Late Antiquity: Perceptions and Practices* (pp. 127–137). Routledge.

12 From *Theoria* to Pilgrimage

Ships, shores and sacred travel around the Mediterranean Sea in Late Antiquity

Amelia R. Brown

This chapter examines the impact of Christianity on sacred aspects of sea travel in Late Antiquity. Combining literary and archaeological evidence, it charts the changes brought about by the new religion to the way travellers experienced spaces connected to their journeys. In general terms, Christian travellers acted in ways that had much in common with those of earlier mariners. Shipbuilding, steering by oars and uses of space onboard ships seem to have remained relatively constant until the ninth or tenth century. Whether en route from east to west, or north to south, late antique travellers faced the same threats as in ancient times: stormy weather, rocky shores or leaky ships' hulls (Casson, 1971). Christians responded to these threats in remarkably similar ways: they still burned incense on altars as part of their prayers aboard ships, and they addressed the heavens with raised arms, from the higher stern area of the ship. At the same time, this era was marked by important changes in the lived religion of those aboard ships and ashore, whether pilgrims on a personal quest or bishops attending a church council (see also Rebecca Sweetman's chapter in this volume). Christians prayed to the One God, Jesus, or Mary. They carried images of saints, instead of images or shrines of the ancient gods. They sailed to visit new sacred sites and to collect physical souvenirs to carry home as spiritual protection.

The first section of the chapter looks into the smallest scale, analysing how the experience of the space of the ship was progressively Christianised from the fourth century to the early seventh century. As I will argue, the addressees of shipboard prayers and rituals changed, as did the iconography of protective amulets carried aboard, even though the *formulae* and the places for seeking assistance continued Greco-Roman religious traditions. I will then consider, in Section 12.2, how Mediterranean sea routes were influenced by Christian pilgrimage, especially in terms of the number and frequency of passengers. Western and Eastern Christians alike went in increasing numbers to eastern ports, from Caesarea south to Alexandria, to visit biblical holy places like the Galilee, Jerusalem and Mt Sinai. Christian pilgrims flowed east to the sites of the Holy Land, and then back home again with portable – and sometimes physical – blessings from the trip (Hunt, 1982). They also helped to shift patterns of land-based coastal devotional activity at established ports like Alexandria, away from traditional sailors' patrons like Isis and Sarapis to new Christian saints and their shrines, such as St Menas, south of the city, and SS Cyril and John at Menouthis/Abukir, which pilgrims visited on their way to the biblical Mt Sinai (Frankfurter, 1998). Constantine's choice of Constantinople as his new capital shifted the routes of Alexandrian grain ships eastwards, and this political change, along with the intensity of ecclesiastical activity around the court in the Bosporus, meant that there were more frequent passages and water stops for Christian shipping along the eastern coastline and islands of the Aegean. I follow Janet Wade in arguing for the importance of the new capital's maritime draw and the importance of the growing number of relics at the City in Christianising traditional

DOI: 10.4324/9780429427152-15

maritime religion there, and adding new Christian pilgrimage routes (Wade, 2014, 2018, 2020; Nadine Viermann, in this volume).

Finally, section 12.3 will consider the consequences of the rise of new maritime patron saints and their sites of worship away from the major centres in the Holy Land, Alexandria or Constantinople – as in the case of St Phocas of Sinope, St Isidore of Chios or St Nicholas of Myra. As I will argue, their images aboard ships and at shrines ashore supplanted ancient civic, regional or Hellenic gods (Vitaliotis, 1997). Through pilgrimage, Christians increased the standing of previously modest cities such as Seleucia on the Calycadnus river for St Thecla in south-east Pisidia and Cilicia (Kristensen, 2017; Cortese, 2022, 62–93). In many ports, new basilica churches were constructed on harbours, and new souvenirs manufactured, while the ancient need continued for warehouses, breakwaters and housing ashore at ports favoured by Christian travellers in Late Antiquity (Constable, 2003, e.g. traces the Greek *pandocheion* through to the Arabic *funduq* as a hostel with commercial functions always needed ashore).

On the largest scale, then, pilgrims played a key role alongside Constantine and the professional sailors who carried them abroad, in both Christian space- and place-making, helping to reshape pagan into Christian sacred landscapes via their travels by sea. Pilgrims drove the Mediterranean-wide interchange of their own personal, local and ecumenical practices of Christianity with those at their destination or stopping points, changing life – and especially its sacred aspects – aboard ship and ashore in Late Antiquity and changing the addressees of prayers of salvation made for safety at sea most of all (Sweetman, 2019, 2022).

12.1 Ships

Ships were mobile spaces: short-term, well-defined communities of professional mariners and passengers aboard what was a floating home, temple/church and *agora* all rolled into one diverse, perilous and often cramped space. Ships brought their own distinctive spatial dynamics to the coastal interplay of pilgrims, harbours and cults in Late Antiquity (an aspect analysed by Rowan Munnery in this volume). They carried cargoes with passengers, and the *ethos* of the passengers' own habits, beliefs, home ports and reasons for pilgrimage. Christian pilgrims came aboard ships with many other non-human "passengers" both large and small, known and unknown: fleas, rats, cats, herbs and bad diseases. They brought along useful knowledge, old and new – like new names for the divine, for example.

The archaeological record for shipwrecks shows that the types of ships, their technological features and the routes they sailed (Constantinople aside) did not change substantially in Late Antiquity. The early Hellenistic Kyrenia ship, a Rhodian merchantman which foundered off the rocky headlands in the straits between Lycia and Cyprus, was comparable in size (approximately 15 metres), cargo and crew to the *Isis*, a North African merchantman of the late fourth century CE wrecked off NW Sicily (Katzev & Swiny, 2022; McCann & Freed, 1994). Both of these merchantmen were on well-travelled ancient sea routes: along the southern Anatolian shores between the Aegean Sea and the Levant, and west of Sicily, between Rome and Carthage. Literary references as well as ceramic evidence confirm these continuities of travel on merchant ships in relation to Christian pilgrimage in Late Antiquity. When St Nicholas of Sion sought a vessel to take him on his second pilgrimage to the Holy Land in the middle of the sixth century, for example, he took an Egyptian ship to Askalon eastwards from Tristomon in Lycia (via the delta of Egypt), and then found a Rhodian ship to carry him back home westwards from Askalon to the ports of southern Lycia. He left the ship at Andriake, the port of Myra, with assistance from his own prayers to ensure the stop was made before Rhodes (*Life* of *Nicholas of Sion*, 27–38).

The *Yassıada* shipwreck went down off the harbour of Halicarnassus just north of Rhodes and carried evidence of Christian merchants and possibly churches trading and sailing between the Black Sea, Constantinople, the islands of the Ionian coast, Cyprus and the Levant. Coins of Emperor Heraclius of 625/6 were found aboard, giving a *terminus post quem*, along with a cargo of 800+ *amphorae*, many marked with crosses or other Christian protective symbols (Bass & van Doorninck, 1982; van Alfen, 1996). The stern area once had a cabin with a galley and above it the raised rear deck with the steering oars, helm and traditional devotional space aboard ship. Tools, kitchenware and other artefacts from the stern were interpreted by the excavators as galley equipment and the property of the captain and crew. The captain was "George the Presbyter (or 'Senior') Ship Captain (*Naukleros*)", who placed his cargo, ship and profession prominently under the protection of his Christian faith, and, as the excavators suggest, was even employed by the Church to transport supplies southwards to the Roman navy front for fighting against the Persians or Arabs. George had engraved his name and titles prominently on a bronze incense *acerra* (box) found in the stern and carried at least two bronze crosses with it (Bass & van Doorninck, 1982; van Alfen, 1996). *Georgiou Presbyterou Nauklerou* with crosses was also inscribed on the largest steelyard carried on the ship, though its weight was in the traditional form of a bust of Athena or Minerva, a formerly pagan image which retained associations of authority and trustworthiness aboard ship and ashore far into Late Antiquity (Wade, 2020).

Incense (with wine) was the most common offering made aboard ancient ships to enlist the favour of the gods, and it seems to have been used by Christians throughout Late Antiquity and into the Middle Ages. This is archaeologically attested in the form of bronze boxes, censers and stone or terracotta portable altars with traces of burning. Van Doorninck (2005) points as well to the historical use of incense to seal business deals both before and after Christianity, both on land and aboard ships. Among the personnel involved with the *Yassıada* ship (whether crew, passengers or even investors) were Christian men named Theodore (a cruciform monogram on a glass medallion) and Ioannis (a cruciform monogram on a lead seal), as well as manufacturers and/or shippers of Christian-inscribed *amphorae*. The *amphorae* mainly came from Chios and Samos just to the north of Halicarnassus (by that point renamed St Peter's City, today Bodrum), and bore inscribed or painted texts and images, including protective crosses, the declaration *Theos Nike* (God Triumphs!), product names such as oil or lentils and more owners' names with crosses or the *chi rho*. Van Doorninck (2005) argued that this was a Church-owned vessel, attached to a fleet to resupply the Roman navy in Cilicia engaged against Muslim Arabs. Whether primarily mercantile or military in purpose, however, this early seventh-century ship certainly attests to the Christianised rituals of captain, crew and passengers – possibly even pilgrims – aboard this medium-sized merchant ship.

We can visualise some of the changes and continuities in the way religion was experienced aboard ships between the early imperial period and the time of St Nicholas of Sion or the sailors of the *Yassıada* ship in the mid-sixth/early seventh century by comparing this information to what is shown on the Torlonia relief (MT430) (Figure 12.1). A marble votive relief (1.22 × 0.75 metres) excavated at Rome's *Portus* in the middle of the nineteenth century, the Torlonia relief was probably dedicated to a god like Liber/Bacchus and/or Neptune/Poseidon in thanks for safe arrival ashore by a *navicularius* of the wine trade returning home to Rome in the early third century. A merchant ship is shown entering *Portus* with its Claudian lighthouse. Two men with a woman in between them gather on the high stern to burn incense on a portable altar, probably as an offering to god(s) in thanks for a safe arrival (Wachsmuth, 1967, 143–150; Casson, 1971, Figure 144; Basch, 1987, 465, Figure 1038; Mott, 1996, 12, Figure 1.2; Tuck, 2008; Bonino, 2015; Feuser, 2015; Settis & Gasparri, 2021, no. 26, 175–178, 310–311). This interpretation of the relief is supported by statues shown ashore: Liber Pater with his panther and Neptune on the

Figure 12.1 The Torlonia relief.

Lighthouse island (both at a larger scale than the ship), along with Water Nymphs (or Danaids), an emperor, a four-elephant chariot-driving emperor on an arch (Domitian?) and two *Genii* on pedestals on the shore of the *Portus*. The letters VL (*Votum Libero* – "Vow discharged for Liber" (?), *CIL* XIV, 2033) along with two images of the Roman she-wolf and twins on the ship's sail, the prominence of Liber's statue at the upper right and other evidence for his shrine on the north-east side of the hexagonal harbour, all suggest that he was the primary deity celebrated in the relief and the focus of the ritual of thanks for return to the safety of his harbour on the ship's stern (Settis & Gasparri, 2021, no. 26, 178; *CIL* XIV, 30; Apuleius, *Met.* 11.16–11.17; Feuser, 2015, 30). The relief also appears to represent a common ritual of thanksgiving aboard Roman imperial merchant ships, and the physical and symbolic images of gods, spirits and even the evil eye, which were addressed ashore by that ritual on the ship as it entered harbour. The *Navicularius'* ship's physical consecration to those gods is also apparent in the sail's imagery, Venus as a wind goddess carved on the stern, images of Victory atop the main mast, and stern, and especially Liber's figure carved on the bow.

Portable devotional images or shrines were also carried on ancient ships, along with altars and incense and are sometimes found today in shipwrecks if they were made of durable materials such as terracotta, lead, bronze, gold or coral (good against drowning). This practice is documented for different periods across antiquity. The *Augustan Comacchio* or *Valle Ponti* wreck from the Po delta, just north-east of Ravenna, was a medium-sized merchant ship carrying *amphorae* and 102 lead ingots up the Adriatic, possibly from the north Aegean (Bode et al., 2021). She also had six miniature temple-shaped shrines made of thin lead sheets aboard, with two-dimensional images of the gods inside showing Aphrodite/Venus, Hermes/Mercury and other gods, and with rings to suspend them as devotional items either from the rafters of the ship's cabins or from the rigging (Parker, 1992, 444).

Such portable images in the form of miniature shrines, gods, spirits or heroes might have been bought at one port, or brought from home, serve aboard the ship as a devotional image for some time and then be dedicated ashore at a different port; alternatively, they might be retained for many successful voyages. Phoenician and Greek sailors since the late Bronze Age had carried statuettes of Smiting Gods (Melqart/Herakles) and Nude Goddesses (Astarte/Aphrodite) from east to west. Bronze statuettes of the Smiting God were found in Late Bronze Age Mycenae, Tiryns and the island of Melos as well as at Archaic Delos, Lindos and Cadiz (Gadir); a bronze and gold nude goddess figurine and a golden nude goddess plaque were found on the Uluburun shipwreck off Lycia from *c.* 1300 BC (Cruz Marín Ceballos, 2011; Brody, 1998; EAM Pi2631 from Mycenae, Stampolidis, 2003, 455–457; Pulak, 1998; Fawcett & Zietsman, 2001).

A Hellenistic mariner of Naucratis spun a story quoted by Athenaeus of an Archaic ancestor buying an Aphrodite figurine at Paphos and then dedicating her in Naucratis when saved from seasickness and storm by her power and the healing smell of myrtle aboard the ship (Athenaeus, 15.675–676). References to this practice of carrying protective images continue throughout the imperial period, with Juvenal mocking the dedicators of *tabellae* to Isis for saving them from shipwreck on the way to Rome, and material evidence like a mould for making silver plaques with an image of Isis *Pelagia* from the third century found in the Athenian Agora (Juvenal, *Sat.* 12.28; Williams, 1985). Popkin (2017) notes that souvenir glass flasks and drinking cups from the port of Puteoli, just west of Naples, were brought aboard ship to advertise the power and grandeur of that city's harbourside shrines for sacred and casual travellers from near and far when they returned home.

For Late Antiquity, Gary Vikan has catalogued comparable material evidence which was carried back from Christian pilgrimages or on subsequent sea voyages. This includes images (i.e. icons) of saints, crosses and illustrated pilgrim flasks or *ampullae* (and other containers), used for carrying protective sacred materials like water, oil or dust – all referred to in Greek as *eulogiai*, "blessings" (Latin *benedictiones*) and often mass-produced of lead or terracotta (Vikan, 2003, 2010). *Ampullae* or stamped terracotta tokens showing ships, or scenes of saints blessing ships, are in the minority of those produced but could be marketed to any visitors by sea eager for a safe return home. Like the terracotta figurines of Aphrodite produced from the Archaic to the Hellenistic period or the early imperial plaques depicting Isis, these Christian pilgrim "souvenirs" or "amulets" could be carried for a short or long stretch of time to bless their bearers. They were subsequently gifted, dedicated or "sacrificed" to a saint at their shrine, used to close off the sacred transaction of a vow, gifted in diplomacy or even left to an heir, for the blessing to continue to operate in the space between the human and divine (Vikan, 2010).

An example is a terracotta pilgrim token stamped with St Elizabeth, the infant St John the Baptist and the sacred cave which saved them, now Ein Kerem west of Jerusalem, which reads "Blessing of the Lord from the Refuge of St Elizabeth". This token is part of treasures including stamped lead pilgrim flasks from the Holy Land, which were collected at St Columbanus's Monastery at Bobbio near Milan in the early seventh century, many of which were buried with Columbanus himself (Bobbio no. 4, Vikan, 2010, 20–22; Filipová, 2015). Portable images like these tokens of blessing and stamped containers for sacred water, oil or dust promoted the fame, status and efficacy of the saint (and place) depicted in helping those who visited their shrines and were often carried great distances homeward or in the course of further business on the sealanes (Vikan, 2010, 23–26, 70).

Martyred, entombed and even living saints could give protection at sea to the person who bore their relics afloat or who collected the water, oil, myrrh, dust, images, amulets or even just dirt from their shrines and then kept or dedicated them ashore, with some carriage by sea and votive practices little changed from Greco-Roman traditions (Krueger, 2010; Vikan, 2010, 15–16). Faith in the efficacy of these practices of carrying, vowing and dedicating pilgrimage materials was partly related to Christian dogma but also influenced by pre-existing maritime religious habits (and beliefs) around the shores of the Mediterranean Sea.

Belief in the power of the *Kyrios* (Lord) Jesus Christ and the One God to effect divine salvation at sea entered Christianity early, as Jesus lived and taught beside the Sea of Galilee and "calmed the waters" in a number of Biblical miracles (Matthew 14:22–33; Mark 6:45–52; John 6:16–21; Wachsmann, 1995). Jesus was then invoked as *Kyrios*, Lord, already a long-standing title for gods and heroes in the ancient Near East, and was believed to grant assistance when asked alongside his mother Mary and a growing array of other saints in answer to sailors' and pilgrims' prayers (Parker, 2017, 154–172). However, stamped terracotta pilgrim tokens, flasks or lead *ampullae* depict Christ's miracles less often than they show the local saints of the shorelines and harbours of the eastern Mediterranean, especially St Menas just south of Alexandria. Among the many lead Holy Land *ampullae* of the sixth century now preserved at St Columbanus in Bobbio, or at the Cathedral Treasury of St John in nearby Monza, most show composite scenes with many figures like the Ascension, or a simple True Cross. Only two of these flasks, made to hold oil from the True Cross, show Jesus saving St Peter by walking on water, and thus seem likely to have promised safety at sea to the bearer (Vikan, 1998, 2003, 247–248; Grabar, 1958, no. 11; Milburn, 1988, 264).

The use of substances collected from saints, whether martyred, entombed or living at the time of the collection, and particularly materials connected with St John are also better represented in the literary evidence for Christian beliefs in salvation at sea. Dust from the tomb of St John the Evangelist (*ho Theologos*) was collected at his acropolis basilica at the port city and provincial capital of Ephesus, and the city was renamed *Hagios Ioannis ho Theologos*, Ayasoluk in Turkish. This dust was carried away from Ephesus specifically to help mariners at sea, replacing benefits previously ascribed to statuettes, silver plaques or pilgrimage to see Artemis/Diana of the Ephesians (Koester, 1995; Bangert, 2010).

Silversmiths of Ephesus who hammered out these protective portable images of Artemis/Diana *Ephesia* were already portrayed by St Luke as threatened by St Paul's teachings at Ephesus in the first century (*Acts* 19:21–40), while the Virgin Mary's house, tomb and basilica church also drew pilgrims alongside those seeking blessings from St John the Evangelist (widely "syncretised" with St John of Patmos, writer of *Revelation*). Galla Placidia dedicated a church at the Roman naval port and imperial capital of Ravenna in thanks to St John the Evangelist for her salvation in a storm at sea in the mid-fifth century and brought his sandal from Ephesus to Ravenna, where it became a protective relic for sailors and embarking pilgrims (Deliyannis, 2010; Sivan, 2011).

In the east, the dust (Syriac *hnana*) which had been blessed by St Simeon Stylites the Elder from atop his tall column in his shrine at Qalat Seman, just inland from Antioch was also carried for protection at sea (and in other situations) by pilgrims who had visited the stylite. Dalrymple (1997, 156–157) quotes Theodoret of Cyrrhus, *History of the Monks of Syria*, on pilgrims "arriving from every side" to see Simeon, including

> inhabitants of the extreme west, Spaniards and Britons and the Gauls . . . Italy. The man became so celebrated in the great city of Rome that at the entrance of all the workshops men have set up small images of him, to provide thereby some protection and safety for themselves.

Vikan (2010, 16) quotes St Simeon the Elder's *Miracle* 71 from his Syriac *Life* (61–72) on the belief that dust from his shrine could calm storms at sea. Episode 235 from the later *Ancient Life* of St Simeon the Younger (also quoted by Vikan, 2010, 16) follows his namesake's benefits in claiming that dust from his shrine sprinkled on a ship and into the sea summoned his spirit to aid the monk Dorotheos when he carried it aboard ship during a winter storm.

There is thus good evidence that Christian pilgrims directly replaced many traditional devotional practices of safe sailing by using incense, carrying images as amulets and making prayers from the stern of the ship, addressing the heavens with their images of saints or representations of the cross aboard. Use of space aboard ships was reshaped, and rituals afloat remained concentrated in the stern, focused on amulets and the heavens, and most common in times of trouble. John Moschus's *Spiritual Meadow*, written *c.* 600, for example, gives two miracles for obtaining fresh drinking water on ships at sea (*Miracles* 173–174, *PG* 87.3, 3041–3044). The eunuch anchorite Theodore on his way back to Constantinople from a Holy Land pilgrimage used prayer and the sign of the cross to bring the rains. In another tale, a *naukleros* sailing from Constantinople to the Holy Land on pilgrimage spent three days in fasting and prayer before summoning the clouds himself. Trombley (2001) summarises similar Christian seafaring and pilgrimage miracles in hagiography of the seventh to ninth centuries. Thirst at sea could be as dangerous as a storm, so bringing fresh water via rain was as important as a following breeze, and available to the pious pilgrim or ship's captain who knew the right prayers, was on the right route and had the proper personal devotion to attract the attention, and hence assistance, of the invoked saint, Jesus Christ or Mary (McCormick, 2001).

12.2 Saviours by sea and ashore

Aboard ship or back ashore, Christian pilgrims also reshaped ancient traditions of invoking saviour gods in order to call on new saviour saints, and the Christian or Jewish One God (for Jewish mariners, see Syn. *Ep.* 4). At the sheltered Bay of Grammata on the northern end of Syros, for example, the inscribed prayers of pre-Christian sacred travellers on *theoria* to Delos or Naxos are joined in the fourth and fifth centuries by Jewish and Christian inscriptions using the same format of Greek address, the same term *euploia* ("good sailing") and the same *tabulae ansatae* but also some new formulas of prayer and many new addressees such as *Kyrios* (The Lord), Mary the *Panagia* or sailors' patrons like St Phocas of Sinope (*IG* XII, 5; Jewish inscriptions in Battinou, 2018, nos. 65–66; Nowakowski, 2017).

Most northern Mediterranean coastlines are rocky and indented, and so many ancient ports located away from silt-bearing river mouths, such as Athens's Piraeus, Mykonos's civic harbour opposite Delos or Syros's Bay of Grammata in between are still in use today. Their ruined and standing ancient structures, rupestral inscriptions and surrounding landscapes represent primary evidence for changes in the choice of protectors by seafarers and pilgrims in Late Antiquity. They are also nodes for tracing the seasonal use of the ancient sea routes in between them and how the ports and harbours have changed since Late Antiquity (Beresford, 2013; Morton, 2001; Semple, 1927, 354, Figure 3) (Figure 12.2).

Dangerous *aktai* (or promontories), along with more welcoming urban ports in between, were often consecrated in antiquity to maritime saviour gods like Poseidon or the Dioscuri with altars, statues or temples. Eventually, the majority of these promontories, and the shelters in between, were reconsecrated to Christian saviours like the *Panagia* Mary or the Archangel Michael. At the north-eastern outlet of the Bosporus, for example, the harbourside *Hieron* (Shrine) of Poseidon on the eastern shore was reconsecrated to the Archangel Michael in the fourth or fifth century (Quirini-Poplawski, 2014). Here this change was encouraged by pilgrims travelling to St Phocas's home port of Sinope, or onwards to his subsidiary shrine and hostel in the Crimea, as well as by the reverse flow of Black Sea, Georgian and Armenian pilgrims southwards to the Holy Land (where their presence is attested by graffiti). Mariners and pilgrims

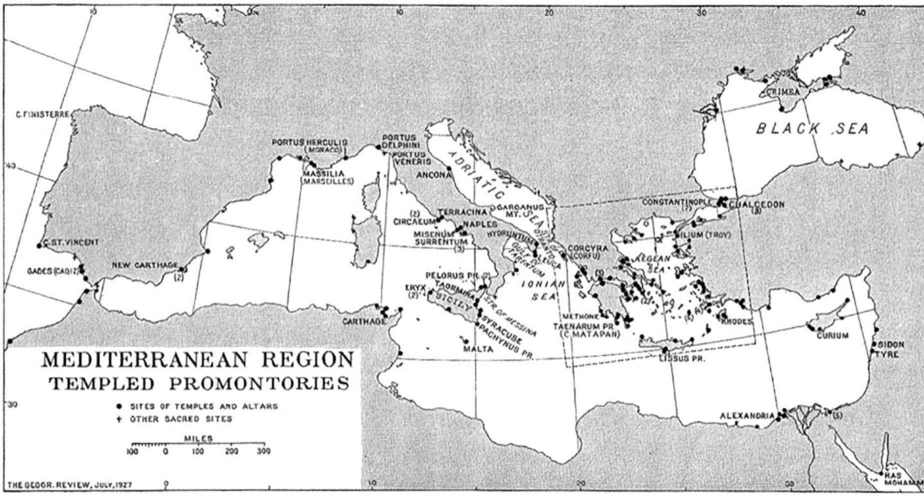

FIG. 3—Map of the Mediterranean region showing distribution of headland sanctuaries identified from classical literature. The dots show sites where altars or temples are known to have existed. Crosses show headlands regarded as sacred but lacking record of altar or temple. See Figure 4 for an enlarged map of the Aegean area. Scale of map, 1:25,000,000.

Use was made of the following authorities: Aeschylus, Anyte, Apollonius Rhodius, Appian, Archilochus, Aristophanes, Aristotle, Athenaeus, Catullus, Cicero, Diodorus Siculus, Euripides, Herodotus, Hesiod, Homer, Homeric Hymns, Horace, Josephus, Juvenal, Livy, Lucan, Macrobius, Ovid, Pausanias, Philostratus, Pliny the Elder, Plutarch, Polybius, Pomponius Mela, Procopius, Propertius, Ptolemy, Scylax, Seneca, Sophocles, Strabo, Suetonius, Tacitus, Thucydides, Vergil, Vitruvius, Xenophon.

Figure 12.2 Templed promontories of the Mediterranean.

Source: From Semple (1927, Figure 3)

together thus created new "local" shipboard and then coastal Christian religious landscapes, which united the seas beyond the horizons of any individual person, ship or even port (Beck, 2020).

On the urban side of coastal saviour gods, a good example of continuity of space with change of cult is at Corinth's two ports. Lechaeum, on the western-facing Isthmus of Corinth, was dug out into an artificial harbour lagoon under Periander *c.* 600 BCE, and it was still bordered by shrines of Poseidon and a venerable sacred grove of Aphrodite under the Roman empire – this is where Plutarch set his *Banquet of the Seven Sages* (Plut. *Mor.* 146D; *SEG* 23, no. 170). Lechaeum harbour was enhanced and Christianised for visiting ships, sailors and pilgrims to Corinth with dredging, moles and a massive Christian basilica, the Lechaion Basilica, in the later fifth to early sixth centuries (Pirenne-Delforge, 1994, 94–96, 434; Brown, 2018, 135–140). Corinth's rocky eastern port at Cenchreae also remained in use throughout Late Antiquity. As the closest harbour to the Panhellenic Isthmian shrine of Poseidon, it had been refurbished during the imperial period to serve both practical and spiritual needs of pilgrims and other travellers with long stone moles, statues and harbourside shrines of Poseidon, Artemis, Aphrodite and Isis (the setting for Apuleius' *Metamorphoses* 11). Temples were replaced with seaside churches: in the case of the former temples of Aphrodite and Isis, one new church on each headland welcomed ships into the harbour in the fifth and sixth centuries (Paus. 2.2.3; Brown, 2018, 135–140; Korka & Rife, 2022).

The Dioscuri, or "Sons of Zeus", heroic horse-riding twins adopted as patrons in places like Samothrace, Delos and Rome, were only replaced by pilgrims and mariners late in Late Antiquity (Walker, 2015; *ID* 1562, 1898; Parker, 2017, 141–2; Bruneau, 1970, 397). They had been worshipped since at least the Archaic poetry of Alcaeus and their *Homeric Hymns* and were said in *HH* 33 to answer prayers made from the highest point of the ship, usually the stern (Alcaeus Frag. 34; *HH* 17, 33). The Homeric poet mentioned a sacrifice of white lambs in their honour, followed by a succinct but vivid narrative of the Dioscuri coming to calm the wind and waves of a winter storm when invoked from the *akroteria* on the stern (*prymne*) of the ship (*HH* 33, lines 10–11). A procession in their honour was still held outside Rome on the Isola Sacra between Portus and Ostia into the late fifth century (Lizzi Testa, 2010, 2016). The harmless coronal discharge which flashes along ships' spars in storms was also widely called by the name of Dioscuri in antiquity and was considered to bring good luck by ancient sailors as a sign that the twins were near them on the seas (Alcaeus Frag. 34; Paus. 1.31, 8.21.4). This phenomenon was also Christianised in Late Antiquity as St Elmo's Fire, after St Erasmus of Formia (later Gaeta), bishop and martyr of this port south of Rome, who by the fifth century was both a patron of sailors and the new namesake of this fortune-bearing light (Damjan, 2004; Vesely, 1972, 1980).

Seasonal festivals in the spring and autumn also held special significance for ancient seafarers leaving and returning to port, while biennial festivals like the Panhellenic Isthmia were in the summer at the heart of the sailing season; somewhat like music festivals today, these recurring events created a larger locality and a feeling of community among the attendees beyond the spaces of individual ships or cities (Beresford, 2013). Christian "Blessing of the Fleet" ceremonies shifted some of these communal gatherings to midwinter Epiphany in the eastern Mediterranean, while Carnival, May Day and especially Easter with Pentecost took on some of the springtime ship-launching traditions of earlier festivals such as the *Dionysia*, *Aphrodisia* or *Navigium* of Isis *Pelagia* (Ap. *Met.* 11 [Isis]; Piacenza Pilgrim, *Travels*, 11 [Epiphany]).

Some cities also marked the "closing" of the sea to smaller boats in the autumn and the return of naval warfleets to shipsheds. This season dedicated to Demeter and Kore for hauling ships ashore was Christianised as a Byzantine (fifth century) and then western medieval (seventh century) Christian feast dedicated to the nativity of the *Panagia* Virgin Mary on 8 September, spreading westwards from Jerusalem's church of St Anne atop the Shepherd's Pool (Aeneas Tacticus, *Poliorcetica* 17.1; Simon, 1983; *LFF*, 2022; Brubaker & Cunningham, 2011).

12.3 Shores and harbours

I conclude with a look at spatial continuities and changes driven by Christian pilgrims coming and going from harbours in Late Antiquity. The origins, growth and sacred significance of Christian pilgrimage to the Holy Land in the late antique period have long been the subject of research (Ousterhout, 1990; Janin, 2002; Kuelzer, 2002; *Piacenza Pilgrim*). The Holy Land was always the main destination for Christian pilgrims, and most arrived there by sea, whether at Alexandria, Caesarea or ports in between. This is the case of the pilgrimages of St Nicholas of Sion, for example, who arrived at the Holy Land via Askalon (the second time via the Egyptian Delta, cf. *Life of Nicholas of Sion*, 9 and 27–35). Souvenirs of such a pilgrimage (or even just the voyage itself) were widely believed to help Christian pilgrims with safe travel later in life (Pryor, 2008). Along the Levantine coast, ports like Gaza, Askalon or the metropolitan bishopric of Caesarea gained new churches for local and pilgrim use, as well as enhanced harbour facilities in Late Antiquity (Limor, 2006). Other ancient, converted or newly built seaside shrines around the eastern Mediterranean littoral also gained new Christian fame for their power and special efficacy for certain professions (such as sailors) and certain situations, such as travel by sea. Continuities and changes in ancient sacred travel routes and facilities ashore are visible in evidence from Late Antiquity, whether or not ancient "pagans" on *theoria* before the third century are dubbed pilgrims or not (Elsner & Rutherford, 2005; Kristensen & Friese, 2017).

Beyond coastal sacred sites of importance in the Old or New Testament or in the *Apocrypha*, efficacious saints for sailors emerged along the newly invigorated ancient north–south sea routes between Alexandria, Cyprus and Constantinople and along the east–west routes from

Rome of SS Peter and Paul to the ports of the Holy Land: Gaza, Askalon and especially Caesarea. The seaside saviour saints St Isidore of Chios and St Phocas of Sinope on the Black Sea were also brought to Constantinople as relics by sailors in the fifth century after their fame spread from their home ports (Brown, 2023; Nowakowski, 2017). Pryor emphasises that most of the traditional ports of call in the eastern Mediterranean remained in use into the Byzantine era, partly through habit and the need for fresh water, and partly through the practical, bodily and spiritual services available there (Pryor, 2008; McCormick, 2001; Pryor & Jeffreys, 2006).

The route between the new capital, Crete, and Alexandria retained many of its traditional ports, those which were islands and in Asia Minor became increasingly important for practical ports of access after the Arab invasions. In the year 949, for example, the Byzantine naval fleet heading south out of Constantinople to retake Crete from the Muslim Arabs watered first at the ancient port of Abydos in the Propontis, then at the island of Chios, and finally at Ephesus (Constantine Porphyrogennetos, *De Cer*. 2.45.664–665; McCormick, 2001, 413–414; Pryor & Jeffreys, 2006). This route of sailing south to the Holy Land for pilgrims from the north and the Latin West continued via Rhodes, the Lycian coast (especially Patara, or Myra's Andriake), and then along the coast of Cyprus (Zavagno, 2017), while other routes from Constantinople stretched north-east into the Black Sea and around the Peloponnese to the Latin West (Morton, 2001).

Alexandria was one of the biggest winners in the growth of Christian pilgrimage ports, as well as one of the earliest harbours to remould the waterfront, shipping facilities and other maritime spaces to cater to the new Christian cult. It was a port for travellers by sea to the Holy Land, who could go east via Mt Sinai after first seeking blessings from the relics of St Mark, St Menas across Lake Maeotis and SS John with Cyrus at Menouthis/Abukir. Mariners, Alexandrians and Egyptians, all gave special credit to St Mark the Evangelist as the founder of Alexandria's cathedral. Local tradition traced his birth to Cyrene in Libya and claimed that he was martyred in Alexandria on 25 April, under Nero, after missionary work in Rome and the East (Peter 1 5:3; 2 Tim 4:11). Venetian merchants and Greek monks in 828 smuggled St Mark's relics out of Alexandria by sea in a cargo of pork (to avoid Muslim customs inspection) and cabbage leaves. As patron saint of Venice from the ninth century onwards, his relics were visited by local sailors, naval forces and plenty of outbound (and returning) pilgrims. He was first kept in the Doge's palace, then in a new ninth-century canal-side basilica with campanile, and since 1094 in San Marco, a copy of the church of the Holy Apostles in Constantinople. San Marco is decorated with mosaics of his miracles, including aiding sailors to arrive safely into the port of Alexandria, where its famed ancient lighthouse, the *Pharos*, still stood in Late Antiquity (Eusebius, *H.E.* 2.14.6, 2.24.1, 2.24.15–2.24.16; Choat, 2012; Bagnall, 2007; Vio, 2003).

Many pilgrims to Alexandria also sailed south across Lake Maeotis to visit the shrine of the martyr St Menas, and collect his holy water as a blessing. Small terracotta pilgrim flasks were made and filled as souvenirs from the fifth to seventh centuries there and have been found at Greek ports like Chios or Thessalonike and at Latin-speaking cities of the Spanish and British coasts, revealing the reach of the pilgrims to the site who returned home with their blessing. Most flasks show St Menas between two camels, and on the flipside sometimes a merchant ship under sail, a blessing of the voyage for mariners who carried the water or for pilgrims on their way home (Figure 12.3: BM 1882,0510.51, *c*. 480–650, with a cross on the bow of the ship; Buckton, 1994; Kaufmann, 1910, pl. 96.12; Grossmann, 1998; Anderson, 2007; Vikan, 2010, 33–34; Bangert, 2010). St Menas was also the dedicant of churches along the Nile, and outside of Egypt at Constantine's harbour of Thessalonike, and at Constantinople. Today the domes of *Hagia* Sophia and St Eirene still show up to sailors from the Bosporus, since they stand on Byzantium's former Acropolis on the ruins of temples of Artemis and Athena. However, St Menas's

Figure 12.3 Pilgrim flask of St Menas.
Source: Courtesy of the British Museum

church was also once visible to sailors at the north-east corner of the city, just where ships turned west into the Golden Horn, and near a statue of Tyche/*Fortuna* (Mango, 2001; Dark & Harris, 2008).

Constantine's successors made a progressive consecration of the ports of Byzantium as Constantinople, a New Rome, with relics of other patrons of pilgrims and sailors like St Isidore of Chios and St Phocas of Sinope, and harbourside churches for the Virgin Mary *Panagia*, St Nicholas and St Artemios (Wade, 2014, 2018, 2020). The relics of the apostles St Andrew (from Patras), St Luke (from Boeotian Thebes) and St Timothy (from Ephesus) were also brought to rest in the Church of the Holy Apostles in Constantinople, alongside the tomb of Constantine, diverting pilgrims to Constantinople from those cities in Greece and Asia Minor (Woods, 1991; Wisniewski, 2019, 22–25).

A martyr saint popularised in Late Antiquity by pilgrims and sailors on the route between Constantinople and the southern Mediterranean, and brought to the Latin West, was St Isidore of Chios, whose first church was built in Chios beside a freshwater spring on the northern side of the harbour, a protected natural bay on the east coast of the island just opposite the densely settled Erythrae/Çeshme Peninsula (S00425 on the Oxford *CSLA*; *BHG* 960). The basilica of St Isidore and its spring, which is today off the narthex, were visited for fresh water, Christian services and spiritual protection by fishermen, sailors, Christian pilgrims and local Chiots from at least the early fifth century into the fourteenth century, when it was damaged by an earthquake, and may have been converted to use as a mosque (Mullen, 2004, 92). St Isidore of Chios, whose name "Gift of Isis" evokes the earlier naval patroness, was said to be a naval officer drowned at the spring in the third century and buried nearby as a Christian martyr (Greg. Tur., *Glory of the Martyrs* 30, 101; Krueger, 2010).

Pilgrims to his *martyrium* from as far as Numidia or as close as the island of Tenos collected and carried away holy water in flasks, or terracotta *eulogiai*, both stamped with an image of the saint: an engraved bronze stamp survived in Thessalonike and is now in the Walters Art Museum, Baltimore (Figure 12.4: inv. no. 54.230, Vikan, 2010, 20–21, Figure 8). St Isidore stands frontally, wearing naval military garb, and (in the imprint) blesses a merchant ship with his lowered left hand while holding a crook (or schematic steering oar) in his right. His name in Greek is given in sixth-century lettering, with a cross at its start and end, and the engraver also fit in an abbreviation for "Jesus Christ", and "Receive the *Blessing*" under the ground line, and a schematic depiction of the *ciborium* over his holy spring, with a cross, a font and what might be a votive ship or lamp (or both) hanging above it (+ ὁ ἅγιος Ἠσιδόρος +/Ἰ(ησοῦ)ς Χ(ριστό)ς/Δέξ(αι) εὐλογί(αν)). The *oikonomos* St Marcian (died *c.* 475), enshrined some of Isidore's relics in Constantinople, on waterfront land reclaimed from the sea on the Golden Horn below the church of

Figure 12.4 Pilgrim stamp of St Isidore.

Source: Courtesy of the Walters Art Museum

St Eirene. Mango (2001) sets this into the context of shoreline reclamation at Constantinople for new churches, harbour facilities and other maritime services, citing Himerius's *Oration* 41 and the *Vita Marciani* (Gedeon, 1899, 275; Snee, 1998). St Isidore had a church in Rome, too, by the Porta Tiburtina leading to the harbours. His relics in Constantinople were translated in 1525 to a new chapel at San Marco Basilica in Venice to protect the Republic's fleets, mariners and embarking pilgrims.

A reliquary for the sailors' patron St Phocas of Sinope was also brought to a new church in Constantinople in Late Antiquity, via the south-west coast of the Black Sea. St Phocas's relic was translated from Sinope to Constantinople *c.* 400, and according to St John Chrysostom, a seaborne procession brought the saint to his new church on the waterfront (John Chrysostom, *Homily on St. Phocas* in Mayer & Neil, 2006; Mayer, 1998). St Phocas had already become a patron and saviour saint for pilgrims and mariners on the routes leading both north-east and south-west from the new imperial capital. His name ("Seal"), like that of St Isidore, had deeper symbolism for seafarers, who respected seals, dolphins and other swimming mammals as impervious to drowning. He was honoured as a martyr, sailor and son of a shipbuilder; some icons show him holding a ship's steering oar (see *Martyrdom of St Phokas*; Nowakowski, 2017). His Syriac *Passio* and a homily of Bishop Asterius of Amasea credited St Phocas with calming the seas, steering, reefing and even keeping drowsy sailors awake on watch (see *Martyrdom of St Phokas* for the Syriac text; Asterius of Amasea, *Homily* 9 [*PG* 40, 308]). Mariners's *graffiti* in the sheltered Bay of Grammata on the northern shore of the island of Syros, as noted earlier, appeal for his help (or give him thanks), alongside older inscriptions invoking Aphrodite *Euploia* ("of Good Sailing"), and contemporary inscriptions from Late Antiquity appealing to Mary or the Apostles, while Jewish merchants from Naxos appeal to their One God as *Kyrios* (Nowakowski, 2017).

A *eulogia* terracotta amulet of St Phocas now in the Hermitage Museum attests in the Greek text and imagery to pilgrims who once collected these tokens on the north shore of the Black Sea to carry with them (as it says on the token) as a mobile "Blessing (*Eulogia*) of St Phocas of the Hospice of Cherson". From excavations at Chersonese, the ancient city of Kherson on the western side of the Crimean peninsula, this token shows that the Tauric Chersonesos remained Greek-speaking in Late Antiquity and became a destination for Christian pilgrims (Figure 12.5: inv. no. X-263, Vikan, 2010, 20–21, Figure 7). The stamped image surrounded by the writing shows St Phocas in a frontal pose, wearing naval uniform, with his arms raised in blessing. Instead of holding a ship like St Isidore, however, he is shown over life size, standing in a merchant ship with a high bow, stern and double steering oars and alongside an oversized incense burner. Besides Sinope and Cherson, shrines of St Phocas catering to sailors were built at other urban ports around the Black Sea before and after his translation to Constantinople, where theophoric city names and harbour temples were also Christianised, for example Apollonia to Sozopolis, now in Bulgaria, and in Late Antiquity "The (Christian) Saviour's City" (Quirini-Poplawski, 2014).

12.4 Conclusion

This chapter has traced religious changes onboard ships, in the adoption of seafaring protectors, and also ashore in Late Antiquity. Sacred travel by sea was changed by Christian pilgrims aboard ships and at shores in the eastern Mediterranean from Greek *theoria*, "going to see god(s)", to the carriage of new images of saints and passage east to new ports of call for new reasons through the seventh century of Late Antiquity. From the early fourth century, the rising tide of Christian pilgrims and other travellers from the new religion, along with Constantine's new

Figure 12.5 Stamped amulet of St Phocas.

Source: Courtesy of the Hermitage Museum

capital city, reshaped some rituals and routes of sacred travel. Christians adopted and modified traditional prayers and ritual actions of ancient sacred travel, such as offering incense, gathering at altars at the stern of the ship closest to the sky and carrying protective amulets aboard ship. Holy Land ports, and especially Alexandria, grew in importance, as did those ports and their saints, on routes to and from Constantinople and near new pilgrimage sites.

Editions and abbreviations

Aeneas Tacticus, *Poliorcetica – Aineias the Tactician: How to survive under siege*, trans., with introduction and commentary, by D. Whitehead (1990). Oxford.

Alcaeus, *Frag.* – West, M. L. (ed. and trans.). *Greek Lyric Poetry: The poems and fragments of the Greek iambic, elegiac, and melic poets (excluding Pindar and Bacchylides) down to 450 B.C.* (1993). Oxford University Press.

Apuleius, *Met.* – Apuleius, *Metamorphoses*, 2 vols, trans. J. A. Hansen (Loeb Classical Library) (1989, 1996). Harvard University Press.

Asterius of Amasea, *Homilies* (*PG* 40, 155–480), ed. C. Datema (1970) *Asterius of Amasea: Homilies I-XIV*. E. J. Brill.

Athenaeus – Athenaeus, *The Learned Banqueters*, ed. and trans. S. D. Olson (Loeb Classical Library) (2007). Harvard University Press.

BHG – Bibliotheca Hagiographica Graeca.

CIL – Corpus Insccriptionum Latinarum.

Constantine Porphyrogenitus, *De Cer*. – *Constantine Porphyrogennetos: The Book of Ceremonies* Byzantina Australiensia 18, trans. A. Moffatt & M. Tall (2012). Brill.

CSLA – The Cult of Saints in Late Antiquity database. http://csla.history.ox.ac.uk

Eusebius, *H.E – Eusebius: The History of the Church*, trans. G. A. Williamson & A. Louth (1989). Penguin.

Greg. Tur., *Glory of the Martyrs – Gregory of Tours, Glory of the Martyrs* (Translated Texts for Historians), trans. R. Van Dam (1988). Liverpool University Press.

HH – Homeric Hymns. In West, M. (ed. and trans.). *Homeric Hymns.Homerica Apocrypha. Lives of Homer* (Loeb Classical Library) (2003). Harvard University Press.

ID – Inscriptions de Délos.

IG – Inscriptiones Graecae.

Juvenal, *Sat*. – Juvenal, *Satires*. In S. M. Braund (ed. & trans.). *Juvenal and Persius* (Loeb Classical Library) (2004). Harvard University Press.

LFF – Lesser Feasts and Fasts 2022. Church Publishing Incorporated.

Life of Nicholas of Sion – Sevcenko, I., & Sevcenko, N. P. (eds.). *The Life of St. Nicholas of Sion* (1984). Holy Cross Press.

Martyrdom of St Phokas – Brock, S. *The Martyrdom of St Phokas of Sinope* (2013). Gorgias Press.

Piacenza Pilgrim – Wilkinson, J. *Jerusalem Pilgrims before the Crusades*. Revised ed. (2002). Warminster.

SEG – Supplementum Epigraphicum Graecum.

Syn. *Ep*. – Lacombrade, C. (ed. & trans.). *Synésios de Cyrène, tome 2* (1978). Les Belles Lettres.

Bibliography

Anderson, W. (2007). Menas flasks in the west: Pilgrimage and trade at the end of antiquity. *Ancient West and East*, *6*, 221–243.

Bagnall, R. S. (2007). *Egypt in the Byzantine world, 300–700*. Cambridge University Press.

Bangert, S. (2010). The Archaeology of pilgrimage: Abu Mina and beyond. In D. M. Gwynn & S. Bangert (Eds.), *Religious diversity in late antiquity* (Late Antique Archaeology 6) (pp. 293–327). Brill.

Basch, L. (1987). *Le musée imaginaire de la marine antique*. Institut hellénique pour la préservation de la tradition nautique.

Bass, G. F., & van Doorninck, F. H., Jr. (Eds.). (1982). *Yassi Ada 1: A seventh-century Byzantine shipwreck*. Texas A&M University Press.

Battinou, Z. (Ed.). (2018). *Corpus Inscriptionum Judaicarum Graeciae: Corpus of Jewish and Hebrew inscriptions from Island and Mainland Greece (end of 4th c. BCE – 15th c.)*. The Jewish Museum of Greece.

Beck, H. (2020). *Localism and the ancient Greek City-state*. University of Chicago Press.

Beresford, J. (2013). *The Ancient Sailing season*. Brill.

Blackman, D., Rankov, B., Baika, K., Gerding, H., & Pakkanen, J. (2013). *Shipsheds of the Ancient Mediterranean*. Cambridge University Press.

Bode, M., Hanel, N., & Rothenhöfer, P. (2021). Roman lead ingots from Macedonia – the Augustan shipwreck of Comacchio (prov. Ferrara, Italy) and the reinterpretation of its lead ingots' provenance deduced from lead isotope analysis. *Archaeological and Anthropological Sciences*, *13*, 163. https://doi.org/10.1007/s12520-021-01430-0

Bonino, M. (2015). *Navi mercantili e barche di età romana*. "L'Erma" di Bretschneider.

Brody, A. J. (1998). *"Each man cried out to his God": The specialized religion of Canaanite and Phoenician seafarers*. Scholars Press.

Brown, A. R. (2018). *Corinth in late antiquity: A Greek, Roman, and Christian City*. I.B. Tauris.

Brown, A. R. (2023). The rites of the (late) ancient mariners: Professional and casual sailors as Christian pilgrims in the late antique Mediterranean. In A. Collar & T. M. Kristensen (Eds.), *Pilgrims in place, pilgrims in motion: Sacred travel in the ancient Mediterranean* (ASMA 15, pp. 199–212). Aarhus University Press.

Brown, A. R., & Smith, R. (2019). Guardian goddess of the surf-beaten shore: The influence of mariners on sanctuaries of Aphrodite in *Magna Graecia*. In S. Blakely & B. J. Collins (Eds.), *Religious convergence in the ancient Mediterranean* (Studies in Ancient Mediterranean Religions 2, pp. 19–41). Scholars Press.
Brubaker, L., & Cunningham, M. (Eds.). (2011). *The cult of the mother of God in Byzantium: Texts and images*. Oxford University Press.
Bruneau, P. (1970). *Recherches sur les cultes de Délos à l'époque hellénistique et à l'époque impériale*. Carl Roebuck.
Buckton, D. (1994). *Byzantium: Treasures of Byzantine art and culture from British collections*. British Museum Press.
Bultrighini, I. (2015). *Demi attici della Paralia*. Carabba.
Casson, L. (1971). *Ships and seamanship in the Ancient world*. Princeton University Press.
Chankowski, V. (2008). *Athènes et Délos à l'époque classique: Recherches sur l'administration du sanctuaire d'Apollon délien*. Ecole Française d'Athênes.
Choat, M. (2012). Christianity. In C. Riggs (Ed.), *The Oxford handbook to Roman Egypt* (pp. 474–489). Oxford University Press.
Christidis, A.-Ph., Arapopoulou, M., & Chrite, M. (2007). *A history of ancient Greek: From the beginnings to late antiquity*. Cambridge University Press.
Constable, O. R. (2003). *Housing the stranger in the Mediterranean world: Lodging, trade, and travel in late antiquity and the middle ages*. Cambridge University Press.
Cortese, A. (2022). *Cilicia as sacred landscape in late antiquity: A journey on the trail of apostles, martyrs and local saints*. Reichert.
Cruz Marín Ceballos, M. (2011). *Cultos y ritos de la Gadir fenicia*. Universidad de Cádiz.
Dalrymple, W. (1997). *From the holy mountain: A journey in the shadow of Byzantium*. Flamingo.
Damjan, D. (2004). St. Erasmus (Lychnid) and St. Thecla (Seleucia): A study of two early Christian cultic centers [MA dissertation, Bilkent University]. Bilkent University Institutional Repository. https://repository.bilkent.edu.tr/items/431daf06-5567-4515-8a14-31fd4c43b624.
Dark, K. R., & Harris, A. L. (2008). The last Roman forum: The forum of Leo in fifth-century Constantinople. *Greek, Roman and Byzantine Studies*, *48*, 57–69.
Deliyannis, D. M. (2010). *Ravenna in late Antiquity*. Cambridge University Press.
Dillon, M. (1997). *Pilgrims and pilgrimage in ancient Greece*. Taylor & Francis.
Elsner, J., & Rutherford, I. (2005). *Pilgrimage in Graeco-Roman and early Christian antiquity: Seeing the gods*. Oxford University Press.
Emlyn-Jones, C. J., & Preddy, W. (2017). *Plato I: Euthyphro; Apology; Crito; Phaedo*. Harvard University Press.
Fawcett, N., & Zietsman, J. C. (2001). Uluburun – the discovery and excavation of the world's oldest known shipwreck. *Akroterion*, *46*, 5–20.
Feuser, S. (2015). Kult und Kultbauten in Hafendarstellungen. Hafendarstellungen in Kult und Kultbauten. In T. Schmidts & M. M. Vučetić (Eds.), *Häfen im 1. Millennium AD: Bauliche Konzepte, herrschaftliche und religiöse Einflüsse* (pp. 35–52). Schnell & Steiner.
Filipová, A. (2015). On the origins of the Monza collection of Holy Land *ampullae*: The legend of Gregory the Great's gift of relics to Theodelinda reconsidered. *Arte Lombarda*, *173*(4), 5–16.
Frankfurter, D. (Ed.). (1998). *Pilgrimage and holy space in late antique Egypt*. Brill.
Gallop, D. (1975). *Plato: Phaedo*. Oxford University Press.
Garland, R. (2001). *The Piraeus: From the fifth to the first century B.C.* Bloomsbury.
Gedeon, M. (1899). *Byzantinon heortologion*. Printed by the author.
Grabar, A. (1958). *Ampoules de Terre Sainte (Monza, Bobbio)*. Klinksieck.
Grossmann, P. (1998). The pilgrimage center of Abu Mina. In D. Frankfurter (Ed.), *Pilgrimage and holy space in late antique Egypt* (Religions in the Graeco-Roman World 134, pp. 281–302). Brill.
Horden, P., & Purcell, N. (2000). *The corrupting sea: A study of Mediterranean history*. Oxford University Press.
Hunt, E. D. (1982). *Holy land pilgrimage in the later Roman empire, AD 312–460*. Oxford University Press.

Janin, H. (2002). *Four paths to Jerusalem: Jewish, Christian, Muslim, and secular pilgrimages, 1000 BC to 2001 CE*. McFarland.

Katzev, S. W., & Swiny, H. W. (Eds.). (2022). *The Kyrenia ship final excavation report, volume I: History of the excavation, amphoras, ceramics, coins and evidence for dating* (The Kyrenia Ship Final Excavation Report Vol. 1). Oxbow Books.

Kaufmann, K. M. (1910). *Die Menasstadt und das Nationalheiligtum der altchristlichen Aegypter in der westalexandrinischen Wüste: Ausgrabungen der Frankfurter Expedition am Karm Abu Mina* (Vol. 1). K.W. Hiersemann.

Koester, H. (1995). *Ephesos, metropolis of Asia: An interdisciplinary approach to its archaeology, religion, and culture*. Trinity Press International.

Konstantinidou, P. S. (1972). *Ho Hagios Spyridōn: Episkopos Trimythountos, ho Thaumatourgos*. Theopress.

Korka, E., & Rife, J. L. (2022). *On the edge of a Roman port: Excavations at Koutsongila, Kenchreai, 2007–2014* (*Hesperia* Suppl. 52). ASCSA.

Kristensen, T. M. (2017). Excavating Meriamlik: Sacred space and economy in late antique pilgrimage. In T. M. Kristensen & W. Friese (Eds.), *Excavating pilgrimage: Archaeological approaches to sacred travel and movement in the ancient world* (pp. 224–244). Taylor & Francis.

Kristensen, T. M., & Friese, W. (Eds.). (2017). *Excavating pilgrimage: Archaeological approaches to sacred travel and movement in the ancient world*. Taylor & Francis.

Krueger, D. (2010). The religion of relics in late antiquity and Byzantium. In M. Bagnoli, H. A. Klein, C. Griffith Mann, & J. Robinson (Eds.), *Treasures of heaven: Saints, relics, and devotion in medieval Europe* (pp. 5–18). Walters Art Museum.

Kuelzer, A. (2002). Byzantine and early post-Byzantine pilgrimage to the Holy Land and to Mount Sinai. In R. Macrides (Ed.), *Travel in the byzantine world* (pp. 149–164). Taylor & Francis.

Limor, O. (2006). "Holy journey": Pilgrimage and Christian sacred landscape. In O. Limor & G. G. Stroumsa (Eds.), *Christians and Christianity in the Holy Land: From the origins to the Latin Kingdoms* (pp. 321–354). Brepols.

Lizzi Testa, R. (2010). *Insula ipsa libanus almae veneris nuncupatur*: Culti, celebrazioni, sacerdoti pagani a roma, tra iv e vi secolo. In G. Bonamente & R. Lizzi Testa (Eds.), *Istituzioni, carismi ed esercizio del potere, IV-VI secolo d.C* (pp. 273–303). Edipuglia.

Lizzi Testa, R. (2016). Vrbs Roma between Pagans and Christians. In M. R. Salzman, M. Sághy, & R. Lizzi Testa (Eds.), *Pagans and Christians in Late antique Rome: Conflict, competition, and coexistence in the fourth century* (pp. 401–410). Cambridge University Press.

Loriaux, R. (1975). *Le Phédon de Platon* (Vol. 2). Editions J. Duculot.

Lovén, B. (2011). *The ancient Harbours of the Piraeus*. Aarhus University Press.

Malkin, I. (2011). *A small Greek world: Networks in the Ancient Mediterranean*. Oxford University Press.

Mango, C. (2001). The shoreline of Constantinople in the fourth century. In N. Necipoğlu (Ed.), *Byzantine Constantinople: Monuments, topography and everyday life* (pp. 17–28). Brill.

Matton, R. (1960). *Corfou*. L'Institut Francais d'Athenes.

Mayer, W. (1998). The sea made holy. The liturgical function of the waters surrounding Constantinople. *Ephemerides Liturgicae, 112*, 459–468.

Mayer, W., & Neil, B. (2006). *The cult of the saints: Select homilies and letters by St John Chrysostom*. St Vladimir's Seminary Press.

McCann, A. M., & Freed, J. (1994). *Deep water archaeology: A late-Roman ship from Carthage and an ancient trade route near Skerki Bank off northwest Sicily* (*JRA* Suppl. 13). Journal of Roman Archaeology.

McCormick, M. (2001). *Origins of the European economy: Communications and commerce, AD 300–900*. Cambridge University Press.

Milburn, R. (1988). *Early Christian art and architecture*. Scolar Press.

Morton, J. (2001). *The role of the physical environment in Ancient Greek seafaring*. Brill.

Mott, L. V. (1996). *The development of the rudder: A technological tale* (Studies in Nautical Archaeology 3). Texas A&M University Press.

Mullen, R. L. (2004). *The expansion of Christianity*. Brill.
Nowakowski, P. (2017). Phokas, rocks near Grammata on the island of Syros (E01232). *Cult of Saints in Late Antiquity*. http://csla.history.ox.ac.uk/record.php?recid=E01232.
Ousterhout, R. (1990). *The blessings of pilgrimage* (Illinois Byzantine Studies 1). University of Illinois Press.
Papadopoulou, C. (2010). Aphrodite and the fleet in classical Athens. In A. C. Smith & S. Pickup (Eds.), *Brill's companion to Aphrodite* (pp. 217–233). Brill.
Parke, H. W. (1977). *Festivals of the Athenians*. Cornell University Press.
Parker, A. J. (1992). *Ancient shipwrecks of the Mediterranean & the Roman provinces*. Archaeopress.
Parker, R. (2017). *Greek Gods abroad: Names, natures, and transformations*. University of California Press.
Paspalas, S. (2008). The Panagia Myrtidiotissa: The changing image of a Kytherian icon. In W. Caraher, L. Jones Hall, & R. S. Moore (Eds.), *Archaeology and history in roman, medieval and post-medieval Greece. Studies on method and meaning in honor of Timothy E. Gregory* (pp. 197–225). Routledge.
Pirenne-Delforge, V. (1980). Épithètes cultuelles et interpretation philosophique: à propos d'Aphrodite Ourania et Pandémos à Athènes. *Antiquité Classique*, *57*, 142–157.
Pirenne-Delforge, V. (1994). *L'Aphrodite grecque: Contribution à l'étude de ses cultes et de sa personnalité dans le panthéon archaïque et classique*. Centre international de l'Étude de la Religion antique.
Popkin, M. L. (2017). Souvenirs and memory manipulation in the Roman empire: The glass flasks of ancient Pozzuoli. In L. Munteán, L. Plate, & A. Smelik (Eds.), *Materializing memories in art and popular culture* (Routledge Research in Cultural and Media Studies 102, pp. 45–61). Routledge.
Potts, J. (2010). *The Ionian Islands and Epirus: A cultural history*. Oxford University Press.
Pryor, J. H. (2008). Ships and seafaring. In E. Jeffreys, J. F. Haldon, & R. Cormack (Eds.), *The Oxford handbook of Byzantine studies* (pp. 482–490). Oxford University Press.
Pryor, J. H., & Jeffreys, E. (2006). *The age of the Dromon: The Byzantine navy ca 500–1204* (The Medieval Mediterranean 62). Brill.
Pulak, C. (1998). The Uluburun shipwreck: An overview. *International Journal of Nautical Archaeology*, *27*(3), 188–224.
Quirini-Poplawski, R. (2014). Seaside shrines in the late mediaeval Black Sea Basin. Topography and selected historical and art historical questions. In M. Bacci & M. Rohde (Eds.), *The Holy Portolano/Le Portulan sacré: The Sacred geography of navigation in the middle ages. Fribourg colloquium 2013/La Géographie religieuse de la navigation au moyen âge. Colloque Fribourgeois 2013*. De Gruyter.
Rutherford, I. (2013). *State pilgrims and sacred observers in Ancient Greece: A study of* Theôriâ *and* Theôroi. Cambridge University Press.
Semple, E. C. (1927). The Templed promontories of the ancient Mediterranean. *The Geographical Review*, *17*, 353–386.
Settis, S., & Gasparri, C. (Eds.). (2021). *The Torlonia Marbles. Collecting Masterpieces*. Electa.
Simon, E. (1983). *Festivals of Attica: An archaeological commentary*. University of Wisconsin Press.
Sivan, H. (2011). *Galla Placidia: The last roman empress* (Women in Antiquity). Oxford University Press.
Snee, R. (1998). Gregory Nazianzen's Anastasia Church: Arianism, the goths, and hagiography. *Dumbarton Oaks Papers*, *52*, 157–186.
Sourvinou-Inwood, C. (1979). *Theseus as son and stepson: A tentative illustration of Greek mythological mentality*. Institute for Classical Studies.
Stamatopoulos, N. (1993). *Old Corfu: History and culture* (3rd ed.). Printed by the author.
Stampolidis, N. C. (Ed.). (2003). *Sea routes . . . From Sidon to Huelva: Interconnections in the Mediterranean (16th-6th c. BC). Catalogue of the exhibit in the Museum of Cycladic Art*. Museum of Cycladic Art.
Sweetman, R. (2019). Islands and resilience: Christianization processes in the cyclades. In M. A. Cau Ontiveros & C. Mas Florit (Eds.), *Change and resilience: The occupation of Mediterranean islands in Late Antiquity* (Joukowsky Institute 9, pp. 193–216). Oxbow Books.
Sweetman, R. (2022). Networking Christians? The spread of Christianity in the Eastern Mediterranean. In A. Collar (Ed.), *Networks and the spread of ideas in the past: Strong ties, innovation and knowledge exchange*. Routledge.

Trombley, F. (2001). Mediterranean sea culture between Byzantium and Islam C. 600-850. In E. Kountoura-Galake (Ed.), The Dark Centuries of Byzantium (7th – 9th c.), pp. 133–169. National Hellenic Research Foundation.

Tuck, S. L. (2008). The expansion of triumphal imagery beyond Rome: Imperial monuments at the harbors of Ostia and Leptis Magna. In R. L. Hohlfelder (Ed.), *The maritime World of Ancient Rome* (Memoirs of the American Academy in Rome Suppl. 6, pp. 325–341). University of Michigan Press.

Tzavella, E. (2012). Urban and rural landscape in early and middle Byzantine Attica (4th–12th c. AD) [Doctoral thesis, University of Birmingham]. UBIRA E Theses. https://etheses.bham.ac.uk/id/eprint/4321/.

van Alfen, P. G. (1996). New light on the 7th-c. Yassı Ada Shipwreck: Capacities and standard sizes of LRA1 Amphoras. *Journal of Roman Archaeology*, *9*, 189–213.

Van Doorninck, F. H. (2005). The ship of Georgios, priest and sea captain: *Yassiada*, Turkey. In G. F. Bass (Ed.), *Beneath the seven seas: Adventures with the institute of nautical archaeology*. Thames & Hudson.

Vesely, J. M. (1972). *Chiesa di S. Erasmo a Formia*. Formia.

Vesely, J. M. (1980). Sant'Erasmo di Formia o di Ochrida? In *Actes du X congress international d'archaeologie chrétienne. Thessalonike 28 Septembre – 4 Octobre 1980* (pp. 681–690). Pontifico Instituto di Archeologia Cristiana.

Vikan, G. (1998). Byzantine pilgrims' art. In L. Safran (Ed.), *Heaven on Earth: Art and the Church in Byzantium* (pp. 229–266). Pennsylvania State University Press.

Vikan, G. (2003). *Sacred images and sacred power in Byzantium*. Routledge.

Vikan, G. (2010). *Early Byzantine Pilgrimage Art*. Dumbarton Oaks Byzantine Collection Publications.

Vio, E. (2003). *St. Mark's: The art and architecture of Church and state in Venice*. Riverside Book Company.

Vitaliotis, Y. (1997). *Journeys on the seas of Byzantium*. Hellenic Ministry of Culture.

Wachsmann, S. (1995). *The sea of Galilee boat: An extraordinary 2000-year-old discovery*. Springer US.

Wachsmuth, D. (1967). *Pompimos ho daimōn: Untersuchung zu den antiken Sakralhandlungen bei Seereisen* [Doctoral thesis, Universität Berlin]. Universität Berlin.

Wade, J. (2014). "Lock up your valuables": Perceptions of sailors and sea-merchants in port cities of late antiquity and early Byzantium. *Journal of the Australian Early Medieval Association*, *10*, 47–75.

Wade, J. (2018). The eternal spirit of Thalassa: The transmission of classical maritime symbolism into Byzantine cultural identity. *Journal of the Australian Early Medieval Association*, *14*, 51–69.

Wade, J. (2020). Athena, patroness of the marketplace: From Athens to Constantinople. In B. Neil & K. Simic (Eds.), *Memories of Utopia: The revision of histories and landscapes in Late Antiquity* (pp. 232–250). Routledge.

Walker, H. J. (1995). *Theseus and Athens*. Oxford University Press.

Walker, H. J. (2015). *The twin horse gods: The Dioskouroi in mythologies of the ancient world*. I. B. Tauris.

Williams, E. R. (1985). Isis Pelagia and a Roman Marble matrix from the Athenian agora. *Hesperia*, *54*, 109–119.

Wisniewski, R. (2019). *The beginnings of the cult of relics*. Oxford University Press.

Woods, D. (1991). The date of the translation of the relics of SS. Luke and Andrew to Constantinople. *Vigiliae Christianae*, *45*, 286–292.

Woolf, G. (2016). Only connect? Network analysis and religious change in the Roman world. *Hélade*, *2*, 43–58.

Woolf, G. (2018–2019). Global deities? Gods on the move in the ancient Mediterranean world. *Bandue*, *11*, 111–128.

Woolf, G. (2020). *The life and death of ancient cities: A natural history*. Oxford University Press.

Young, P. H. (2005). The Cypriot Aphrodite cult: Paphos, Rantidi, and Saint Barnabas. *Journal of Near Eastern Studies*, *64*, 23–44.

Zavagno, L. (2017). *Cyprus between Late Antiquity and the early Middle Ages (ca. 600–800): An island in transition*. Routledge.

Part 4

Changing spaces

13 Clean death or messy resilience?

Forum Traiani and *Forum Romanum* as activity spaces during the sixth century

Christina Videbech[1]

In 500 CE, King Theodoric (454–526 CE) visited the *Forum Romanum* during his *adventus* (Anon. Val. 65–67). The ceremony positioned Theodoric as the heir of Rome's grand emperors – the continuator of Roman values. Around 50 years later – Theodoric's *adventus* was likely a distant memory by then – someone threw a dinner party and dumped the leftovers in the *Atrium Vestae* (Filippi, 2001, 601). No sources reveal the reason for the dinner or its participants. Still, as discussed later, the tableware's quality indicates that they were part of Rome's large well-off but unnamed population. Easily overshadowed by historical events, the historical sources for their daily activities are often absent when the history of Rome's *fora* is written. Nevertheless, scholars increasingly study people who were not rulers, aristocrats, or bishops (e.g. Knapp, 2011; Keddie, 2019). Following research focusing on the monuments' users rather than the monuments themselves (Christie, 2020; see also the chapters of Will Bowden, Rubina Raja, Michelle L. Berenfeld, and Carlos Machado in this volume), this chapter examines who used the *Forum Romanum* (including its adjoining monumental areas along the *Via Sacra* and the *Via Nova* (Figure 13.1)), and the *Forum Traiani* (Figure 13.2) during the sixth century. The so-called *Athenaeum* of Hadrian is also considered, in conjunction with the *Forum Traiani*, as excavations have demonstrated their topographical link (e.g. Galli, 2013) (Figure 13.3). Actors like rulers and aristocrats naturally feature prominently in what follows. However, the chapter also examines evidence of possible non-aristocratic activities, arguing that well-off but non-aristocratic actors increasingly shaped the sixth-century *fora*. Furthermore, it analyses the consequences this had on the *fora*'s development and whether labelling this development as decline is appropriate. In that way, this chapter answers the concerns raised by Bowden in his chapter of this volume. It focuses on people over monuments and a broader range of Rome's actors in spaces which are traditionally studied as spaces of the elite.

To address these questions, the chapter analyses evidence of possible activities in the *fora* recorded in the archaeological, written, and epigraphical sources. This is possible due to scholars increasingly prioritising the sites' later phases (e.g. Bauer, 1996; Kalas, 2015; Meneghini, 2009; Packer, 1997). First, the evidence of actions initiated by emperors and aristocrats is discussed in its historical context. Subsequently, evidence of the well-off inhabitants is examined in their historical and social contexts. Finally, the combined evidence is discussed based on collective memory and community resilience theory. This theoretical combination ensures a cultural and social focus stressing human agency. Resilience theory seeks to understand an ecological or cultural system's capacity to deal with change and reinvent itself (Holling & Gunderson, 2002; Walker & Salt, 2012). Systems develop in an adaptive cycle, alternating between

DOI: 10.4324/9780429427152-17

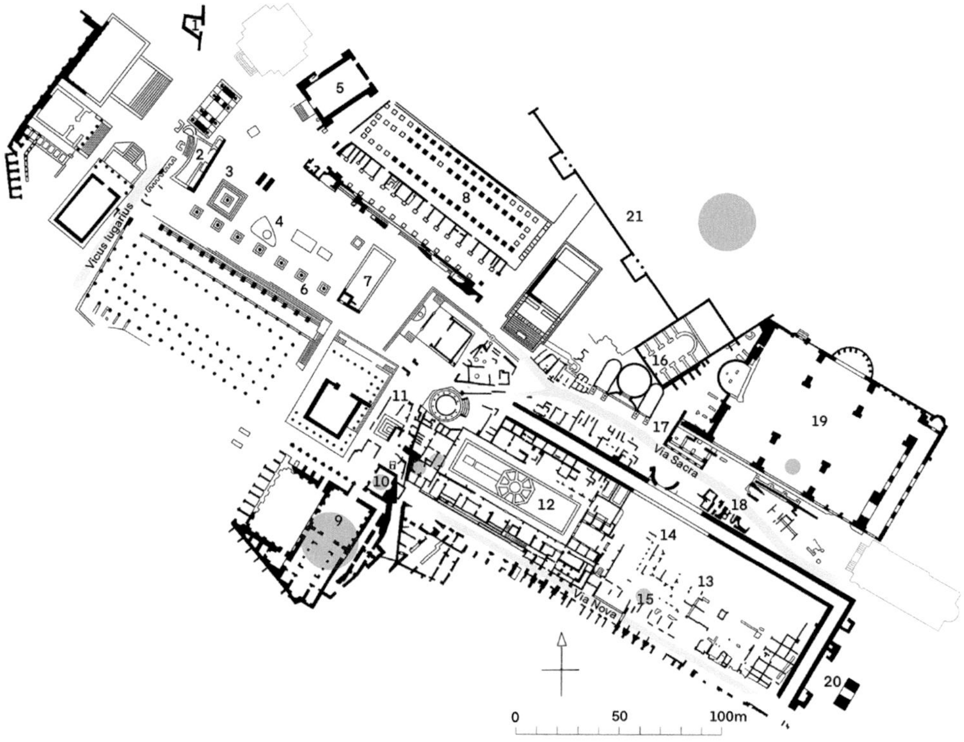

Figure 13.1 The *Forum Romanum* area with locations and monuments mentioned in the text marked.

Note: Circles: Sixth-century graves.

Key to the map – 1: *Carcer*, 2: Western *rostra*, 3: Column of Phocas, 4: *Lacus Curtius*, 5: *Curia*, 6: The Seven Column Monument, 7: Eastern *rostra*, 8: *Basilica Aemilia*, 9: *Santa Maria Antiqua*, 10: The Oratory of the Forty Martyrs, 11: *Lacus Iuturnae*, 12: *Atrium Vestae*, 13: *Horrea Vespasiani*, 14: *Horrea Vespasiani, hor.* b, 15: Room XII,1, 16: The Church of St. Cosmas and Damian, 17: 'Medieval porticus', 18: Fountain, 19: *Basilica Nova*, 20: *Arcus Titi*, 21: *Forum Pacis*. The *Atrium Libertatis, Secretarium Senatus, Ad Palmam, Porticus Curva,* and *Aedes Jani* are not marked due to their uncertain locations.

Source: Drawn and adapted by Vandel and Videbech from maps in Palombi (1988, Figure 1); Coarelli (2014, 73); Carandini (2012, add. tables 19); Coarelli (2008, 56–57); and 'Ground map Santa Maria Antiqua – Rome' by Afernand74 (Wikimedia Commons).

periods of resource accumulation and windows of transformation (Figure 13.4) (Holling & Gunderson, 2002, 25–62; Redman & Kinzig, 2003). Problems arise only if a system is too rigid to adapt. The theory has proved its ability to spark interesting discussion in late Roman archaeology (Malmberg et al., 2024). It is thus suitable for interpreting evidence of both positive and negative transformations spurred on by the sixth century's political, social, and environmental events. The challenges that hit Rome during this period are critical in understanding the *fora*'s development, as they became drivers for human action. Because of them, windows of opportunity to transform the *fora* appeared. Based on community resilience theory, which stress the role of human actions as drivers of resilience, this chapter focuses primarily on the roles humans played to ensure – and inhibit – resilience (Berkes & Ross, 2013). Thus, Rome's human actors – even its non-aristocratic inhabitants – played an active role in this process. They were not just bystanders to uncontrollable events.

The way different social agents used the *fora* – two of Rome's most important memory spaces – was shaped by tradition and the historical perception of the spaces. Thus, the attitudes

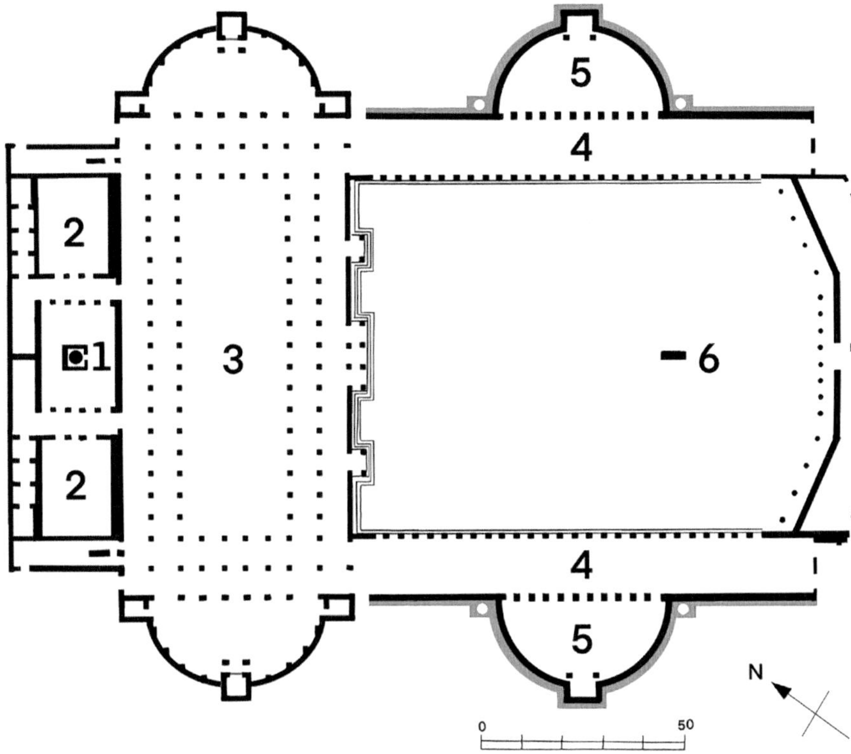

Figure 13.2 The *Forum Traiani*.

Note: Key to the map – 1: The Column of Trajan, 2: Library, 3: *Basilica Ulpia*, 4: Porticus, 5: *Exedrae*, 6: *Equus Traiani*, 7: Forum entrance and connection to the *Forum Augustum*.

Source: Drawn and adapted by Vandel and Videbech from maps in Meneghini and Santangeli Valenzani (2007) and Bianchi and Meneghini (2021, Figure 20).

of Rome's actors towards their mnemonic function could either contribute to or inhibit the *fora*'s resilience, depending on how they acted on it. Collective memory theory helps examine this aspect. Collective memory is a shared memory that provides members of a group with a perceived mutual identity based on a mythological or historical past (Halbwachs, 1950). Aided by rituals, traditions, and the retelling of stories, memory can be stored in mnemonic vessels such as monuments (Halbwachs, 2008, 117–164). Such vessels prolong memory's life beyond individual lifetime by eliciting human remembrance (e.g. Hirst & Manier, 2008, 184–189). In many ways, they work as when one makes a mark in a calendar to remember an appointment. However, unlike marks in a calendar, collective memory is continually subject to reorganisation, manipulation, and inaccurate recall shaped by the people who remember it (Halbwachs, 2008, 129–130). Thus, while regimes might try to portray it that way, no homogeneous memory exists. Collective memory is always firmly connected and shaped by its individual 'users'. The combination of resilience and collective memory theory enables an examination of how political, social, and environmental events affected the *fora*'s mnemonic meaning and how the *fora*'s use mirrored this meaning. Additionally, it helps analyse how the *fora* and their use reflected changes in social behaviour and self-understanding as they were transformed to fit the world view and needs of rulers, aristocrats, Christians, and Rome's well-off inhabitants. Possibly,

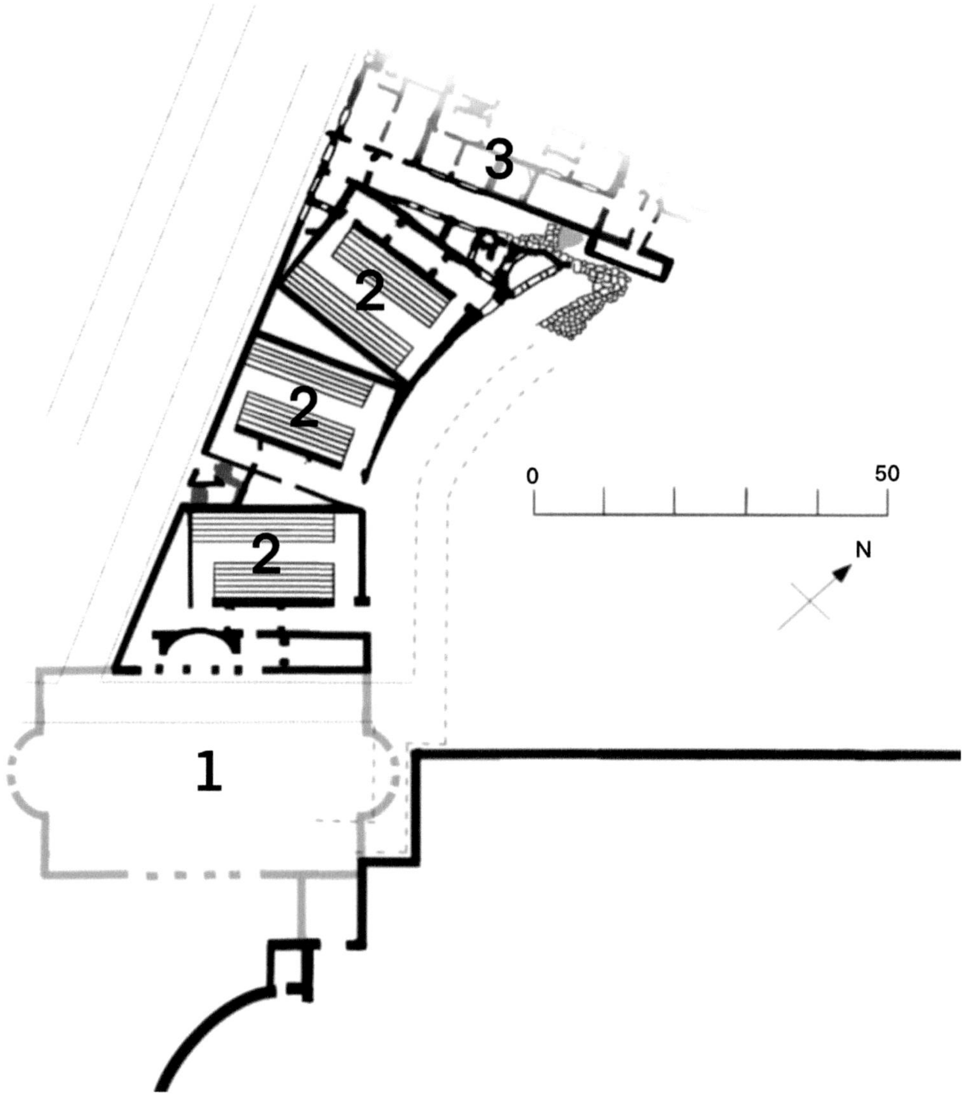

Figure 13.3 The so-called *Athenaeum* area.

Key to the map – 1: Traces of a possible structure suggested as the *Schola Fori Traiani* in Carandini, 2012, 2: *Auditoria*/ so-called Hadrian's *Athenaeum*, 3: Residential areas.

Source: Drawn and adapted by Vandel and Videbech from maps in Carandini (2012, tab. 54) and Bianchi and Meneghini (2021, Figure 20)

Theodoric's *adventus* was one of the last grand ceremonies of the *Forum Romanum*. Yet the forum was not abandoned, as the dinner party illustrates. Meanwhile, the sources are strikingly silent on what happened in the *Forum Traiani*.

13.1 A Gothic restoration?

Cassiodorus's *Variae* contains several letters in which Ostrogothic rulers – Theodoric especially – take action to maintain and protect Rome's buildings (e.g. 3.30; 3.31; 7.15; Fauvinet-Ranson,

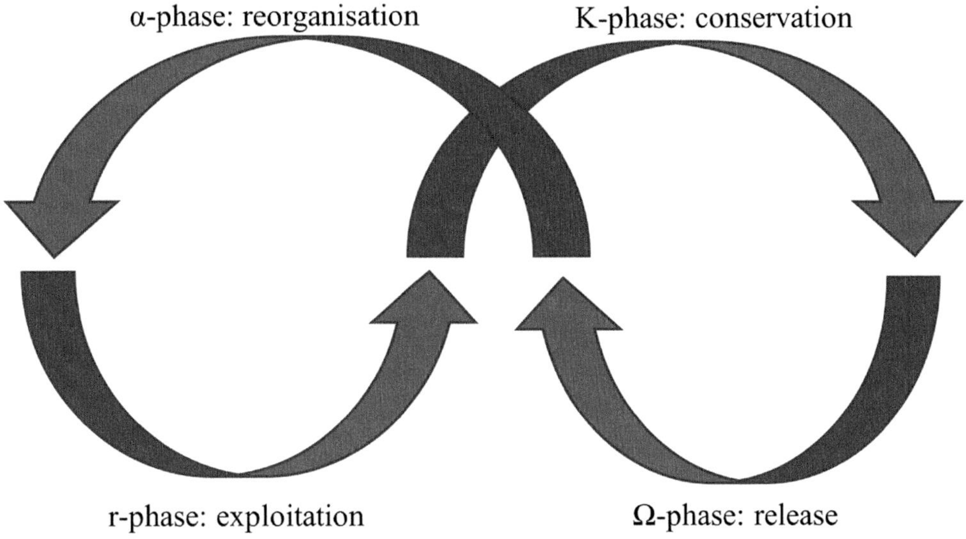

Figure 13.4 The adaptive cycle.

Note: The cycle has four phases: In the r-phase natural capital is accumulated, resulting in the K-phase, in which the production and storage of energy and material are maintained at high levels. This phase's instability brings eventual collapse, resulting in the Ω-phase, which releases the accumulated capital. Finally, the α-phase results in the system organising into an alternate state.

Source: Drawn by Videbech based on Holling and Gunderson (2002, Figure 2–1).

2006, 102–105, 135–141). Although Cassiodorus (*c*. 485–583 CE) was not unbiased and notoriously highlighted the continuity of the Gothic rule (Bjornlie, 2013, 7–33), the need for restorations in Rome fitted well with Theodoric's wish to display himself as Rome's restorer. Being visible in Rome's daily life, the *fora* were potentially powerful places for Theodoric to show this restoration (La Rocca, 1993). As clusters of mnemonic vessels, they visualised a shared Roman past, strengthening a feeling of togetherness (Häkli & Paasi, 2003, 142; Machado, 2021, 652–661). This function must have been important in a traditional, yet continuously evolving, society. Therefore, one would not be surprised to find evidence of traditional ruler activities in the *Forum Traiani*, encouraged or carried out by the Ostrogothic rulers.

Liebs (1985, 215–216) demonstrated how imperial pronouncements, letters, and laws were published on bronze plaques in the forum as late as 451. This practice made Arnold (2017, 115–116) suggest, based on the *Anonymus Valesianus*, that Theodoric followed earlier traditions and published a speech here. Unfortunately, nothing in the text supports this. While, as Arnold rightly points out, there was precedent for publishing ruler pronouncements in the *Forum Traiani*, the text Arnold cites only mentions that the plaques should be put up in public, not in this specific forum. As other places – including the *Forum Romanum* – were used for this purpose, it cannot be taken for granted that Theodoric had the *Forum Traiani* in mind. In fact, no sources account for activities in this forum even remotely connected to an Ostrogothic ruler. Still, Cassiodorus insisted that '*Traiani forum vel sub assiduitate videre miraculum est*' (*Var*. 7.6). He also listed the forum together with other great monuments, suggesting its continued historical importance (*Chron*. a. 95). It was still remembered as the legendary Trajan's burial place, tied to Roman history and supremacy (Cass., *Chron*. a. 120).

Turning to the *Forum Romanum*, unequivocal evidence of Gothic rulers directly using or influencing the space is also sparse, despite Cassiodorus's claims that they wanted to maintain

Rome's buildings. In practice, Gothic use of the *Forum Romanum* was limited to a few symbolic events. Theodoric likely used the *Curia*, as was the custom, when he spoke to the Senate during his *adventus* in 500 (Fraschetti, 1999, 194–197). Subsequently, he presented himself *ad Palmam* (Anon. Val. 65–67; Cass., *Chron*. a. 500). This location is debated, but as it was used for speeches during this time, Liverani's suggestion that it was connected to the *rostra* seems plausible (Liverani, 2007, 177–180). Following tradition, *ad Palmam* and the *Curia* were likely similarly used during the other possible Gothic *adventūs*, but the sources do not mention this explicitly (Eutharic, 518: Anon. Val. 80; Cass., *Chron*., a. 518. Theodahad, 535: Cass., *Var*. 10.30). Furthermore, Eutharic (*c*. 480–522 CE) granted honours *in curiam* in 519 (Cass. *Chron*., a. 519). While we cannot be sure that *in curiam* refers to the Curia building and not the Senate's presence (discussed later), this event must be considered a possible activity performed by an Ostrogothic actor.

Turning to the archaeological evidence, the *Forum Romanum* was traditionally a place for imperial and aristocratic self-promotion through sculpture. However, as elsewhere, the statue habit was waning in the forum (Machado, 2010, 239–243). No statues from the Gothic period have been found. Still, literary sources indicate an Ostrogothic intention of maintaining the statue habit. Procopius indirectly account for several statues of Theodoric erected in Rome as he accounts for the accusation that Boethius's widow destroyed some of them (*Bell*. 7.20.29). Isidore of Seville also mentioned a statue of Theodoric being erected (*Hist. Goth. Wand. Sue*. 1.59). Possibly, some of these statues were erected in the *fora*, but it cannot be confirmed archaeologically. The inscription *CIL* VI, 1795/*LSA*-1479, found in the *Via Sacra*, has been interpreted as evidence of a statue of Theodoric (Johnson, 2016, 357) or Ostrogothic restorations in the area (Arnold, 2014, 226). Nevertheless, as it only reads '*Salvo d(omino) n(ostro) [Theode]/ rico re[ge glorio]/siss[imo –]*', neither of the interpretations is justified. The inscription does not prove that a Gothic king directly shaped the forum. Still, it demonstrates how an unknown activity was traditionally celebrated with an inscription during the Gothic rule. Another piece of evidence is Theodahad's wish to restore the famous elephant statue group in the *Via Sacra* in 535/536, mentioned by Cassiodorus (*Var*. 10.30). As a minimum, Theodahad's wish indicates a Gothic intention of directly shaping the area. Yet no sources account for Theodahad (*c*. 480–536 CE) acting on it.

Despite the limited evidence for Gothic rulers shaping the *Forum Romanum*, Theodoric's *adventus* illustrates how it should be equipped for symbolic ceremonies, underscoring important continuities in the use of Rome's spaces during the Gothic rule. This is further underlined by the *Variae*, indicating that the Gothic rulers encouraged restoration, as discussed later (Fauvinet-Ranson, 2006, 267–274). Nevertheless, we should note that the *Edictum Theodorici*, an updated law collection probably composed on Theodoric's initiative, unlike previous laws, never treated the maintenance of monuments (Lafferty, 2010, 339–345, 349). Possibly, the Gothic rulers had limited time and funds to spend on a city that, although symbolically important, was not their capital. Possibly, they had local actors who were happy to aid them, making it unnecessary for the rulers themselves to take action. As the above evidence shows, the Gothic rulers were rarely physically present in Rome. Their 'Gothic restoration' would therefore benefit greatly from having allies in Rome who could act out their wishes – allies like Rome's aristocrats and the increasingly powerful pope. It is to the evidence of these people's actions we now turn.

13.2 Christian allies

During the sixth century, pagan religion was possibly not completely gone from the *Forum Romanum*. According to Procopius (*Bell*. 1.25.18–1.25.25), someone opened the doors of the

Aedes Jani as late as 537. Moreover, in 494, Pope Gelasius complained that people celebrated the *Lupercalia* (*Coll. Avell.* 100). Exactly when this ended is unknown. Possibly, it continued into the sixth century, becoming increasingly based on tradition rather than religion, as Machado argued (2019, 162–165). Still, Christianity became increasingly prominent in the area. The *Forum Romanum*'s memory vessels were gradually linked with Christian myths and martyrs, evolving the area's collective memory (Videbech, 2024). The *Acta Restituti* (*AASS* Mai VII, 11–13) accounts how Restitutus was eaten by dogs *ad Palmam*. Additionally, the water collected in the impressions of Peter's knees in the *Via Sacra* was believed to possess healing powers (Greg. Tour., *Glor. Mart.* 27). This accumulation of Christian myths in the area could be an early sign of the forum being a site of Christian pilgrimage (Videbech, 2024). Simultaneously, Christians possibly began to make their mark on the forum's monuments as the structure, which would later be converted into the church of *Santa Maria Antiqua*, was decorated with Christian wall paintings, possibly as early as the first half of the sixth century (before the consecration of the church: Bordi, 2021, 389–394).

In 526–530, Pope Felix IV consecrated a pre-existing building as the Church of St Cosmas and Damian (*Lib. Pont.* 56). It was the first church with access to the forum area. While the interior was adapted for the new purpose, attested by brick stamps dated to Theodoric's reign, there is no evidence for changes in its façade (Bauer, 1996, 51–57; Krautheimer et al., 1937, 137–143; Steinby, 1986, 140). It thus did not change the area's old aesthetics and the pope likely consecrated the church with the Ostrogothic ruler's acceptance. Adapting an existing building would fit perfectly with the Ostrogothic restoration agenda. Innovation was wrapped in tradition, making it acceptable even to the most traditional Romans. Thus, the church could signal cooperation between the increasingly powerful pope and the mostly Arian rulers (Vitiello, 2004, 119). The pope became an ally in preserving the forum, especially as Christian processions to the Church of Santa Maria Maggiore and St Peter's Basilica took their starting point in the Church of St Cosmas and Damian in 590 (Greg. Tour., *Hist. Franc.* 10.1) and 603 respectively (Greg. Mag., *App.* 9). Simultaneously, it is possible that the church's dedication to two Eastern saints was a gesture to the Eastern Emperor, who officially was Rome's ruler. In the nearby *Forum Traiani*, however, no evidence of Christian activities or myths dating to the sixth century is recorded.

13.3 Aristocratic allies

Other potentially important allies in the Gothic restoration were Rome's aristocrats, and while evidence of restorations initiated by Gothic rulers is sparse, restorations carried out by aristocrats are attested in the *Forum Romanum*. One inscription mentioning the *Atrium Libertatis* indicates restoration and the placement of a statue here in 491/518 by the city prefect (*CIL* VI, 1794; Fraschetti, 1999, 202–205). Scholars debate the *atrium's* location as it is not attested archaeologically, but most agree on its functional connection to the *Curia*. Bauer (1996, 12–14) summarises the evidence. Equally debated is the location of the *Secretarium Senatus*, also mentioned in the inscription and possibly repaired simultaneously (Fraschetti, 1999, 202–205). It was likely adjacent to the *Atrium Libertatis* and the *Curia*. Lavan (2021, vol. 2, 755–758) summarises the debate.

Brick stamps, dated to Theodoric's reign, are generally accepted by scholars to indicate a restoration of the *Curia* (Steinby, 1986, 128, 140, 2001, 138–139). In themselves, the brick stamps do not prove that aristocrats initiated the restorations. However, they could be connected to some unspecified work on the *Curia* – possibly a restoration – undertaken by the praetorian prefect Ioannes in 507/527 (Cass. *Var.* 9.7). The inscription '*Reg[nante d(omino) n(ostro)*

Theoderico? ---] =/ris Fla[vi–]/curia[---]' could also be from this restoration, but its fragmentary state complicates the interpretation, especially as the word *curia* lacks context (Lavan, 2021, vol. 2, 755). Furthermore, its initiator is not mentioned in the inscription.

Though the evidence is sparse, it indicates that even if restoration was part of Theodoric's political agenda, it was primarily aristocrats who executed it. Yet the line between the agency of Gothic rulers and the aristocracy is blurred (Salzman, 2021, 246–253). *CIL* VI, 1794 – mentioned earlier – illustrates this. While the inscription mentions the city prefect as the initiator, it also honours Theodoric by mentioning him. Additionally, the aristocrats who executed the restorations were often royal officials. Whether these people acted on their own or on the ruler's initiative is the question. It is unlikely that they would go directly against the king's wishes. There must therefore have been a limit to the actions these officials could perform without conferring with the king. Thus, royal agency would always be present indirectly in the actions of aristocrats, who, following earlier practice, acted as the ruler's representatives (e.g. Machado, 2019, 27–61). Moreover, the *Variae* suggests that the Gothic rulers – residing in Ravenna – had people in Rome reporting back to them. Letters like *Var.* 2.36 illustrates how Theodoric was eager to micromanage Italy's cities. Unfortunately, this proved hard in practice as letters expressing his frustration with unauthorised reuse, misuse of official funds, and people acting against his orders show (*Var.* 2.34; 3.31; 7.13). The *Variae* evidence is thus ambiguous, as it both illustrates a Gothic wish to micromanage Rome and how, despite this, Gothic control of the city was weaker than the kings should have liked.

Letter *Var.* 9.7's wording also suggests that the already-mentioned *Curia* restoration by Ioannes was initiated by himself. This indicates that aristocrats initiated restorations, even if the final authorisation was the king's. Later, Athalaric would use this restoration to justify the promotion of Ioannes's son. Evidently, royal agency could assert itself as positive reinforcement, encouraging other aristocrats to act similarly. Another letter (*Var.* 4.30) also indicates that a patrician initiated an adaptation of the *Porticus Curva*, but only after Theodoric gave his permission. Meanwhile, a letter to another aristocrat from the same king strongly suggested that he should restore the Theatre of Pompey with royal funds (*Var.* 4.51). Though not written explicitly, the suggestion resembles an order. The evidence thus points to Gothic rulers continuing an already well-established practice: instead of restoring monuments themselves, they encouraged aristocrats to do it for them – sometimes directly, sometimes indirectly. It is thus hard to identify whether Rome's aristocrats acted on their own initiative. Still, it seems possible that aristocratic activities, including those of officials, were, to some extent, initiated by the officials themselves. Likely this tendency would increase when rulers had other things to deal with during the period. Thus, keeping the evidence's ambiguity in mind, the evidence discussed here was assessed to be at least partially based on aristocratic agency.

All this considered, Rome's aristocracy was probably inclined to restore the *Forum Romanum* as they likely still used it. In 522, Boethius (*c.* 480–524 CE) read a panegyric to Theodoric *in curia* in connection with his sons being appointed consuls (Boet., *De cons. Phil.* 2,3,8). In 533, Cassiodorus read a panegyric to Amalasuntha in an unmentioned location (*Var.* 11.1). Furthermore, according to Cassiodorus, Rome's *vicarius* was present in the *Aula Libertatis* (*Var.* 6.15), and notaries frequented the *Curia Libertatis* (*Var.* 6.16), while *Var.* 5.21, 8.10, and 9.25.3 indicate that the *Atrium Libertatis* and the *Curia Libertatis* were used for senate meetings. Unfortunately, *curia* and the different *libertatis*-combinations could refer to both the assembled Senate, the *Curia*, or the *Atrium Libertatis*, making it hard to establish the activities' precise location (Fraschetti, 1999, 188–194). Additionally, phrases like *in curia/in curiam* could refer to the *Curia* or the Senate's presence. This potential presence is not enough to link the events with the building, as the locations of senate meetings were traditionally chosen based on purpose and practicality (Talbert, 1984, 113–114, 118–119). The same was the case in 438 when the Theodosian Code was promulgated in the house of the aristocrat Anicius Achillius Glabrio Faustus (*Gesta Senatus* 1; Matthews, 2000, 32–35). Therefore, the connection between these events and the *Curia* building cannot be presupposed, although at least some of them likely occurred here.

Meanwhile, the evidence suggests that the aristocracy's interest in the *Forum Traiani* waned as they sought new arenas for their self-representation (see below). Previously, statues honouring the emperor and Rome's aristocracy were erected in the forum (Chenault, 2012; Weisweiler, 2012), but there is no evidence of this practice continuing during the Ostrogothic rule. Packer (1997) suggested based on an archaeological report from 1931 and Lanciani's folio 112 (Buonocore, 1997, 214; Ricci, 1931) that the city prefect repaired the forum around 508 (pp. 8–10). However, neither of these sources provides any real evidence to support such a suggestion. Scholars like Marrou (1932) and Galli (2013, 64–67) have argued that the *Forum Traiani* and the nearby *auditoria* were places for literary education during the fourth and fifth centuries. While this chapter does not aim to discuss their arguments, it should be noted that no sources point to such activities occurring during the Ostrogothic rule. Yet this rule would soon be over as the Eastern Emperor decided to reconquer Rome – a decision which would impact the two *fora* and their actors.

13.4 Gothic Wars, Eastern Emperors, and fleeing aristocrats

The Gothic Wars (535–554 CE) hit Rome hard. Procopius accounts for both famine and despair (*Goth.* 7.17). Unfortunately, having won the war, Justinian (482–565 CE) seemed less eager to maintain Rome. After a flood, plague came to Rome during the 590s (Greg. Tour., *Hist. Franc.* 10.1). Scholars debate its exact demographic impact, but it likely resulted in a decline in taxpayers (Meier, 2016; Sessa, 2019, 236–237). Therefore, it would be understandable if the Eastern emperors did not prioritise maintaining a city most never visited. Their disinterest in Rome is underlined by Gregory the Great's continual complaints of imperial troops neglecting Rome (Greg. Mag., *Ep.* 2.38; 5.36; 8.22; 9.240). Nevertheless, the Pragmatic Sanction explicitly mentions how the *Forum Romanum* should be maintained, apparently demonstrating an intention to maintain Rome (Just., *Nov.* 802, App. 7.25). Unfortunately, it also states that such maintenance should be preserved with traditional funds and taxes, placing the responsibility on the senatorial aristocracy and the decreasing group of taxpayers. Consequently, the lack of sources accounting for imperial actions in the *Forum Traiani* is hardly surprising. One possible exception is the so-called *Athenaeum*, which was converted into a mint. Its *auditoria* were stripped from their marble decorations, and over a hundred furnaces were installed (Serlorenzi & Egidi, 2013, 194–195). The quality of the work led Serlorenzi and Egidi (2013, 194–195) to suggest that it was an official mint.

Despite the aforementioned claims of the Pragmatic Sanction, imperial interference is equally hard to trace in the *Forum Romanum*. In 565–578, the Church of Santa Maria Antiqua was installed in a possible vestibule of the Palatine Palace (Krautheimer et al., 1959, 249–270). Although Hurst (2021) pointed out inconsistencies in the excavation reports, the scholarly consensus is that it dates to the sixth century, based on coin finds and the building's sixth-century Christian paintings. On the basis of the *Liber Pontificalis*'s silence on the church and its consecration while Narses, commander of Justinian's army in Italy, lived in the Palatinc Palace, Bordi (2021, 395–397) suggested that it was either an imperial dedication or Narses's project. Coates-Stephens argued similarly (2006, 158), pointing out that the dedication to Maria indicates a Byzantine connection as she featured prominently in Eastern orthodoxy. While the argument builds on speculation, it is intriguing due to the *Liber Pontificalis*'s well-attested tendency to omit monuments not erected on papal initiative (Dey, 2021, 72). Maskarinec (2018, 38–39) further suggested that the church was used for imperial administration but admitted it was an argument *ex silentio*. If the church is evidence of imperial interference in the forum, it is a rare example.

Even the Column of Phocas, dating to 608, can be questioned as an imperial initiative. It was a rededication of an earlier monument to Emperor Phocas by the Italy-based exarch (*CIL* VI, 1200; Giuliani & Verduchi, 1987, 174–177). As Coates-Stephens (2006, 150–154) argues, the column illustrates how the forum was still symbolically important. Yet the only actor explicitly mentioned in its inscription is the exarch, not the emperor (Coates-Stephens, 2011, 401). Again, evidence of direct imperial agency is hard to establish. Still, it cannot be excluded that the *fora* were officially imperially controlled spaces. Machado (2019, 66) showed how imperial approval was previously required for aristocratic building projects in Rome. As Salzman (2021, 291) argues, the Pragmatic Sanction indicates that Rome's monuments were imperial property even after the Gothic Wars. Initiatives like the Column of Phocas likely needed imperial approval in some shape or form or – at the very least – an emperor who did not actively act against them. While this was not necessarily different from the situation in the *fora* during previous centuries and the Gothic rule, something had changed by the end of the sixth century. Rulers had previously counted on aristocrats to aid them in making the imperial presence visible by dedicating and restoring monuments in the emperor's name. This practice was now challenged. While emperors were relatively safe from the Gothic Wars in Constantinople, Rome's aristocracy was not.

According to Procopius, the wars resulted in Rome's senators having no funds to maintain monuments (*Bell.* 8.22.1–8.22.16). Moreover, sources suggest that many aristocrats were executed during the war's final years and that a significant number of the survivors – including Cassiodorus – fled Rome (Salzman, 2021, 267). The Lombard invasion did not improve the situation (Barnish, 1998, 140–151). While the Senate – last mentioned in 603 – was not dissolved, the relocation of many senators must have been a severe blow (Greg. Mag., *App.* 8). Nevertheless, Salzman (2021, 267–269) shows how some aristocrats returned to Rome. In other parts of the city, maintenance is evident, resulting from cooperation between local aristocrats and the Byzantine administration (Dey, 2021, 73–74). Thus – despite Procopius's claims – some aristocrats had funds for maintenance. In the case of the two *fora*, however, with the exception of the Column of Phocas, sources of their aristocratic use disappear almost entirely. The *Curia* was last referred to as the senate house in 536–539 by Procopius as he described the *Aedes Jani* (*Bell.* 5.25.19). On the basis of this description, Bartoli (1963, 72) and Fraschetti (1999, 197) argued for the building's continuous use as senate house. Fraschetti also suggested that the *Curia* kept its function after it became a church (Fraschetti, 1999, 198–199). Unfortunately, no sources indicate this. Likewise, Coates-Stephens (2011, 401–403) argued for the continued ceremonial use of the forum during the late sixth century. While the argument is appealing, it is only based on circumstantial evidence.

Meanwhile, Venantius Fortunatus twice mentions poetry readings in the *Forum Traiani*, echoing its former use for literary education (*Carm.* 3.18.7–3.18.8; 7.8.25–7.8.26). His accounts are usually taken at face value. However, his value as a witness to events in Rome is limited, as nothing points to Venantius ever having visited Rome. Besides, he was trained in rhetoric and grammar, and his authorship illustrates his knowledge of classical poetry (George, 1995, xx–xxi). I would therefore argue that his *Carmina* could be literary *topoi*, a rhetorical tool used by ancient authors to show learning. Instead of choosing contemporary Christian poets, Venantius referred to Homer and Virgil, the forefathers of antique poetry, underlining his poems' historicising tone. This reference demonstrated Venantius's learnedness, indicating that he was name-dropping rather than accounting for actual events. Even if we, despite the above, trust Venantius's accounts, *Carm.* 3.18.7–3.18.8's wording suggests that the practice of reading poetry in the *Forum Traiani* decreased during the sixth century compared to earlier centuries – maybe even that it had ended. It is also interesting to note that Venantius, when he needed a rhetorical education, travelled to Ravenna, not Rome (Salzman, 2021, 285).

To summarise, the evidence of imperial and aristocratic activities in the two *fora* is sparse during the last part of the sixth century. Even when evidence of imperial self-presentation is evident, as in the case of the Column of Phocas, aristocrats seemed to be the instigators. This does not exclude that imperial agency was enforced indirectly. However, as aristocrats struggled due to the Gothic Wars and began to increasingly use Rome's other available spaces, their presence and the activities they performed on behalf of the emperor declined in the two *fora*. It is clear that part of this argument is built on the absence of evidence and that the fragmentary state of the sources available to modern scholarship can partially explain this lack of evidence. Nevertheless, as discussed later, the tendency is so consistent that this explanation seems unsatisfactory.

13.5 The well-off

Having thus examined the development of the *fora*'s use by Rome's elite actors, it is time to turn to this chapter's other protagonists, namely Rome's well-off inhabitants. As the evidence for the actions of rulers and aristocrats waned in the *Forum Romanum*, an increasing amount of evidence indicate the growing presence of the well-off in the area's daily life. These signs have eluded scholars due to the difficulties inherent in the study of this group, most notably the issue of its very definition. In this chapter, the well-off include actors recorded in the sources who cannot be identified as popes, aristocrats, or rulers. Consequently, their identification is often uncertain. By the sixth century, they were predominantly Christian and likely had different ethnicities, as Rome had attracted people from all over the empire over time. Some would have identified as Roman, most were likely Roman citizens. Some might have been *liberti*, others freeborn. Their economic background was diverse but secure. They were not aristocrats; they had no direct political influence but could influence politics indirectly and shape their community. People belonging to this group could be priests, scribes, nuns, traders, artisans, wealthy widows, and doctors, among others.

In the *fora*, most traces for the activities of the well-off are archaeological. Unfortunately, this evidence is problematic since, as Keddie (2019, 198–99, 201–2) argued, wealth does not determine social class. Aristocrats could be relatively poor, and freedmen could reside in rich houses. Consequently, it is often hard to establish if the evidence is left behind by aristocrats or the well-off, as discussed later. In the case of the *Forum Traiani*, unfortunately, this is a non-issue as neither group's activities are detectable in the sources. Fortunately, this is not the case in the *Forum Romanum*. Scattered small finds provide glimpses into the day-to-day use of the area.

One reason for visiting the forum was to collect water. Two (possibly) medieval wells, made of reused building materials, have been found in the *Via Sacra*, although the area's complex excavation history makes a precise dating hard (Le Pera & D'Elia, 1986, 245–246). Another fountain in the *Via Sacra* was possibly used during Late Antiquity, as indicated by late antique restorations (Palombi, 1988, 85–97). The find of water jugs and other locally produced ceramics also shows that people continued to collect water at the *Lacus Iuturnae* (Steinby, 2012, 58–59). Unfortunately, precise datings are hard to establish as Boni did not record the stratigraphy of this evidence (Boni, 1901, 97–103). Late antique coins – often with no recorded context – were found scattered in the same area (Örmä, 1989, 2012b). Örmä (1989, 315) suggested that they were part of an early seventh-century deposit. I find it equally likely that they were dropped when people collected water. Other signs of continuous activities are the presence of medieval mortars (Burton Brown, 1904, 18) and lamps dated up to the seventh century (Örmä, 2012a).

Boni (1904, 582) also reported finding several Constantinian and proto-Byzantine coins in connection with the *Lacus Curtius*, which traditionally had been used by Rome's inhabitants to toss coins for the emperor's health (Suet., *Aug*. 57.1). Boni's coins indicate a possible late antique continuation of this popular tradition. Sadly, the lack of precise dating of the coins prohibits establishing if it continued during the sixth century. Another intriguing evidence group consists of 77 game boards incised on the area's monuments and pavement (Trifilò, 2012). These games indicate that people played games in the forum. Due to their generic design the game boards are hard to date. However, they could have been used during the sixth century as games dating to this century have been found in Eastern cities (Lavan, 2021, vol. 1, 352, 361).

Written sources rarely account for Rome's well-off inhabitants. Consequently, the letter *Var*. 4.30 must be considered as it often features prominently in scholarship on the late *Forum Romanum* and possibly indirectly indicates non-aristocratic use. As mentioned earlier, Ostrogothic rulers handed over public buildings to the care of Rome's aristocracy. Never are non-aristocrats recorded to have acquired monuments. Nevertheless, this could be an unforeseen consequence of the Ostrogothic practice. In *Var*. 4.30, Theodoric granted a patrician permission to build *fabricae* connected with the *Porticus Curva* around 506/511. The letter also accounted for private dwellings in and around the porticus. While it cannot be ruled out, nothing suggests that the *fabricae* and dwellings were intended for the aristocracy. One could speculate how their use was delegated to well-off inhabitants while aristocrats collected rent from shopkeepers and tenants. Unfortunately, the *Porticus Curva*'s possible connection to the *Forum Romanum* is debated as it has not been possible to identify its archaeological remains with certainty (Marchetti Longhi, 1949). Bauer's suggestion (1983) that it was identical to the *Porticus Absidata*, located behind the *Forum Transitorium*, is the most widely accepted suggestion. Therefore, *Var*. 4.30 does not definitively prove that shops and dwellings were constructed in the forum area during the Gothic rule, but in an area immediately adjacent to it. Yet archaeological evidence indicating such a development is present.

By the sixth century, the *Atrium Vestae* had lost its function as the Vestals' residence. Scholars have suggested that it became lodgings for visitors to the imperial court from the last part of the fourth century due to its topographical connection to the Palatine Palace (Bossi, 2017, 753; Filippi, 2001, 600–601). A large coin hoard deposited in the house in 472 possibly supports this hypothesis, as wealthy people must have been present. Still, we should be cautious to accept this suggestion. While it is possible to determine that wealthy people inhabited the space, the evidence is insufficient to determine that they were imperial guests. During the sixth century, parts of the house were adapted to what could be interpreted as small-scale manufacturing areas (Bossi, 2017, 742). Simultaneously, several rooms were divided into smaller spaces with heating systems and brick floors (Filippi, 2001, 600–601). On the house's top floor, the bath area was divided into smaller rooms with basins (Bossi, 2017, 741–743). While establishing a possible small-scale production area does not rule out an interpretation of the house as an official lodging place, placing manufacturing areas close to apartments meant for official guests seems odd. It would, however, fit well with the house being turned into dwellings for less grandiose – but wealthy – inhabitants. Possibly, these people belonged to Rome's well-off inhabitants as expensive materials such as marble were prominent in the restructuring.

The adjacent building, often identified as the *Horrea Vespasiani* (Papi, 1996), has traditionally been interpreted as a space for storage or a fish market due to its design and its several basins (Ippoliti, 2017, 757–776). Between the fourth and sixth centuries, the *horrea* were divided into smaller units (Ippoliti, 2017, 767–768, 774–775). In the fourth century, the courtyard of *Horreum* A was divided into two parts by a wall. Passage between the two parts of the courtyard was made possible by at least one entrance. In the same century, a structure, which has been interpreted as a shop counter was erected in room 118 of *Horreum* B. The floor in the courtyard was restored and a further fish basin was erected. All this has been interpreted as the building's continued use as a place for trade (Ippoliti, 2017, 774–775). The restructuring of the building complex continued during the fifth and sixth centuries. A pillar was added to the entrance of *Horreum A* and *B*'s layout was changed. Furthermore, the pavement was restored, and a new drainage system channelling rainwater to the courtyard's centre was established. A wall was erected, dividing the courtyard and removing one of the basins' stairs. Despite this, the basins were still functional, meaning that the *horrea* could still function as a fish market (Ippoliti, 2017, 775). A further basin was even constructed in *Horreum A* (Ippoliti, 2017, 768). In Room XII,1, large amounts of glass, vessels, and fireplaces indicate a sixth-century glass production (Storozhylova, 2019, 19–22).

Filippi (2001, 601) has suggested that the rework of the *horrea* was a way to adapt it to residential use. Unfortunately, the evidence is too ambiguous to confirm this interpretation, and no literary or epigraphical sources are available to help. Consequently, the evidence is hard to interpret, and the initiators of the adaptations remain unknown. Still, the establishment of the glass workshop in the *horrea* resembles the development of the adjacent *Atrium Vestae*, possibly indicating a general theme in the area's development (see Carlos Machado in this volume for similar developments in the Campus Martius). Similarly, the nearby so-called medieval porticus in the *Via Sacra*, constructed with reused building materials, was possibly used for trading. It is dated later than 283, but a more precise dating is impossible due to its complicated excavation history (Le Pera & D'Elia, 1986, 249–252).

Meanwhile, the *Basilica Aemilia* was traditionally used for trading and financial activities in the early empire (Freyberger et al., 2007, 523). Brick stamps dating to Theodoric's reign indicate a sixth-century restoration (Bauer, 1996, 35; Steinby, 1986, 128, 140). Three floors – located at the basilica's entrance and in two of its *tabernae* – were also restored. On the basis of the mosaic style, they have been dated between the (possibly late) sixth and seventh centuries (Guidobaldi & Guidobaldi, 1983, 264–277, 353). Their restoration could indicate that the basilica was used for trading, but no other traces of trade are recorded (Coates-Stephens, 2006, 163). Alternatively, the floors and the brick stamps could be interpreted as the basilica being turned into a private house (Mortera, 2017) or administrative offices (Coates-Stephens, 2011, 389–390). While the lack of inscriptions recording the restorations does not exclude aristocratic actors, the evidence points to minor restorations focusing on keeping the building functional rather than showcasing political symbolism. It illustrates the need to keep the space functional in the day-to-day life of Rome's inhabitants.

A similar development could be argued for the postholes from possible market stalls found around the Seven Column Monument, the Column of Phocas and the *rostra* (Giuliani & Verduchi, 1987, 173, 187). They have been dated to the sixth/seventh century, but recently scholars suggest a dating around 1000 as fixed market stalls were uncommon in Italy during the sixth century (Goodson, 2021, 132–134). Nevertheless, examples likely dating between the fourth and the seventh centuries are documented in the Eastern Empire. Occasionally paired with *topos* inscriptions, their identification as shops is easier (Lavan, 2021, vol. 1, 375–379). Furthermore – if we accept the postholes as evidence of trade – it fits well with the evidence found in the forum's north-western area. Here a coin weight and 205 bronze coins, dating between the second and the late sixth centuries, were found in a way that suggested them having been kept together in a container (Molinari, 2004, 204, 206; Paganelli, 2004, 180, 182). Paganelli (2004, 182) suggested that this container burned during a possible late sixth-/early seventh-century fire – attested by large amounts of charcoal. The latest coin was minted between 565 and 578, giving the fire's *terminus post quem*. As the coins were stored together with a coin weight, it is plausible they came from a shop or a trader.

Additionally, lime deposits and spoliation of the pavement are detectable in the same area (Maetzke, 1991, 85–87, 191; Paganelli, 2004, 180–183). Reuse activities are also attested at the eastern *rostra*, where a lead workshop was incorporated. Nearby an iron foundry was established. Both date to the sixth/seventh centuries (Giuliani & Verduchi, 1987, 165–166 and Figure 264). A third workshop was also installed during the sixth century in the *rostra* (Giuliani & Verduchi, 1987, 165–166). Based on the find of a partially recut marble block found alongside charcoal, marble pieces, animal bones, and ceramics, it could be a workshop specialising in reusing building materials (Lavan, 2021, vol. 2, 885). This would fit with the evidence of spoliation from the nearby area. However, as Coates-Stephens (2011, 392) argues, it could also be a workshop attending repairs on the area's buildings. No matter their precise function, the workshops' presence is evidence that the forum's monumental space was now included in the daily life of Rome's inhabitants. Moreover, regardless of whether the workshops were connected to spoliation, they were themselves reusing an old monument, namely the *rostra*.

According to Cassiodorus (*Var*. 3.49.3), ruins were an unwanted element in the cityscape. Consequently, dilapidated structures were adapted for reuse or had their building material used to restore other complexes (Cass., *Var*. 2.7; Fauvinet-Ranson, 2006, 64–65, 121). If the ruler granted his permission, reuse was defendable – even appreciated (La Rocca, 2014, 16–25). According to Cassiodorus (*Var*. 7.13), unsolicited reuse was another matter, causing '*publicus dolor*'. Nevertheless, we have reason to doubt if ordinary Romans were outraged by reuse. Scholars often use the account of the riot against Lampadius (Amm. Marc. 27.3.8–27.3.10) to exemplify Rome's inhabitants reacting violently to reuse. Yet people were not outraged by the reuse but by Lampadius's failure to pay for it. Likewise, no angry reactions to the forum workshops are recorded – possibly because they were approved by the ruler or because Rome's inhabitants did not mind. Perhaps they even benefitted from the reuse as the workshops would provide work for non-aristocratic inhabitants, even if aristocrats administrated it. No matter the case, the workshops of the *rostra* indicate that Rome's sixth-century actors had a pragmatic attitude towards the historical monuments.

The forum area was also adapted in other ways during this period. The sixth-century *Edictum Theodorici* illustrates how burials '*intra urbem Romam*' were officially banned even at that date (*Edict. Theod.* 111). Even so, dated stratigraphically to this century, the first cemetery linked with the *Forum Romanum* was founded near St Cosmas and Damian in the *Forum Pacis* (Capponi & Ghilardi, 2002). A grave in the *Basilica Nova* dates to around the same time (Fea, 1821, 30) (Figure 13.1). Identifying the buried people is impossible, but the graves' design, and the scanty grave goods do not suggest aristocratic burials but rather 'ordinary' inhabitants of Rome. This development continued after the Gothic Wars. Graves, dating to the late sixth century, was found in the *Atrium Vestae* and the *Horrea Vespasiani* (Filippi, 2001, 601; Meneghini & Santangeli Valenzani, 1993, 105; Storozhylova, 2019, 17–18). As in the case of the other graves, these graves were probably not aristocratic either. A cemetery was also founded in connection with *Santa Maria Antiqua*, dated based on the funerary inscriptions of the goldsmith Amantius (*ICUR* n.s. I, 1403, 572 CE) and Ypolita (*ICUR* n.s. I, 1404, 582/602 CE), who either herself or through her family belonged to the '*arte argent[ar]i*' (Coates-Stephens, 2021, 200). Both belonged to Rome's well-off inhabitants. The inscriptions were found in the connected Oratory of the Forty Martyrs (Gulowsen, 2001, 79–81). Scholars debate whether they were initially put there or reused in the chapel (Augenti, 2001, 163–168). Coates-Stephens (2021) argues for the first option based on their stratigraphy, suggesting that the oratory was initially a guild chapel dedicated by Amantius (pp. 197–200). If so, it was the first building founded by a well-off inhabitant in the area. Sadly, other graves, potentially dating to the cemetery's early phase, have been disturbed (Tea, 1937, 243–244).

Briefly leaving the *Forum Romanum*, two grave inscriptions are also recorded in the *Forum Traiani*. One is in Greek and dated to the so-called Christian period by Paribeni (1933, 501, n. 203) with no further elaboration on why or what this entails. Another inscription is in Latin and dated to 550–649 based on the dating formula (Ricci, 1994, 183–184, n. 9; Paribeni, 1933, 500, n. 197). None of them has any explicitly Christian symbols or phrases. The associated graves have not been found, and no information on their context is provided. Therefore, they do not prove that graves were present in the *Forum Traiani* nor the well-off's presence. Possibly, the inscriptions were simply reused in the area's Medieval houses.

Thus, we finally arrive at the dinner party mentioned in the introduction. Its leftovers – animal bones with butcher marks, fragments of imported *sigilata*, glass, lamps, kitchen vessels, and African *amphorae* – were buried in the *Atrium Vestae* around 550 (Filippi et al., 2004). Similar deposits of amphorae and dinner service from the end of the sixth/beginning of the seventh centuries were dumped in the forum's north-western part (Paganelli, 2004, 180, 182). Neither purpose nor participants of these dinners are known. They could be connected to the funeral rites of the area's buried or private parties held by local residents. In either case, their participants would belong to the well-off, which would fit with the nature of the leftovers. Still, it cannot be outruled that the leftovers could come from a dinner held in a nearby aristocratic house, meaning they had been carried a certain distance from that house. While possible, it seems more logical that the leftovers came from the forum area, so one would not have to carry them for long. All the evidence of the forum's sixth-century use indicates its users' pragmatism. Above, it was argued how well-off inhabitants likely lived in the same house as the leftovers were found. The most logical solution therefore seems to be that the house's inhabitants threw a party and discarded the leftovers in a nearby unused room. Thus, while one could look to other parts of Rome to find the dinner party's participants, it is unnecessary.

13.6 Clean death or messy resilience?

Having examined the evidence of the two *fora*'s actors, the quantitative difference in the evidence is striking. As Christianity was increasingly intertwined with everything from trade to politics, the *fora*'s mnemonic vessels proved that Roman culture was fundamentally intact. Inspired by Nora's influential contribution to collective memory theory, it is tempting to suggest that the *fora* were simply great museums, so-called *lieux de mémoire* (Nora, 1989). Influenced by this theory, Hölkeskamp and Stein-Hölkeskamp (2006) edited a volume in which the *Forum Romanum* feature prominently as an example. However, in describing the *lieux de mémoire*, Nora stressed their crystallisation and irrelevance to real, living memory. They were kept artificially alive by organising what was once living memory (Nora, 1989, 7, 12, 19). They were not wholly dead but not living or evolving either. They existed in the absence of real *milieux de mémoire*, which integrated memory and everyday experience (Nora, 1989, 1, 11–12). Consequently, the data from the period strongly suggest that *Forum Romanum* was not a *lieu de mémoire*. It was a lively space adapted to the changing times. Christian wall paintings, churches, legends, and possible pilgrims can be detected. These activities adapted the space to Rome's role as a Christian city, proving its flexibility and causing it to stay resilient and relevant to Rome's inhabitants (Videbech, 2024). *Forum Romanum*'s long history, combined with the Roman ideal of traditionality (e.g. Hekster, 2020a, 2020b), could hinder this process. Nevertheless, no sources attest to Rome's inhabitants objecting to the change. The *Forum Romanum* had countless times transformed according to Rome's political climate and changing power players. Rome's actors were used to such transformation adding new layers to the space's memory. The city's gradual Christianisation became a window of opportunity and reinvention of the forum.

One consequence of this process was the introduction of graves in the area. While no satisfactory explanation has been found for their presence (Costambeys, 2001, 169–172), I would problematise the notion that graves equal the decline of the monuments they occupy. Instead, graves could contribute to keeping the forum relevant as people would come there to visit them. The same argument goes for the market stalls, workshops, and dwellings in and between the monuments. Such adaptations of the forum would ensure that the space was lively and relevant to Rome's inhabitants. The gradual appearance of graves, shops, and legends in the *Forum Romanum* might seem messy. One could argue that they were an expression of the area's decline compared to earlier times. However, we must question such judgemental interpretations. Modern perceptions of a clean ancient city were not the reality. Not during the sixth century, not before. Local actors continually formed and re-formed Rome, ensuring its overall resilience. Although the evidence of possible non-aristocratic activities is often hard to interpret, it still deserves scholarly attention. In the end, the forum's daily use kept it alive. Studies suggest that collective memory needed more than society's top to survive (e.g. Davis, 2007). Epic tales of past rulers and heroes kickstarted the space's importance, but they were insufficient to maintain it. To the well-off, the *fora* were likely historical places, sparking pride, but before anything else, the *Forum Romanum* had a function. Rome's inhabitants traded there, played games, and interacted. They went there to worship God, bury their dead, and throw dinner parties. Possibly, they lived there. All of this is visible in the above data.

Does this mean aristocrats no longer attended the forum during the sixth century? Probably not. As is clear from the evidence examined earlier, they likely continued using the forum during the Gothic rule. Their withdrawal from the forum seems to have been a gradual process, spurred on by events like the Gothic Wars and the changing political situation. After the wars, sources of the forum's aristocratic use are sparse. This is not the same as saying they were not present in Rome anymore. Evidence from other sites suggests that they sought new political and religious arenas. Thus, in 603, the Senate and clergy received the imperial portraits in the Lateran (Greg. Mag., *App.* 8), and both emperors and aristocrats increasingly donated gifts in St Peter's (Greg. Mag., *Ep.* 11.26; *Lib. Pont.* 54, 55, 58, 61). Meanwhile, documented building activities in the *Forum Romanum* focus on practical needs to keep them functioning or adapting them for daily activities such as workshops, trading, and living.

Studying *Forum Romanum*'s resilience, we must consider who benefitted from it and what value is attributed to this. Sources like Cassiodorus want us to believe that Rome's resilience was tied to the aristocracy. Salzman recently convincingly argued the Senate's importance in Rome's resilience and how its demise was a significant turning point in Rome's history (Salzman, 2021). I agree on the Senate's importance. As one of the most prominent *fora* actors for centuries, its demise was a tipping point in their development. However, Salzman leaves little room for non-aristocratic actors as contributors to Rome's resilience. This chapter suggests that the focus of rulers and aristocrats on other spaces became a window of opportunity for the well-off actors (Michelle L. Berenfeld discusses a similar development in the area of the Caelian Hill during the fifth century in her chapter in this volume). While it is easy to devaluate evidence of spoliation and rubbish dumps as signs of decline, we should acknowledge it proving that people attended the area. The evidence demonstrates how trade, graves, possible dwellings, and dinner parties existed simultaneously in the same space. Written elite sources, like the *Edictum Theodorici*, tell us this should not occur. Nonetheless, it did. The examined evidence thus suggests that while rulers could make laws, ultimately, Rome's inhabitants would approach the forum pragmatically, adapting it when necessary. The evidence of non-aristocratic activity is complicated, leaving more questions than answers, but no inscriptions or written sources link the evidence with aristocratic actors. Moreover, the evidence does not suggest grand symbolic restorations but rather minor adaptations to keep the buildings functioning. Admittedly, this chapter's argument is based partially *ex silentio* and therefore nothing can be definitively concluded. Still, I tentatively suggest that the evidence of the *Forum Romanum* points to ordinary people gradually adapting the forum to their daily needs and that Rome's historical forum was resilient due to this. Yet, as is often the case with systems' resilience, it came at a cost (Walker & Salt, 2012, 21–22). In the eyes of posterity, the price was the forum's aesthetic. The sixth-century forum user apparently did not mind.

Nora's theory on the *lieux de mémoire* does not apply to the *Forum Romanum*. The *Forum Traiani* is another matter. Here, the memory process working so well for the *Forum Romanum* apparently went into overdrive. The sources suggest that collective memory became a hindrance – not an aid – in the forum's resilience. Community resilience theory stresses the importance of actors believing change is desirable (Berkes & Ross, 2013). In the case of the *Forum Traiani*, it was not. The forum has no signs of restoration, political use, new statues, workshops, religious use, graves, private houses, or non-aristocrats. While some of this absence of evidence can surely be attributed to the loss of evidence during early excavations not focused on late antique layers or later spoliation, the picture of the forum's use (or lack thereof) during the sixth century is strikingly consistent, spanning all our sources. Instead of recording activities, sources highlight its splendour and historical importance, pointing to a wish to preserve the memory vessel precisely as it had always been.

While spoliation occurred in Rome's other *fora* in the sixth century, none is found in this forum (Kinney, 2013, 265–266). All sources point to it being well preserved. Nevertheless, as Rome changed, the *Forum Traiani* lost its relevance in everyday life. The possible poetry readings mentioned by Venantius are the only evidence of activity during the later sixth century, and their credibility is questionable. Even if they did occur, they were aimed at Rome's remaining elites and not the rest of the city's inhabitants. It was an elitist event in an elitist place, contrasting the daily life in the *Forum Romanum*. The insistence on keeping the *Forum Traiani* unchanged made it unadaptable, caught in a rigidity trap (Holling & Gunderson, 2002, 96). Even the connected *auditoria* seem to have been separated from the forum as they were adapted to the needs of Rome's actors and transformed into a mint.

The *Forum Traiani*'s less accessible architecture, compared to the *Forum Romanum*, possibly accelerated the process. Largely closed off by walls, the *Forum Traiani* became a place existing parallel to the city rather than as an integrated part. Its purpose became to crystallise the past. Paraphrasing Nora (1989, 12), the nostalgic persistence of memory replaced living memory. The *Forum Traiani* became the objectified fear of forgetting – a *lieu de mémoire* – detached from daily life. Was this what Cassiodorus alluded to as he continued his description of its splendour '*sed numquid per ea vivitur aut corporis salus aliqua inde delectatione recreatur*' (*Var*. 7.6)? Ultimately, neither rulers' wishes to restore, nor aristocratic restorations ensured the *fora*'s survival. Only people using them actively renewed their relevance. The *Forum Traiani* reached a tipping point. There was no way back as it became superfluous in the evolving city. Keeping the forum untainted and clean in its historicity – the exact thing meant to preserve it – became its downfall.

13.7 Conclusion

The *Forum Romanum*'s sixth-century development could be labelled as 'decline'. Yet the above examination suggests a more complex situation. It indicates that the forum was adapted to the changing times and Christianity, maintaining its relevance to Rome's actors. Rulers expressed the wish to maintain Rome, and sometimes aristocrats executed their will. However, what started as a top-down process was continued by other strata of Roman society. Changes were coming, accelerated by wars, natural disasters, and tiny bacteria. Considering the emperors' problems, they did not prioritise the funds necessary to uphold the symbolic spaces. Had *Forum Romanum*'s adaptation only been a top-down process, it would likely have dwindled. The forum would have been abandoned as rulers and aristocrats concentrated on other spaces, including Rome's Christian basilicas. This did not happen as Rome's well-off seized the space for their purposes, demonstrating how they could deal with change as windows of opportunity. Contrastingly, the *Forum Traiani* was shaped by an elitist wish for it to stay unchanged, forever crystallising Roman memory. Rome's inhabitants accepted this, taking their businesses elsewhere. The forum became a ghost of the past, fantastically irrelevant in its splendour. It ended up just as dead as the emperor buried there. In the sources, the *Forum Traiani* appears pristine and clean, starkly contrasting the messy *Forum Romanum*, shaped by adaptation. Ironically, this 'messiness', which ensured its continued relevance and resilience, has often been deemed a symptom of the forum's decline. Maybe it is time to reconsider.

Note

1 I wish to thank Professor Mark Humphries for commenting on an early draft. I would also like to thank Dr Christian Thrue Djurslev, who was a great help in analysing Venantius Fortunatus's poems. Dr Carlos Machado and Professor Rebecca Sweetman also deserve sincere thanks for letting me be a part of this publication. I also want to thank the anonymous reviewers, who helped me improve the chapter immensely. However, I especially wish to thank Professor Simon Malmberg, who took the time to discuss every nerdy detail and challenge my conclusions. Thank you.

Editions and abbreviations

AASS – *Acta Sanctorum*

Amm, Marc – Ammianus Marcellinus, *Res Gestae*. In *Ammianus Marcellinus*, 3 vols. (Loeb Classical Library), trans. J. C. Rolfe (1982). Harvard University Press.

Anon., Val. – *Anonymus Valesianus*. In *Ammianus Marcellinus*, 3 vols. (Loeb Classical Library), trans. J. C. Rolfe (pp. 508–569) (1982). Harvard University Press.

Boet., *De cons. Phil*. – *Boethius, Consolations of Philosophy* (Loeb Classical Library), trans. H. F. Stewart and E. K. Rand (1966). Harvard University Press.

Cass., *Chron*. – Cassiodorus, *Chronica*. In *Chronica Minora 2* (MGH AA 11.2), ed. T. Mommsen (pp. 109–161) (1894). Weidmann.

Cass., *Var*. – Cassiodorus, *Variae*. In *Cassiodori Senatoris Variae* (MGH AA 12), ed. T. Mommsen (1894). Weidmann. Translated in M. Shane Bjornlie (2019). *The Variae. The Complete Translation*. University of California Press.

CIL – *Corpus Inscriptiones Latinarum*.

Coll. Avell. – *Collectio Avellana*. In *Epistulae Imperatorum Pontificum*, 2 vols. (CSEL 35), ed. O. Günther (1895). Verlag der Österreichischen Akademie der Wissenschaften.

Edict. Theod – *Edictum Theodorici*. In *Monumenta Germaniae Historica. Leges Nationum Germanicarum* 5, ed. F. Bluhme (1889) (pp. 149–170). Hahn. Repr. P. L. Falaschi (1966). *Edictum Theoderici Regis Italiae*. Giuffrè.

Gesta Senatus – *Gesta Senatus Romani de Theodosiano publicando*. In *Theodosiani Libri XVI cum constitutionibus Sirmondianis*, ed. T. Mommsen (pp. 1–4) (1905). Weidmann.

Greg. Mag., *App*. and *Ep*. – *S. Gregorii Magni, Registrum Epistolarum*, 2 vols. (CSSL 140–140a), ed. D. Norberg (1982). Brepols. Translated in J. R. C. Martyn (2004). *The Letters of Gregory the Great* (3 vols.). Pontifical Institute of Medieval Studies.

Greg. Tour., *Glor. Mart* – Gregory of Tours, *Liber in Gloria Martyrum*. In *Gregory of Tours: The Glory of the Martyrs* (TTH), trans. R. Van Dam (1988). Liverpool University Press.

Greg. Tour., *Hist. Franc.* – Gregory of Tours, *Historia Francorum*. In *Gregorii Episcopi Turonensis Libri Historiarum X* (MGH Script. R. Mer. 1.1), ed. B. Krusch & W. Levison (1951). Hahn. Translated in L. Thorpe (1974) *Gregory of Tours: The History of the Franks*. Penguin Books.

ICUR n.s. I – Inscriptiones Christianae Urbis Romae, nova series, vol. 1: Inscriptiones incertae origines, ed. A. Silvagni (1922). Ex officina libraria Doct. Befani.

Isidore of Seville, *Hist. Goth. Wand. Sue.* – Isidore of Seville, *Historia Gothorum Wandalorum Sueborum*. In Chronica Minora saec. IV, V, VI, VII (MGH AA 11.2), ed. T. Mommsen (1894) (pp. 241–303). Weidmann. Translated in G. Donini & G. B. Ford. *Isidore of Seville's History of the Kings of the Goths, Vandals, and Suevi* (1966). Leiden: BRILL.

Just., *Nov.* – Justinian, *Novellae*. In *Corpus Iuris Civilis*, vol. 3, ed. P. Krüger (1895). Weidmann. Translated in D. J. D. Miller & P. Sarris (2018) *The Novels of Justinian. A Complete Annotated English Translation*. Cambridge University Press.

Lib. Pont. – *Liber Pontificalis*. In *Le Liber Pontificalis*, 2. Vols., ed. L. Duchesne (1886). Ernest Thorin. Translated in R. Davis (1989). *The Book of Pontiffs: The Ancient Biographies of the First Ninety Roman Bishops to AD 715* (TTH). Liverpool University Press.

LSA – *The Last Statues of Antiquity* online database. http://laststatues.classics.ox.ac.uk

Procopius, *Bell.* – Procopius, *De Bellis*. In *The Wars,* 5 vols. (Loeb Classical Library), trans. H.B. Dewing (1914–1928). Harvard University Press. Revised and modernised in H. B. Dewing & A. Kaldellis. *The Wars of Justinian* (2014). Hackett Publishing Company.

Suet., *Aug.* – Suetonius, *De Vita Caesarum. Divus Augustus*. In *Suetonius, Lives of the Caesars*, 2 vols. (Loeb Classical Library), trans. J. C. Rolfe (1950–1951). Harvard University Press.

Venantius Fortunatus, *Carm.* – Venantius Fortunatus, *Carmina*. In *Venanti Honori Clementiani Fortunati presbyteri Italici opera poetica* (*MGH AA* 4.1), ed. Friedrich Leo (1881). Weidmann, trans. J. George. *Venantius Fortunatus: Personal and Political Poems* (TTH) (1995). Liverpool University Press.

Bibliography

Arnold, J. J. (2014). *Theoderic and the Roman imperial restoration*. Cambridge University Press.

Arnold, J. J. (2017). Theodoric and Rome: Conquered but unconquered. *Antiquité Tardive*, *25*, 113–126.

Augenti, A. (2001). *Il Palatino nel medioevo: Archeologia e topografia (Secoli VI-XIII)*. L'Erma Di Bretschneider.

Barnish, S. J. B. (1998). Transformation and survival in the Western senatorial aristocracy, c. A.D. 400–700. *Papers of the British School at Rome*, *56*, 120–155.

Bartoli, A. (1963). *Curia Senatus. Lo scavo e il restauro*. Instituto di Studi Romani.

Bauer, F. A. (1983). Porticus Absidata. *Römische Mitteilungen*, *90*, 111–184.

Bauer, F. A. (1996). *Stadt, Platz und Denkmal in der Spätantike*. von Zabern.

Berkes, F., & Ross, H. (2013). Community resilience: Toward an integrated approach. *Society and Natural Resources*, *26*(1), 5–20.

Bianchi, E., & Meneghini, R. (2021). L'architettura del Foro di Traiano a nord della Basilica Ulpia. *Bullettino della Commissione Archeologica Comunale di Roma*, *122*, 183–202.

Bjornlie, M. S. (2013). *Politics and tradition between Rome, Ravenna and Constantinople*. Cambridge University Press.

Boni, G. (1901). Il sacrario di Giuturna. In *Notizie degli scavi di Antichità: Anno 1901* (pp. 41–144). Col R. Istituto diArcheologia e Storia dell'Arte.

Boni, G. (1904). Foro Romano. In *Atti del Congresso Internazionale di Scienze Storiche (Roma, 1–9 Aprile 1903) V, sez. IV* (pp. 493–584). Tipografia della R. Accademia dei Lincei.

Bordi, G. (2021). The apse wall of Santa Maria Antiqua (IV-IX centuries). In E. Rubery, G. Bordi, & J. Osborne (Eds.), *Santa Maria Antiqua: The Sistine Chapel of the early middle ages* (pp. 387–422). Harvey Miller Publishers.

Bossi, S. (2017). Aedes, atrium e horreum di Vesta. In A. Carandini, P. Carafa, M. T. D'Alessio, & D. Filippi (Eds.), *Santuario di Vesta, pendice del Palatino e Via Sacra* (pp. 725–754). Edizioni Quasar.

Buonocore, M. (1997). *Appunti di topografia romana nei Codici Lanciani della Biblioteca Apostolica Vaticana: Vol. II*. Biblioteca Apostolica Vaticana.

Burton Brown, E. (1904). *Recent excavations in the Roman Forum, 1898–1904*. Murray.

Capponi, M., & Ghilardi, M. (2002). Scoperta, nel Templum Pacis, di un'area sepolcrale probabilmente contemporanea alla fondazione dei SS. Cosma e Damiano. In F. Guidobaldi & A. G. Guidobaldi (Eds.), *Ecclesiae Urbis* (Vol. 1, pp. 733–756). Pontificio Istituto di Archeologia Cristiana.

Carandini, A. (Ed.). (2012). *Atlante di Roma Antica* (2 vols.). Mondadori Electa.

Chenault, R. (2012). Statues of senators in the Forum of Trajan and the Roman forum in late antiquity. *Journal of Roman Studies*, *102*, 103–132.

Christie, N. (2020). From royalty to refugees: Looking for the people in reconstructing urban change in late antique Italy. In A. Carneiro, N. Christie, & D. Blasco (Eds.), *Urban transformations in the Late Antique West: Materials* (pp. 323–342). Imprensa da Universidade de Coimbra.

Coarelli, F. (2008). *Guida archeologica di Roma*. Mondadori.

Coarelli, F. (2014). *Rome and environs: An archaeological guide*. University of California Press.

Coates-Stephens, R. (2006). Byzantine building patronage in post-reconquest Rome. In M. Ghilardi, C. J. Goddard, & Porena (Eds.), *Les cités de l'Italie tardo-antique (IVe-VIe siècle)* (pp. 149–166). École Française de Rome.

Coates-Stephens, R. (2011). The Forum Romanum in the Byzantine period. In O. Brandt & P. Pergola (Eds.), *Marmoribus Vestita* (2 vols., pp. 385–408). Pontificio Istituto di Archeologia Cristiana.

Coates-Stephens, R. (2021). The 'oratory of the forty Martyrs'. In E. Rubery, G. Bordi, & J. Osborne (Eds.), *Santa Maria Antiqua: The Sistine Chapel of the early middle ages* (pp. 195–211). Harvey Miller Publishers.

Costambeys, M. (2001). Burial topography and the power of the Church in fifth- and sixth-century Rome. *Papers of the British School at Rome*, *69*, 169–189.

Davis, J. L. (2007). Memory groups and the state. In N. Yoffee (Ed.), *Negotiating the past in the past* (pp. 227–256). University of Arizona Press.

Dey, H. (2021). *The making of medieval Rome*. Cambridge University Press.

Fauvinet-Ranson, V. (2006). *Decor civitatis, decor Italiae: Monuments, travaux publics et spectacles au VIe siecle*. Edipuglia.

Fea, C. (1821). *Sopra le Terme Tauriane, il Tempio di Venere e Roma, il Foro di Domiziano e d'Augusto*. Francesco Burlié.

Filippi, D. (2001). L'Atrium Vestae in età Tardoantica. In M. S. Arena, P. Delogu, L. Paroli, M. Ricci, L. Saguì, & L. Vendittelli (Eds.), *Roma dall'Antichità al Medioevo. Archeologia e Storia* (pp. 599–601). Electa.

Filippi, D., Ricci, G., Giuseppe, H. D., Capelli, C., & Delussi, F. (2004). La Casa delle Vestali: Un immondezzaio di VI secolo d.C. In L. Paroli & L. Vendittelli (Eds.), *Roma dall'Antichità al Medioevo II* (pp. 164–179). Ministero per i Beni e le Attività Culturali Soprintendenza Archeologica di Roma.

Fraschetti, A. (1999). *La Conversione. Da Roma pagana a Roma cristiana*. Laterza.

Freyberger, K. S., Ertel, C., & Bitterer, T. (2007). Neue Forschungen zur Basilica Aemilia auf dem Forum Romanum. *Römische Mitteilungen*, *113*, 493–552.

Galli, M. (2013). L'Athenaeum di Adriano. *Bolletino di Archeologia On Line. Direzione Generale per Le Antichitá*, *IV*, 60–71.

George, J. (1995). *Venantius fortunatus: Personal and political poems*. Liverpool University Press.

Giuliani, C. F., & Verduchi, P. (1987). *L'Area centrale del Foro Romano*. Olschki.

Goodson, C. (2021). *Cultivating the city in Early Medieval Italy*. Cambridge University Press.

Guidobaldi, F., & Guidobaldi, A. G. (1983). *Pavimenti marmorei di Roma dal IV al IX secolo*. Pontificio Istituto di Archeologia Cristiana.

Gulowsen, K. (2001). The oratory of the Forty Martyrs: From imperial ante-vestibule to Christian room of worship. In J. R. Brandt & O. Steen (Eds.), *Imperial art as Christian art, Christian art as imperial art, acta ad Archaeologiam et Artium Historiam Pertinentia* (Vol. 15, pp. 77–91). Bardi Editore.

Häkli, J., & Paasi, A. (2003). Geography, space and identity. In J. Öhman & K. Simonsen (Eds.), *Voices From the North: New Trends in Nordic Human Geography* (pp. 141–155). Ashgate.

Halbwachs, M. (1950). *La mémoire collective*. Les Presses Universitaires de France.

Halbwachs, M. (2008). *La topographie légendaire des Evangiles en Terre sainte*. Presses Universitaires de France.

Hekster, O. (2020a). Ruling through religion? Innovation and tradition in imperial representation. In R. Dijkstra (Ed.), *The early reception and appropriation of the Apostle Peter (60–800 CE)* (pp. 26–40). Brill.

Hekster, O. (2020b). Anchoring political change: Adaptive government in the classical world. *Journal of Comparative Politics*, *13*, 99–107.

Hirst, W., & Manier, D. (2008). Towards a psychology of collective memory. *Memory*, *16*(3), 183–200.

Hölkeskamp, K.-J., & Stein-Hölkeskamp, E. (Eds.). (2006). *Erinnerungsorte der Antike*. Beck C. H.

Holling, C. S., & Gunderson, L. H. (2002). Resilience and adaptive cycles. In L. H. Gunderson & C. S. Holling (Eds.), *Panarchy: Understanding transformations in human and natural systems* (pp. 25–62). Island Press.

Hurst, H. (2021). The early Church of Santa Maria Antiqua. In E. Rubery, G. Bordi, & J. Osborne (Eds.), *Santa Maria Antiqua: The Sistine Chapel of the early middle ages* (pp. 161–176). Harvey Miller Publishers.

Ippoliti, M. (2017). Horreum Vestae (75–600 d.C). In A. Carandini, P. Carafa, M. T. D'Alessio, & D. Filippi (Eds.), *Santuario di Vesta, pendice del Palatino e Via Sacra* (pp. 757–775). Edizioni Quasar.

Johnson, M. J. (2016). Art and architecture. In J. Arnold, S. Bjornlie, & K. Sessa (Eds.), *A companion to Ostrogothic Italy* (pp. 350–389). Brill.

Kalas, G. (2015). *Restoration of the Roman forum in late antiquity*. University of Texas Press.

Keddie, A. (2019). *Class and power in roman Palestine*. Cambridge University Press.

Kinney, D. (2013). Spoliation in medieval Rome. In S. Altekamp, C. Marcks-Jacobs, & Seiler (Eds.), *Perspektiven der Spolienforschung 1* (pp. 261–285). De Gruyter.

Knapp, R. (2011). *Invisible Romans*. Harvard University Press.

Krautheimer, R., Corbett, S., & Frazer, A. K. (1937). *Corpus Basilicarum Christianarum Romae. Vol. I*. Pontificio istituto di Archeologia Cristiana.

Krautheimer, R., Corbett, S., & Frazer, A. K. (1959). *Corpus Basilicarum Christianarum Romae. Vol. II*. Pontificio istituto di Archeologia Cristiana.

Lafferty, S. D. W. (2010). Law and society in ostrogothic Italy: Evidence from the *Edictum Theoderici*. *Journal of Late Antiquity*, *3*(2), 337–364.

La Rocca, C. (1993). Una Prudente Maschera "Antiqua". La Politica Edilizia di Teoderico. In *Atti del XIII Congresso internazionale di studi sull'Alto Medioevo. Milano, 2–6 novembre 1992* (pp. 451–515). Centro Italiano di Studi sull'alto Medioevo.

La Rocca, C. (2014). *Mores tuos fabricae loquuntur*. Building activity and the rhetoric of power in Ostrogothic Italy. *Haskins Society Journal*, *26*, 1–49.

Lavan, L. (2021). *Public space in the late Antique City* (2 Vols.). Brill.

Le Pera, S. B., & D'Elia, L. (1986). "Sacra via": Note topografiche. *Bullettino della Commissione Archeologica Comunale di Roma*, *91*(2), 241–262.

Liebs, D. (1985). Promulgationslokale im Spätantiken Rom. In J. A. Ankum, J. E. Spruit, & F. B. J. Wubbe (Eds.), *Satura Roberto Feenstra: Sexagesimvm qvintvm Annvm Aetatis Complenti ab Alvmnis Collegis Amicis Oblata* (pp. 215–228). Universitätsverlag Fribourg.

Liverani, P. (2007). Osservazione sui Rostri del Foro Romano in età Tardoantica. In A. Leone, D. Palombi, & S. Walker (Eds.), *Res bene gestae. Ricerche di storia urbana su Roma Antica in onore di Eva Margareta Steinby* (pp. 169–193). Edizioni Quasar.

Machado, C. (2010). Public monuments and civic life: The end of the statue habit in Italy. In Delogu & S. Gasparri (Eds.), *Le trasformazioni del V secolo. L'Italia, i barbari e l'Occidente romano* (pp. 238–257). Brepols.

Machado, C. (2019). *Urban space and aristocratic power in Late Antique Rome*. Oxford University Press.

Machado, C. (2021). Statue habit and statue culture in Late Antique Rome. *Journal of Roman Archaeology, 34*, 632–666.
Maetzke, G. (1991). La struttura stratigrafica dell'area nordoccidentale del Foro Romano come appare dai recenti interventi di scavi. *Archeologia Medievale, 18*, 43–200.
Malmberg, S., Seland, E., & Aavitsland, K. (Eds.). (2024). *City, hinterland and environment: Urban resilience during the first millennium transition. Acta ad Archaeologiam et Artivm historiam Pertinetia* 34 (n.s. 20). Bardi Edizioni.
Marchetti Longhi, G. (1949). Senatus ad Palmam, Porticus Curva e Porticus Absidata. *Rendiconti della Pontificia Accademia Romana di Archeologia, 25–26*, 183–229.
Marrou, H.-I. (1932). La vie intellectuelle au Forum de Trajan et au Forum d'Auguste. *Mélanges d'Archéologie et d'Histoire, 49*, 93–110.
Maskarinec, M. (2018). *City of Saints: Rebuilding Rome in the early middle ages*. University of Pennsylvania Press.
Matthews, J. F. (2000). *Laying down the law*. Yale University Press.
Meier, M. (2016). The 'Justinianic Plague': The economic consequences of the pandemic in the Eastern Roman Empire and its cultural and religious effects. *Early Medieval Europe, 24*(3), 267–292.
Meneghini, R. (2009). *I Fori Imperiali e i Mercati di Traiano: Storia e Descrizione dei Monumenti alla Luce degli Studi e degli Scavi*. Istituto Poligrafico dello Stato.
Meneghini, R., & Santangeli Valenzani, R. (1993). Sepolture intramuranee e paesaggio urbano a Roma tra V e VII secolo. In L. Paroli & P. Delogu (Eds.), *La storia economica di Roma nell'alto Medioevo alla luce dei recenti scavi archeologici* (pp. 89–111). All'Insegna del Giglio.
Meneghini, R., & Santangeli Valenzani, R. (2007). *I Fori Imperiali. Gli scavi del Comune di Roma (1991–2007)*. Viviani.
Molinari, M. C. (2004). Le monete dell'ambiente D. In L. Paroli & L. Vendittelli (Eds.), *Roma dall'Antichità al Medioevo II* (pp. 204–219). Ministero per i Beni e le Attività Culturali Soprintendenza Archeologica di Roma.
Mortera, A. (2017). Trasformazioni del paesaggio urbano nell'area del Foro Romano alle soglie del Medioevo: Il caso della Basilica Aemilia. In *III Ciclo di Studi Medievali. Atti del Convegno 8–10 settembre 2017 (Firenze)* (pp. 343–364). Nume.
Nora, P. (1989). Between memory and history: Les lieux de mémoire (M. Roudebush, Trans.). *Representations, 26*, 7–24.
Örmä, S. (1989). Le monete del deposito Boni. In E. M. Steinby (Ed.), *Lacus Iuturnae I* (pp. 303–316). De Luca.
Örmä, S. (2012a). Lucerne. In E. M. Steinby (Ed.), *Lacus Iuturnae II.2.* (pp. 137–142). Institutum Romanum Finlandiae.
Örmä, S. (2012b). Monete. In E. M. Steinby (Ed.), *Lacus Iuturnae II.2.* (pp. 305–307). Institutum Romanum Finlandiae.
Packer, J. E. (1997). *The forum of Trajan in Rome: A study of the monuments* (3 vols.). University of California Press.
Paganelli, M. (2004). Area N-O del Foro Romano: L'ambiente D nell'alto Medioevo. In L. Paroli & L. Vendittelli (Eds.), *Roma dall'Antichità al Medioevo II* (pp. 180–203). Ministero per i Beni e le Attività Culturali Soprintendenza Archeologica di Roma.
Palombi, D. (1988). Contributo alla topografia della Via Sacra, dagli appunti inediti di Giacomo Boni. *Topografia Romana. Ricerche e Discussioni, 10*, 77–97.
Papi, E. (1996). Horrea vespasiani. In E. M. Steinby (Ed.), *Lexicon Topographicum Urbis Romae, vol. 5 (H-O)* (pp. 49–50). Edizioni Quasar.
Paribeni, R. (1933). Iscrizioni dei Fori Imperiali. In G. Bardi (Ed.), *Notizie degli scavi di Antichità: Vol. IX* (pp. 431–523). Col R. Istituto diArcheologia e Storia dell'Arte.
Redman, C., & Kinzig, A. (2003). Resilience of past landscapes: Resilience theory, society, and the longue durée. *Conservation Ecology, 7*(1), 1–19.
Ricci, C. (1931). Esplorazione archeologica delle Cantine a Macel de' Corvi. *Bulletino della Commissione Archaeologica Comunale di Roma, 59*, 117–22.

Ricci, G. C. (1994). Tagliamonte, Iscrizioni cristiane nelle Collezioni Comunali. Inediti e revisioni. *Bullettino della Commissione Archeologica Comunale di Roma*, *96*, 179–194.

Salzman, M. R. (2021). *The falls of Rome: Crises, resilience, and resurgence in Late Antiquity*. Cambridge University Press.

Serlorenzi, M., & Egidi, R. (2013). L'Athenaeum di Adriano. Storia di un edificio dalla fondazione al XVII secolo. Note conclusive. *Bolletino di Archeologia On line. Direzione Generale per le Antichitá*, *IV*, 192–198.

Sessa, K. (2019). The new environmental fall of Rome: A methodological consideration. *Journal of Late Antiquity*, *12*(1), 211–255.

Steinby, E. M. (1986). L'industria laterizia di Roma nel Tardo Impero. In A. Giardina (Ed.), *Società romana e impero tardoantico* (Vol. 2, pp. 99–164). Laterza.

Steinby, E. M. (2001). La Cronologia delle 'Figlinae' Tardoantiche. In M. Cecchelli (Ed.), *Materiali e Tecniche dell'Edilizia Paleocristiana a Roma* (pp. 127–150). Edizione De Luca.

Steinby, E. M. (Ed.). (2012). *Lacus Iuturnae II.1. Saggi degli anni 1982–85*. Institutum Romanum Finlandiae.

Storozhylova, Y. (2019). *Provenance and archaeometric analysis of late Roman glass from the Palatine Hill* [Master's thesis, The Sapienza University of Rome].

Talbert, R. J. A. (1984). *The senate of imperial Rome*. Princeton University Press.

Tea, E. (1937). *La basilica di Santa Maria Antiqua*. Vita e Pensiero.

Trifilò, F. (2012). *The Forum Romanum games project. Preliminary project report (20/03/2012)* (Report). The University of Kent. www.romansociety.org/Portals/0/Forum_Games_Project_01.pdf

Videbech, C. (2024). Christians, memory, and resilience at the Late Antique Forum Romanum. In S. Malmberg, E. Seland, & K. Aavitsland (Eds.), *City, hinterland and environment: Urban resilience during the first millennium transition. Acta ad Archaeologiam et Artivm historiam Pertinetia* 34 (n.s. 20). Bardi Edizioni.

Vitiello, M. (2004). Teoderico a Roma. Politica, amministrazione e propaganda nell 'adventus' dell'anno 500. *Historia*, *53*, 73–120.

Walker, B., & Salt, D. (2012). *Resilience practice: Building capacity to absorb disturbance and maintain function*. Island Press.

Weisweiler, J. (2012). From equality to asymmetry: Honorific statues, imperial power, and senatorial identity in Late-Antique Rome. *Journal of Roman Archaeology*, *25*, 319–350.

14 Lived spaces in late antique and early Islamic Jerash – old questions in the light of new evidence

Rubina Raja

Late antique and early Islamic city spaces in the Near East have over the past decades become the focus of intensified research, as archaeological work has developed and these often-complex spaces have been given more attention in archaeological fieldwork. However, research on these periods has often been ignored when reconstructing the bigger picture, while structures from the Roman period have attracted most attention due to their often monumental character and due to the traditional focus on the study of the incorporation of the East into the Roman Empire. As Bowden's chapter in this volume shows, this bias can also be observed for other areas of the ancient world. Furthermore, large-scale excavations of the late nineteenth and early twentieth centuries often cleared the post-classical phases without leaving much documentation about the late antique – apart from the churches – and Islamic periods. This contribution takes its point of departure in the published material from the Danish-German Jerash Northwest Quarter Project and revisits that evidence to extrapolate what we can learn about the lived spaces – the spaces inhabited by the society in Gerasa in the late antique and early Islamic periods – through this scattered and complex, but also crucial, evidence. The central questions are: what is the impact on our understanding of urban space in Gerasa/Jerash of this (slowly) growing archaeological interest for post-classical periods? And also, what has this shift in archaeological focus revealed about that society and its use and reuse of the spaces of the city?

Gerasa, also known as Antiochia on the Chrysorrhoas in the Roman period and Jerash/Jarash from the early Islamic period onwards, is one of the Near Eastern Decapolis cities, famous for its monumental ruins, mostly of the Roman and late antique periods (Figure 14.1) (Lichtenberger & Raja, 2015c, 2018f, 2020c; Raja, 2012). Gerasa was "rediscovered" and placed correctly on the map of the region by Jasper Ulrich Seetzen in 1806 (Seetzen, 1854). In the years thereafter its ruins were visited by numerous Western travellers on their Grand Tours (Lichtenberger, 2002). In the second and third centuries CE, Gerasa covered about 85 hectares and was circumvallated by city walls of about 4 kilometres in length (Raja, 2012, 137–190). The city held all the monuments one would expect of a middle-sized city of the Roman period in the East, including theatres, a basilica, numerous sanctuaries, a *macellum* and a colonnaded street 1.2 kilometres long (Figures 14.2 and 14.3). The city was even embellished with a small hippodrome (Kehrberg, 2017), a monument that was not very common in urban landscapes of the Near East. From the late fifth century onwards, numerous churches were also constructed in and around the city (Crowfoot, 1931, 1938, Michel, 2001, 224–274). The city also flourished in the early Islamic period until it was hit by the devastating earthquake of January 749 CE (Lichtenberger & Raja, 2019a, 2019c, Marco et al., 2003; Tsafrir & Foerster, 1992). The city contracted considerably afterwards, but activity can still be traced – not to the extent of earlier periods, and evidence for substantial habitation only picks up again in the middle Islamic period

DOI: 10.4324/9780429427152-18

Figure 14.1 View of the south part of Gerasa with the Oval Piazza, the Sanctuary of Zeus Olympios, the South Theatre as well as the hippodrome.

Source: Author's image

(Blanke, 2022; Kalaitzoglou, 2018; Lichtenberger et al., 2021; Lichtenberger & Raja, 2016, 2018c, 2018d, 2018e; Rattenborg & Blanke, 2017).

The site was the focus of much research since archaeological exploration began in 1907, when soundings and smaller trial excavations were undertaken by a German team (Mortensen, 2017, 2018; Lichtenberger & Raja, 2020c). They did not publish their findings to any great extent, and the excavation archive and some material is still in the *Archiv der Zentrale* at the main seat of the *Deutsches Archäologisches Institut* in Berlin. For many years Gottlieb Schumacher's (1902) plan was the most precise map of the site's topography and its then-known complexes (Figure 14.2). Extensive archaeological exploration at Gerasa was only initiated in the 1920s by first a joint American and British expedition and later an exclusively American mission (Kraeling, 1938), which focused almost exclusively on the Roman and Christian monuments of the city. World War II had put a temporary halt to archaeological work in the region, and it was only with the extensive Jerash Archaeological Project in the 1980s that research at the site was commenced again, with a set of large-scale excavation projects, many of which are still awaiting final publication (Zayadine, 1986; also see *Syria. Archéologie, art et histoire*, vol. 66). Over the course of the past decades, archaeological research at the site has continued in a set of projects focusing on various areas of the city and on particular chronological phases (Lichtenberger & Raja, 2021a; Seigne, 1985, 1997, 2018; Blanke, 2016, 2018, 2022; historiographic overview of archaeological research at the site, Lichtenberger & Raja, 2020b, 2020c; Brody & Raja, 2022, for a new project on the archives at Yale University Art Gallery).

While much has been published about the site, most research has focused on the Roman-period development and monuments, as is often the case with classical sites in the Near East

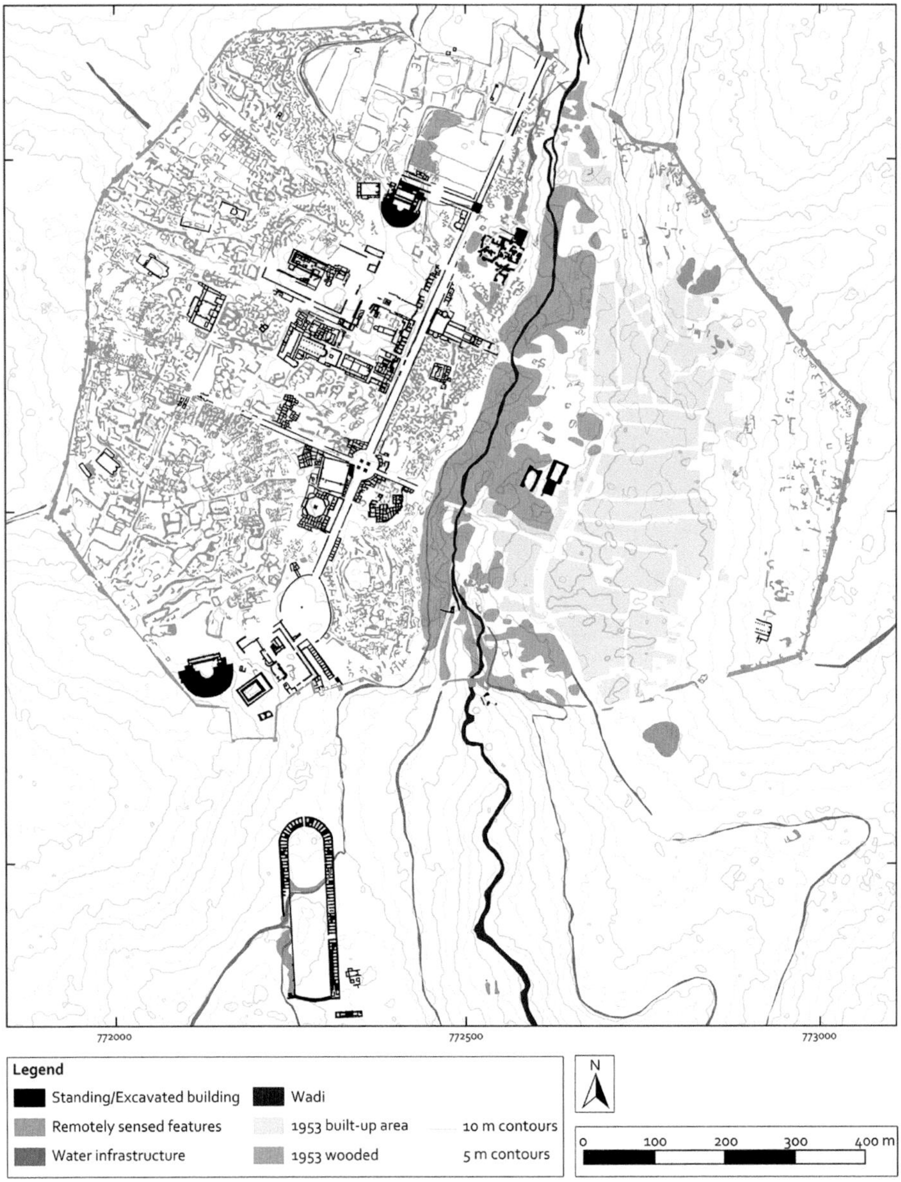

Figure 14.2 The most up-to-date map of Gerasa was done within the framework of the Danish-German Jerash Northwest Quarter Project.

Source: Courtesy of the Danish-German Jerash Northwest Quarter Project

(Lichtenberger & Raja, 2018a, 2018c, 2019b, 2020b; Blanke, 2022). Only over the past four decades has research picked up on the post-classical and Islamic periods, and these can now begin to be seen in a different light (Barnes et al., 2006; Blanke, 2016, 2018; Blanke et al., 2007, 2010; Blanke & Walmsley, 2010; Rattenborg & Blanke, 2017; Walmsley, 2018). Following these recent developments, the aim of this chapter is to highlight new conclusions drawn from already-published evidence.

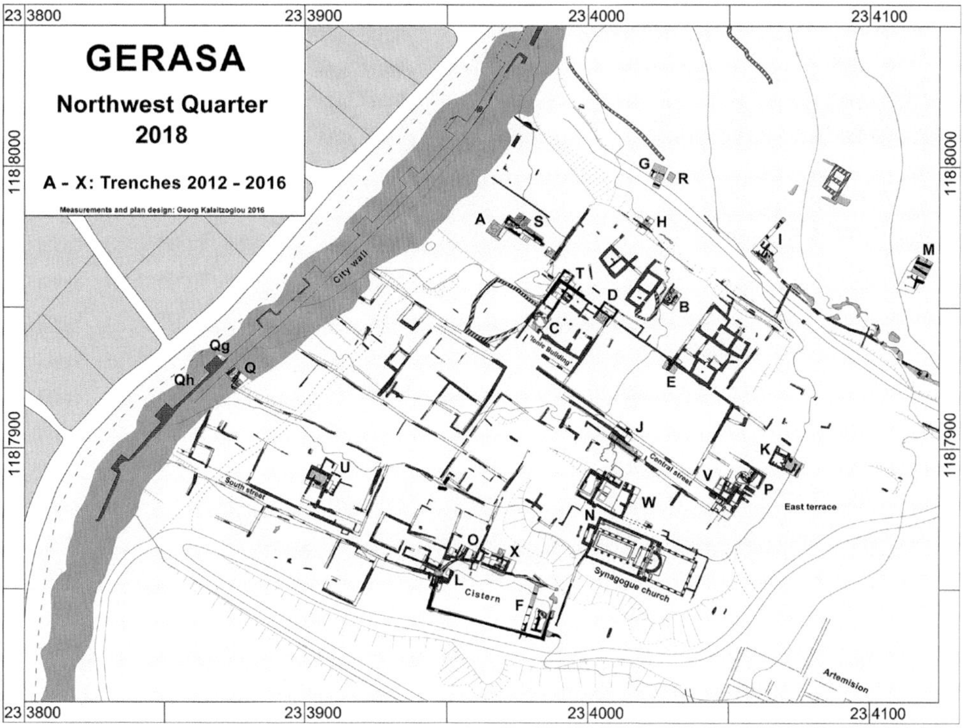

Figure 14.3 Map of the Northwest Quarter in Gerasa with all trenches excavated between 2012 and 2016 marked with letters.

Note: These trenches are referred to by their letters in the text of this contribution, but a figure reference is not given every time.

Source: Courtesy of the Danish-German Jerash Northwest Quarter Project

14.1 Late antique and Islamic archaeology at Jerash/Jarash in the light of new research

When the Jerash Archaeological Project started in the 1980s, with its greater interest for the late antique and Islamic remains, one of the issues encountered was the clearance of extensive post-antique phases by the large-scale excavations of the early twentieth century (Figure 14.4) (Kraeling, 1938). A great deal of knowledge was lost during these clearings: their scale was comparable to the brutal clearings by French archaeologists in Palmyra in the late 1920s and early 1930s of, for example, the Sanctuary of Bel, when an entire village was demolished and the locals were forcibly moved to new housing away from the ancient site (Raja, 2022a, 124; Raja et al., 2021; Baird et al., 2023). In the two volumes that came out of the Jerash Archaeological Project, four contributions in the first volume address post-antique findings (Clark, 1986; Gawlikowski, 1986; Gawlikowski & Musa, 1986; Schaefer & Faulkner, 1986), and in the second volume three contributions concern post-antique periods (Naghawi, 1989; Rasson & Seigne, 1989; Watson, 1989), while a further one includes observations on the post-antique phases of the hippodrome (Ostrasz, 1989). Although post-antique material was considered in a reasonable number of studies, none of them presented a synthesis of post-antique Jerash. The

Figure 14.4 Large-scale clearings undertaken by the British-American team in the vicinity of the Sanctuary of Artemis.

Note: The temple is visible to the top right in the photo.

Source: Courtesy of Yale University Art Gallery

articles in these volumes are in general project-driven and do not anchor the archaeological findings in an overarching narrative of the city or the region.[1] So even after the Jerash Archaeological Project, the publication edited by Kraeling and published in 1938 still provided the best overview of the site's development until the Arab invasion (Crowfoot, 1938). This imbalance has begun to be redressed with the increased focus on (particularly) early Islamic finds of the Islamic Jarash Project (directed by Alan Walmsley), and more recently of the late antique Jarash Project, directed by Louise Blanke (Blanke, 2016, 2018, 2022; Blanke & Walmsley, 2022). In fact, more recent contributions have sought to sum up some of the evidence for the late antique and early Islamic periods, presenting a more general picture (Blanke & Walmsley, 2022; Rattenborg & Blanke, 2017).

In the past decade, late antique and early Islamic evidence from the Northwest Quarter has been published in great detail, adding further information about this part of the city, an approximately 4 hectares large area. This evidence includes several complexes with oil presses (Kalaitzoglou et al., 2013, 63–68, 2015, 19–26, 2022a, 20–26), a cistern which was used as a production and habitation space during the late antique period (Kalaitzoglou et al., 2014, 22–29, 2015, 31–37, 2022a, 33–40, 2022b, 169–175), a building with mosaic floors (called the Mosaic Hall) that was connected to the so-called Synagogue Church (Kalaitzoglou et al., 2022a, 26–33, 2022b, 163–169), as well as a set of early Islamic houses, which were all destroyed in the earthquake that hit the city in January 749 (Kalaitzoglou et al., 2015, 26–31, 2022a, 40–46, 2022b, 153–163). Furthermore, numerous articles on various findings and complexes have been published (Lichtenberger & Raja, 2021a for a bibliography), and until now four volumes of the final publications have appeared (Lichtenberger & Raja, 2020a, 2020d, 2021b, 2022a). These publications have addressed all periods encountered in the archaeological evidence from the Northwest Quarter of the city – including those of the post-classical periods – allowing researchers to access all data from these excavations. All these spaces were in use one way or another and underwent several transformations between the late Roman period and the date when the earthquake struck. Being dynamic spaces, where neighbourhoods could be massively changed for various reasons that might not always be clear to us, they all raise questions about lived spaces in the late antique and early Islamic city.

So where does that leave us, when we want to investigate late antique and Islamic Jerash in terms of its spaces – the lived spaces of the city? Such a study will necessarily be influenced by the lack of publications of some of the excavations already undertaken, and it is difficult to conduct comparative studies of various spaces within the city due to the unequal standards of publication of the evidence. One solution for this situation is to focus on lived spaces, while accepting that we are only ever able to address specific snapshots of a larger and complex picture. It has certainly become clear, over the past decades, that urban life in late antique and Islamic Jerash was richer in terms of material culture than earlier thought. But how must that, in turn, change our view of these periods in relation to the until now known lived spaces in the city, such as that of the early Islamic mosque for example?

14.2 Tracing lived spaces in Jerash/Jarash

Since 2012, the term "Lived Religion", in the sense of the experience and expression of religion seen from the perspective of the individual or group who participated in it – and not through an institutional lens – has become a concept which methodologically and theoretically has pushed forward agendas beyond the history of religion, with deep implications for archaeology as well (McGuire, 2008; Raja & Rüpke, 2015). Now the concept, with its emphasis on the dimension of what is *lived*, has also entered the realm of everyday life in antiquity and beyond, urging us not only to take closer look at cities, their layouts and single complexes but also to address in much greater detail the fine-grained evidence from which we might tease much more knowledge and thus broaden our understanding of the people and the organisation of urban space (Latzer-Läsar et al., 2020; Rau & Rüpke, 2020; Raja & Sindbæk, 2018, 2020).

The concept of "Lived Religion" was originally developed for the study of contemporary religion by Meredith McGuire (2008). This approach does not take its point of departure in religious organisations or belief systems, but rather it begins with the individual's use of religion, taking its cue from everyday experiences, practices, expressions and interactions, which can in turn be connected to religion. In this way, "religion" is to be understood as a range of experiences and actions, as well as beliefs (Raja & Rüpke, 2015; Rüpke et al., 2015). The term "Lived Space" can, in principle, mean the same, but it covers a much wider range of everyday life, including realms beyond religious behaviour (Raja, 2022b; Raja & Rüpke, 2021a). While we might say that this is what we always study in archaeology and ancient history, namely the people of the past, we often tend to forget to put the individual at the centre of the narrative, and we often lose ourselves in the description of the evidence, in its recording and contextualisation, without even mentioning the people who interacted with the materials we find (Raja & Rüpke, 2021b; see Bowden's remarks in this volume). Productive new discussions of lived spaces, which feed into the approach advocated in this chapter, are also found in other contributions in this volume, including those by Michelle L. Berenfeld, Carlos Machado and Christina Videbech. They, however, all focus on the late antique period in different spaces in the city of Rome.

14.3 Lived spaces and oil presses: Roman and late antique agricultural production complexes and urban spatial reorganisation

The hinterland of Jerash was fertile, and numerous oil presses were found, attesting to oil production both inside and outside the city (Kalaitzoglou et al., 2013, 2022a, 2022b; Lichtenberger et al., 2013 also for further references). Some of these presses were visible in photos from the excavations published in the volume edited by Kraeling (1938), and others have been published by later teams working at the site, including the French mission (Seigne, 2013). It is not surprising that Jerash would have had numerous oil presses. Oil manufacturing was one of the core productions in ancient Mediterranean cities, Jerash being no exception.

Two complexes with oil presses were excavated in the Northwest Quarter. The first complex was excavated in an area that had suffered from modern disturbances and where a large architectural element with altar iconography was visible above ground (Trench B: Kalaitzoglou et al., 2013, 63–68; Lichtenberger & Raja, 2015c, 491–493; Lichtenberger et al., 2013, 12–13). The Roman monumental architectural piece originated from a sacred space, and it had been reused as part of the late Roman press (Lichtenberger & Raja, 2015a, 2022b). At some point in the late antique period, the complex in which the press stood was entirely filled in with a dense fill consisting almost only of late Roman and late antique ceramic fragments, among which a large number of fragments from Jerash Bowls were present. The second complex was found in an underground cave (Trench J: Kalaitzoglou et al., 2015, 19–26, 2022a, 20–26; Lichtenberger et al., 2015a, 46–48), with an oil press. When the central street in the Northwest Quarter was laid out, in the late Roman period, the northern entrance to the cave, including a staircase to the complex, was blocked with a dense fill consisting of a lot of ceramic waste, much alike the situation in trench B.[2] The southern entrance remained open for some time until the Mosaic Hall was constructed in the sixth century (see below).

It is not possible to say why exactly these two complexes fell out of use, based on the archaeological evidence, and we can only speculate about the reasons for it. This abandonment consisted of a massive reorganisation of space on the hill, possibly caused by earthquake destruction and/or the lack of water resources in the Northwest Quarter, caused by broader changes to the overall water supply of the city, indicated by the backfilling of a large cistern on the top of the hill in the fourth century (trenches A and S; Kalaitzoglou et al., 2013, 58–63, 2022b, 134–141; Lichtenberger & Raja, 2015b).[3] A large-scale transformation was undertaken, and it is clear that such a project would have required organisation, most likely also people in charge of the task and manpower to move filling material and implement the closing of the complexes. Furthermore, questions pertaining to the provenance of the filling material (which consisted largely of broken ceramic vessels, many of which were unused) come to the forefront, as well as questions about who was allowed to gather such filling material and who needed to have given their consent for it to be used. What remains entirely clear, though, is that this massive reorganisation of at least parts of the Northwest Quarter would have caused immense changes to this lived space, most likely turning it from an area dominated by production complexes to one which came to hold buildings (at least partly) connected to the Late Roman army, underlined by the inscriptions in the Mosaic Hall (Haensch et al., 2016), and, in later periods, also domestic houses. This invasive, and profound, change to the space would have involved people: people changing places, people moving materials, people being part of entirely reorganising part of their urban environment in a profound way. It would also have involved production places being moved and therefore also the people who worked there.

14.4 Lived spaces and water infrastructure: organisation and reorganisation of cistern spaces

Already at the time of the American-British excavations, a large cistern on the south slope of the Northwest Quarter was surveyed and drawn (Figure 14.5), and numerous areas were excavated in and around its vicinity during the Danish-German Jerash Northwest Quarter Project. This was a rock-cut cistern, measuring 40 × 18 metres, the largest found in Gerasa until now (Trenches F, L, O and X; Kalaitzoglou et al., 2014, 22–29, 2015, 31–37, 2022a, 33–40, 2022b, 169–175; Lichtenberger et al., 2014, 43, 2015a, 49). As far as the published evidence indicates, it was most likely constructed during the Trajanic–Hadrianic period, during a period of expansion of the city (Lichtenberger et al., 2015a). At this point in time, the Northwest Quarter played a prominent role in respect of the organisation of the water supply for this part of the city and most likely for some of the lower-lying areas as well. After the main cistern on the top of the hill was closed off, our structure – located much further down the hill on its south slope – might have become the most important water resource in the Northwest Quarter. The cistern was used for several centuries, and it is clear from the archaeological examinations that it underwent several phases of repair before it fell out of use in Late Antiquity (Lichtenberger et al., 2015b; Lichtenberger & Raja, 2015c). After it stopped being used as a water reservoir, various buildings were constructed inside it, and an underground cistern, partly natural, was used for water supply in a now much smaller space. In the early Islamic period, the cistern was abandoned as a location for habitation and production, and it was entirely filled up.

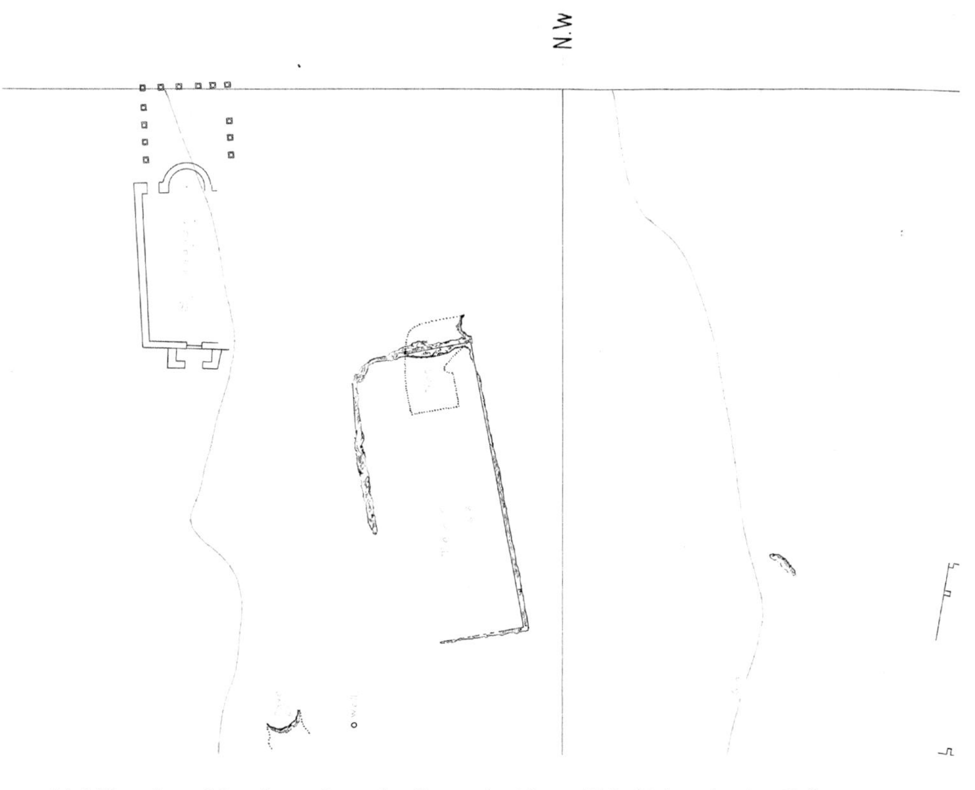

Figure 14.5 Drawing of the cistern from the Gerasa Archive at Yale University Art Gallery.

Source: Courtesy of Yale University Art Gallery

This extremely rudimentary construction, a rock-cut rectangular structure tells us about urban development and lived spaces in a longue durée perspective. Not only did the cistern stay in use as a water reservoir for hundreds of years, but it was afterwards turned into a space for habitation and production: a fact which in many ways does not agree with the ordinary view of the ancient city as a place where certain activities were confined to certain spaces. Furthermore, the cistern's life underlines the flexible way in which spaces were adapted according to changing circumstances as people continued the need to use that space. The cistern's various phases show that space within the city was used to the maximum capacity, and that it had to be adapted for a variety of purposes, in every possible way. It remained functional as a lived space with complexes for domestic and productive use even when it went out of use *as a cistern* (possibly because the rock-cut bottom partly collapsed over a natural cave) – most likely because there still was some access to water resources through the now-underground cistern installed in the natural cave underneath it. The space was filled up in the early Islamic period, at some point before the devastating earthquake of 749 CE, and at that point in time it was obviously not a desirable place for neither habitation nor production. We can only speculate, but maybe water was not available at this point anymore. However, as in the case of the oil presses, this reorganisation of the space would have required time and energy to move material and ensure proper in-filling. In terms of reorganisation of space in the Northwest Quarter in general, one might wonder whether the in-filling of the cistern might be contemporary with the early Islamic conversion of the Mosaic Hall (discussed later) into a different use and the building of large private houses on the so-called East Terrace (see below). From the Roman through the late antique and into the early Islamic period, the area seems to have profoundly changed character more than once.

14.5 Lived spaces and religion in Late Antiquity: the Mosaic Hall, the so-called Synagogue Church and early Islamic reuse

The so-called Synagogue Church was excavated in the 1920s within three weeks and was only published in a short report (Crowfoot, 1938, 216–220; Crowfoot & Hamilton, 1929; see also Dvorjetski, 2005; Michel, 2001, 251–155). The original excavators were interested first and foremost in understanding the church and the synagogue phases (Crowfoot, 1931; see also Crowfoot, 1938) and not in the earlier or later phases. In the short publications of the Synagogue Church, it is stated that the synagogue was built into parts of a late Roman period (third-fourth century) house with a courtyard lined with columns, but there are no proper plans of these phases in the publication and no in-depth descriptions of the house phases (Crowfoot, 1938, 216–220; Crowfoot & Hamilton, 1929). No hard evidence gave a date for the construction of the synagogue, and it was assumed by the excavators that it was constructed at some point in the fifth century according to the style of the mosaics found and building techniques employed. Since repairs to the mosaic decorations of the synagogue can be detected, and the building held several dedicatory inscriptions as well, it was also assumed that it was in use for quite some time. The date of the church construction is also not entirely secure, but a dedicatory inscription placed in front of the choir screen carries a date equating to 530/531 and may, due to its location, have been the main dedicatory inscription, providing a date for the building (Haensch et al., 2016; Lichtenberger & Raja, 2018b).

In the sixth century, a hall with mosaic inscriptions dating to, respectively, 576 and 591 CE was constructed south of the cave complex with the oil press (see above; Trenches N and W: Kalaitzoglou et al., 2022a, 26–33, 2022b, 163–9). The Mosaic Hall, as it is called, was connected to the Synagogue Church by way of a staircase that led to the *atrium* in front of the church. The hall was roughly rectangular and measured 18 × 10 metres (Haensch et al., 2016; Lichtenberger & Raja, 2018b). The dedicatory inscriptions inform us about ecclesiastic individuals living in Gerasa in the sixth century, some of which were connected to the Roman army, namely to the special unit of the *Electi Iustiniani*, which is also mentioned in the inscriptions (Haensch et al., 2016). The Mosaic Hall must have had an ecclesiastical function, which is confirmed by the dedicatory inscriptions, its attachment to the church and the plan of other churches from the site and the region. The hall was financed and decorated by the *Electi Iustiniani*, who are only known from these inscriptions. In the early Islamic period, the layout of the hall was changed, and a set of dividing walls was set neatly into the already-existing complex, partly on top of one of the dedicatory inscriptions. This new phase of reuse continued until the earthquake of 749, when the complex collapsed entirely.

According to Flavius Josephus, Gerasa had a Jewish community before the First Jewish War, and evidence of Jewish presence has been discussed in scholarship (Lichtenberger & Raja, 2018b; Raja, 2017). The late antique synagogue, however, underlines that a Jewish community remained present in the city in later periods as well. It is not possible to say what led to the conversion of the synagogue into a church, nor whether the synagogue was still in use when it was converted. It is unclear if the Jewish community was forcefully expelled from their synagogue, for example. The early Islamic period's use of the Mosaic Hall, which has now been documented in detail, underlines the fact that the Northwest Quarter underwent profound changes in this period, and that earlier buildings were reused and reframed to suit the needs of the new community in the early Islamic period. It is likely that they also used the church next door, but since this was not documented during the early twentieth-century excavations, it remains a speculation. The use and reuse of the Mosaic Hall underlines the profound changes that the Northwest Quarter underwent, and it shows that the lived experience of this space was changed over time – from being a religious space belonging to a Jewish community to being connected to a late antique special Roman army unit and, later, serving the early Islamic community. Together, the history of use, reuse and reorganisation of the Roman house, turned synagogue and then church, and that of the Mosaic Hall, from being connected to the church to its early Islamic phase, show that the lived experiences within these spaces could be changed profoundly and used in entirely different contexts from that of their original purpose, depending on the group or part of society which took over the space.

14.6 Lived spaces and domestic architecture: prime location real estate in the early Islamic period

Apart from the changes to the Mosaic Hall, domestic architectural complexes of the early Islamic period were located in several places in the quarter, both on the so-called East Terrace, at the top of the hill, and on its southern slope (Lichtenberger et al., 2016; Lichtenberger & Raja, 2017). The north slope of the hill does not have traces of early Islamic domestic architecture. On the East Terrace, a set of trenches was excavated between 2014 and 2016 (trenches K, P and V; Kalaitzoglou et al., 2015, 26–31, 2022a, 40–46, 2022b, 153–163). These all yielded substantial evidence for domestic structures of the early Islamic period underlining that life in the Northwest Quarter was thriving in this period in general. The houses were destroyed by the earthquake of 749, and the area was left untouched until the middle Islamic period. The trenches on the East Terrace held parts of two domestic buildings: a two-storey house inhabited at the time of the earthquake was partly excavated in trench K (Lichtenberger et al., 2016). Parts of a large house that was undergoing major renovations at the time the earthquake hit were found in trenches P and V (Lichtenberger & Raja, 2017). The houses do not seem to have any pre-Islamic phases, and the use of at least large parts of the East Terrace date to the early Islamic period. While the area at the top of the Northwest Quarter is dominated by middle Islamic remains dating to between eleventh and thirteenth centuries, it is clear that earlier structures had been in place as well (Lichtenberger & Raja, 2016, 2018c, 2018d; Kalaitzoglou, 2018). Evidence for earlier occupation was found underneath middle Islamic structures in trenches C, D, E and T, occupation which had come to an end with the 749 earthquake (Kalaitzoglou et al., 2013, 68–75, 2014, 13–22, 2022b, 141–147). Sections of an early Islamic-period kitchen, dated through ceramic and coin finds as well as 14C dates were excavated in trench D on the north side of the middle Islamic wall, outside the middle Islamic building. A bottle-shaped cistern in trench C was excavated. This cistern had only been filled in during the middle Islamic period. The south slope of the Northwest Quarter also seems to have been an area in which a lot of reorganisation and construction was going on in the early Islamic period. In trench U, parts of a residential area of the early Islamic period were found (Kalaitzoglou et al., 2022b, 147–153). These domestic structures had undergone several alterations, from their construction until the earthquake of 749.

All these partly excavated early Islamic houses yield crucial information about everyday life. Their construction phases inform us that massive building activity was going on in the period after the Arab invasion and that a community seems to have prospered in the Northwest Quarter until the earthquake struck. The contents of the houses, in particular of the house in trench K, show that the inhabitants had a wide range of objects at their disposal – objects of a quality that are seldom found (or published from excavations) in urban contexts of this period (Barfod et al., 2016; Larsen et al., 2016; Lichtenberger et al., 2016). This does not necessarily mean that the same situation was not encountered in different areas of Jerash or the region in general.

Evidence for a profound reorganisation of the space on the hill in the late antique and early Islamic periods is abundantly present. This reorganisation seems to have mainly consisted of changes not only to already-existing spaces but also to the layout of entirely new monumental domestic complexes. This prominent hill, the highest point within the city wall, seems to have been a hub for activity in this period, which underlines its importance within the broader early Islamic city, even though we do not know the layout of the city as a whole for this period.

14.7 Conclusion: changing lived spaces

The case studies discussed earlier illustrate how scholarly focus on the changing lived spaces of Gerasa in the late antique and early Islamic periods can yield much new information from the already-published material – despite the fact that the evidence is complex and scattered across an approximately 4-hectare-large area. The areas of the Northwest Quarter examined by the project underwent a series of profound changes throughout our period, underscoring society's evolving priorities – possibly as a result of natural catastrophes or of a slow change in water supply, but also as a result of different groups taking over spaces of others, as in the case of the Synagogue Church and the Mosaic Hall. These changes can sometimes be connected directly to historical evidence (the inscriptions in the Mosaic Hall), but usually we must tease out potential reasons for spatial reorganisations from complex archaeological evidence. As outlined, a major change seems to have taken place in the early part of Late Antiquity (fourth century CE), which might have been spurred by changes in the water supply to this part of the city. Later changes include the presence of the sixth-century military unit, which most likely took over the building of an earlier synagogue. In the early Islamic period, further profound changes to the use of the space happened, with a focus on the development of domestic architecture. All these changes are reflections of lived spaces, spaces which were inhabited by the people of Gerasa/Jerash. They were the ones – no matter whether they belonged to the military, such as the *Electi*

Iustiniani, the local Jewish society, production workers or families and communities living in the area – who took part in changing these spaces over time.

The wider implications of the results from Gerasa/Jerash are that we should begin to look at urban societies in the Near East in a much more diachronic perspective than hitherto done and that only through such a longue durée perspective will we be able to contextualise more firmly the changing lived spaces in urban settings.

Notes

1 The most prominent attempts at narratives, which also, to some extent, include the evidence from Jerash, are the publications by Walmsley (1995; 1996; 1997; 2000; 2005; 2007; 2011; 2018).

2 For methodological analyses of more than 800,000 pieces of ceramics found in the excavations across five campaigns, see Romanowska et al. (2022); Romanowska et al. (2018); and Romanowska et al. (2021).

3 For an overview of water supply in Gerasa/Jerash, see the works by Boyer (2016a; 2016b; 2018; 2019; 2022); see also Kristiansen & Stott (2020) and Stott et al. (2018).

Bibliography

Albrecht, J., Degelmann, C., Gasparini, V., Gordon, R., Patzelt, M., Petridou, G., Raja, R., Rieger, A.-K., Rüpke, J., Sippel, B., Urciuoli, E. R., & Weiss, L. (2018). Religion in the making: The lived ancient religion approach. *Religion*, *48*(4), 568–593.

Baird, J., Kamash, Z., & Raja, R. (2023). Knowing Palmyra. Mandatory production of archaeological knowledge. *Journal of Social Archaeology*, *23*(1), 76–98. https://doi.org/10.1177/14696053221144013

Barfod, G. H., Larsen, J. M., Lichtenberger, A., & Raja, R. (2016). Revealing text in a complexly rolled silver scroll from Jerash with computed tomography and advanced imaging software. *Nature Scientific Reports*, *5*, Article 17765. https://doi.org/10.1038/srep17765

Barnes, H., Blanke, L., Damgaard, K., Simpson, I., Sørensen, M. L., & Walmsley, A. (2006). From 'guard house' to congregational mosque: Recent discoveries on the urban history of Islamic Jarash. *Annual of the Department of Antiquities of Jordan*, *50*, 285–314.

Blanke, L. (2016). Late antique Jarash project. *American Journal of Archaeology*, *120*(4), 643–645.

Blanke, L. (2018). Abbasid Jerash reconsidered: Suburban life in Jerash's southwest district over the longue durée. In A. Lichtenberger & R. Raja (Eds.), *The archaeology and history of Jerash: 110 years of excavations* (pp. 39–48). Brepols.

Blanke, L. (2022). (Re)constructing Jarash: History, historiography and the making of the ancient city. In E. K. Fowden, S. Cagaptay, E. Zychowicz-Coghill, & L. Blanke (Eds.), *Cities as palimpsests? Responses to Antiquity in eastern Mediterranean urbanism* (pp. 351–370). Oxbow Books.

Blanke, L., Damgaard, K., Simpson, I., & Walmsley, A. (2007). From bathhouse to congregational mosque: Further discoveries on the urban history of Islamic Jarash. *Annual of the Department of Antiquities of Jordan*, *51*, 177–197.

Blanke, L., Lorien, P. D., & Rattenborg, R. (2010). Changing cityscapes in central Jarash: Between late Antiquity and the Abbasid period. *Annual of the Department of Antiquities of Jordan*, *54*, 311–327.

Blanke, L., & Walmsley, A. (2010). Islamic Jerash project. *American Journal of Archaeology*, *114*, 517–519.

Blanke, L., & Walmsley, A. (2022). Resilient cities: Renewal after disaster in three late antique towns of the East Mediterranean. In J. Martinez Jiménez & S. Ottewill-Soulsby (Eds.), *Remembering and forgetting the ancient city* (pp. 69–108). Oxbow Books.

Boyer, D. D. (2016a). Aqueducts and birkets: New evidence of the water management system servicing Gerasa (Jarash), Jordan. In R. A. Stucky, O. Kaelin, & H.-P. Mathys (Eds.), *Proceedings of the 9th International Congress on the Archaeology of the Ancient Near East, 9–13 June 2014, Basel, Vol. 3* (pp. 517–531). Harrassowitz Verlag.

Boyer, D. D. (2016b). Recent advances in understanding the water delivery system to Gerasa of the Decapolis. In G. Wiplinger (Ed.), *De Aquaeductu atque Aqua Urbium Lyciae Pamphyliae Pisidiae: The legacy of Sextus Julius Frontinus. Tagungsband des Internationalen Frontinus-Symposiums Antalya, 31. Oktober – 9. November 2014* (pp. 143–154). Peeters.

Boyer, D. D. (2018). The role of landscape in the occupational history of Gerasa and its hinterland. In A. Lichtenberger & R. Raja (Eds.), *The archaeology and history of Jerash: 110 years of excavations* (pp. 59–86). Brepols.

Boyer, D. D. (2019). *An analysis of the historical water management system to Gerasa in the period 100 BC to CE 700* [Doctoral dissertation, University of Western Australia].

Boyer, D. D. (2022). *Water management in Gerasa and its hinterland: from the Romans to AD 750 (Jerash Papers 10)*. Turnhout: Brepols

Brody, L., & Raja, R. (2022). The Gerasa archives at the Yale University art gallery: Exploring the archives of the Anglo-American excavations 1928–1930 and 1930–1934. In O. Bobou, A. C. Miranda, & R. Raja (Eds.), *Archival historiographies: The impact of 20th century legacy data on archaeological investigations* (pp. 55–82). Brepols.

Clark, V. A. (1986). The church of Bishop Isaiah at Jerash. In F. Zayadine (Ed.), *Jerash archaeological project: Vol. I. 1981–1983* (pp. 303–342). Department of Antiquities of Jordan.

Crowfoot, J. W. (1931). *Churches at Jerash: A preliminary report of the joint Yale-British expeditions to Jerash, 1928–1930*. Council of the British School of Archaeology in Jerusalem.

Crowfoot, J. W. (1938). The Christian churches. In C. H. Kraeling (Ed.), *Gerasa: City of the Decapolis: An account embodying the record of a joint excavation conducted by Yale University and the British School of Archaeology in Jerusalem (1928–1930), and Yale University and the American Schools of Oriental Research* (pp. 171–262). American Schools of Oriental Research.

Crowfoot, J. W., & Hamilton, R. W. (1929). The discovery of a synagogue in Jerash. *Palestine Exploration Quarterly*, *61*(4), 211–219.

Dvorjetski, E. (2005). The synagogue-church at Gerasa in Jordan: A contribution to the study of ancient synagogues. *Zeitschrift des Deutschen Palästina-Vereins*, *121*, 140–167.

Gasparini, V., Patzelt, M., Raja, R., Rieger, A.-K., Rüpke, J., & Urciuoli, E. (Eds.). (2020a). *Lived religion in the ancient Mediterranean world: Approaching religious transformations from archaeology, history and classics*. De Gruyter.

Gasparini, V., Patzelt, M., Raja, R., Rieger, A.-K., Rüpke, J., & Urciuoli, E. (Eds.). (2020b). Pursuing lived ancient religion. In V. Gasparini, M. Patzelt, R. Raja, A.-K. Rieger, J. Rüpke, & E. Urciuoli (Eds.), *Lived religion in the ancient Mediterranean world: Approaching religious transformations from archaeology, history and classics* (pp. 1–8). De Gruyter.

Gawlikowski, M. (1986). A residential area by the South Decumanus. In F. Zayadine (Ed.), *Jerash archaeological project: Vol. I. 1981–1983* (pp. 107–136). Department of Antiquities of Jordan.

Gawlikowski, M., & Musa, A. (1986). The church of Bishop Marianos. In F. Zayadine (Ed.), *Jerash archaeological project: Vol. I. 1981–1983* (pp. 137–162). Department of Antiquities of Jordan.

Haensch, R., Lichtenberger, A., & Raja, R. (2016). Christen, Juden und Soldaten im Gerasa des 6. Jahrhunderts. *Chiron*, *46*, 177–204.

Kalaitzoglou, G. (2018). A middle Islamic hamlet in Jerash: Its architectural development. In A. Lichtenberger & R. Raja (Eds.), *Middle Islamic Jerash (9th century – 15th century): Archaeology and history of an Ayyubid-Mamluk settlement* (pp. 97–116). Brepols.

Kalaitzoglou, G., Lichtenberger, A., & Raja, R. (2013). Preliminary report of the second season of the Danish-German Jarash Northwest Quarter Project 2012. *Annual of the Department of Antiquities of Jordan*, *57*, 57–79.

Kalaitzoglou, G., Lichtenberger, A., & Raja, R. (2014). The Danish-German Jarash North-West Quarter Project 2013: Preliminary field report. *Annual of the Department of Antiquities of Jordan*, *58*, 11–37.

Kalaitzoglou, G., Lichtenberger, A., & Raja, R. (2015). Preliminary report of the fourth season of the Danish-German Jerash Northwest Quarter Project 2014. *Annual of the Department of Antiquities of Jordan*, *59*, 11–43.

Kalaitzoglou, G., Lichtenberger, A., Möller, H., & Raja, R. (with Bangsgaard, P., Eger, C., Merkel, S., Prange, M., Springer, A., Polla, S., & Larsen, J. M.). (2022a). Preliminary report of the fifth season of the Danish-German Jarash Northwest Quarter Project 2015. *Annual of the Department of Antiquities of Jordan*, *60*, 17–129.

Kalaitzoglou, G., Lichtenberger, A., Möller, H., & Raja, R. (with Eger, C.). (2022b). Preliminary report of the sixth season of the Danish-German Jarash Northwest Quarter Project 2016. *Annual of the Department of Antiquities of Jordan*, *60*, 131–226.

Kehrberg, I. (2017). Ancient Burials at the Hippodrome of Gerasa/Jarash. *Annual of the Department of Antiquities of Jordan*, *58*, 181–213.

Kraeling, C. H. (Ed.). (1938). *Gerasa, city of the Decapolis: An account embodying the record of a joint excavation conducted by Yale University and the British School of Archaeology in Jerusalem (1928–1930), and Yale University and the American Schools of Oriental Research*. American Schools of Oriental Research.

Kristiansen, S. M., & Stott, D. (2020). Mapping Jerash by remote sensing. In A. Lichtenberger & R. Raja (Eds.), *Environmental studies, remote sensing, and modelling: Final publications from the Danish-German Jerash Northwest Quarter Project I* (pp. 157–172). Brepols.

Larsen, J. M., Lichtenberger, A., Raja, R., & Gordon, R. L. (2016). An Umayyad period magical amulet from a domestic context in Jerash, Jordan. *Syria. Archéologie, art et histoire*, *93*, 369–86.

Latzer-Läsar, A., Raja, R., Rüpke, J., & Urciuoli, E. (2020). Intersecting religion and urbanity in Late Antiquity. In A. Latzer-Läser & E. Urciuoli (Eds.), *Urban religion in Late Antiquity* (pp. 1–14). De Gruyter.

Lichtenberger, A. (Ed.). (2002). *U. J. Seetzen: Unter Mönchen und Beduinen. Reisen in Palästina und angrenzenden Länders 1805–1807*. K. Thienemanns Verlag.

Lichtenberger, A., Lindroos, A., Raja, R., & Heinemeier, J. (2015a). Radiocarbon analysis of mortar from Roman and Byzantine water management installations in the Northwest Quarter of Jerash, Jordan. *Journal of Archaeological Science: Reports*, *2*, 114–127.

Lichtenberger, A., Peterson, A., Polla, S., Raja, R., Springer, A., Stukenbrok, H., & Ting, C. (2021). New insights from Middle Islamic ceramics from Jerash. *Journal of Islamic Archaeology*, *8*(1), 53–86.

Lichtenberger, A., & Raja, R. (Eds.). (forthcoming). *By the Gold River: Gerasa through the eyes of 19th and early 20th century visitors*. Brepols.

Lichtenberger, A., & Raja, R. (2015a). An architectural block with altar-iconography from the North-west Quarter of Jerash. *Levant*, *47*(1), 112–130.

Lichtenberger, A., & Raja, R. (2015b). Intentional cooking pot deposits in late Roman Jerash (Northwest Quarter). *Syria: Archéologie, art et histoire*, *92*, 309–328.

Lichtenberger, A., & Raja, R. (2015c). New archaeological research in the Northwest Quarter of Jerash and its implications for the urban development of Roman Gerasa. *American Journal of Archaeology*, *119*(4), 483–500.

Lichtenberger, A., & Raja, R. (2016). Ǧeraš in the middle Islamic period: Connecting texts and archaeology through new evidence from the Northwest Quarter. *Zeitschrift des Deutschen Palästina-Vereins*, *132*, 63–81.

Lichtenberger, A., & Raja, R. (2017). Mosaicists at work: The organisation of mosaic production in Early Islamic Jerash. *Antiquity*, *91*(358), 998–1010.

Lichtenberger, A., & Raja, R. (Eds.). (2018a). *The archaeology and history of Jerash: 110 years of excavations*. Brepols.

Lichtenberger, A., & Raja, R. (2018b). From synagogue to church. The appropriation of the synagogue of Gerasa/Jerash under Justinian. *Jahrbuch für Antike und Christentum*, *61*, 85–100.

Lichtenberger, A., & Raja, R. (Eds.). (2018c). *Middle Islamic Jerash (9th century – 15th century): Archaeology and history of an Ayyubid-Mamluk settlement*. Brepols.

Lichtenberger, A., & Raja, R. (2018d). Middle Islamic Jerash through the lens of the longue durée. In A. Lichtenberger & R. Raja (Eds.), *Middle Islamic Jerash (9th century – 15th century): Archaeology and history of an Ayyubid-Mamluk settlement* (pp. 5–36). Brepols.

Lichtenberger, A., & Raja, R. (2018e). The middle Islamic period in northern Jordan and its pottery: Ceramics in context. In A. Lichtenberger & R. Raja (Eds.), *Middle Islamic Jerash (9th century – 15th century): Archaeology and history of an Ayyubid-Mamluk settlement* (pp. 1–4). Brepols.

Lichtenberger, A., & Raja, R. (2018f). A view of Gerasa/Jerash from its urban periphery: The Northwest Quarter and its significance for the understanding of the urban development of Gerasa from the Roman to the Early Islamic period. In A. Lichtenberger & R. Raja (Eds.), *The archaeology and history of Jerash: 110 years of excavations* (pp. 143–166). Brepols.

Lichtenberger, A., & Raja, R. (2019a). 18. Januar 749 n. Chr: Ein Erdbeben zerstört die Stadt Gerasa in Jordanien. *Antike Welt*, *1*(2019), 4.

Lichtenberger, A., & Raja, R. (Eds.). (2019b). *Byzantine and Umayyad Jerash reconsidered: Transitions, transformations, continuities*. Brepols.

Lichtenberger, A., & Raja, R. (2019c). Defining borders: The Umayyad-Abbasid transition and the earthquake of CE 749 in Jerash. In A. Lichtenberger & R. Raja (Eds.), *Byzantine and Umayyad Jerash reconsidered: Transitions, transformations, continuities* (pp. 265–286). Brepols.

Lichtenberger, A., & Raja, R. (Eds.). (2020a). *Environmental studies, remote sensing, and modelling: Final publications from the Danish-German Jerash Northwest Quarter Project I*. Brepols.

Lichtenberger, A., & Raja, R. (Eds.). (2020b). *Hellenistic and Roman Gerasa: The archaeology and history of a Decapolis City*. Brepols.

Lichtenberger, A., & Raja, R. (2020c). Late Hellenistic and Roman Antiochia on the Chrysorrhoas, also called Gerasa: A reappreciation of the urban development in the light of the findings of the Danish-German Jerash Northwest Quarter Project (2011–17). In A. Lichtenberger & R. Raja (Eds.), *Hellenistic and Roman Gerasa: The archaeology and history of a Decapolis city* (pp. 7–54). Brepols.

Lichtenberger, A., & Raja, R. (Eds.). (2020d). *Metal finds and coins: Final publications from the Danish-German Jerash Northwest Quarter Project II*. Brepols.

Lichtenberger, A., & Raja, R. (2021a, 22 August). *Danish-German Jerash Northwest Quarter Project – Bibliography*. Figshare.

Lichtenberger, A., & Raja, R. (Eds.). (2021b). *Glass, lamps, and Jerash Bowls: Final publications from the Danish-German Jerash Northwest Quarter Project III*. Brepols.

Lichtenberger, A., & Raja, R. (2022a). *Architectural elements, wall paintings, and mosaics: Final publications from the Danish-German Jerash Northwest Quarter Project IV* (Vols. I – II). Brepols.

Lichtenberger, A., & Raja, R. (2022b). A monumental architectural limestone block with altar iconography. In A. Lichtenberger & R. Raja (Eds.), *Architectural elements, wall paintings, and mosaics: Final publications from the Danish-German Jerash Northwest Quarter Project IV: Vol. I. Architecture and building ceramics* (pp. 65–74). Brepols.

Lichtenberger, A., Raja, R., Eger, C., Kalaitzoglou, G., & Sørensen, A. H. (2016). A newly excavated private house in Jerash. Reconsidering aspects of continuity and change in material culture from Late Antiquity to the Early Islamic Period. *Antiquité Tardive*, *24*, 317–359.

Lichtenberger, A., Raja, R., & Stott, D. (2019). Mapping Gerasa: A new and open data map of the site. *Antiquity*, *93*(367), E7.

Lichtenberger, A., Raja, R., & Sørensen, A. H. (2013). Preliminary registration report of the second season of the Danish-German Jarash Northwest Quarter Project 2012. *Annual of the Department of Antiquities of Jordan*, *57*, 9–56.

Lichtenberger, A., Raja, R., & Sørensen, A. H. (2014). The Danish-German Jarash Northwest Quarter Project 2013: Preliminary registration report. *Annual of the Department of Antiquities of Jordan*, *58*, 39–103.

Lichtenberger, A., Raja, R., & Sørensen, A. H. (with Bangsgaard, P., Eger, C., & Jensen, P. F.). (2015b). The Danish-German Jerash Northwest Quarter Project 2014: Preliminary registration report. *Annual of the Department of Antiquities of Jordan*, *59*, 45–131.

Lichterman, P., Raja, R., Rieger, A.-K., & Rüpke, J. (2017). Grouping together in lived ancient religion. *Religion in the Roman Empire*, *3*(1), 3–10.

Marco, S., Hartal, M., Hazan, N., Lev, L., & Stein, M. (2003). Archaeology, history, and geology of the A.D. 749 earthquake, dead sea transform. *Geology*, *31*(8), 665–668.

McGuire, M. B. (2008). *Lived religion: Faith and practice in everyday life*. Oxford University Press.

Michel, A. (with Duval, N., & Piccirillo, M.). (2001). *Les églises d'époque byzantine et umayyade de la Jordanie (Provinces d'Arabie et de Palestine), V^e^– VIIIe siècle: Typologie architecturale et aménagements liturgiques*. Brepols.

Mortensen, E. (2017). Archive archaeology: rediscovering Jerash in the digital archives. In A. Lichtenberger & R. Raja (Eds.), *Gerasa/Jerash: From the urban periphery* (pp. 27–31). Fællestrykkeriet, AUTRYK.

Mortensen, E. (2018). The early research history of Jerash: A short outline. In A. Lichtenberger & R. Raja (Eds.), *The archaeology and history of Jerash: 110 years of excavations* (pp. 167–186). Brepols.

Naghawi, A. (1989). Umayyad filses minted at Jerash. *Syria. Archéologie, art et histoire*, *66*, 219–222.

Ostrasz, A. A. (1989). The hippodrome of Gerasa: A report on excavations and research 1982–1987. *Syria. Archéologie, art et histoire*, *66*, 55–77.

Raja, R. (2012). *Urban development and regional identity in the eastern Roman provinces, 50 BC – CE 250: Aphrodisias, Ephesos, Athens, Gerasa*. Museum Tusculanum Press.

Raja, R. (2017). Zeus Olympios, Hadrian and the Jews of Antiochia-on-the-Chrysorrhoas-formerly-called-Gerasa. In E. Minchin & J. Wade (Eds.), *Text and the material world: Essays in honour of Graeme Clarke, studies in Mediterranean archaeology, PB185* (pp. 171–195). Astrom Editions.

Raja, R. (2022a). *Pearl of the desert: A history of Palmyra*. Oxford University Press.

Raja, R. (2022b). Urban transformations seen through the lens of urban religion. In A. Cortese & G. Fioratto (Eds.), *Urban space between the Roman age and late antiquity: Continuity, discontinuity and changes* (pp. 132–147). Schnell und Steiner.

Raja, R., & Rüpke, J. (2015). Appropriating religion: Methodological issues in testing the 'lived ancient religion' approach. *Religion in the Roman Empire*, *1*(1), 11–19.

Raja, R., & Rüpke, J. (2021a). Introduction: Urban archaeology. *Religion in the Roman Empire*, *7*(2), 217–220.

Raja, R., & Rüpke, J. (Eds.). (2021b). *An Urban Archaeology of Ancient Religion, Religion in the Roman Empire, 7.2*. Mohr Siebeck.

Raja, R., & Sindbæk, S. M. (2018). Urban network evolutions. Exploring dynamics and flows through evidence from urban contexts. In R. Raja & S. M. Sindbæk (Eds.), *Urban network evolutions. Towards a high-definition archaeology* (pp. 13–18). Aarhus Universitetsforlag.

Raja, R., & Sindbæk, S. M. (2020). Urban networks and high-definition narratives: Rethinking the archaeology of urbanism. *Journal of Urban Archaeology*, *2*, 173–186.

Raja, R., Steding, J., & Yon, J.-B. (2021). *Excavating Palmyra: Harald Ingholt's excavation diaries. A transcript, translation, and commentary* (Vols. 1–2). Brepols.

Rasson, A.-M., & Seigne, J. (1989). Une citerne byzantine-omeyyade sur le Sanctuaire de Zeus. *Syria. Archéologie, art et histoire*, *66*, 117–152.Rattenborg, R., & Blanke, L. (2017). Jarash in the Islamic Ages (*c.* 700–1200 CE): A critical review. *Levant*, *49*(3), 312–332.

Rau, S., & Rüpke, J. (2020). Religion and urbanity: Reciprocal formations. In S. Rau & J. Rüpke (Eds.), *Religion and urbanity online*. De Gruyter. https://doi.org/10.1515/urbrel.13230336

Romanowska, I., Brughmans, T., Bes, P., Carrignon, S., Egelund, L., Lichtenberger, A., & Raja, R. (2022). A study of the centuries-long reliance on local ceramics in Jerash through full quantification and simulation. *Journal of Archaeological Method and Theory*, *29*, 31–49.

Romanowska, I., Brughmans, T., Lichtenberger, A., & Raja, R. (2018). Urban networks seen through ceramics: Formal modelling approaches to pottery distribution in Jerash. In R. Raja & S. M. Sindbæk (Eds.), *Urban network evolutions: Towards a high-definition archaeology* (pp. 131–138). Aarhus University Press.

Romanowska, I., Lichtenberger, A., & Raja, R. (2021). Trends in ceramic assemblages from the Northwest Quarter of Gerasa/Jerash, Jordan. *Journal of Archaeological Science: Reports*, *36*, Article 102778. https://doi.org/10.1016/j.jasrep.2020.102778

Rüpke, J., Feldmeier, R., King, K. L., Raja, R., Reed, A. Y., Riedweg, C., Schwartz, S., Smith, C., & Vinzent, M. (2015). Editorial: Religion in the Roman empire. *Religion in the Roman Empire*, *1*(1), 1–7.

Schaefer, J., & Faulkner, B. K. (1986). An Umayyad potters' complex in the North Theatre in Jerash. In F. Zayadine (Ed.), *Jerash archaeological project: Vol. I. 1981–1983* (pp. 411–460). Department of Antiquities of Jordan.

Schumacher, G. (1902). Dscherasch. *Zeitschrift des Deutschen Palästina-Vereins*, *25*, 109–177.

Seetzen, U. J. (with Kruse, F.). (1854). *Reisen durch Syrien, Palästina, Phönicien, die Transjordan-Länder, Arabia Petraea und Unter-Aegypten: Vol. I.* Verlag Bei G. Reimer.

Seigne, J. (1985). Le Sanctuaire de Zeus à Jerash: Éléments de chronologie. *Syria. Archéologie, art et histoire*, *62*(3), 287–295.

Seigne, J. (1997). De la grotte au périptère: Le Sanctuaire de Zeus Olympien à Jerash. *Topoi*, *7*(2), 993–1004.

Seigne, J. (2013). Des jardins, des oliviers et de la vigne aux portes de la cité. À propos de quelques installations agricoles de Gerasa de la Décapole. *Topoi. Orient-Occident. Supplément 12, 2013. Villes et campagnes aux rives de la Méditerranée ancienne. Hommages à Georges Tate*, 407–420.

Seigne, J. (2018). Pourquoi Hadrien a-t-il passé l'hiver de 129/130 à Gerasa? In A. Lichtenberger & R. Raja (Eds.), *The archaeology and history of Jerash: 110 years of excavations* (pp. 207–213). Brepols.

Stott, D., Kristiansen, S. M., Lichtenberger, A., & Raja, R. (2018). Mapping an ancient city with a century of remotely sensed data. *Proceedings of the National Academy of Sciences of the United States of America*, *115*(24), E5450–E5458.

Tsafrir, Y., & Foerster, G. (1992). The dating of the 'earthquake of the sabbatical year' of 749 CE in Palestine. *Bulletin of the Schools of Oriental and African Studies*, *55*, 231–235.

Walmsley, A. (1995). Tradition, innovation, and imitation in the material culture of Islamic Jordan: The first four centuries. *Studies in the History and Archaeology of Jordan*, 5, 657–668.

Walmsley, A. (1996). Byzantine Palestine and Arabia: Urban prosperity in Late Antiquity. In N. Christie & S. T. Loseby (Eds.), *Towns in transition: Urban evolution in Late Antiquity and the early Middle Ages* (pp. 126–158). Scolar Press.

Walmsley, A. (1997). Land, resources and industry in early Islamic Jordan (7th-11th century): Current research and future directions. *Studies in the History and Archaeology of Jordan*, *6*, 345–351.

Walmsley, A. (2000). Production, exchange and regional trade in the Islamic East Mediterranean: Old structures, new systems? In I. L. Hansen & C. Wickham (Eds.), *The long eighth century: Production, distribution and demand* (pp. 265–344). Brill.

Walmsley, A. (2005). The village ascendant in Byzantine and early Islamic Jordan: Socio-economic forces and cultural responses. In J. Lefort, C. Morrisson, & J.-P. Sodini (Eds.), *Les Villages dans l'Empire byzantin, IVe – XVe siècle* (pp. 511–522). Lethielleux.

Walmsley, A. (2007). *Early Islamic Syria*. Bloomsbury Academic.

Walmsley, A. (2011). Trends in the urban history of eastern Palaestina Secunda during the late antique – Early Islamic transition. In A. Borrut (Ed.), *Le Proche-Orient de Justinien aux Abbassides: peuplement et dynamiques spatiales. Actes du colloque 'Continuités de l'occupation entre les périodes byzantine et abbasside au Proche-Orient, VIIe – IXe siècles', Paris, 18–20 octobre 2007* (pp. 271–284). Brepols.

Walmsley, A. (2018). Urbanism at Islamic Jerash: New readings from archaeology and history. In A. Lichtenberger & R. Raja (Eds.), *The archaeology and history of Jerash: 110 years of excavations* (pp. 241–256). Brepols.

Watson, P. M. (1989). Jerash Bowls: Study of a provincial group of Byzantine decorated fine ware. *Syria. Archéologie, art et histoire*, *66*, 223–261.

Zayadine, F. (Ed.). (1986). *Jerash archaeological project: Vol. 1. 1981–1983*. Department of Antiquities of Jordan.

15 "Places of continuity"

Before, around, and after the end of the Roman villas in Central Italy

Enrico Zanini

This chapter aims to identify and explore vertical lines of continuity that cross the well-established horizontal lines of chronological discontinuity in Late Antiquity. In other words, I will try to define which elements may have caused a differential lifespan between different settlements and living spaces that appear to be homologous and located in the same territory. My intent is therefore to outline a more articulated perception of the evolution of the landscape that occurs in different times and modalities, according to the nature of the human settlements and their ability to interact with the surrounding territory on various levels: administrative, economical, and functional.

My analysis originates from the ongoing research on the site of Vignale (near Piombino, in Tuscany), where the archaeological investigation revealed – as we will see – the presence of a large Roman villa that was built between the first century BCE and the first century CE in an area inhabited since the Etruscan period. Unlike many other Roman villas in the same area, the settlement of Vignale shows an unexpected continuity of life, substantially preserving its shape until the early Middle Ages. In this long span of time, different functions were contemporary or sequentially associated to the basically unaltered physical structure of the villa, giving rise to an overall phenomenon, which might be defined as "structural resilience". The physical structure of the Roman villa of Vignale apparently survived the collapse of the Roman world, and the site was abandoned only in the medieval period, when the community that occupied it moved to a castle located on the top of the adjacent hill. However, human occupation of the area did not cease with the disruption of the medieval castle at the end of the fourteenth century and, after a period of apparent abandonment, Vignale became relevant again as a "central place" in the geography of that territory from the seventeenth to early twentieth centuries, this time assuming the shape of a large estate, hosting some 40 families of peasants, with a large mansion house.

Reflecting on Vignale's history in the perspective of the *longue durée* (or even in a "medium durée"), we can assume that, in terms of interactions between a single human settlement and the rest of the world, we should consider two different types of relationship, each characterized by a different temporality. The first one – that we could define as macro-economic – is the link connecting a single site with the main systems of the world in which it is inserted: the economy, administration, defence, communication, and religious organization. This kind of interaction tends to change following the transformation of those general macro-economies and the changes in the role of our settlement inside them; such inter-related development tends to unfold in the short to medium term in its duration. The second kind of relationship could be defined as micro-ecological, since it deals with the network of connections that a single human settlement, once established in a pre-existing landscape, builds with that landscape itself. This is tendentially a *longue durée* relationship because the human settlement becomes the anthropic

DOI: 10.4324/9780429427152-19

part of the "natural" landscape, assuming those characters of resilience and inertia that characterize nature by definition.

Most interesting, in this perspective, is that a single settlement might act as a point of contact between these two dimensions: the general macro-economic and the local micro-ecological ones. This is particularly the case when the settlement is originally the product of an economic choice made by a member of the upper classes, as in the case of a large Roman villa, and/or when the settlement is connected with a major element of infrastructure, like a road or a port. In these cases, an external – and therefore macro-economic – choice changes the previous natural landscape, adding a new element that tends to become permanent in the landscape itself, thus creating a permanent – or at least a long lasting – connection between the local micro-ecology and the regional/general macro-economies.

15.1 The vantage point

The archaeological site of Vignale is located on the Etruscan coast, and it was already recognized as the site of a Roman settlement by 1830 (Figure 15.1). The first excavations were carried out by the Grand Duke of Tuscany, Leopold II, as a part of reclamation works in the coastal swamp of the Maremma. On that occasion, extensive remains of a building were unearthed and identified – mostly correctly – as a Roman bath complex which was clearly part of a larger settlement. This was only partially excavated at the time: the *thermae* and part of the complex into which they were inserted were carefully recorded in an archaeological map of the area (Zanini, 2019a).

On the basis of entries written by the Grand Duke himself in his personal diary (De Ruggiero, 2016; Pesendorfer, 1987), it is clear that a large figurative and polychrome floor mosaic was brought to light at the time. However, despite its outstanding quality and the attention shown

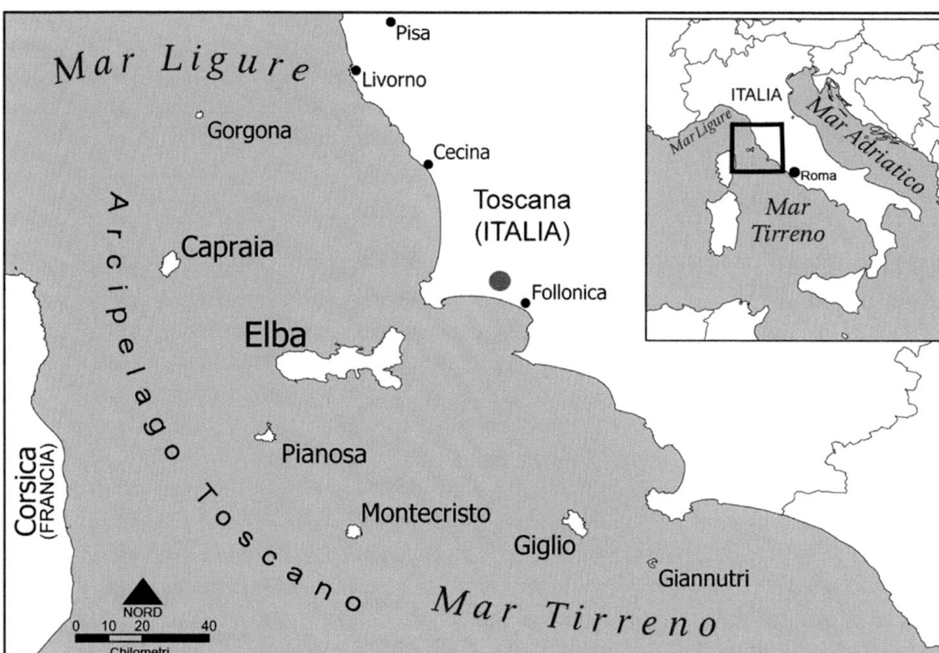

Figure 15.1 Location of the archaeological site of Vignale in the spatial context of coastal Etruria.

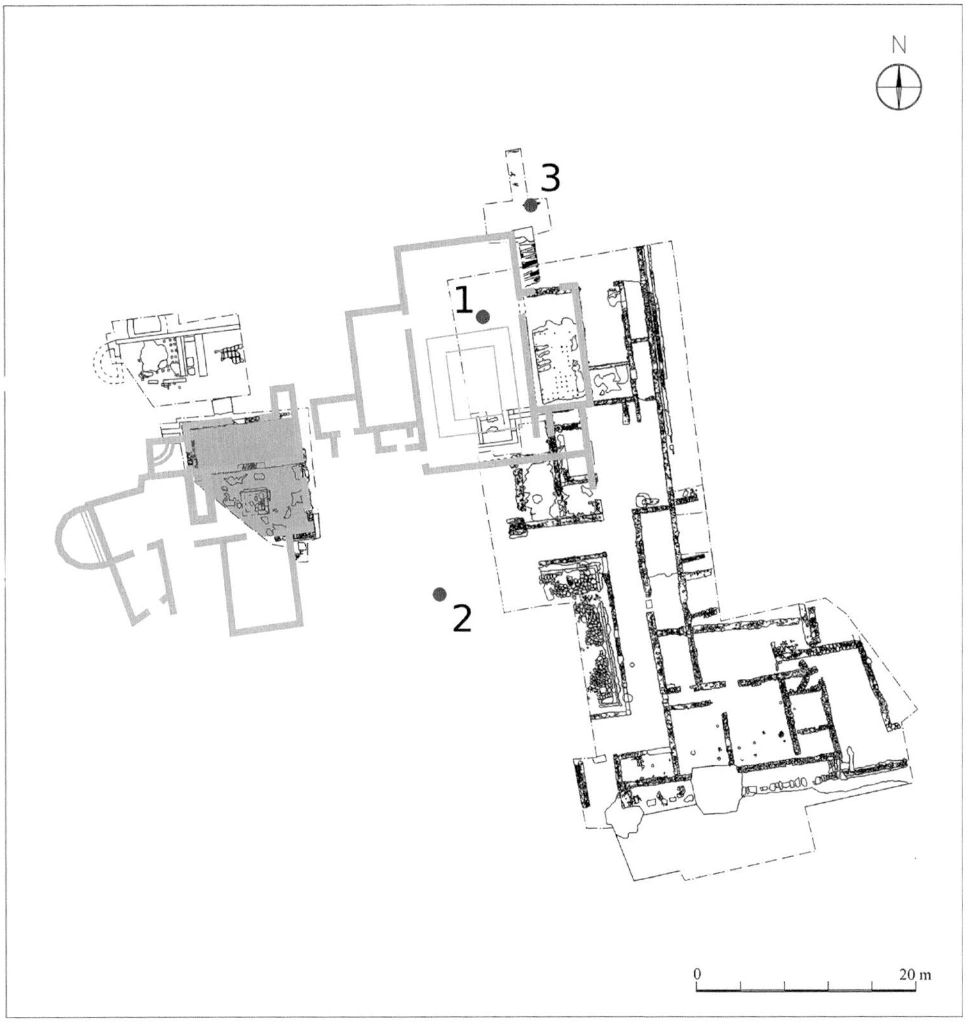

Figure 15.2 Archaeological plan of the excavated structures with, superimposed, the sketch plan of the excavation made by Leopold the II.

Note: The grey area indicates the position of late antique floor mosaic; the black dots indicate the radiocarbon-dated burials (numbers refer to Figure 15.6).

by Leopold II, it was (oddly) neither published nor documented in any detail. A few years later, the mosaic was covered with a solid shed to protect it, and the excavated remains of the surrounding site became part of one of the first examples of an archaeological park in Italy (Zanini, 2019a). This small park remained in use up to an undefined moment in the second half of the nineteenth century, when the external remains were buried again. A few decades later, possibly at the end of the same century or even at the beginning of the twentieth century, the floor mosaic was hidden below a rustic floor, and the shed that covered it was converted into a shelter for agricultural machinery (Giorgi & Zanini, 2015). This was used for several decades as a working place, while the memory of the late antique mosaic underneath its pavement faded, in spite of the fact that the field was known as the "field of the mosaic". The shed was finally demolished

Figure 15.3 Drone view of the archaeological site.

Note: (a) Ancient Via Aurelia path; (b) villa/mansio; (c) late antique bath complex; (d) late antique floor mosaic; (e) Roman bath complex; (f) amphora and roof tiles kilns; (g) ancient lagoon (?, the identification is probable but not 100% certain).

in 1960 and the field was ploughed in order to plant a vineyard (Dallai et al., 2003): the ploughs largely destroyed what was below the soil, sparing only the mosaic, which was protected by the solid foundations of the nineteenth-century shed.

In the following decades, sequential ploughing brought to light a large amount of pottery, collected mainly by looters and just sporadically by agents dedicated to the protection of archaeological remains. A renewed archaeological investigation, started in 2004 and still ongoing (Giorgi, 2016; Giorgi et al., 2005; Zanini, 2011; Zanini & Giorgi, 2014), allowed us to piece together the whole set of events, to bring again to light the famous mosaic and to identify the buildings already excavated by Leopold II. Most importantly, it allowed us to contextualize all these elements in their spatial setting, placing their relationship with the surrounding territory within a long-term dynamic, on the one side, and with the economic and administrative macrostructures of the Roman, late antique, and the early medieval worlds, on the other.[1] Nearly 20 years of excavation have allowed us to appreciate the complexity of the settlement and its several phases of life, which can only be presented here just in a synthetic form (Figures 15.2 and 15.3).

15.2 A (un)typical Tyrrhenian scenario

This part of the Maremma was probably already populated in prehistoric and protohistoric times, while the presence of a continuous settlement in the Etruscan and Hellenistic periods is attested by finds revealed by the agricultural ploughing, including sherds of *bucchero*, Attic red-figure pottery, Corsican pottery, and a bronze *fibula* datable to the seventh and sixth centuries BCE (Dallai et al., 2003, 290); no building structures relating to this period have been unearthed so far. The history of the site changed drastically between the third and second centuries BCE, coinciding with the construction of the *Via Aurelia*, which crossed this part of Tyrrhenian Etruria and whose route very likely ran only some 50 metres away from the site itself.

The foundation of the first settlement in Vignale during the Roman period seems to be directly connected to the construction of this new road, probably correlated with the presence of kilns exploiting the widespread accessibility of clay, water, and wood for the production of roof tiles, amphorae, and plain pottery. The construction of these kilns also laid the foundation for the establishment of a consolidated tradition of productive activities that has survived down to our time (Zanini, 2019b).

During the late Republican era, this first productive nucleus gave way to a new set of buildings: four rooms in a row, opening onto a courtyard paved with gravel, were built at that time. They were probably part of a first villa, which was expanded in different and luxurious forms during the following decades and centuries. It seems that the villa evolved into a new complex of buildings in the Augustan period, covering a surface of approximately 3 hectares. It consisted of a main residential unit (*pars urbana*), a productive area (*pars rustica*), and a series of open spaces and gardens. At that time, these buildings were probably arranged on the boundaries of a coastal lagoon which offered favourable micro-climatic conditions, together with a good economic potential represented by fishing or fish farming and transport facilities for the tiles, amphorae, and plain pottery produced in the kilns. The latter activity is archaeologically attested by a new and larger set of pottery kilns that was erected on the lagoon's edge at the time of Augustus or a few decades later, under the ownership or perhaps the management of an Antiochus, whose stamps appear very frequently on tiles coming from our site. Antiochus might have been a freedman connected with a branch of the *gens Fulvia*, known in Rome for their entrepreneurial activities in this sector (Giorgi, 2019).

The extension of the villa, by far the largest in this area (Cucini, 1985, 290–295), and its unusual position between the lagoon and the *via Aurelia* – instead of on a more regular and favourable location on the slope of the nearby hill – suggests that several functions were associated with this settlement already in this initial phase: (1) an upper-class residence, testified to by the quality of what is left of the pavements and the wall facing, both of them characterized by the use of valuable marbles; (2) one or more productive functions (pottery making and fishing); and (3) probably a resting place and horse change facility for travellers using the *via Aurelia* on behalf of the government (Giorgi, 2016).

These multiple functions are the key to understanding another peculiarity of the Roman settlement of Vignale, which is its long lifespan. The second and third centuries CE are traditionally seen as the beginning of the decline – in number, geographical distribution, and economic prosperity – of the wine-producing *villae* in Tyrrhenian Etruria (Cambi, 1993). In contrast, Vignale seems to have remained an active centre during this period, as attested to by coin finds. Despite the damage produced by modern looters operating with metal detectors, coins were retrieved in large amounts, proving the existence of commercial activities involving the use, and therefore the loss, of coins, along the entire imperial age until at least the fifth century CE.

A second marker of the continuity of life and operation during the same period is given by the evident traces of structural renovations: namely, the refurbishment of the pavement in one of the yards with peristyle during the first half of the second century CE, and, at approximately the same time, the building of new rooms subdividing another former courtyard. Towards the south, very probably along a *diverticulum* connecting the villa to the main route of the *via Aurelia*, a new bath complex was built. Its location, in isolation at some 30 metres away from the main buildings, suggests that it was related to the function of road station that at least a part of the villa could have assumed in this period (as discussed in Giorgi & Zanini, 2019).

15.3 A late antique floor mosaic with a long history

The area of Vignale continued to be occupied in Late Antiquity, as testified by the installation of the mosaic floor originally revealed by Leopold II in the nineteenth century. The date of the mosaic is clearly indicated by its style and manufacturing technique and was confirmed during our excavation by the discovery of a coin of Constantine the Great in the bedding of the floor (Figure 15.4).

Unfortunately, almost nothing is known about the topographic and stratigraphic contexts of this mosaic, because of the processes of the 1830s excavations and later preservation attempts. Until now, it has been impossible to determine its original function, which could be connected either to a late antique bath complex or to a large dining room and a new phase of the villa, developed through the renovation of the original complex (Giorgi & Zanini, 2018). The topographical location favours the first hypothesis, since the mosaic is in close proximity with the contemporary bath building, partially excavated already in nineteenth century and now most extensively investigated during our research (Figure 15.5). However, the iconography suggests that it might have belonged to a large reception hall, as its main theme – the personification of

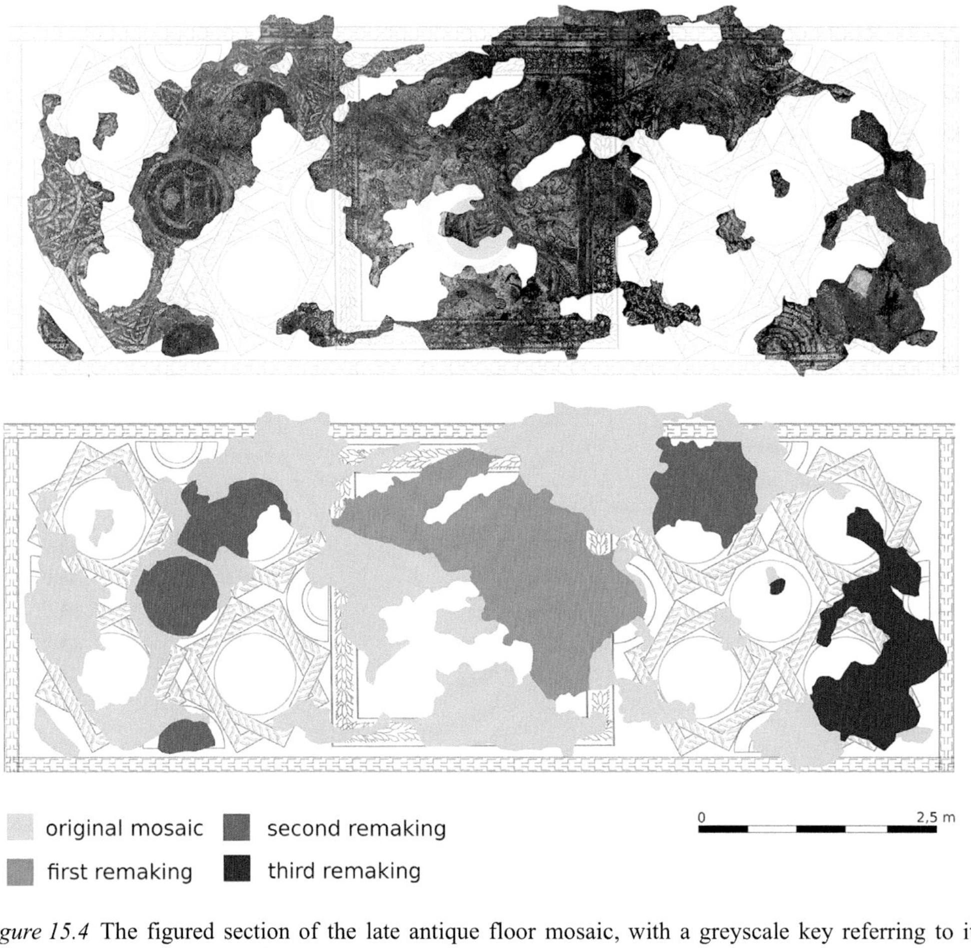

Figure 15.4 The figured section of the late antique floor mosaic, with a greyscale key referring to its distinct phases of life.

Figure 15.5 The figured section of the late antique floor mosaic in the topographical context of late antique bath complex.

cyclic Time (*Aion/Saeculum/Annus*) surrounded by the four Seasons – is often found in large dining rooms of late antique luxurious residences in North Africa (Parrish, 1984; Dunbabin, 1991). In fact, the style and technical quality of Vignale's mosaic suggests the direct involvement of African manufacturers, possibly using material imported from that same region since, after the first petrographic analyses, the mosaic tiles did not originate from local quarries – as should be expected, given that those quarries were widely exploited in Roman times.

Regardless of which topographical-functional context will be confirmed for the mosaic and the space it decorated in our future research, the continuous life of the human settlement of Vignale is indisputable. In case we are dealing with a dining room, it would indicate that in the first half of the fourth century the villa of Vignale was the seat of an upper-class aristocrat, rich enough to build a prestigious residence and pay for an engaging mosaic with a sophisticated iconographic theme. The personification of cyclic Time is depicted seated on a celestial globe, a very rare and meaningful representation that cannot be discussed here, but that suggests a direct connection to the highest rank of the late antique aristocracy (Zanini, 2018). Alternatively, if the mosaic belonged to an isolated bath complex, new and possibly more interesting lines of research will open up. A key focus would be the identification of the individual with interest in investing important resources to fund a building of this kind, presumably dependent on the road station that was still active at that time in Vignale.

Our mosaic has a life history, perhaps a long one, indicated by at least three successive phases of repair that changed its original composition. Unfortunately, we are still not able to assign a precise date to each of these interventions, but we can make a few considerations based on clues that emerged in our study. The first and most important repair occurred very probably sometime after the first damage, as demonstrated by the use of new mosaic tiles instead of a simple recovery of the original ones, which were evidently lost. This could suggest that the mosaic was abandoned for a period after a large disruption that affected more or less half of its central figurative field (Giorgi & Zanini, 2015). The changes weren't limited to the dimensions and the quality of the *tesserae* employed but extended also to the style and iconography of the figurative scene: the general structure was maintained (the portrayal of the Seasons on the corners of the central field recalls the original representation faithfully), while the background was drastically – and somewhat inexplicably – modified. The original white background with isolated *xenia*, which would have been very easy to restore, was instead replaced with a decidedly more complicated motif of vegetable spirals with pairs of birds. Even more extensive was the redesign of the central part of the figurative section, where the original personification of Time – a scantily dressed young man seated on a celestial globe – was transformed into the portrait of an actual human, rather than a personification: a fully dressed man, who remained seated on the roughly restored globe, whose head appears arranged in an incongruous position in relation to the chest. Since it is difficult to conceive such a resounding mistake on this very crucial figure – moreover made by artisans who proved to be excellent in repairing or reorganizing the remaining parts of the same mosaic – the change in the main figure has to be intentional and intended as a form of re-semanticization of the whole mosaic, maybe due to its insertion in a new context, whether residential or not.

An elaborate series of stratigraphic observations, whose discussion is out of the scope of this chapter, suggests that this drastic makeover of the floor mosaic is contemporary with an enlargement of the late antique bath complex, where a new system of semicircular heated pools, coated with fine marbles slabs was created. This is also likely contemporary to the reorganization of the living spaces of the Roman *villa*. This new late antique – or even early medieval – luxurious residence is marked by the creation of a new large hall, arranged in the space previously occupied by a peristyle, and the extensive remaking of the pavements in the surrounding rooms with high-quality floors: polychrome mosaics; *opus sectile*; and *signinum*, the latter decorated with marble insertions (Bueno, 2011). This phase, in part exposed by the nineteenth-century excavation, was unfortunately largely destroyed by ploughing. However, on the basis of a thorough work of reconstruction starting with the few remaining pieces of evidence, converging clues from stratigraphy and finds suggest that this overall remaking of the villa should be provisionally dated to between the end of the fifth and the beginning of the sixth centuries (pending the potential discovery of further elements that may confirm or contradict such a hypothesis). The same date is also suggested by the iconography of the first remaking of the mosaic.

The stratigraphic sequence of the making and remaking of the Seasons mosaic continues with two further extensive interventions, both of them recognizable from the different techniques and materials employed. The second renovation (hence the third phase in the mosaic life) involved mainly the two side-carpets, where the *clipei* inserted into interlaced frames and originally decorated with personifications were replaced with aniconic black-and-white motifs, employing larger tiles. Finally, a third and final renovation was characterized by the sole use of even larger black tesserae, organized in order to fill up the gaps produced by the passage of time, both in the original mosaic and in its renovations. It should also be noted that, during the excavations of areas beyond those that were disturbed in the nineteenth century, mosaics with the same materials and decorative patterns were also found beyond the large Seasons mosaic. This fact proves that we are not dealing with modern restoration work (a possibility that always has to be kept in mind) and forces us to consider the time span in which these renovations occurred, as well as their spatial and functional contexts.

As observed earlier, the more recent phases of the villa (probably dating from the fifth to sixth centuries) have been almost completely lost, but significant traces of structural modification and changes in function have been found in the area of the late antique bath complex, the only section where the more recent stratigraphy is preserved in a readable manner. On the basis of the data collected in the current excavation and its comparison with the map drafted during the Leopold II excavation, we can hypothesize a prolonged use of the late antique *thermae*, even beyond the lifespan of the villa, possibly within different spatial setups (Figure 15.2). This is an additional factor to be considered when attempting to investigate about the entity, public or private, which was able to guarantee the continuity of function of such a large and articulated settlement in a still undefined moment of the early Middle Ages.

15.4 Burials and bishops in the early Middle Ages

In addition to the stratigraphic data concerning the bath house and the villa, it is possible to draw on two pieces of evidence: the first coming from the burials found within the excavation area, and the second consisting of a number of archival documents. Apparently, Vignale was the burial site of a fairly large community for centuries (Figure 15.6). We have so far found three burial nuclei: the first and most ancient one consists of tombs unearthed in the area of the villa already explored by Leopold II and then heavily damaged by ploughs, a fact that hampered us from obtaining a precise identification of their topographic and stratigraphic contexts. One of these tombs we were able to fully analyse was without grave goods, but it can be dated to the seventh century based on radiometric analysis[2]; a similar date (i.e. the end of the sixth century or the beginning of the seventh century) can be suggested for three bronze belt buckles found in the 1980s in a ploughed field and very probably coming from destroyed burials (Dallai et al., 2003, 302–310).

A second group of burials was located in one of the corners of the portico surrounding a stone-paved yard which is supposed to have been the central structure of the southern and non-residential part of the villa. This sector probably functioned as a resting place (*mansio*) for people moving along the *via Aurelia*, both in Roman imperial times and in Late Antiquity. In this case the radiometric analysis of one of the skeletons indicated a date between the end of the eighth and the ninth centuries, with the possibility to extend this period until the tenth century. The tomb is located in the stratigraphy that lies below the collapse of the ceiling, and this would imply that in this late period the portico itself was still standing – at least partially – and that the surrounding structure was still functioning in one way or another.

The third group of tombs can be dated even later in time. This is an extended burial ground which has, so far, been investigated only by means of an exploration trench: inside it, we found some 40 simple pit graves, without any goods, but arranged on at least two overlapping levels. The tombs are located inside what was originally the garden of the Roman villa, fenced by a rectangular ambulacrum, and the trench we excavated represents a random sample of the entire burial site; the high density of graves suggests that the overall number of the burials must be in the hundreds. This indicates that this space was operating as a burial place for a long period of time and for a rather numerous population; after this, it is even more notable that the only radiometric dating obtained so far reports a date between the eleventh and the twelfth centuries. This dating is surprising because, until our investigation, it was commonly assumed that by the early Middle Ages the lagoon of Vignale had transformed into the swamp that survived until the

Figure 15.6 The radiocarbon-dated burials.

Note: Numbers refer to the position in Figure 15.2: (1) seventh century; (2) ninth century; (3) eleventh to twelfth centuries.

nineteenth-century land reclamation, and this should make it unlikely that such an inhospitable area could host a large and enduring human settlement. More credibly, our burial site could be instead associated with the presence of a church, which might have served as a burial point for many different small villages distributed in the surrounding area.

This hypothesis is also supported by literary sources, in particular a series of archival documents preserved in the *Codex Diplomaticus Langobardiae* (i.e. the Lombard Diplomatic Code) and a series of documents related to the history of the Duchy of Lucca. These documents pertain to the activities of Lucca's bishops in the Cornia valley between the second half of the eighth century and the year 1000, that is more or less the time span indicated by the radiocarbon dating of the burials found in Vignale. Starting from the 770s (*CDL*, II, 239, 240), the Langobardic bishops used the church of San Vito in Cornino as a "logistical base" for the administration of their properties in the Val di Cornia district, as indicated in the signing formula of a number of donations, exchanges, and other property documents. From at least 906 onwards (*MDL*, V, 3, 1092), the same place is defined as a *curtis* (i.e. the economic and management centre of a large early medieval rural estate), and a church named San Vito in Cornino is mentioned in two interesting documents dating back to 970 and 974, respectively. In the first one, an act is stipulated in *Loco Cornino, Villa qui dicitur S. Vito*, and associates the church's hagionym (saint's name) to an open settlement (*villa*), which might be the heir of the structure we are currently excavating at Vignale (*MDL*, V, 3, 1416). In the second document, the same place is described as located in *chomitato et territorio Populonense*, and it is stated that San Vito owns a series of land plots and rural houses, and that it adjoins a *via publica* (most probably identifiable as what survived of the ancient *via Aurelia/Aemilia Scauri*) and a watercourse, the Botrangolo canal, that today runs in close proximity to Vignale (*MDL*, V, 3, 1449). Merely six years later, the toponym Vignale appears for the first time in a document from 980 that was signed *in loco et finibus Cornino ubi dicitur Viniale* (*MDL*, V, 3, 1517), while in a document from 996, the expression changes into *loco Cornino juxta Turre qui dicitur Sancto Vito* (*MDL*, V, 3, 1705). In the last document of the series, dated back to the year 1000, the church appears again (*loco ubi dicitur ad Sancto Vito*) and its location in proximity to a *via publica* is reaffirmed (*MDL*, V, 3, 1751).

These documents are not totally conclusive because the church and the *curtis* of San Vito are not clearly associated with the toponym of Vignale. However, a simple evaluation of the probabilities seems to point to a continuity of function at the same site during this time, albeit with different modalities: at first as a church and its *villa* annex, then as a structured *curtis*, and finally as a *turris* (Ceccarelli Lemut, 1985). Of course, while there is no definitive proof that the literary data refer to the site of Vignale, the balance of probabilities suggests that it does, given that the descriptions made offer a credible image when paralleled with the continuous occupation shown in the archaeology.

15.5 Looking for an explanation

While it is important to try to identify the structures being excavated with the *curtis* of Cornino, it is also important to note the contribution this work is making to our understanding of the territory across Late Antiquity and into the early Middle Ages. In Vignale, we observe something that is very different from the well-consolidated interpretative model that outlines the "end of the villas" – both as structural decay of the single settlements and as an overall crisis of the economic system based on them – as the crucial division between the Roman and the early medieval worlds (Brogiolo et al., 2005; Francovich & Hodges, 2003).

At Vignale, the Roman villa does not disappear as happened to its counterparts on the Tyrrhenian coast, such as the case of Settefinestre, whose interpretation (Carandini, 1985) has deeply marked the subsequent historiographical debate on the fate of villa economy. Our villa is subject to neither a secondary and parasitic occupation, according to the "*squatting model*" that was largely adopted in previous literature (Lewit, 2003) nor the mutilation of its structures, which were instead maintained, and even reorganized and improved until a very late time. At present, it is not yet possible to claim if this is an isolated phenomenon or if a similar evolution can be seen elsewhere. In the first case, Vignale would be a *hapax* that should be explained; in the second one, it would be simply a rare case of extraordinary visibility of a more common trend that can only be followed here, thanks to its rich archaeological and extra-archaeological set of documentation. In both cases, the evidence indicates the need for a reassessment of our understanding of the evolution of space not just at Vignale but elsewhere too.

The first element to take into consideration is the strong topographical and functional connection between our site and the ancient *via Aurelia*. This connection may have played a crucial role in at least three aspects. First of all, it now seems correct to conclude that the road maintained its original functionality far longer than has previously been claimed. The hypothesis of the abandonment of the *via Aurelia* following the devastations perpetrated by the Visigoths at the beginning of the fifth century rests essentially on the contemporary testimony of Rutilius Namatianus. In a passage of the poetical account of his travel towards Gaul (*De reditu*, 1.39–1.41), he claims to have chosen to travel by sea because "the fields of Tuscia and the Via Aurelia had been devastated by the Goths". However, Rutilius himself, in a subsequent passage of the same text, indirectly testifies to the practicability of travelling along the *via Aurelia* and also the continuous functioning of the *cursus publicus* in the area. He reports having received news of the appointment of his friend Rufius Volusianus as a new *Praefectus Urbi* when he arrived at Populonia (Rut. Nam., *De reditu* 1.415–419). Since Rufius Volusianus was appointed after the departure of Rutilius from Rome, the speed with which the news reached him suggests that the system of transmission of official information was still active and this indirectly proves the survival of the network of *mutationes/stationes/mansiones* that supported it.

Additional indicators confirm the continued use of the *via Aurelia* at least throughout the fifth century. For example, another well-known passage of the same *De reditu* (1.453–490) praises the splendour of the residence of another upper-class aristocrat, Caecina Decius Acinatius Albinus (PLRE II, Albinus 7), whose *villa maritima* arose nearby at Vada Volterrana, controlling its saltworks – the latter represented a stable source of income until at least the end of the eighth century. It is difficult to imagine an aristocrat as Caecina Albinus, who had previously served as *Praefectus Urbi*, having his residence in an area that was isolated and cut off from any terrestrial communication and connected to the outside world only by sea (Dallai et al., 2003, 302–310). The presence of upper-class late antique villas in this region is, after all, confirmed by recent archaeological investigations for instance at San Vincenzino (Donati, 2012) and in this scenario Vignale would also be included. In other words, if the Visigoths did not cause the disruption of the *via Aurelia*, and assuming that the Gothic War had only a limited impact on this territory (Zanini & Celani, 2020), we may suppose that the structure and/or the function of the ancient road survived for a longer period, until the times of the Lombard bishops of Lucca.

A second and more general consideration suggests the continuity of function of the road: such relevant infrastructural elements, once settled and established, will tend to survive after the disappearance of the original reason for their construction. Two of the main communication routes to and from Rome contemporary to the *via Aurelia*, that is, the *via Flaminia* and the *via Appia*, continued in use during the early Middle Ages, as demonstrated by the many episodes of the Byzantine–Lombard conflict that occurred along the trail of the *Flaminia* (Menestò, 1999), or the established use of the *Appia*, and then of the road towards Calabria, during the Italian expedition of Costans II, in the second quarter of the seventh century (Corsi, 1983). In a similar way, our road could also have remained in use during the early Middle Ages, although with a different organization and under a new and undefined authority responsible for its maintenance. Lucca's bishops still preserved their large properties in Val di Cornia, and it is therefore probable that the old Roman road was used as their point of access to the territory, since there was no other connection, by land or sea, between Lucca and this part of Maremma. By this time, the ancient Roman road had long since lost its original function, that being a major piece of service infrastructure inside the macro-economic system of the Roman world. However, this does not necessarily imply that it had lost its physical shape; the road, its route, and its paved sections remained integrated into the landscape, albeit the memory of its original role and function was largely lost; in other words, they could now be interpreted like a relic of a past anthropic addition to the natural landscape.

The permanence of a few physical remains of the road – or, at least, the memory of their existence – can be traced in the toponym "*via della silice*", deriving from the presence of large paving stones, which is recorded on different nineteenth-century maps, not far from Vignale (Dallai et al., 2003, 299–300). Sections of this same road, close to Vignale, were still visible at the end of the seventeenth century (Fiorelli, 1880, 172). Furthermore, the original route of the ancient *via Aurelia* was still well defined from the juridical point of view in the first half of the nineteenth century. When Grand Duke Leopold II built the new *via Regia* (royal road) between Pisa and Grosseto, a number of land plots along the old *via Emilia* (to be read as the ancient

Aemilia Scauri/Aurelia) were transferred from public to private property, marking the demise of public control over the route of the ancient road. From this point of view, the relics of the ancient *via Aurelia* continued to play an important role in local life, although its original function of integrating the areas it crossed to the broader Roman imperial economy was lost at the end of antiquity. The ancient road trail became a specific feature of that landscape, which was able to condition how that landscape "worked", allowing travellers to reach Vignale more rapidly and easily than other places and therefore making it an "ideal" place from which to administrate that part of the territory. In a *longue durée* perspective, this worked fine whatever the power – public or private – that exerted its control over that specific space.

The last aspect of our analysis concerns the complex network of relationships between the road and the territory in which it was built. A Roman road is a multidimensional entity: it is certainly a physical structure that crosses an area with the main purpose of improving the mobility of people and goods in the same area, but it is also a physical connection between that area and the cities the road was originally conceived to join together, in our case Rome and Pisa. In other words, the road establishes a network of micro- and macro-economic relations with different territories, which do not necessarily disappear when the road loses its original use. The construction of a new road always implies the creation of a new spatiality in an area, something we could define as a permanent or semi-permanent transformation of its local microecology.

As such, the construction of a new road can be compared to that of an aqueduct, something that many ancient authors described as the creation of an artificial river (e.g. Gregory of Nazianzus, *Oration* 33.6). Aqueducts were capable of changing the life of a city or a portion of territory by bringing water to an area where this had been unavailable, therefore fostering the establishment of new settlements as well as new functional, productive, and economic relationships between different areas of the same territory (Giorgi, 2021). In many respects, a road operates in the same way, building new relationships between parts of a space and reorienting its connectivity with distant areas and their macro-economies. Such relationships and connections tend to auto-regenerate in different ways, and thus to survive even when the original road disappears, simply by creating the need of the reconstruction of a new infrastructure, maybe with a different form but still capable of fulfilling the same role. On these bases, it is possible to talk about "infrastructural resilience", that is, the survival of an economic network that may involve the continuous regeneration of its originating element, even though this will assume different and adapted forms.

Vignale appears to be a place of infrastructural resilience, given that, after its origins in connection with the construction of the *via Aurelia* and then its long life next to the same road, it was able to evolve into a *central place* of a micro-territory. As in the somewhat parallel case of the road station of Santa Cristina in Caio, not far from Siena (Bertoldi & Valenti, 2016), the central place of Vignale was eventually able to outlive the Roman road (Zanini, 2016). Only later in time, possibly in the thirteenth century, was the human settlement definitely disconnected from the trail of the ancient road and receded back to one of the surrounding hills, assuming the shape of a medieval castle. However, after the early abandonment of the castle, at the very beginning of the fifteen century, the original place of settlement regenerated again – some two centuries later – as a modern farm, halfway between the hill and the lowland. In this long-term perspective, the movement from the plain to the hilltop is more a parenthesis than a definite break that marks the end of antiquity and the beginning of the Middle Ages. The human settlement was a shifting one: born as a "product" of the new road in the third or second century BCE, it moved up to the hills only many centuries after the end of the Roman world; it then returned to the plain as soon as possible and became, in turn, a landmark for the building of a new road, more than

2,000 years after the original one. From this point of view, it seems to me that one could speak about "functional resilience", a specific character that ensured that our territory took a different trajectory as compared to surrounding areas.

15.6 Structural resilience?

Together with this complex form of infrastructural resilience of the road and the human landscape created by its construction and function, a second form of resilience seems to be recognizable, one that I would call "structural". This was represented by the physical structure of the *villa-mansio*, which became a stable part of the landscape, beyond the economic and functional reasons that propelled its formation many centuries before.

Some parts of the villa's physical structure – according to our observations, mainly the late antique bath complex, and maybe the courtyard of the *mansio* and surrounding buildings – provided suitable rooms and spaces for the duties related to the presence of the later church of San Vito (if it was really located in this area) or any other centre of the *curtis*. In this case too, the new public and private powers that over time succeeded each other in the administrative and economic control of this part of Tuscany could have reused the structures built many centuries before, in a completely different context, maintaining their original function or adapting it to the new needs that arose with the new times.

The micro-economic relationship established between the *villa-mansio* and the surrounding territory certainly played a crucial role in favouring these two forms of resilience. Born in close functional relationship with the *via Aurelia*, the villa of Vignale, and then the large and complex settlement that over centuries developed around it, represented the point of contact between this micro-territory and the large macro-economic and administrative system of the Roman world, at least until the fifth century CE. In this respect it is easy to imagine that our site developed the connotation of a *central place* on a local and subregional scale, becoming a physical point of reference for the social organization and the population of the surrounding area.

Such connotations of being a *central place*, solidly supported by the persistence of the built structures in the landscape, survived in common perception even when its original system had long disappeared and allowed Vignale to outlive the disappearance of the Roman world and the "chaotic" formation of the early medieval Tuscan landscape (Valenti, 1996), surviving until the formation of the landscape of castles in the central centuries of the Middle Ages. The construction of the medieval castle on the hill surrounding our site during the thirteenth century, and then, after its abandonment, the birth of the sole residential and productive pole of this area with the settlement of the Franceschi family during the second half of the seventeenth century in the farm of Vignal Nuovo (i.e. the original nucleus of a still active farm centre) represent further stages of an extraordinary long path of continuity in the role of this site against its surrounding natural and anthropic space (Zanini, 2019b).

A further aspect, of no less importance, is that the structural and functional continuation of the settlement of Vignale could be due to its persistent connection with public power, albeit in different forms over the time. This is a line of research that still needs to be fully developed, but it seems reasonable to suggest that the lands where the Roman villa was built during the first century BCE were among those requisitioned following the Sullan wars and that, from that moment on, they effectively became part of the public trust to be assigned to senatorial families that sustained the new statal power (synthesis in Celuzza, 1994) or, perhaps, as a fiscal property. This status could have been maintained with the progressive changing of the holders of public power (Augenti & Terrenato, 2000): first the imperial administration, until its crisis in the fifth century; after that, the new post-Roman aristocracy of the Ostrogothic period; then again, the

Lombard bishops from the last quarter of the sixth century on. This line of continuity, precisely in this area of coastal Tuscany, is still waiting to be fully explored and it is nowadays the subject of a rich debate among historians and early medieval archaeologists (Bianchi & Collavini, 2018; Chirico & Citter, 2018; Bianchi, 2022). Our archaeological investigation in Vignale might offer new and "fresh" archaeological data to reassess the issue.

Notes

1 The archaeological research in Vignale is carried out by the Department of History and Cultural Heritage of the University of Siena, under the joint scientific direction by Dr Elisabetta Giorgi and myself, after a research permission by the Italian Ministery of Culture.

2 All radiometric analyses discussed here were made possible, thanks to the generous support of Richard Hodges and Giovanna Bianchi, within the European project NEUMed.

Editions and abbreviations

CDL – *Codex Diplomaticus Langobardiae* (*Historiae Patriae Monumenta* 13) (1873). Stamperia Reale.

MDL – *Memorie e documenti per servire all'istoria del Ducato di Lucca*, ed. D. Barsocchini, (1841). Felice Bertini.

Rut. Nam., *De reditu* – Rutilius Namatianus, *De Reditu Suo*. In *Rutilius Namatianus, Sur son retour*, ed. E. Wolff (2007). Les Belles Lettres.

Bibliography

Augenti, A., & Terrenato, N. (2000). Le sedi del potere nel territorio di Volterra: una lunga prospettiva (secoli VII a.C.-XIII d.C.). In G. P. Brogiolo (Ed.), *II Congresso nazionale di archeologia medievale: Musei Civici, Chiesa di Santa Giulia: Brescia, 28 settembre-1 ottobre 2000* (pp. 298–303) . All'Insegna del Giglio.

Bertoldi, S., & Valenti, M. (2016). Santa Cristina in Caio a Buonconvento (Siena): Diacronia di un central place. In P. Basso & E. Zanini (Eds.), *'Statio Amoena'. Sostare e vivere lungo le strade romane tra antichità e alto medioevo* (pp. 185–196). Archaeopress.

Bianchi, G. (2022). *Archeologia dei beni pubblici. Alle origini della crescita economica in una regione mediterranea (secc. IX-XI)*. All'Insegna del Giglio.

Bianchi, G., & Collavini, S. (2018). Public estates and economic strategies in Early Medieval Tuscany: Towards a new interpretation. In G. Bianchi & R. Hodges (Eds.), *Origins of a new economic union (7th–12th centuries), Preliminary results of the nEU-Med project: October 2015-March 2017* (pp. 147–159). All'Insegna del Giglio.

Brogiolo, G. P., Chavarría Arnau, A., & Valenti, M. (A c. Di). (2005). *Dopo la fine delle ville: Le campagne dal VI al IX secolo 11. Seminario sul tardo antico e l'alto Medioevo, Gavi, 8–10 maggio 2004*. SAP Società Archeologica.

Bueno, M. (2011). *Mosaici e pavimenti della Toscana: II secolo A.C.-V secolo D.C.* Quasar.

Cambi, F. (1993). Paesaggi d'Etruria e di Puglia. In A. Giardina & A. Schiavone (Eds.), *Storia di Roma 3.2* (pp. 229–254). Einaudi.

Carandini, A. (1985). *Settefinestre: Una villa schiavistica nell'Etruria romana*. Panini.

Ceccarelli Lemut, M. L. (1985). Scarlino: Le vicende medievali fino al 1399. In R. Francovich (Ed.), *Scarlino I. Storia e territorio* (pp. 19–74). All'Insegna del Giglio.

Celuzza, M. G. (1994). Etruria romana. In *Enciclopedia dell'arte antica, classica e orientale. Secondo supplemento (1971–1994), Vol. II* (pp. 549–554). Istituto dell'Enciclopedia Italiana.

Chirico, E., & Citter, C. (2018). I beni pubblici e della corona dall'Impero romano ai Longobardi: Il caso di Roselle (Grosseto). In C. Giostra (Ed.), *Città e campagna: Culture, insediamenti, economia (secc. VI-IX)* (pp. 97–120). SAP Società Archeologica.

Corsi, P. (1983). *La spedizione italiana di Costante II*. Patron.

Cucini, C. (1985). Topografia del territorio delle valli del Pecora e dell' Alma. In R. Francovich (Ed.), *Scarlino I. Storia e territorio* (pp. 147–320). All'Insegna del Giglio.

Dallai, L., Patera, A., Shepherd, E. J., & Zanini, E. (2003). Il Vignale ritrovato. In C. Mascione & A. Patera (Eds.), *Materiali per Populonia 2* (pp. 281–313). All'Insegna del Giglio.

De Ruggiero, A. (2016). *Leopoldo II granduca di Toscana. I viaggi, i documenti e la bonifica in Maremma*. Aska Edizioni.

Donati, F. (2012). *La villa romana dei Cecina a San Vincenzino (Livorno): Materiali dello scavo e aggiornamento sulle ricerche*. Felici Editore.

Dunbabin, K. M. D. (1991). Triclinium and Stibadium. In W. J. Slater (Ed.), *Dining in a Classical Context* (pp. 121–148). University of Michigan Press.

Fiorelli, G. (1880). Notizie degli scavi di antichità comunicate dal Socio G. Fiorelli al Presidente nel mese di novembre 1880. *Notizie degli Scavi di Antichità, 6*, 168–174.

Francovich, R., & Hodges, R. (2003). *Villa to Village: The Transformation of the Roman Countryside in Italy, C. 400–1000*. Bristol Classical Press.

Giorgi, E. (2016). La villa-mansio di Vignale: Vivere e viaggiare nell'Etruria costiera tra il I e il V secolo d.C. In P. Basso & E. Zanini (A c. Di), *Statio Amoena. Sostare e vivere lungo le strade romane tra antichità e alto medioevo* (pp. 173–183). Archaeopress.

Giorgi, E. (2019a). Appunti per ripensare la cronologia dei bolli laterizi di Marco Fulvio Antioco a Vignale. In M. Modolo, S. Pallecchi, E. Zanini, & G. Volpe (Eds.), *Una lezione di archeologia globale. Studi in onore di Daniele Manacorda* (pp. 335–339). Edipuglia.

Giorgi, E. (2019b). Gestire l'acqua, costruire il paesaggio. L'impatto ecologico dell'acquedotto romano-bizantino di Gortina sull'ambiente naturale e l'insediamento umano. *Annuario della Scuola Archeologica Italiana di Atene, 99*(1), 443–461.

Giorgi, E. (2021). Vignale tra archeologia, comunità, territorio: istantanee e riflessioni su un progetto in corso. *Rassegna di Archeologia, 29*, 275–295.

Giorgi, E., Patera, A., & Zanini, E. (2005). Il sito del Vignale tra storia, ricerca e valorizzazione. In *I segni dell'uomo, Atti del Convegno (Livorno 2004)* (pp. 115–127). Pacini Editore.

Giorgi, E., & Zanini, E. (2015). Il mosaico trasformato: Un pavimento di una villa tardoantica nella Toscana costiera. *Arte Medievale, 9*, 277–296.

Giorgi, E., & Zanini, E. (2018). Una residenza aristocratica nella Tuscia tardoantica e un mosaico pavimentale di complessa interpretazione. In I. Baldini Lippolis & C. Sfameni (Eds.), *Abitare nel Mediterraneo tardoantico* (pp. 365–374). Edipuglia.

Giorgi, E., & Zanini, E. (2019). Vignale (Piombino). Le terme di una villa/mansio nel tempo, tra antichità e alto medioevo (?). In M. Medri & A. Pizzo (Eds.), *Le terme pubbliche nell'Italia romana (II secolo a.C. – fine IV secolo d.C.). Architettura, tecnologia e società* (pp. 492–509). Romatre Press.

Lewit, T. (2003). Vanishing villas: What happened to élite rural habitation in the West in the 5th-6th c.?. *Journal of Roman Archaeology, 16*, 260–274.

Menestò, E. (1999). *Il corridoio bizantino e la via amerina in Umbria nell'Alto medioevo/a cura di Enrico Menesto*. Centro italiano di studi sull'Alto Medioevo.

Parrish, D. (1984). *Season Mosaics of Roman North Africa*. G. Bretschneider.

Pesendorfer, F. (Ed.). (1987). *Il governo di famiglia in Toscana: Le memorie del granduca Leopoldo II di Lorena, 1824–1859*. Sansoni.

Valenti, M. (1996). La Toscana tra VI e IX secolo. Città e campagna tra fine dell'età tardoantica ed altomedioevo. In G. P. Brogiolo (Ed.), *La fine delle ville romane: Trasformazioni nelle campagne tra tarda antichità e alto medioevo* (pp. 81–106). All'Insegna del Giglio.

Zanini, E. (2011). Vignale 2004–2010. Ridefinizioni progressive di un progetto di ricerca archeologica. *Materiali per Populonia, 10*, 263–274.

Zanini, E. (2016). Qualche appunto per un'archeologia contestuale delle stazioni di sosta nel mondo romano e tardoantico. In P. Basso & E. Zanini (Eds.), *Statio Amoena. Sostare e vivere lungo le strade romane tra antichità e alto medioevo* (pp. 71–79). Archeopress.

Zanini, E. (2018). Il Tempo seduto sul globo celeste. Appunti in divenire su un tema iconografico tardoantico. In S. Pedone & A. Paribeni (Eds.), *'Di Bisanzio dirai ciò che è passato, ciò che passa e che sarà'. Scritti in onore di Alessandra Guiglia* (pp. 751–764). Bardi Edizioni.

Zanini, E. (2019a). Leopoldo II di Toscana, il mosaico di Vignale e un progetto di parco archeologico nella Maremma dell'800. In M. Modolo, S. Pallecchi, E. Zanini, & G. Volpe (Eds.), *Una lezione di archeologia globale. Studi in onore di Daniele Manacorda* (pp. 115–121). Edipuglia.

Zanini, E. (2019b). Trama e ordito di un territorio: Un esperimento tra (micro) Big History e archeologia pubblica a Vignale di Maremma (Toscana). *European Journal of Post-Classical Archaeologies*, *9*, 303–325.

Zanini, E., & Celani, J. (2020). Archeologia della guerra greco-gotica: prolegomeni a una ricerca in corso. In J. M. Macias Solé, A. Ribera Lacomba, & M. Rosselló (a cura di), *Recintos fortificados en época visigoda: historia, arquitectura y técnica constructiva* (pp. 11–24). Istitut Català d'Arqueologia Clàssica.

Zanini, E., & Giorgi, E. (2014). Dieci anni di ricerche archeologiche sulla mansio romana e tardoantica di Vignale: Valutazioni, questioni aperte, prospettive. *Rassegna di Archeologia*, *24*(2), 23–42.

16 Lived space and social change in late antique Rome

The Campus Martius[1]

Carlos Machado

According to Procopius, Rome was nearly razed to the ground and turned into pasture for sheep in 546 CE. After ten years of war against the troops sent by Justinian to reconquer Italy, the Ostrogothic king Totila had decided to destroy the city to show his displeasure with the Romans and because it was impossible to protect it from a besieging enemy. Hearing about this, the eastern general Belisarius wrote a letter to his Ostrogothic rival pleading that the king be merciful. The city's magnificent monuments were testimony of its unique history and greatness, and therefore a treasure that belonged to all; destroying them would bring more infamy than military advantage. Accordingly, the city was spared (Proc. *Wars*, 7.22.6–7.22.17). As late as the middle of the sixth century, the city's monumental apparatus was still seen as a living symbol of the glory of Rome and its civilisation (Pani Ermini, 1999, 35–37; Wallace-Hadrill, 2022, 27–31). Belisarius and Totila followed a long tradition of seeing the city through its buildings and adornments. Two centuries earlier, when narrating the arrival of Constantius II (in 357), Ammianus Marcellinus listed the monuments that awed the emperor in a superlative language that characterised the largest city of the ancient world as 'the home of empire and of every virtue' (16.10.13: 'imperii virtutumque omnium larem').

To this day, the monumental spaces of the city and the texts that described them have remained a privileged medium for approaching late antique Rome. The evolution of the *Urbs* from imperial capital to Christian centre and its impact on the urban fabric and the people who inhabited it has recently attracted renewed attention (Machado, 2019; Salzman, 2021; Wagner, 2021; Dey, 2021). Rejecting the old debate between continuity or catastrophe, scholars have adopted different perspectives, looking at the imperial, senatorial, and ecclesiastical elites and the structures associated with them as a way of understanding their role in the transformation of the city.[2] Building on the accomplishments of previous generations (Pietri, 1976 and Krautheimer, 1980 remain fundamental), these works have shown how social, economic, political, and religious processes redefined the government of Rome, fostering the development of new architectural forms and spaces.

This chapter reconsiders these transformations from a different perspective, focusing on a specific part of Rome to examine the relationship between material and social changes as a way of understanding how urban space was redefined between the late third and the late sixth centuries CE. As the contributions to this volume illustrate, 'space' has been approached by late antique scholars through a variety of approaches.[3] In this chapter, I will examine the formative role played by urban space in late antique Rome's social life, and how different social dynamics gave shape to it in turn (a perspective argued by Lefebvre, 1991; Machado, 2019, 18–22). To do this, I will focus on Campus Martius, a region that played a central role in the urban development and image of the city across antiquity. This area roughly corresponds to the Augustan *Regio IX, Circus Flaminius* – a floodplain extending along the left bank of the Tiber at the foot

DOI: 10.4324/9780429427152-20

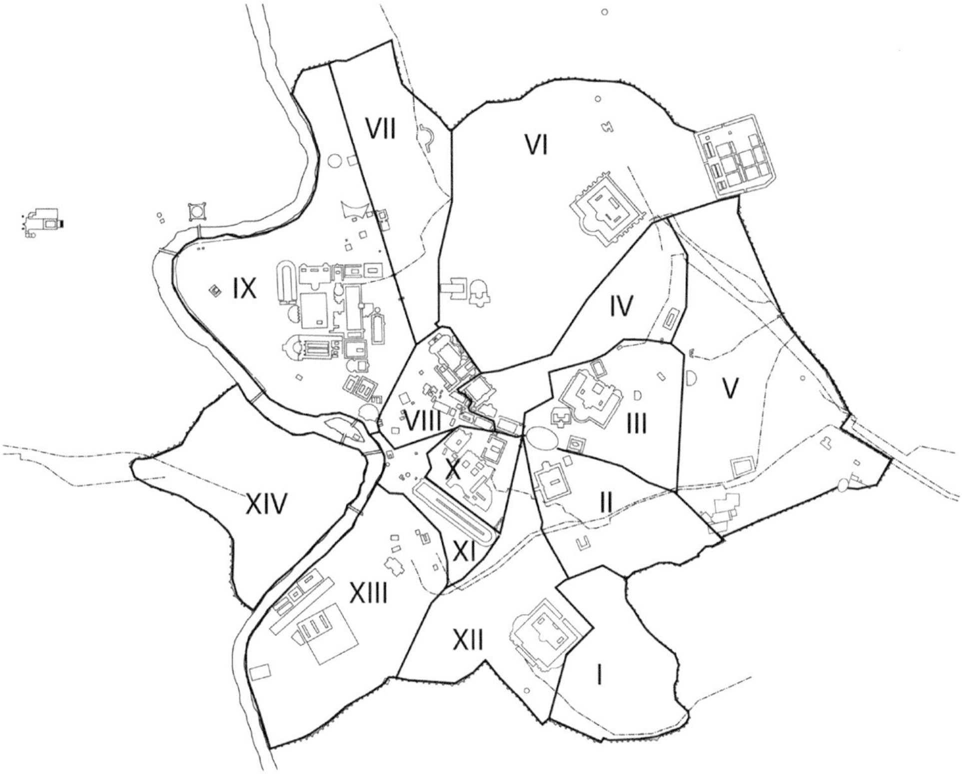

Figure 16.1 Rome's administrative regions.

Source: Adapted by Mary Woodcock Kroble and Carlos Machado from R. Meneghini and R. Santangeli Valenzani (2004). Produced with the authors' kind permission

of the Pincian, Quirinal, and Capitoline hills (Figure 16.1). Developments in neighbouring areas of the city, most notably *Regio VII* (*Via Lata*) and *Regio XIV* (*Transtiberim*), will also be considered for full context.

In spite of its importance, the Campus Martius was never the subject of comprehensive excavations like the Roman and Imperial *Fora* and has not attracted comparable attention among late antique scholars (e.g. Bauer, 1996; Kalas, 2015; see Christina Videbech in this volume). Almost nothing is known about the area next to the bend of the Tiber, to the West of Piazza Navona/stadium of Domitian, for example. And yet, recent archaeological work connected to the building of the new line C of the metro has added precious information to what is known through both discoveries made in previous centuries and the survival of ancient structures like the Pantheon (Egidi et al., 2010; Filippi, 2015a). Much of this information and its contribution for our understanding of the late antique topography of the area has been synthesised in the work of Spera (2014), which remains an invaluable demonstration of the continued importance of this region. As a result, the Campus Martius offers an unequalled wealth of information about various forms of use (monumental, residential, religious, and economic) by different social groups. An examination of this part of the city allows us to consider the making of late antique Rome from the ground up, from the perspective of its space and its inhabitants. Moreover, as Michelle L. Berenfeld demonstrated in her chapter, focusing on a specific neighbourhood allows us to see the connection between agents and processes that would otherwise remain disconnected in the analysis.

To consider these different elements, this chapter will be divided into four sections: after briefly characterising some of the more general elements that defined its landscape (section 1), we will look at how imperial and ecclesiastical elites used the Campus as a form of expressing power (section 2). We will then consider the development of a topography of dependence that bound the people who lived and worked in the Campus to the city's social and economic elite (section 3), before considering the impact of these developments on the lived experiences and daily struggles of subaltern groups (section 4). As we will argue, the area played a unique part in the making of late antique Rome, being a dynamic and contested space for social interaction.

16.1 A changing landscape

The late antique history of the Campus Martius started in 271, when Emperor Aurelian began works for building his wall around Rome, setting off a series of changes that would have important ramifications for the city (Dey, 2011; Coates-Stephens, 2012). For the entirety of its 19-kilometre circuit, the wall cut through previously existing neighbourhoods, demolishing buildings and blocking streets. By creating a massive physical barrier between the northern part of the Campus Martius and its outskirts, the wall redefined the landscape and topography of the area. It reoriented movement between the inside and the outside of the city, directing travellers through the *Porta Flaminia* into the *via Lata* (Figure 16.2: 1) or through one of its smaller gates. Six of these *posterulae* have been identified along the river wall, at least three of them corresponding to docking structures (Corvisieri, 1878; De Caprariis, 1999, 220–221; Dey, 2011, 194–195). These changes impacted the street system of the area (Dey, 2011, 304–309), a development with important economic and social consequences. This is shown by an inscription that records the arrival of wine from northern Italy at the *Ciconiae*, wharves located near the bend of the Tiber (Figure 16.2: A). The wine was then transported along the so-called *via Recta* to the temple of the Sun on the *via Lata* (Figure 16.2: B), where it was stored and distributed to the people (*CIL* VI, 1785=*EDR*111565; Chastagnol, 1960, 324). As De Caprariis observed, the nearby Monte Citorio is an artificial hill made of discarded pottery that attests to the economic dynamism of this zone (De Caprariis, 1999, 231).

The construction of the Aurelian wall, and especially the quantity of debris it generated, contributed to a different – and more profound – development in the history of Rome's landscape: the rise in the ground level of the city. From its very earliest days, the city's ground level was raised by floods, earthquakes, fires, as well as demolition and building campaigns (Manacorda, 2022). This was particularly true of the Campus Martius, an area prone to floods until the beginning of the twentieth century (Aldrete, 2007, 33–39). Between the fourth and the end of the sixth centuries, floods are recorded for the years 371, 398, 411, 555, 570, and 589, and (although sources rarely specify it) it is certain that our area was badly affected (Aldrete, 2007, 243). Ammianus Marcellinus (29.6.17–29.6.18) noted that the Tiber 'covered nearly everything', causing famine to occur in 371, and Claudian described the river running between the roofs of buildings in 398 (*De Bel. Gild.* 41–43). A similar picture is presented by Gregory of Tours, who mentions a flood that destroyed churches and warehouses in the late sixth century, causing famine and plague (*H.F.* 10.1). As the river receded after every flood, buildings were left damaged, structures abandoned, and even new routes for circulation opened.[4] Natural disasters offered building opportunities as much as suffering. This is seen in the case of the great conflagrations that destroyed large parts of the area from time to time, as in 64 and 80 CE (Wiseman, 1993, 223–224). Coarelli convincingly argued that the Diocletianic works recorded

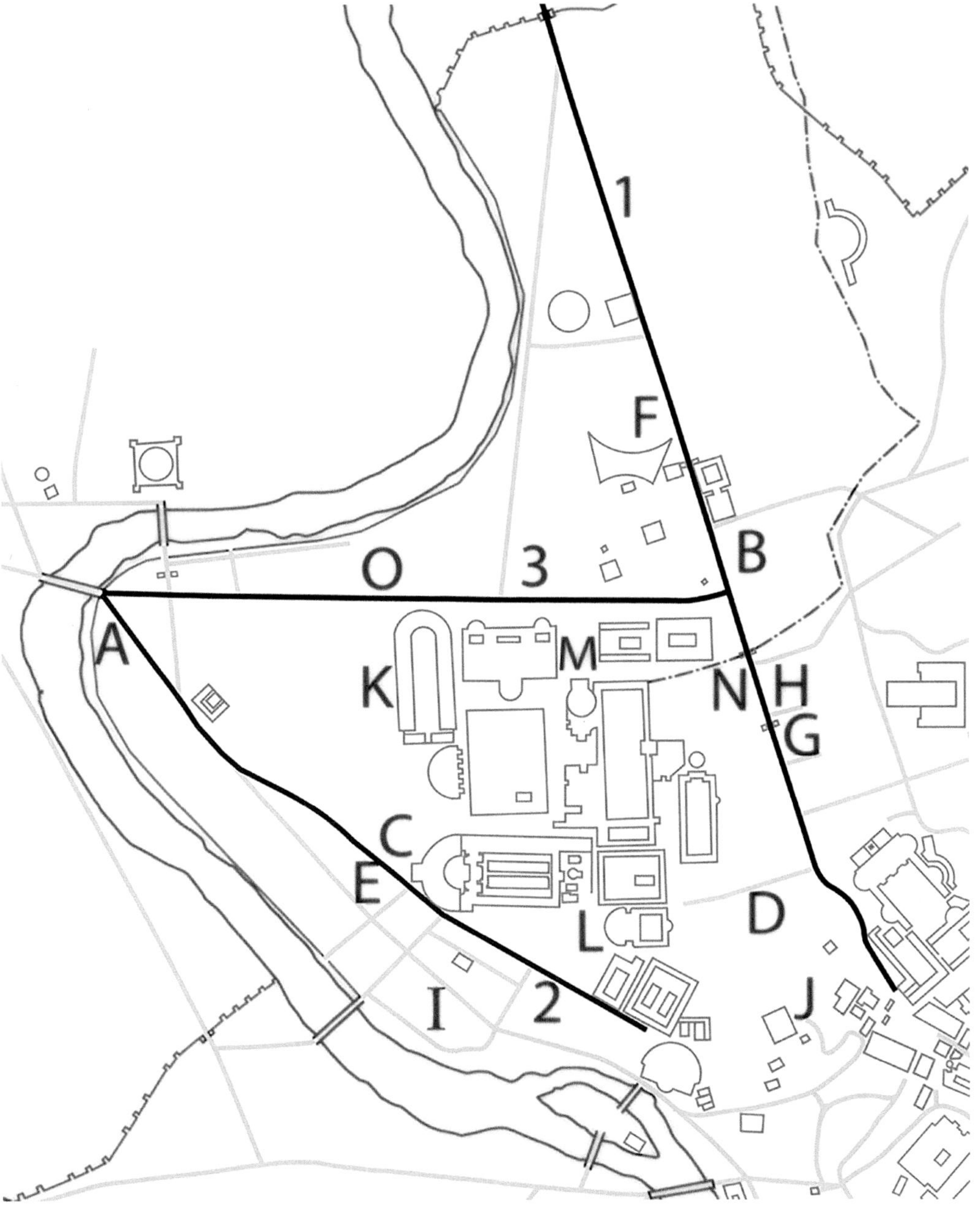

Figure 16.2 The Campus Martius, with structures cited in the text.

Key for the map: 1. *Via Lata/Via Flaminia*; 2. *Via Triumphalis*; 3. *Via Recta*; A. *Ciconiae*; B. Temple of the Sun; C. Theatre and porticus of Pompey; D. San Marco; E. San Lorenzo in Damaso; F. S. Lorenzo in Lucina; G. S. Marcello al Corso; H. *Insulae* off via Lata; I: *Insulae* at S. Paolo alla Regola; J. *Insulae* nea *Ara Coeli*; K. Stadium of Domitian; L: *Crypta Balbi*; M. Pantheon; N: Column of Marcus Aurelius; O. Forum of Sibidius.

Source: Adapted by Mary Woodcock Kroble from R. Meneghinin and R. Santangeli Valenzani (2004), produced with the authors' kind permission

in the *Chronographer of 354*, including on buildings in the Forum as well as on the complex of Pompey (Figure 16.2: C) and on the sanctuary of Isis in the Campus Martius, were carried out in response to the fire of 283 (Coarelli, 1999a, 23).

The raising of ground levels had a clear impact on this part of Rome. As Manacorda (2022, 39) noted, the level of the *via Lata* went from 12 metres above sea level in 220 BCE to 13.15 in Late Antiquity. Sediari collected data from excavations in different parts of the Campus, observing that the ground around the *Ara Pacis*, on the northern end of the *via Lata*, had gone from 10.7 metres above sea level at the time of Augustus to 12.7 in the second century. Changes were more dramatic in the South: the area of San Paolo alla Regola, near the theatre of Pompey, was raised from 10.45 metres above sea level at the end of the first century to 15.75 in the fourth, and the nearby Palazzo della Cancelleria went from 9.64 metres above sea level in the late Republic to 14.9 at the time of Constantine (Sediari, 1997, 237–247). It is in the context of these broader developments, which redefined the landscape of the Campus Martius at its most basic level, that we can consider political, religious, and social developments, and how the very identity of the area was refashioned in Late Antiquity.

16.2 The imperial Campus Martius

During the final decades of the Republic, powerful generals and politicians like Pompey, Caesar, and Agrippa used the Campus Martius as a space for defining their position in the city and before the *populus Romanus*. Temples, theatres, and porticoes were among the many types of monuments used to characterise their builders as magnificent patrons, adding to the complex of myths and cults that had populated the area since the origins of the city (Coarelli, 1997; Pentiricci, 2009). During the centuries that followed, emperors added to the area a dizzying array of entertainment structures, porticoes, religious buildings, and honorific monuments that confirmed the status of Rome and its rulers as the centre of the world (Figure 16.2) (Boatwright, 1987; Favro, 1996; Albers, 2013). The ideological character of the area and its association with the ruling court continued into the late antique period. Even the Aurelian wall, in essence a defensive structure, might have served as a symbol of imperial power (Dey, 2011, 116–137). To the monuments that lined both sides of the *via Lata* (Figure 16.2: 1), Aurelian added the already-mentioned temple of the Sun. Close to it was the so-called arco di Portogallo, an arch possibly built during the reign of Gallienus (253–268) or Aurelian himself, and whose remains were still visible in the sixteenth century (Torelli, 1993a). Further down the same road, a monumental arch was added to celebrate Emperors Diocletian and Maximian, celebrating the anniversary of their joint reign and perhaps their victory over usurpers in Britain (Torelli, 1993b).

The tetrarchs left their mark on this part of the city when they restored the theatre and porticus of Pompey (Figure 16.2: C): two inscriptions found in the area record the rebuilding of the complex and its dedication to Jupiter and Hercules Augusti, celebrating the association of the rulers with the divine (*CIL* VI, 255=*LSA*-1506; *CIL* VI, 256=*LSA*-1528, from 301–302). The theatre continued to attract imperial attention around the end of the fourth and the start of the fifth centuries, when Emperors Arcadius and Honorius sponsored extensive restorations, rebuilding the outer circuit and the interior of the structure (*CIL* VI, 1191=*EDR*111507).[5] Honorius was celebrated as the 'defender of Roman dominion and freedom' and 'unconquered ruler' in the same area with a statue dedicated by the urban prefect Aurelius Anicius Symmachus in 418–420 (*CIL* VI, 1193=*LSA*-1310).[6] A century later, the Ostrogothic king Theoderic requested that the patrician Quintus Aurelius Memmius Symmachus restore the theatre, offering to cover the expenses for the work (Cass. *Var*. 4.51), attesting to the enduring significance of Pompey's monument in the area.

Besides buildings, imperial *adventus* and triumphal processions were another privileged form of interaction between Romans and their rulers, and they too left an imprint in the Campus Martius (Liverani, 2007, 2016). The *via Triumphalis* (Figure 16.2: 2) was the traditional route followed by returning armies in triumph since the Republic and continued in use during our period (Pentiricci, 2009, 16–25). At the northern end of the road, the Senate dedicated a triumphal arch decorated with statues of Arcadius, Honorius, and Theodosius II, possibly also connected with the *via Triumphalis* (*CIL* VI, 1196=*LSA*-1311). Not far from it, an archway marked the end of the recently built *porticus Maximae* by Emperors Gratian, Valentinian II, and Theodosius I in 379–383 towards the *pons Aelius* (*CIL* VI, 1184=*LSA*-2668). These works might be related to a passage in a sermon of Augustine, in which he mentions a visit of Honorius to St Peter in the Vatican during his stay in the city in 403–404 (Aug., *Cum pagani ingrediderentur*, 26; Fraschetti, 1999, 244–247; Liverani, 2007). A new ceremonial route was developed with the Christianisation of the city, connecting its traditional imperial heart to the developing centre of Christianity at the other side of the Tiber on the Vatican. The expanding Christian topography was in this way integrated into the old forms of celebrating imperial rule.

This combination of continuity and adaptation can also be seen in the case of religious structures. As mentioned earlier, Aurelian built the temple of the Sun, and Diocletian and Maximian restored the sanctuary of Isis. Besides these two works, one of which was a restoration, there is limited evidence for official intervention on temples in the area, even from local authorities. Two inscriptions record the restoration of the temple of Apollo (probably Apollo Sosianus) by the urban prefect Memmius Vitrasius Orfitus in 357–359 (*CIL* VI, 45=*EDR*142032 and *AE* 2013, 142=*EDR*142033). Ammianus mentions that, a few years later, the prefect Claudius Hermogenianus Caesarius restored the porticus connected to the temple of *Bonus Eventus* (29.6.19). Like in other parts of the city, Christian buildings represented the main innovation to the area's religious topography (Cecchelli, 1998; Spera, 2014). Bishop Mark (336) built a basilica near the *Pallacinae*, close to the start of the *via Lata* (*Lib. Pont.* I, 202) (Figure 16.2: D). Not far from it, towards the Quirinal hill in *Regio* VII, Bishop Julius (337–352) built a basilica near the forum of Trajan, according to the Liberian Catalogue (*Lib. Pont.* I, 9). Bishop Damasus (366–384) built a basilica dedicated to St Laurence on his property behind the theatre of Pompey (*Lib. Pont.* I, 212) (Figure 16.2: E). Although known through archaeology, there is no evidence for the precise date or identity of the founders of the basilica later known as S. Lorenzo in Lucina, built after 350 (Brandt, 2012, 31), and the *titulus Marcelli*, on the *via Lata*, which is datable to the late fourth and early fifth centuries (Episcopo, 2003, 84–85) (Figure 16.2: F and G). As a result, by 400, the Campus Martius combined a strong Christian identity with its imperial associations. From the perspective of the imperial and episcopal elites, this was a key space for the representation of power, whether political or religious. As such, these privileged agents shaped the way we still see the area today, through churches and imperial monuments. However, although addressed to the Roman populace in general, these developments were closely connected to a rather narrow group. We need to consider a different (or messier, to echo Christina Videbech in this volume) type of use of space if we want to come closer to an appreciation of the deeper changes that were redefining this part of the city.

16.3 A social topography

Over the years, a large number of *insulae*, multi-storeyed apartment blocks, have been identified in different parts of the Campus Martius. As excavations have shown, the church of San Lorenzo in Lucina was built over a third-century *insula* just off the *via Lata* (Boman, 2012). A few metres further South, excavations in the early twentieth century revealed a group of six *insulae* of Hadrianic date, separated by paved streets (Figure 16.2: H). At least three of these structures had porticoes facing the street at the front, and *tabernae* around the ground floor (Figure 16.3) (Cavallero, 2011). Further South, a large complex of Domitianic date was found on via di San Paolo alla Regola, near the porticus of Pompey (Quilici, 1986–1987) (Figure 16.2: I). As the evidence available indicates, from the middle of the first century onwards, the Campus Martius was transformed by the intensive process of construction of multi-functional and multi-storeyed buildings that involved other parts of the city as well (De Laine, 2018). These contained utilitarian spaces like *tabernae* on the ground floor, often a larger apartment on the first floor, and smaller units on the upper storeys, famously of poor condition (Packer, 1971; Fogagnolo, 2022). A good illustration of this type of building is the structure on the slope of the Capitoline hill, near the staircase of the basilica of S. Maria in Ara Coeli – probably the eastern wing of a larger complex (Figure 16.2: J). The building, of which five storeys survive, had *tabernae* equipped with mezzanines on the ground floor, a courtyard, and progressively smaller apartments as one moves upwards (Packer, 1968–1969; Ippoliti, 2015). Rather than resting in splendid isolation, the imperial monuments of *Regio IX* were surrounded by a densely concentrated population, with the structures built for housing it.

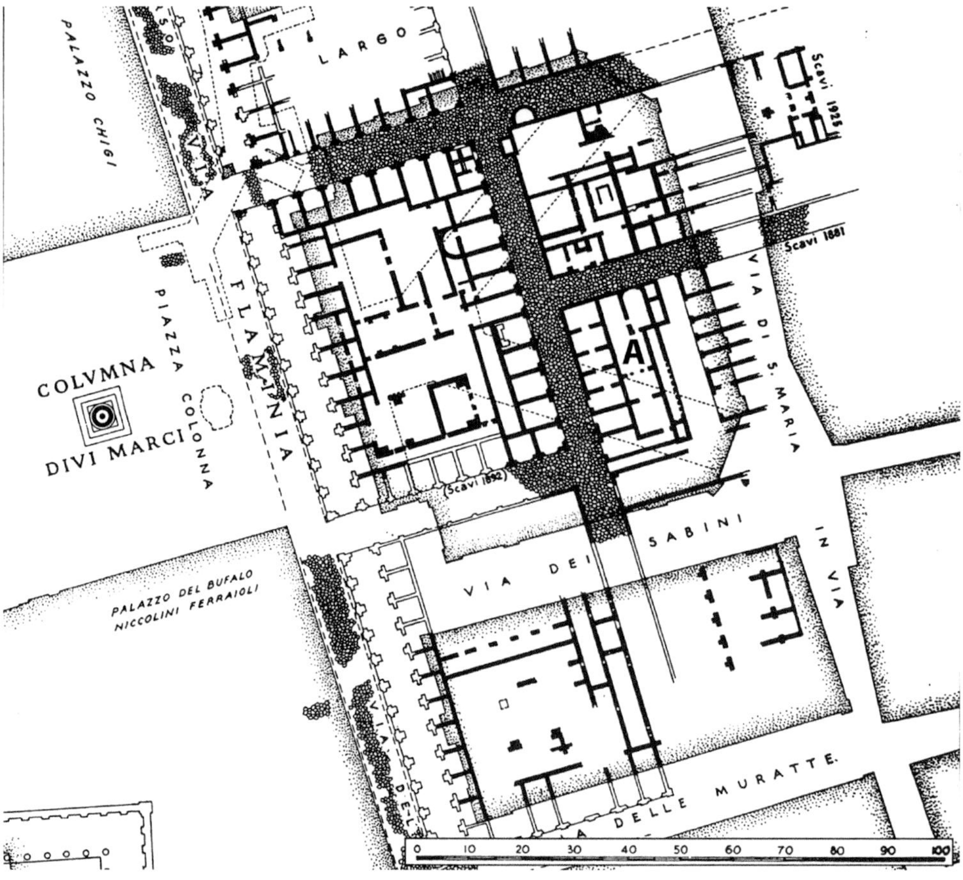

Figure 16.3 The *insulae* off *via Lata*.

Source: Adapted by Mary Woodcock Kroble from Meneghini (1999), with the author's permission.

Since at least the time of Augustus, the margins of the Tiber were developed as a harbour area, part of a complex system of quays, docks, wharves, and warehouses that stretched across the city (Castagnoli, 1980; Muzzioli, 2018; Malmberg, 2021). As the requirements of the imperial capital and of its population grew, *horrea* (warehouses) were constructed at different points, and the Campus Martius was no exception (e.g. D'Alessio, 2017, 511, 513, 520). Between the second and third centuries, a number of these structures were built along the *via Lata*, most likely servicing the docks by the Tiber (De Caprariis, 1999, 221–225; Machado, 2019, 49). The importance of this type of structure for the topography of the area is indicated by the so-called *Regionary Catalogues*, two lists (the *Notitia* and the *Curiosum*) of monuments compiled region by region around the middle of the fourth century. In this case, both lists record 25 *horrea* in *Regio IX* (p. 557), more than *Regio XIV Transtiberim* (with 22 *horrea*: p. 564) – an area more often associated with this type of structure. It is easy to see how the supply of grain, wine, and oil served the function of celebrating the power of the city's generous imperial patrons. But other forms of social relations also took place in these structures, whether we are talking about apartments or warehouses. Most scholars agree that apartment units in multi-storeyed buildings were rented by their inhabitants (Purcell, 1999, 151; De Laine, 2018, 318). Whether the poorer residents were able to pay the full rent or not, the fact remains that living in Rome represented, to a large part of the population, being dependent on a property owner.

This landscape of dependence experienced important developments in Late Antiquity, which can be traced in the material record. During the fourth century, in different parts of the city, *insulae* were converted at least in part into private *domus*, some with marks that distinguished them as elite residences (Machado, 2012; Fogagnolo, 2022). A good illustration of this phenomenon is the complex on the *vicus Caprarius* (*Regio VII*), where the ground floor of a multi-storeyed building was converted into a house, decorated with mosaic pavement, marble revetment on the walls, and statues. The survival of part of the staircase leading to the first floor raises the possibility that the apartments above remained in use as residence (Insalacco, 2005). A similar process can be seen in one of the *insulae* excavated in via in Arcione, a few hundred metres to the East (Lissi Carona, 1985). On the other side of the Tiber, in Trastevere, the excavation of the building known through an inscription as the *insula Bolani* revealed that the complex was being converted into a *domus* by the early fifth century, when works were interrupted and the structure was abandoned (Fogagnolo, 2013).

We can gain a better understanding of this process of conversion by considering a case for which we have more contextual information, the *domus* identified as belonging to the senatorial family of the Artemii in the fourth century, as indicated by an inscribed *fistula aquaria* (Meneghini, 1999). The house was built adapting the ground floor of one of the already-mentioned *insulae* off the *via Lata* (Figure 16.3). It was a sophisticated building, entered through a vestibule, with a room giving access to a courtyard from where an apsidal reception room could be reached (Figure 16.4). A long but narrow nymphaeum with niches for sculpture occupied one of the sides of the house. We cannot be certain about the type or extent of the influence that might have been exerted by a wealthy family over its immediate neighbours, especially in such a densely settled neighbourhood. It is possible, furthermore, that the house was rented to a different family, who in this case were dependent on the Artemii. In any case, it is likely that a family of senatorial status would possess enough social, political, and economic capital to offer support and protection to its neighbours.

The social power of members of the Roman elite was not merely an abstract sociological concept, but a reality expressed in the topography of the city. Their houses were not isolated points in the map of the city, but a central element in what could be an extensive and varied portfolio of economic and social investments (Machado, 2019, 201–230). When the aristocratic

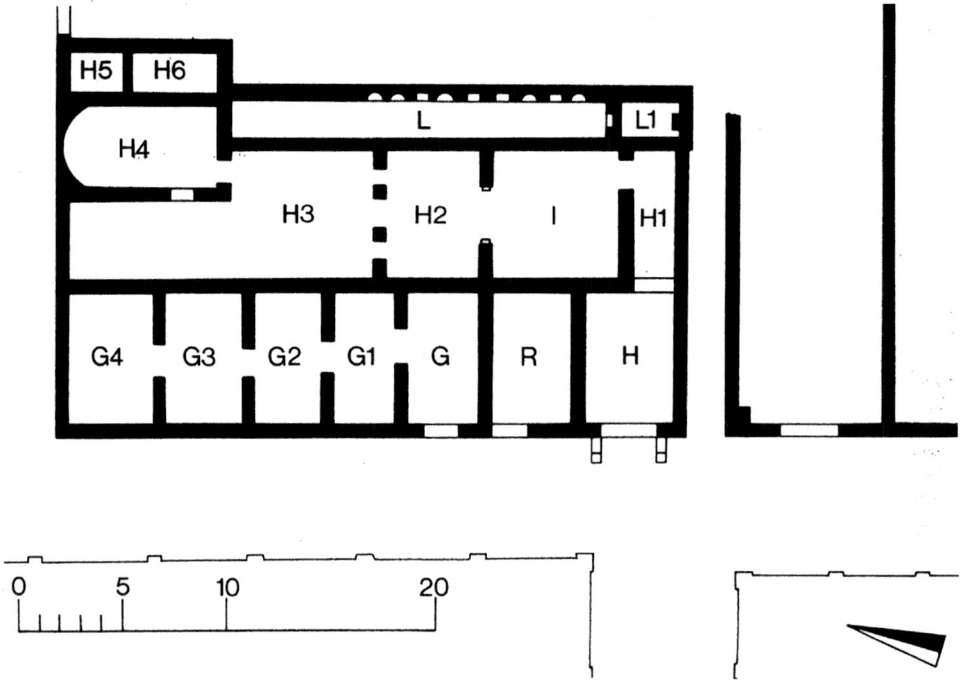

Figure 16.4 The *domus* of the Artemii.

Source: From R. Meneghini (1999). Reprinted with permission

lady Vestina donated properties to endow the basilica of Saints Gervasius and Protasius on the Quirinal, in the early fifth century, she presented (among other properties) six *domus*, two *balnea* (baths), and a *pistrinum* (a bakery), all in the same neighbourhood (*Lib. Pont.* I, 221–222). Writing in the late sixth century, Bishop Gregory I mentioned the house of a former cleric, located near the baths of Agrippa in the Campus Martius, left to the church along with a *pistrinum* and a *salganum* (a food shop) nearby (Greg. I, *Ep.* 6.42; 9.137). The presence of aristocrats and their properties in a specific neighbourhood might help us understand Ammianus's account of an attack by the lowly populace (*plebs infima*) against the house of the urban prefect Ceionius Rufius Volusianus in 365, which would have destroyed it if not for the help from the members of his household and his neighbours, *familiares* and *vicini* (27.3.8). The presence of wealthy houses, and the spatial distribution of their owners' property portfolios, contributed to the dominance of elite Romans over their neighbours. In this context, the adaptation (even when only partial) of *insulae* into *domus* had important consequences for the social topography of such a densely inhabited area.

The evolving landscape of dependence of the Campus Martius also impacted the world of labour, as we can see in the case of warehouses and the guilds that operated in them. Roman *collegia* and *corpora* traditionally sought to establish close links with powerful patrons. Patronage links gained a new significance in Late Antiquity, when aristocrats became more directly involved in the administration of the city – including the *annona*, the imperial system of food supply (Machado, 2019, 47–49). A number of inscribed bases of honorific statues, usually set up in the *atria* of the houses of their patrons, record the patronage relation between aristocrats and different guilds, attesting to the close connection between personal power, domestic spaces, and associations of workers.[7] This is clearly expressed in the monument dedicated by the *mensores* (grain measurers) of Portus to their patron Ragonius Vicentius Celsus, prefect of the *annona* in 389 (*CIL* VI, 1759=*LSA*-1464; Ciambelli, 2019). The fact that this base was first seen on the Tiber Island in the fifteenth century adds to its significance, as no aristocratic *domus* are known in the area. It is more likely that the monument was originally dedicated in one of the margins of the Tiber, in Trastevere or in the Campus Martius. Either way, its most likely original context was one of the many warehouses that occupied this area, possibly an administrative room or a courtyard, where his clients laboured under the cold and hard likeness of their patron.

If the identification of the provenance of Celsus's statue is correct, it illustrates the growing presence of aristocrats in the physical structures connected with the supply of the city – and, as a consequence, on the routine of the *collegia* and *corpora* that worked there. Since the Republican period, wealthy Romans had invested on the building and management of warehouses.[8] This presence became more intense from the fourth century onwards, as aristocrats often held offices that involved supervising work carried out in these structures, which made the threat of irregularities more frequent (Machado, 2019, 54–57). During the fourth and fifth centuries, the imperial court remained concerned with this presence, issuing legislation that tried to prevent powerful Romans from attempting to take over *horrea* illegally (e.g. *Cod. Theod.* 15.1.12; 14.15.4; 15.1.4). However, the situation had already changed by the beginning of the sixth century, as suggested by a letter of Theoderic to the prefect, Argolicus, in which the court acknowledged the rights of those taking over warehouses (Cass. *Var.* 3.29).

The growing weight of relations based on socio-economic dependence might also help us understand a more elusive process of spatial change, the adaptation of previously existing structures to productive use.[9] There is good evidence for the presence of workshops in the Campus Martius in earlier periods. We are informed, for example, about the *corinthiarius de theatro Balbi* Lucius Aufidius Aprilis, a trader or a worker (probably both) specialising in Corinthian bronze working, based in the vicinity of the *Crypta Balbi* during the last decades of the first century (*AE* 1977, 25=*EDR*079145). Towards the margin of the Tiber, excavations at Largo Perosi near modern-day via Giulia revealed a space for manufacturing, as indicated by the type of pavement (*cocciopesto* in certain rooms, *bipedales* or travertine in others) and the presence of a water basin, also datable to the end of the first century (Filippi, 2015b, 419).[10]

The most important change taking place in Late Antiquity was the new relationship between industrial use and its physical context. Sometime between the fourth and the fifth centuries, part of the *imma cavea* of the stadium of Domitian (now Piazza Navona – Figure 16.2: K) was used as a marble workshop, as indicated by the number of carefully cut fragments of coloured marble found in the area (Molinari, 2014, 264). Towards the middle of the fifth century, the exedra of the *Crypta Balbi* was adapted for the production of glass, as indicated by the installation of an oven as well as by the large number of glass fragments found (*c.* 2500) (Saguì, 1993, 114) (Figure 16.2: L). The most eloquent example of this process is the so-called *Athenaeum*, a building formed by two halls of Hadrianic date excavated on piazza Madonna del Loreto, next to the Forum of Trajan, that was still in use in the fifth century (Egidi, 2010). The structure was spoliated and converted into a large workshop in the sixth century, when it was equipped with furnaces of varying sizes and types for work on iron and copper (as indicated by the structure of the furnaces and the analysis of the residues found in them), possibly for the mint (Serlorenzi et al., 2011; Molinari, 2015, 623). The use of public structures would not have happened without the knowledge of public authorities, which suggests that the agents involved were either well connected or representatives of the government.

Industrial structures also had an impact on the fate of residential buildings, especially towards the end of our period. These were more opportunistic conversions of spaces, more likely under the initiative of the owners of the *insulae* where they took place, without the involvement of public authorities. *Insulae*, as noted earlier, were traditionally connected to economic activities, usually in their ground-floor *tabernae* with their mezzanines. This use seems to have become more intensive from the fifth century onwards, when different types of space were adapted for productive use.[11] An *insula* excavated on Piazza Venezia at the beginning of the *via Lata* had its ground floor adapted for the workshop of a coppersmith in the early sixth century (Serlorenzi, 2010, 133). At San Paolo alla Regola, a large amount of animal bones was found discarded in one of the the *insulae* during the sixth century (as indicated by the stratigraphy of the area), suggesting the activity of a butcher (Quilici, 1986–1987, 260 and 303). The ground floor of the *insula* by the *Crypta Balbi*'s exedra, where a mithraeum had been installed in the third century, was converted into a workshop in the late sixth century. This was recently identified as a bakery, consisting of one room with a large circular oven (similar to those in other *pistrina*) and another with a smaller structure that might have been used for kneading the bread (Serlorenzi et al., 2016). In the seventh century, one of the buildings excavated underneath the modern Galleria Rinascente, on via del Tritone (*Regio VII*), was adapted for a lime kiln, equipped with two furnaces, at a time when the other structures in the area were in the process of abandonment (Acampora, 2017, 155 and 157). These operations required substantial investment, making structural adaptations to pre-existing rooms, building furnaces and ovens, and securing access to water. Rather than a sign of decay, they were part of new economic strategies of the elite. Like the conversion of *insulae* into *domus* that took place at an earlier period, these also impacted the relationship between property owners, tenants, and workers, redefining the geography of the area. At the same time, the spatial context of these structures (as well as the political and legal relations in which they were involved) helped to shape the social interactions taking place in the Campus Martius, making it more openly marked by social dependence and economic interests. It was through space, and the built environment that shaped it that social hierarchies were established and consolidated.

16.4 Space and social conflict

The topographic and economic changes that redefined Campus Martius in Late Antiquity had a direct impact on the social conditions faced by the different groups living and working there, going well beyond the elite. The production of space expresses unequal relations of power, in which the actions of a few play a dominating role. The visibility of large monuments and inscriptions on marble naturally leads scholars to focus on emperors, bishops, and the senatorial elite, while the silence of our sources – or the outright hostility, as in the case of Ammianus Marcellinus – seems to confirm this bias. However, as Yeoh (2003, 17) observed with regard to modern colonial cities, 'various elements of the built environment were differentially perceived and interpreted by different communities based on their own values, priorities, and resources'. What we need to consider now is how other social dynamics might have contributed to the definition of Campus Martius as a lived space. As I will argue, the workers and subaltern groups that frequented this area helped to shape its social configuration as a complex topography that combined sites of oppression, exploitation, and – inevitably – resistance.

Labour and exploitation were an essential aspect of how space was experienced. This is illustrated by a curious passage of the *Ecclesiastical History* of Socrates, written in Constantinople in the early fifth century. After narrating Theodosius I's victory over the usurper Maximus in the West, Socrates describes the moral decay that corrupted the city of Rome and the emperor's initiatives to correct it (*H.E.* 5.18). He mentions the existence, in the city, of bakeries that were built underneath taverns that kept prostitutes as a form of luring in innocent visitors, who would then be kidnapped and forced to work in the horrendous conditions of the *pistrinum* below. Although probably a fabrication narrated to highlight the Christian character of the Eastern Emperor, the passage stresses the violence involved in this type of work. Writing over two centuries earlier, Apuleius had already referred to the harsh conditions of the people (and animals) working in bakeries, poorly dressed, chained, whipped and beaten blue (*Met.* 9.12–9.13). That two authors so far apart (chronologically, geographically, and culturally) could make use of the same *topos* to tell very different stories only shows that this was a powerful image, more compelling because it was acceptable to readers. The Campus Martius was certainly an important *locus* for poor and exploited workers like these, being equipped with 20 *pistrina* according to the *Regionary Catalogues* (p. 557). The catalogues do not record information about how these structures were managed, nor whether they were publicly or privately owned – or both, as Carrié suggested (Carrié, 1975, 1064–1068; Ceparano, 1998). Socrates refers to their owners using the Latin term *mancipes*, which is specific enough to suggest that contractors played an important role in their management. In fact, it is possible that over time the area saw the establishment of a growing number of smaller, privately owned and run *pistrina*, like the one near the *Crypta Balbi* (discussed earlier).

Late antique Romans were well aware of the exploitative nature of the Campus Martius – at least well enough to integrate it into the Christian tradition concerning the martyrdom of Bishop Marcellus (d. 306/309). According to the legend, the bishop was captured at the time of Emperor Maximian and sentenced to forced labour looking after the animals at the *Catabulum*, the headquarters of the pack animal drivers (the *catabolenses*), located on the *via Lata* where the *titulus Marcelli* was later built (*Passio S. Marcelli*, 22 (*AASS*, Ian. II); Coarelli, 1993). ‘Naked but for a hair-shirt, he died in servitude to animals’, in the words of his biographer in the *Liber Pontificalis* – a description that resonates with the references to bakers in the accounts of Apuleius and Socrates of Constantinople.[12] Both the text of the *Passio* and the Life in the *Liber Pontificalis* are datable to the middle of the sixth century, which means that the story must have been taking form already (or at least) in the decades after the building of the church (late fourth or early fifth century).[13] Although we should not accept Christian legends uncritically, the tight chronology between the supposed events, their monumentalisation in the form of a church in the centre of Rome, and the consolidation of the legend in the form known to us suggest that the association between the bishop’s death, the *Catabulum*, and the history of exploitation of the area was very plausible to contemporaries.

It is worth remembering that the Campus Martius was the location of other types of structures where men (and possibly women) were employed in ‘servitude to animals’. An inscribed architrave found reused in the steps leading to the Pantheon (Figure 16.2: M) records a ‘[*stabul*]*um vehiculorum*’ (or possibly ‘[*catabul*]*um*’) restored by the urban prefect Rufius Valerius Messala around the middle of the fifth century. The *Regionary Catalogues*, furthermore, record the presence of the four stables of the circus factions (Coarelli, 1999b, 339–340).[14] The *Catalogues* also mention the existence of the *Trigarium*, a space used for practice by the circus factions in the imperial period (p. 555), but there is no real evidence regarding the continuity of its use or its precise location (Coarelli, 1999c).

Workers who did not have the protection of a guild were probably more exposed to harsher conditions, and the same is true of those who could not afford rent in an *insula*. Although finding complete destitution in the material and literary records is notoriously difficult (as Bowden’s chapter points out), it appears that this period saw a growing presence of informal and precarious forms of residence in the Campus Martius. This is indicated by a law of Arcadius and

Honorius from the late fourth century, regulating the building of *tuguria* – poor dwellings or huts – in the area:

> If any person should attempt to locate a hovel or a hut (*casas seu tuguria*) in the Campus Martius, We command that, by the decree of the Illustrious prefect, he shall be stripped of all his resources and delivered into exile in perpetuity.[15]

A *tugurium* was a specific term, used for poor rural dwellings 'more fitting for rural protection than for housing', usually equipped with a tiled roof, in the words of the jurist Pomponius.[16] Ammianus uses the word to describe the huts where Limigantes soldiers tried (in vain) to hide after their defeat to Constantius II, even though these had thatched roofs (17.13.13), and Paulinus of Nola complains of the presence of two such structures by the basilica of St Felix in Cimitile, made of wood, eyesores 'conveniently' destroyed by a terrible fire (*Carm.* 28.60–28.103). Whether urban or rural, tiled or thatched, what unites these references is the fact that a *tugurium* is a very poor and precarious form of habitation.

The presence of *tuguria* in the Campus Martius should not come as a surprise. Temporary and makeshift dwellings could be a reasonable response to the combination of lack of housing and financial means, and the proximity of economic opportunities. Crucially, there is earlier evidence for the presence of various types of humbler habitation in the Campus Martius, specifically two inscriptions of Severan date, found near the column of Marcus Aurelius (Figure 16.2: N). The first records the letter of the imperial freedman and caretaker of the column, Adrastus, asking permission to build a house in the area (*CIL* VI, 1585a=*EDR*111372); the second records the response of the imperial administration, allowing him to build a hut (*cannaba*) using 'all the tiles and building material from suitable cottages, likewise huts (*de casulis item cannabis*) and (other) buildings'.[17]

What is more revealing is the fact that the law of Arcadius and Honorius is not concerned with the presence of the *tuguria* themselves, but with the agents who 'attempted to locate' them in this part of the city. The threat of punishing such people with the confiscation of all their possessions and their expulsion from the city suggests, in fact, that these were not the poor men and women who inhabited these huts, but those who built (and possibly rented) them.[18] Huts could be illegally built as an opportunistic investment by wealthier Romans to take advantage of those who could not afford an apartment in an *insula* but still needed to be near the economic vitality of this area to make a living. These should be seen, in this sense, as a rational strategy for those who wanted (or needed) cheaper housing in a context of exploitation. The late antique Campus Martius was made of poverty and splendour. Imperial propaganda, patronage, and exploitation shaped the experiences of those who frequented it, and it is in this context that we should see the appearance of *tuguria* in this neighbourhood.

A similar set of issues is suggested by another law, which – although not explicitly referring to Campus Martius – throws important light over the social relations that were taking place there:

> If there should be any persons who adopt the profession of mendicancy and who are induced to seek their livelihood at public expense, each of them shall be examined. The soundness of body and the vigour of years of each one of them shall be investigated. In the case of those who are lazy and not to be pitied on account of any physical disability, the necessity shall be placed upon them that the zealous and diligent informer shall obtain the ownership of those beggars who are held bound by their servile status, and such informer shall be supported by the right to the perpetual colonate of those beggars who are attended by only the liberty of

> their birth rights, provided that the informer should betray and prove such sloth. The masters of the slaves shall be entitled to an unimpaired right of action against those persons who perchance furnished either harborage to fugitives or the advice to adopt the profession of mendicancy.[19]

Addressed by Emperors Gratian, Valentinian II, and Theodosius I to the urban prefect Severus in 382, the law was specifically concerned with those who, in spite of their health and youth, used begging as a form of evading work. As it specifies, the same individuals who denounced the 'illegitimate' beggars would gain control over them under the condition of perpetual colonate and, in the case of slaves, their owners were able to sue those who helped them escape. Rather than showing concern for the plight of the poor, the law is an instrument of social control – an expression of, but also a weapon in, social struggles. It draws an opposition between the condition of beggars, who were free from any bonds to masters or patrons precisely because they were too poor, and the fate of those who were forced to work because they were healthy enough for it. Bishop Marcellus, who died slaving away in the *Catabulum*, would remind us that the line between health and work was precisely the problem, in fact.

The social diversity that characterised Campus Martius and especially the prominent role played by subaltern groups in the area was also related to what was, already by the fourth century, a complex network of spaces for sociability and welfare. These were places where acquaintances were made, social links were established, and ideas exchanged in informal contexts. The area counted with two large imperial *thermae*, the baths of Nero (rebuilt by Alexander Severus), and the baths of Agrippa, which were restored as late as 344.[20] To these should be added the 63 *balnea* recorded in the *Regionary Catalogues* (p. 557). These were possibly privately owned structures like the *tenebrosae balnea* described by Martial in this same area (1.59.3; 2.14.11–2.14.13; Rodriguez Almeida, 1989, 243–249), or even facilities associated with a specific guild or neighbourhood, like the baths of the *Cisiarii* (coachmen) in Ostia (Pavolini, 2006, 55–56). In the Campus Martius, the excavation of the stables of the red faction, in Largo Perosi, revealed next to it the existence of a sizeable bathing structure (approximately 714 square metres) built in the first and restored in the fourth century that might have been related to those workers (Filippi, 2015b, 411–414). The *Catalogues* also record fountains (*lacos*: 120) and shrines (*aediculae*: 35), spaces where different social groups met each other every day (p. 557). Although numbers gathered in the *Regionary Catalogues* should be treated with great care, it should be noted that these are not particularly high when compared to other regions of the city and are consistently repeated in both *Catalogues*, the *Curiosum* and the *Notitia*.

It is likely that the physical condition and actual number of functioning structures used for these social interactions were impacted by lack of repair and abandonment in Late Antiquity, but the period also witnessed the appearance of new ones. We know, for example, that Acilius Glabrio Sibidius established a small forum in the late fourth century next to the so-called *via Recta* (Figure 16.2: O), the route used by the *Falancarii* to transport wine from the *Ciconiae* to the temple of the Sun. This is recorded by a group of statues dedicated by Sibidius's son, the three times prefect of Rome Anicius Acilius Glabrio Faustus (*CIL* VI, 1678=*LSA*-1393; *CIL* VI, 1767=*LSA*-1466; *CIL* VI, 37119=*LSA*-1577), celebrating his family connections in a space where workers associated with the *annona* could gather (Machado, 2019, 49–51). The development of Christianity offered new opportunities in this sense. Not only because churches are essentially spaces for social interaction, but also because Christian elites were engaged in the establishment of a network of social assistance and welfare (Santangeli Valenzani, 1996–1997; Dey, 2008; Brown, 2012). Lay benefactors were involved in the construction of *xenodochia*, structures for the assistance of the very poor: the general Belisarius founded one such institution on the *via Lata*, and it is possible that the *xenodochium Aniciorum* mentioned by Bishop Gregory I in the late sixth century had already been founded as an institution for the assistance of the poor in the fifth century.[21]

The foundation of churches and *tituli* was associated with the development of institutionalised help, a process that was well established by the middle of the fifth century, during the episcopate of Leo I (Salzman, 2017, 74–77; Machado, 2022, 247). In a series of sermons on the Collections, Leo urged the faithful to bring their donations to their churches, from where those resources could be distributed to the poor (*Serm*. 23.3; 25.2). The large number of people in need that surrounded churches is suggested by an anecdote preserved in the work of Priscus, according to which Emperor Anthemius tried (in vain) to hide from his enemies by mingling with the beggars that crowded the area of S. Crisogono in Trastevere, after being defeated by Ricimer in the civil conflict of 472 (Priscus, Frag. 64.1=John of Antioch, Frag. 209.1). Christian spaces played, in this way, a crucial role in the socio-economic life of the various neighbourhoods of late antique Rome, and the Campus Martius was no exception to this trend. It is worthy of note, in fact, that the area counted with two foundations associated with the deacon and martyr Laurence, celebrated among Roman Christians for his care of the poor – the modern-day churches of San Lorenzo in Lucina and San Lorenzo in Damaso. Together with the *titulus* of Marcellus (a bishop who died due to over-exploitation at the *Catabulum*), the Christian topography of the Campus Martius acquired a distinctly popular flavour in the fourth and fifth centuries.

Exploitation, dependence, and assistance did not stop the space of late antique Rome from being also a site for resistance. Ammianus (28.4.29) noted the association between spaces for sociability and interaction and (in his opinion) the lowly character of the Roman people: 'You may see many groups of them gathered in the fora, the cross-roads, the streets, and their other meeting-places, engaged in quarrelsome arguments with one another, some (as usual) defending this, others that'.[22] Magalhães de Oliveira (2012, 125–155) showed the centrality of these spaces for the development of horizontal bonds of social cohesion in the cities of Late Antiquity, and the importance these played in the development of subaltern identities and forms of social action. Ammianus also mentions the *tabernae* and the spots shaded by the awnings of theatres as spaces where the crowds of lowest condition and greatest poverty (*turba vero imae sortis et paupertinae*) spent their time. 'Or', he added, 'which is the favourite among all amusements, from sunrise until evening, in sunshine and in rain, they stand open-mouthed, examining minutely the good points or the defects of the charioteers and their horses'.[23] Crowded with theatres, *tabernae*, the stadium, as well as the stables of the circus factions and the *Trigarium*, the Campus Martius was a fertile space for the establishment of horizontal bonds between different subaltern groups, whether guild members, independent workers, or beggars.

The structures that celebrated the glory of Rome and the magnificence of its rulers and that defined the monumental topography of Campus Martius were thus incorporated into a different logic, one that threatened the very social order they were meant to celebrate, and members of the Roman elite were aware of this fact. This is demonstrated by a letter of Sidonius Apollinaris, prefect of Rome in 468, to his friend Campanianus, in which he thanks his friend for introducing him to the prefect of the *annona*, the officer responsible for the supply of Rome, and talks about his fear of unrest caused by famine:

> However, I should wish you also to recommend to his vigilance my own self, that is to say, the upholding and defence of my reputation. For I am afraid that the uproar of the theatre-benches may sound the cry of 'starvation in Rome', and that the general famine may be put down to my luckless management.[24]

The association between theatre and *horrea*, in fact, points to the Campus as the area where these risks were most obvious. Entertainment buildings and food supply, political propaganda and popular dissent were complementary and inextricable faces of this area. It was precisely by bringing together such a wide social diversity, with its antagonistic interests and needs, that the Campus Martius played such a central role in late antique Rome.

16.5 Conclusion

> To see again the lost landscapes of a Rome that does not exist anymore we must therefore close our eyes and look at them with the eyes of the mind, inspired by research and imagination.[25]

By the end of the sixth century, the Campus Martius still boasted of a dazzling collection of monuments. Columns, temples, and porticoes built by generals and emperors since the end of the Republic now coexisted with churches and Christian spaces sponsored by members of a Christian elite. As we saw above, this part of Rome remained an important space for the celebration of power, whether of emperors or gods. The old *Regio IX* illustrates how space synthesises multiple historical periods and temporalities by bringing them together and putting them in relation with each other. Physical structures with different origins and histories coexisted side by side in the Campus Martius, embodying what scholars call the 'contemporaneity of the non-contemporaneous', 'a diversity of temporal strata which are of varying duration'.[26] (Koselleck, 2004, 95). Tetrarchic emperors, members of the Theodosian dynasty, Ostrogothic kings, and bishops paraded through the streets of the area, in a continuous recombination of past glories and present ambitions.

To look at the late antique Campus Martius from the perspective of continuity or decline obscures the fact that change had been one of its defining characteristics since its early days. This is not only because geography and natural catastrophes contributed to the almost continuous raising of its ground level but because Romans unmade it and remade it in the course of their daily activities. This is more spectacularly the case with the building of the Aurelian wall, a massive physical structure that changed the economic and social geography of the area as well as its appearance. It is also the case, however, with the spoliation of many sites across the area, where pieces of marble decoration were removed, stored, and prepared either for reuse in (usually) Christian buildings or to be burned for the production of lime.[27] The adaptation of *insulae* – first into *domus* in the fourth century, and later into industrial facilities – should be seen as part of this same process of re-creation of space as a part of social life.

These developments highlight an important characteristic of space, the fact that its history can only be apprehended in the *longue durée*, in the sense defined by Braudel (1958). In spite of important discontinuities like the building of the Aurelian wall, the most consequential developments taking place in the Campus Martius are associated with the long history of investments by patrons in the form of warehouses, apartment blocks, and sociability spaces. The *insulae* on via di San Paolo alla Regola are an exceptional example of this: built in the late first century, refurbished for residential use in the fourth, and used by a butcher in the sixth, they remained in use until the second half of the twentieth century, when they were identified as an *ancient* building, cleared, and converted into a protected site (Quilici Gigli, 1986–1987). This long transformative process of the urban fabric was associated with a qualitative change in social relations, in the form of an evolving landscape of dependences that entailed violence, exploitation, and social exclusion. Christian legends took shape and circulated in this setting, fostering new pious foundations like the *titulus* of Marcellus on the *via Lata*. Space is, after all, a product and a producer of social synthesis, gathering of different social groups in the same place, from emperors and senators to Christian preachers, workers, and the destitute and infirm. It is in space that these different social categories interact, where patronage links and labour relations, social control and discipline, riots and demonstrations take place. It is through these interactions, conversely, that space is rebuilt and given new meaning.

To a very large extent, the Campus Martius illustrates the changes that redefined late antique Rome as a whole: the continued interest of its rulers; the growing presence of the Church; the involvement of the local elite; as well as the process of restoration, abandonment, spoliation, and adaptation of early imperial structures (Dey, 2021). Considering a specific part of the city, in this sense, allows us to see how these different processes are integrated and the impact they had on the people who lived through them. What makes the Campus Martius particularly important for our purposes is its unique character, the intensity with which processes observable elsewhere impacted this area and the society that frequented it. As the history of the Campus Martius in Late Antiquity shows, the inhabitants of Rome reacted in creative ways to their own history, remaking the city in the process.

Notes

1 Versions of this chapter were presented in Tampere, Pisa, Oxford, and Rome. I would like to thank the audiences at these lectures, and especially Robert Coates-Stephens, for their suggestions. I would like to thank particularly Laura Acampora, Henrik Boman, and Simone Ciambelli for sharing their publications, and Rebecca Sweetman and Lucy Grig for comments. This chapter is belatedly dedicated to Norberto Luiz Guarinello, who inspired many of the ideas presented here, on the occasion of his retirement.
2 But see Bowden, in this volume, for the limits of this perspective.
3 See discussion in the Introduction, and also Tuori (2020). This is well represented by the ongoing series 'Studies in Roman Space and Urbanism', directed by Ray Laurence and published by Routledge.
4 This is best demonstrated by the fate of the *porticus Minucia* and the appearance, in the early fifth century, of a street that would cut through the walls of the complex: Manacorda (2001, 47–48).
5 'D(omini) N(ostri) Arcadius et Honorius [invicti et] / perpetui Augg(usti) theatrum Pompei [collapso] / exteriore ambitu magna etiam [ex parte] / interior[e] r[uen]te convulsum [ruderibus] / subductis et excitatis invice[m fabricis] / [novis restituerunt]'.
6 'Romani orbis libertatisque / custodi, d(omino) n(ostro) Honorio, pio, Aug(usto) / atque invicto principi. / Aur(elius) Anicius Symmachus, v(ir) c(larissimus), / praef(ectus) urb(i), iterum vice / sacra iudicans, devotus / numini maiestatiq(ue) eius, / dicavit'. Cyriacus of Ancona reported seeing the base in the porticus in the fifteenth century.
7 For example, for Lucius Aradius Valerius Proculus, *CIL* VI, 1690=*LSA*-1396 and *CIL* VI, 1693=*LSA*-1399 by the *suarii*, and *CIL* VI, 1692=*LSA*-1398 by the *pistores*; for Attius Insteius Tertullus, *CIL* VI, 1696=*LSA*-1401 by the *magnarii*; for Memmius Vitrasius Orfitus, *CIL* VI, 1739=*LSA*-1441 by the *pistores*, *CIL* VI, 1740=*LSA*-1442 by the *navicularii*, *CIL* VI, 1741=*LSA*-1443 by the *susceptores Ostiensium sive Portuensium*, and *CIL* VI, 1742=*LSA*-1444 by the *corpus omnium mancipum*.
8 See *LTUR* 3 (1996), s.v. 'horrea', 37–51, and *LTUR* 1 (1993), s.v. 'cella', 256–259, for more examples.
9 See, for earlier periods, Morel (2001), and especially Coarelli (2015) and Panella (2015). I am not referring to agriculture, which was present in this area before its urbanisation in the late Republic.
10 The date was indicated by the discovery of an inscription from the Flavian period buried beneath the structure.
11 We cannot discard the possibility that earlier periods also saw this intensive use but are poorly documented as a result of continuous occupation and clearing.
12 *Lib. Pont.* I, 164: 'Qui tamen in servitio animalium nudus amicto cilicio defunctus est'.
13 See Episcopo (2003, 84–85) for the building, and p. 16 for the text; see also Lapidge (2018, 390–398).
14 See, for the stables of the circus factions, Coarelli (1999b); Pentiricci (2009, 50–55); Filippi (2015b). For the *[stabul]um vehiculorum*, see *AE* 2006, 187=*EDR*033301, with discussion in Orlandi et al. (2006).

15 *Cod. Theod.* 14.14.1: 'Eos, qui in campo Martio casas seu tuguria conlocare temptaverint, sententia viri inl(ustris) p(rae)f(ecti) spoliatos omnibus facultatibus tradi in perpetuum exilio praecipimus'.
16 *Dig.* 50.16.180 (Pomponius 30 ad Sabinus): 'Tugurii appellatione omne aedificium, quod rusticae magis custodiae convenit quam urbanis aedibus, significatur. Ofilius ait tugurium a tecto tamquam tegularium esse dictum, ut toga, quod ea tegamur'. See also *Dig.* 8.3.6.1.
17 *CIL* VI, 1585b=*EDR*111374: 'Tegulas omnes et inpensa(m) / de casulis item cannabis / et aedificiis idoneis'. See Moore (2012) for discussion.
18 This is also suggested by the choice of the verb *conlocare* to describe the action punished by the law, which could also be translated as 'to rent'.
19 Cod. Theod. 14.18.1 (transl. Pharr): 'Cunctis adfatim, quos in publicum quaestum incepta mendicitas vocabit, inspectis exploretur in singulis et integritas corporum et robur annorum, adque ea inertibus et absque ulla debilitate miserandis necessitas inferatur, ut eorum quidem, quos tenet condicio servilis, proditor studiosus et diligens dominium consequatur, eorum, vero, quos natalium sola libertas prosequatur, colonatu perpetuo fulciatur quisquis huiusmodi levitudinem prodiderit ac probaverit, salva dominis actione in eos, qui vel latebram forte fugitivis vel mendicitatis subeundae consilium praestiterunt'. See, for this law, the discussion and bibliography cited in Machado (2022, 239–240).
20 On the baths of Nero, see Ghini (1999a, 60–62); on the baths of Agrippa, see Ghini (1999b, 40–42), with *CIL* VI, 1165=*EDR*111337 for the fourth-century restoration.
21 On Belisarius, see *Lib. Pont.* I, 296; on the *xenodochium Aniciorum* and its founder, see Santangeli Valenzani (1996–1997, 205–207) and Machado (2019, 188–189), but see Coates-Stephens (Forthcoming) for doubts concerning this identification.
22 Amm. Marc. 28.4.29 (transl. Rolfe): 'et videre licet per fora et compita et plateas et conventicula, circulos multos collectos in se controversis iurgiis ferri, aliis aliud, fit, defendentibus'.
23 Amm. Marc. 14.6.25 (transl. Rolfe): 'aut quod est studiorum omnium maximum ab ortu lucis ad vesperam sole fatiscunt vel pluviis, per minutias aurigarum equorumque praecipua vel delicta scrutantes'.
24 Sid. Ap. *Ep.* 1.10.2 (transl. Anderson): 'sed et tu vigilantiae suae me, id est famae meae statum causamque commenda. vereor autem ne famem populi Romani theatralis caveae fragor insonet et infortunio meo publica deputetur esuries'.
25 Manacorda (2022, 23): 'Per rivedere i paesaggi perduti di una Roma che non c'è più dobbiamo allora chiudere gli occhi e guardarli con gli occhi della mente, alimentata dallo studio e dalla fantasia'.
26 Koselleck (2004, 95); for a discussion of the rich history of this issue in Marxist thought (more applied to combined social formations), see Morfino (2017).
27 The temple of via delle Botteghe Oscure, dismantled and used for the storage of coins in the fifth century, is a good example of this process: Manacorda & Zanini (1997, 273–274). The fact that the area was later identified as *Calcarario*, referring to the lime kilns, is worthy of note: Manacorda (2001, 52).

Editions and abbreviations

AASS – *Acta Sanctorum*

AE – *L'Anée Épigraphique*

Amm, *Marc.* – *Ammianus Marcellinus*, 3 vols. Trans. J. C. Rolfe (Loeb Classical Library) (1982).

Apuleius, *Met.* – *Apuleius, Metamorphoses*, 2 vols. Trans. J. A. Hansen (Loeb Classical Library) (1989, 1996).

Augustine, *Cum pagani ingredidenrentur*. In F. Dolbeau (ed.) (1991), Nouveaux sermons de saint Augustin pour la conversion des païens et des donatistes. *Revue des Études Augustiniennes*, 37, 37–78.

Cass., *Var.* – *Cassiodori Senatoris Variae*, ed. T. Mommsen (Monumentae Germaniae Historica. Auctores Antiquissimi 12) (1894).

CIL – *Corpus Inscriptionum Latinarum.*

Claudian, *De bello Gild.* – *De bello Gildonico.* In M. Platnauer. Trans.), *Claudian*, vol. 1 (Loeb Classical Library), 98–137 (1922).

Cod. *Theod.* – *Codex Theodosianus*. In T. Mommsen & P. Meyer et al. (eds.), *Theodosiani Libri XVI cum Constitutionibus Sirmondianis*. Trans. C. Pharr, *The Theodosian Code and Novels, and the Sirmondian Constitutions* (1905, 1952).

Dig. – *The Digest of Justinian*, 4 vols., ed. T. Mommsen & P. Krueger, trans. A. Watson (1985).

EDR – *Epigraphic Database Roma* (www.edr-edr.it/default/index.php).

Greg. I, *Ep*. – *S. Gregorii Magni, Registrum Epistolarum*, 2 vols., ed. D. Norberg (Corpus Scriptorum Series Latina 140) (1982).
Greg. Tur, *H.F.* – Gregory of Tours, *Historia Francorum*, ed. B. Krush & W. Levison (Monumenta Germania Historica, Scriptores Rerum Merovingicarum, 1.1) (1951).
Leo, *Serm*. – Leo I, *Sermons* in J. Leclercq & R. Dolle (ed. and trans.), *Léon le Grand. Sermons*, 3 vols. (Sources Chrétiennes) (1947–1973).
Lib. Pont. – *Le Liber Pontificalis*, 3 vols., ed. L. Duchesne (1886).
LSA – *The Last Statues of Antiquity* (http://laststatues.classics.ox.ac.uk).
Martial, Epigrams – *Martial, Epigrams, v. 1: Spectacles, Books 1–5*, trans. D. R. Shackleton Bailey (Loeb Classical Library) (1993).
Paul. Nola, *Carm*. – *Sancti Pontii Meropii Paulini Nolani Carmina*, ed. G. Hartel (Corpus Scriptorum Ecclesiasticorum Latinorum 30) (1894).
Priscus – R. C. Blockley (ed. and trans.) (1983), *The Fragmentary Classicising Historians of Later Roman Empire: Eunapius, Olympiodorus, Prischus and Malchus*, 2, 221–400.
Proc., *Wars* – Procopius, *The Wars*, 5 vols., trans. H. B. Dewing (Loeb Classical Library) (1954).
Regionary Catalogues – *Notitia Urbis Regionum XIV*. In H. Jordan (Ed.), *Topographie der Stadt Rom im Alterthum*, vol. 2 (pp. 539–574). Weidmann (1871).
Sid. Apoll., *Ep*. – Sidonius Apollinaris, *Epistulae*, in W. Anderson (transl.), *Sidonius: Poems and Letters*, 2 vols. (Loeb Classical Library) (1956).
Socrat., *HE* – Socrates, *Historia Ecclesiastica*. In P. Maraval & P. Périchon, trans., *Socrate de Constantinople, Histoire Ecclésiastique*, 4 vols. (Sources Chrétiennes) (2004–2007).

Bibliography

Acampora, L. (2017). Frequentazione e abbandoni dal IV all'VIII secolo. In M. Baungartner (Ed.), *Roma Rinascente. La città antica tra Quirinale e Pincio* (pp. 147–159). De Luca Editori D'Arte.
Albers, J. (2013). *Campus Martius. Die urbane Entwicklung des Marsfeldes von der Republik bis zur mittleren Kaiserzeit* (Studien zur antiken Stadt 11). Dr Ludwig Reichert Verlag.
Aldrete, G. (2007). *Floods of the Tiber in Ancient Rome*. The Johns Hopkins University Press.
Bauer, F. A. (1996). *Stadt, Platz und Denkmal in der Spätantike*. P. von Zabern.
Boatwright, M. T. (1987). *Hadrian and the City of Rome*. Princeton University Press.
Boman, H. (2012). A third century *insula* beneath the basilica of San Lorenzo in Lucina. In O. Brandt (Ed.), *San Lorenzo in Lucina. The Transformations of a Roman Quarter* (pp. 81–122). Swedish Institute in Rome.
Brandt, O. (2012). La chiesa di San Lorenzo in Lucina e il quartiere preesistente: nuove osservazioni. In H. Brandenburg & F. Guidobaldi (Eds.), *Scavi e scoperte recenti nelle chiese di Roma. Atti della giornata tematica dei Seminari di Archeologia Cristiana* (Sussidi di antichità cristiane 24) (pp. 11–31). Pontificio Istituto di Archeologia Cristiana.
Braudel, F. (1958). Histoire et Sciences sociales: la longue durée. *Annales. Economies, sociétés, civilisations*, 13e (4), 725–753.
Brown, P. (2012). *Through the Eye of a Needle: Wealth, the Fall of Rome, and the Making of Christianity in the West, 350–550 AD*. Princeton University Press.
Carrié, J.-M. (1975). Les distributions alimentaires dans les cités de l'empire romain tardif. *Mélanges de l'école française de Rome*, 87(2), 995–1101.
Castagnoli, F. (1980). Installazioni portuali a Roma. In J. H. D'Arms & E. C. Kopf (Eds.), *The Seaborne Commerce of Ancient Rome: Studies in Archaeology and History* (Memoirs of the American Academy in Rome, 36) (pp. 35–42). University of Michigan Press.
Cavallero, F. (2011). La *via Lata* e i suoi quartieri abitativi in età adrianea. In A. Lo Monaco & V. Nicolucci (Eds.), *La Galleria di Piazza Colonna* (pp. 177–186). Umberto Allemandi & C.
Cecchelli, M. (1998). L'edilizia cristiana tra tardoantico e medioevo. Da piazza del Gesù a piazza della Cancelleria. In M. G. Cimino & M. N. Santi (Eds.), *Corso Vittorio Emanuele II tra urbanistica e archeologia. Storia di uno sventramento* (pp. 51–62). Electa.

Ceparano, M. L. (1998). I *pistrina* nei Regionari di IV secolo. *Mélanges de l'école française de Rome – Antiquité*, 110(2), 917–927.

Chastagnol, A. (1960). *La préfecture urbaine a Rome sous le Bas-Empire*. Presses Universitaires de France.

Ciambelli, S. (2019). Ragonius Vincentius Celsus e i *mensores* di Portus. In T. Gnoli (Ed.), *Aspetti di tarda antichità* (pp. 257–277). Pàtron Editore.

Coarelli, F. (1993). Catabulum. In E. M. Steinby (Ed.), *Lexicon Topographicum Urbis Romae* (Vol. 1: A-C, p. 256). Quasar.

Coarelli, F. (1997). *Il Campo Marzio. Dalle origini alla fine della Repubblica*. Quasar.

Coarelli, F. (1999a). L'edilizia pubblica a Roma in età tetrarchica. In W. V. Harris (Ed.), *The Transformations of URBS ROMA in Late Antiquity* (Journal of Roman Archaeology Supplementary Series 33) (pp. 23–33). Journal of Roman Archaeology.

Coarelli, F. (1999b). Stabula IIII Factionum. In E. M. Steinby (Ed.), *Lexicon Topographicum Urbis Romae* (Vol. 4: P-S, pp. 339–340). Quasar.

Coarelli, F. (1999c). Trigarium. In E. M. Steinby (Ed.), *Lexicon Topographicum Urbis Romae* (Vol. 5: T-Z, pp. 89–90). Quasar.

Coarelli, F. (2015). Le attività artigianali nella Roma di età imperiale: fonti letterarie e fonti epigrafiche. In A. Molinari, R. Santangeli Valenzani, & L. Spera (Eds.), *L'Archeologia della produzione a Roma (Secoli V-XV)* (pp. 119–124). École française de Rome/EDIPUGLIA.

Coates-Stephens, R. (2012). The Walls of Aurelian. In R. Behrwald & C. Witschel (Eds.), *Rom in der Spätantike. Historische Erinnerung im städtische Raum* (pp. 83–109). Franz Steiner Verlag.

Coates-Stephens, R. (forthcoming). *Statues After the End of Sculpture: The Statue World of Early Medieval Rome (AD 400–1000)*.

Corvisieri, C. (1878). Delle Posterule tiberine tra la Porta Flaminia ed il Ponte Gianicolense. *Annali della Società Romana di Storia Patria*, 1, 79–121, 137–171.

D'Alessio, M. T. (2017). Regio IX. Circus Flaminius. In A. Carandini (Ed.), *The Atlas of Ancient Rome. Biography and Portraits of the City. 1. Texts and Images* (pp. 493–541). Princeton University Press.

De Caprariis, F. (1999). I porti della città nel IV e V secolo d.C. In W. V. Harris (Ed.), *The Transformations of URBS ROMA in Late Antiquity* (Journal of Roman Archaeology Supplementary Series 33) (pp. 217–234). Journal of Roman Archaeology.

De Laine, J. (2018). Insulae. In C. Holleran & A. Claridge (Eds.), *A Companion to the City of Rome* (pp. 317–323). Wiley Blackwell.

Dey, H. (2008). *Diaconiae, xenodochia, hospitalia* and Monasteries: 'Social security' and the meaning of monasticism in early medieval Rome. *Early Medieval Europe*, 16(4), 398–422.

Dey, H. (2011). *The Aurelian Wall and the Refashioning of Imperial Rome, AD 271–855*. Cambridge University Press.

Dey, H. (2021). *The Making of Medieval Rome. A New Profile of the City, 400–1420*. Cambridge University Press.

Egidi, R. (2010). L'area di Piazza Venezia. Nuovi dati topografici. In R. Egidi, F. Filippi, & S. Martone (Eds.), *Archeologia e Infrastrutture. Il tracciato fondamentale della linea C della metropolitana di Roma: primi indagini archeologiche* (pp. 93–124). Leo S. Olschki.

Egidi, R., Filippi, F., & Martone, S. (Eds.). (2010). *Archeologia e Infrastrutture. Il tracciato fondamentale della linea C della metropolitana di Roma: primi indagini archeologiche*. Leo S. Olschki.

Episcopo, S. (2003). *Il* Titulus Marcelli *sulla Via Lata. Nuovi studi e ricerche archeologiche (1990–2000)* (Tardoantico e Medioevo 6). Palombi Editori.

Favro, D. (1996). *The Urban Image of Augustan Rome*. Cambridge University Press.

Filippi, F. (Ed.). (2015a). *Campo Marzio. Nuove Ricerche. Seminario di Studi*. Quasar.

Filippi, F. (2015b). Il quartiere dello *stabulum Factionis Russatae* nel Campo Marzio occidentale. Rapporto preliminare sull'indagine di scavo di largo Perosi (2009–2014). In F. Filippi (Ed.), *Campo Marzio. Nuove Ricerche. Seminario di Studi* (pp. 393–442). Quasar.

Fogagnolo, S. (2013). Testimonianze del sacco del 410 in un cantiere edilizio a Trastevere (Conservatore di San Pasquale Baylon). In J. Lipps, C. Machado, & P. von Rummel (Eds.), *The Sack of Rome in 410 AD* (pp. 151–161). Reichert Verlag.
Fogagnolo, S. (2022). L'analisi dell'edilizia residenziale tra vecchi dati e nuove interpretazioni. In A. Pizzo & R. Montalbano (Eds.), *Tra le pendici del Quirinale e il Campo Marzio, in memoria di Emilio Rodriguez Almeida* (pp. 193–213). Arbor Sapientiae.
Fraschetti, A. (1999). Veniunt modo reges Romam. In W. V. Harris (Ed.), *The Transformations of URBS ROMA in Late Antiquity* (Journal of Roman Archaeology Supplementary Series 33) (pp. 235–248). Journal of Roman Archaeology.
Ghini, G. (1999a). Thermae Neronianae/Alexandrinae. In E. M. Steinby (Ed.), *Lexicon Topographicum Urbis Romae* (Vol. 5: T-Z, pp. 89–90). Quasar.
Ghini, G. (1999b). Thermae Agrippae. In E. M. Steinby (Ed.), *Lexicon Topographicum Urbis Romae* (Vol. 5: T-Z, pp. 60–62). Quasar.
Insalacco, A. (2005). *Le città dell'acqua. Archeologia sotterranea a Fontana di Trevi.* Electa.
Ippoliti, M. (2015). Le due *insulae* presso la scalinata di S. Maria in *Aracoeli*: analisi archeologica e ricostruzione architettonica. *Scienze dell'Antichità*, 21(1), 181–199.
Kalas, G. (2015). *The Restoration of the Roman Forum in Late Antiquity. Transforming Public Space.* University of Texas Press.
Koselleck, R. (2004). *Futures Past. On the Semantics of Historical Time*. Columbia University Press.
Krautheimer, R. (1980). *Rome. Profile of a City, 312–1308*. Princeton University Press.
Lapidge, M. (2018). *The Roman Martyrs: Introduction, Translations, Commentary*. Oxford University Press.
Lefebvre, H. (1991). *The Production of Space*. Blackwell Publishing.
Lissi Carona, E. (1985). Un complesso edilizio tra via in Arcione, via dei Maroniti e vicolo dei Maroniti. In *Roma. Archeologia nel centro, vol. 2* (pp. 360–365). De Luca Editore.
Liverani, P. (2007). Victors and pilgrims in Late Antiquity and the early middle ages. *Fragmenta*, 1, 83–102.
Liverani, P. (2016). Roma tardoantica come spazio della rappresentazione trionfale. In F. Goldbeck & J. Wienand (Eds.), *Der römische Triumph in Prinzipat und Spätantike* (pp. 487–510). De Gruyter.
Machado, C. (2012). Aristocratic houses and the making of late antique Rome and Constantinople. In L. Grig & G. Kelly (Eds.), *Two Romes. Rome and Constantinople in Late Antiquity* (pp. 136–158). Oxford University Press.
Machado, C. (2019). *Urban Space and Aristocratic Power in Late Antique Rome, AD 270–535*. Oxford University Press.
Machado, C. (2022). Looking for the poor in late antique Rome. In F. Carlà-Uhink, L. Cecchet, & C. Machado (Eds.), *Poverty in Ancient Greece and Rome: Discourses and Realities* (pp. 232–257). Routledge.
Magalhães de Oliveira, J. C. (2012). *'Potestas Populi'. Participation populaire et action collective dans les villes de l'Afrique romaine tardive (ver 300–430 apr. J.-C.)* (Bibliothèque de l'Antiquité Tardive, 24). Brepols.
Malmberg, S. (2021). Understanding Rome as a Port City. In F. Vermeulen & A. Zuiderhoek (Eds.), *Space, Movement and the Economy in Roman Cities in Italy and Beyond* (pp. 315–372). Routledge.
Manacorda, D. (2001). *Crypta Balbi. Archeologia e storia di un paesaggio urbano*. Electa.
Manacorda, D. (2022). *Roma Il racconto di due città*. Carocci.
Manacorda, D., & Zanini, E. (1997). Il tempio di via delle Botteghe Oscure: tra stratigrafia, topografia e storia. *Ostraka*, 6(2), 249–293.
Meneghini, R. (1999). La *domus* degli *Artemii*. *Bullettino della Commissione Archeologica Comunale di Roma*, 100, 229–234.
Meneghini, R., & Santangeli Valenzani, R. (2004). *Roma nell'alto medioevo*. Libreria dello Stato.
Molinari, A. (2014). Gli scavi al n. 62 di Piazza Navona, tra 'microstorie' e 'grandi narrazioni' (secoli V-XV). In J.-F. Bernard (Ed.), *'Piazza Navona, ou Place Navone, la plus belle et la plus grand': du stade de Domitien à la place moderne, histoire d'une évolution urbaine* (pp. 263–274). École française de Rome.

Molinari, A. (2015). La produzione artigianale a Roma tra V e XV secolo. Riflessioni sui risultati di uno studio archeologico sistematico e comparato. In A. Molinari, R. Santangeli Valenzani, & L. Spera (Eds.), *L'Archeologia della produzione a Roma (Secoli V-XV)* (pp. 613–635). École française de Rome/ EDIPUGLIA.

Moore, D. (2012). A note on *CIL* VI.1585a-b and the role of Adrastus, *procurator* of the Column of Marcus Aurelius. *Zeitschrift für Papyrologie und Epigraphik*, 181, 221–229.

Morel, J.-P. (2001). Artisanat et manufacture à Rome (Ier s. av. n. è. – II s. de n. è.). *Pallas*, 55, 243–263.

Morfino, V. (2017). On non-contemporaneity: Marx, Bloch, Althusser. In V. Morfino & P. Thomas (Eds.), *The Government of Time. Theories of Plural Temporality in the Marxist Tradition* (pp. 117–147). Brill.

Muzzioli, M. P. (2018). Le *ripae* di Roma in età imperiale: qualche evidenza dalla pianta marmorea severiana. *RIPARIA*, 4, 28–45.

Orlandi, S., Panciera, S., & Virgili, P. (2006). Attività edilizia monumentale nel centro di Roma nel V sec. d.C. A proposito di una nuova iscrizione de prefetto urbano Rufius Valerius Messala. In M. Ghilardi, C. Goddard, & P. Porena (Eds.), *Les cités de l'Italie tardo-antique (IVe-Ve siècle). Institutions, économie, société, culture et religion* (123–136). École française de Rome.

Packer, J. E. (1968–1969). La casa di via Giulio Romano. *Bullettino della Commissione Archeologica Comunale di Roma*, 81, 127–148.

Packer, J. E. (1971). *The Insulae of Imperial Ostia* (Memoirs of the American Academy in Rome, 31). University of Michigan Press.

Panella, C. (2015). Roma imperiale come centro produttivo: le evidenze archeologiche. In A. Molinari, R. Santangeli Valenzani, & L. Spera (Eds.), *L'Archeologia della produzione a Roma (Secoli V-XV)* (pp. 97–118). École française de Rome/EDIPUGLIA.

Pani Ermini, L. (1999). Roma da Alarico a Teoderico. In W. V. Harris (Ed.), *The Transformations of URBS ROMA in Late Antiquity* (Journal of Roman Archaeology Supplementary Series 33) (pp. 35–52). Journal of Roman Archaeology.

Pavolini, C. (2006). *Ostia* (Guide Archeologiche Laterza). Laterza.

Pentiricci, M. (2009). Il settore occidentale del Campo Marzio tra l'età antica e l'altomedioevo. In C. Frommel & M. Pentiricci (Eds.), *L'Antica basilica di San Lorenzo in Damaso. Indagini archeologiche nel Palazzo della Cancelleria, vol. 1: Gli Scavi* (Monumenta Sanctae Sedis 5.1) (pp. 15–75). De Luca Editori d'Arte.

Pietri, C. (1976). *Roma Christiana: recherches sur l'Église de Rome, son organisation, sa politique, son idéologie de Miltiade à Sixte III (311–440)*, 2 vols. École française de Rome.

Purcell, N. (1999). The populace of Rome in Late Antiquity: Problems of classification and historical description. In W. V. Harris (Ed.), *The Transformations of URBS ROMA in Late Antiquity* (Journal of Roman Archaeology Supplementary Series 33) (pp. 135–161). Journal of Roman Archaeology.

Quilici, L. (1986–1987). Via di S. Paolo alla Regola. Scavi e recupero di edifici antichi e medioevali. *Notizia degli scavi di antichità*, 40, 175–416.

Rakob, F. (1987). Die Urbanisierung des nördliche Marsfeldes. Neue Forschungen im Areal des Horologium Augusti. In *L'Urbs: espace urbain et histoire (Ier siècle av. J.-C. – III siècle ap. J.-C.* (pp. 687–712). École française de Rome.

Rodriguez Almeida, E. (1989). Due note marzialiane: I 'balnea quattuor in campo' e le 'sellae paterclianae' subcapitoline. *Mélanges de l'École française de Rome*, 101(1), 243–254.

Saguì, L. (1993). Produzione vetrarie a Roma tra tardo-antico e alto medioevo. In P. Delogu & L. Paroli (Eds.), *La Storia economica di Roma nell'alto Medioevo alla luce dei recenti scavi archeologici* (pp. 113–136). All'Insegna del Giglio.

Salzman, M. (2017). From a classical to a Christian City. Civic evergetism and charity in late antique Rome. *Studies in Late Antiquity*, 1(1), 65–85.

Salzman, M. (2021). *The Falls of Rome. Crises, Resilience, and Resurgence in Late Antiquity*. Cambridge University Press.

Santangeli Valenzani, R. (1996–1997). Pellegrini, senatori e papi. Gli *xenodochia* a Roma tra il IV e il IX secolo. *Rivista dell'Istituto Nazionale d'Archeologia e Storia dell'Arte*, s. 3, 19–20, 203–226.

Sediari, M. (1997). La topografia della *Regio IX* di Roma in età severiana. *Bullettino della Commissione Archeologica Comunale di Roma*, 98, 215–248.

Serlorenzi, M. (2010). Testimonianze medievali nei cantieri di Piazza Venezia. In R. Egidi, F. Filippi, & S. Martone (Eds.), *Archeologia e Infrastrutture. Il tracciato fondamentale della linea C della metropolitana di Roma: primi indagini archeologiche* (pp. 131–165.). Leo S. Olschki.

Serlorenzi, M., Carrera, F. M., & Colantonio, S. (2016). *Crypta Balbi*. Un panificio di età tardoantica. *Bullettino della Commissione Archeologica Comunale di Roma*, 117, 289–293.

Serlorenzi, M., Ricci, G., de Luca, I., Anguilano, L., La Salvia, V., & Marani, F. (2011). Il contesto archeologico dell'Athenaeum di Adriano tra tardoantico ed alto medioevo: una possibile zecca di VI-fine VII/prima metà VIII secolo? In G. Pardini, N. Parise, & F. Marani (Eds.), *Numismatica e Archeologia. Monete, stratigrafie e contesti. Dati a confronto* (pp. 369–403). Quasar.

Spera, L. (2014). Trasformazioni e riassetti del tessuto urbano nel Campo Marzio centrale tra tarda antichità e medioevo. *Mélanges de l'école française de Rome – Moyen Âge* [online], 126(1). https://doi.org/10.4000/mefrm.1971.

Torelli, M. (1993a). Arco di Portogallo. In E. M. Steinby (Ed.), *Lexicon Topographicum Urbis Romae* (Vol. 1: Α-C, pp. 77–79). Quasar.

Torelli, M. (1993b). *Arcus Novus*. In E. M. Steinby (Ed.), *Lexicon Topographicum Urbis Romae* (Vol. 1: A-C, pp. 101–102). Quasar.

Tuori, K. (2020). Spatial theories and the study of Ancient (Roman) urbanism: Review article. *Arctos*, 54, 357–378.

Wagner, H. (2021). *Das spätantike Rom und die stadtrömische Senatsaristokratie (395–455 n. Chr.)*. De Gruyter.

Wallace-Hadrill, A. (2022). The cities of Cassiodorus: The resilience of urban values. In J. Martínez Jiménez & S. Ottewill-Soulsby (Eds.), *Remembering and Forgetting the Ancient City* (pp. 23–44). Oxbow.

Wiseman, T. P. (1993). Campus martius. In E. M. Steinby (Ed.), *Lexicon Topographicum Urbis Romae* (Vol. 1: A-C, pp. 220–224). Quasar.

Yeoh, B. (2003). *Contesting Space in Colonial Singapore. Power Relations and the Urban Built Environment*. Singapore University Press.

Index